Recent Advances in Applied Nonlinear Dynamics with Numerical Analysis

Fractional Dynamics, Network Dynamics, Classical Dynamics and Fractal Dynamics with their Numerical Simulations

INTERDISCIPLINARY MATHEMATICAL SCIENCES*

Published

Vol. 5: The Hilbert–Huang Transform and Its Applications
eds. Norden E. Huang & Samuel S. P. Shen

Vol. 6: Meshfree Approximation Methods with MATLAB
Gregory E. Fasshauer

Vol. 7: Variational Methods for Strongly Indefinite Problems
Yanheng Ding

Vol. 8: Recent Development in Stochastic Dynamics and Stochastic Analysis
eds. Jinqiao Duan, Shunlong Luo & Caishi Wang

Vol. 9: Perspectives in Mathematical Sciences
eds. Yisong Yang, Xinchu Fu & Jinqiao Duan

Vol. 10: Ordinal and Relational Clustering (with CD-ROM)
Melvin F. Janowitz

Vol. 11: Advances in Interdisciplinary Applied Discrete Mathematics
eds. Hemanshu Kaul & Henry Martyn Mulder

Vol. 12: New Trends in Stochastic Analysis and Related Topics:
A Volume in Honour of Professor K D Elworthy
eds. Huaizhong Zhao & Aubrey Truman

Vol. 13: Stochastic Analysis and Applications to Finance:
Essays in Honour of Jia-an Yan
eds. Tusheng Zhang & Xunyu Zhou

Vol. 14: Recent Developments in Computational Finance:
Foundations, Algorithms and Applications
eds. Thomas Gerstner & Peter Kloeden

Vol. 15: Recent Advances in Applied Nonlinear Dynamics with Numerical Analysis:
Fractional Dynamics, Network Dynamics, Classical Dynamics and Fractal Dynamics with Their Numerical Simulations
eds. Changpin Li, Yujiang Wu & Ruisong Ye

*For the complete list of titles in this series, please go to
http://www.worldscientific.com/series/ims

Interdisciplinary Mathematical Sciences – Vol. 15

Recent Advances in Applied Nonlinear Dynamics with Numerical Analysis

Fractional Dynamics, Network Dynamics, Classical Dynamics and Fractal Dynamics with their Numerical Simulations

Editors

Changpin Li
Shanghai University, China

Yujiang Wu
Lanzhou University, China

Ruisong Ye
Shantou University, China

World Scientific

NEW JERSEY • LONDON • SINGAPORE • BEIJING • SHANGHAI • HONG KONG • TAIPEI • CHENNAI

Published by

World Scientific Publishing Co. Pte. Ltd.
5 Toh Tuck Link, Singapore 596224
USA office: 27 Warren Street, Suite 401-402, Hackensack, NJ 07601
UK office: 57 Shelton Street, Covent Garden, London WC2H 9HE

British Library Cataloguing-in-Publication Data
A catalogue record for this book is available from the British Library.

Interdisciplinary Mathematical Sciences — Vol. 15
RECENT ADVANCES IN APPLIED NONLINEAR DYNAMICS WITH NUMERICAL ANALYSIS
Fractional Dynamics, Network Dynamics, Classical Dynamics and Fractal Dynamics with Their Numerical Simulations

ISBN 978-981-4436-45-8

Printed by FuIsland Offset Printing (S) Pte Ltd Singapore

Professor Zhong-hua Yang

Foreword

Almost no one doubts that Dynamics is always an exciting and serviceable topic in science and engineering. Since the founder of dynamical systems, H. Poincaré, there have been great theoretical achievements and successful applications. Meanwhile, complex and multifarious dynamical evolutions and new social requests produce new branches in the field of dynamical systems, such as fractional dynamics, network dynamics, and various genuine applications in industrial and agricultural production as well as national construction.

Although fractional calculus, in allowing integrals and derivatives of any positive real order (the term "fractional" is kept only for the historical reasons) even complex number order, has almost the same history as the classical calculus, fractional dynamics is still in the budding stage. As far as we know, the beginning era of fractional dynamics very possibly originates from a paper on the Lyapunov exponents of the fractional differential systems published in Chaos in 2010. On the other hand, there have existed a huge number of publications in network dynamics albeit it appeared in 1990's. Besides, network dynamics has penetrated into various sources and more and more theories and applications will be prominently emerged.

With the rapid developments of the nonlinear dynamics, this volume timely collects contributions of recent advances in fractional dynamics, network dynamics, fractal dynamics and the classical dynamics. The contents cover applied theories, numerical algorithms and computations, and applications in this regard. First chapter contributes to surveys on Gronwall inequalities where the singular case has been emphasized which are often used in the fractional differential systems. In the second chapter, recent results of existence and uniqueness of the solutions to the fractional differential equations are presented. In the next chapter, the finite element method and calculation for the fractional differential equations are summarized and introduced. In following three chapters, the numerical method and calculations for fractional differential equations are proposed and numerically realized, where the fractional step method, the spectral method, and the discontinuous finite element method, are used to solve the fractional differential equations, respectively. In the seventh chapter, recent results on the asymptotic expansion of a singularly perturbed problem under curvilinear coordinates are shown with the aid of classical

Laplace transformation. Chapter 8 contributes to investigating the typically dynamical numerical solver-incremental unknowns methods under the background of alternating directional implicit (ADI) scheme for a heat conduction equation. Chapter 9 generalizes the sharp estimates of the two-dimensional problems to the stability analysis of three-dimensional incompressible Navier-Stokes equations solved numerically by a colocated finite volume scheme. In the tenth chapter, numerical algorithms for the computation of certain symmetric positive solutions and the detection of symmetry-breaking bifurcation points on these or other symmetric positive solutions for p-Henon equation are studied. In the following chapter, recent results of block incremental unknowns for solving reaction-diffusion equations are presented. Chapters 12, 13 and 19 contribute to network dynamics, where the models and synchronization dynamics are introduced and analyzed in details. Chapter 14 focuses on chaotic dynamical systems on fractals and their applications to image encryption. Chapter 15 makes contribution to the generation of the planar crystallographic symmetric patterns by discrete systems invariant with respect to planar crystallographic groups from a dynamical system point of view. Chapter 16 investigates the complicated dynamics of a simple two-dimensional discrete dynamical system. Chapter 17 discusses the bifurcations in the delayed ordinary differential equation and the next chapter introduces the numerical methods for the option pricing problems.

We are very grateful to all the authors for their contributions to this volume. We specially thank Ms Tan Rok Ting for her sparing no pains to inform us, replying to us and explaining various details regarding this edited volume. The mostly mentionable question is that the published year of this book happens to be the year of Professor Zhong-hua Yang's 70th birthday. We are privileged and honored to dedicate this edited book to Professor Zhong-hua Yang, our teacher and life-long friend. CL acknowledges the financial support of the National Natural Science Foundation of China (grant no. 10872119), the Key Disciplines of Shanghai Municipality (grant no. S30104), and the Key Program of Shanghai Municipal Education Commission (grant no. 12ZZ084).

Changpin Li, Shanghai University
Yu-jiang Wu, Lanzhou University
Ruisong Ye, Shantou University

May 28, 2012

Preface

This festschrift volume is dedicated to Professor Zhong-hua Yang on the occasion of his 70th birthday.

Zhong-hua Yang was born on October 5, 1942, in Shanghai, China. He graduated in 1964 from Fudan University, a prestigious university in China. After graduation, he was recruited to Shanghai University of Science and Technology (now is called Shanghai University), as a faculty member at the Department of Mathematics.

In 1982, Yang published his first research paper and in the same year he went to California Institute of Technology as a senior visiting scholar to work with the famous mathematician, Professor H.B. Keller, for advanced studies on theory and computation of bifurcation.

Two years later, he returned to Shanghai University of Science and Technology, where he spent twenty years as a faculty member. He has published 70 articles ranging in computational and applied mathematics, especially in computation of bifurcation. In 1989, he was promoted to full professor and appointed as associate director of Department of Mathematics at the university. In 1995, he was appointed as an advisor of the graduated students for Ph.D. degree.

In 1988, 1992 and 1998, he was granted the Science and Technology Progress Award for three times by Ministry of Education, China. In the 1990's, Yang worked on bifurcation computation and applications for nonlinear problems, one of the projects in National "Climbing" Program. He has received special government allowance from the State Council of China since 1992. He was then awarded Shanghai splendid educator in 1995.

In 1996, he moved to Shanghai Normal University, and acted as vice dean of the School of Math. Science (1997-2002). His academic positions and responsibilities also include: Editor of the journal *Numerical Mathematics: A Journal of Chinese Universities (English Series)*, Council member of Shanghai Mathematics Society and reviewer for Mathematical Reviews.

In 2007, his book *Nonlinear Bifurcation: Theory and Computation* was published, which was the first monograph on bifurcation computation in China.

Postgraduate Students under Zhong-hua Yang's supervision

Jizhong Wang (M. Sc., Shanghai University of Science and Technology, 1991)
Ning Ji (M. Sc., Shanghai University of Science and Technology, 1994)
Ruisong Ye (Ph. D., Shanghai University, 1995)
Yujiang Wu (Ph. D., Shanghai University, 1997)
Changpin Li (Ph. D., Shanghai University, 1998)
Wei Zhou (M. Sc., Shanghai Normal University, 2001)
Ying Zhu (M. Sc., Shanghai Normal University, 2002)
Qian Guo (Ph. D., Shanghai University, 2003)
Bo Xiong (Ph. D., Shanghai Normal University, 2004)
Junqiang Wei (M. Sc., Shanghai Normal University, 2004)
Yezhong Li (M. Sc., Shanghai Normal University, 2005)
Quanbao Ji (M. Sc., Shanghai Normal University, 2006)
Xia Gu (M. Sc., Shanghai Normal University, 2006)
Hailong Zhu (M. Sc., Shanghai Normal University, 2007)
Jian Shen (M. Sc., Shanghai Normal University, 2007)
Zhaoxiang Li (Ph. D., Shanghai Normal University, 2008)
Xiaojuan Xi (M. Sc., Shanghai Normal University, 2008)
Yuanyuan Song (M. Sc., Shanghai Normal University, 2008)

Publications Since 1982

Books

(1) Introduction to Numerical Approximation (with De-ren Wang), Higher Education Press, Beijing 1990, in Chinese.
(2) Nonlinear Bifurcation: Theory and Computation, Science Press, Beijing 2007, in Chinese.

Selected Papers

(1) Yang, Z. H. (1982). Continuation Newton method for boundary value problems of nonlinear elliptic differential equations, (in Chinese) *Numer. Math. J. Chinese Univ.* **4**, pp. 28–37.
(2) Yang, Z. H. (1984). Several abstract iterative schemes for solving the bifurcation at simple eigenvalues, *J. Comput. Math.* **2**, pp. 201–209.
(3) Yang, Z. H and Keller, H. B. (1986). A direct method for computing higher order folds, *SIAM J. Sci. Statist. Comput.* **7**, pp. 351–361.
(4) Yang, Z. H and Keller, H. B. (1986). Multiple laminar flows through curved pipes, *Appl. Numer. Math.* **2**, pp. 257–271.
(5) Yang, Z. H. (1987). Folds of degree 4 and swallowtail catastrophe. Numerical

methods for partial differential equations (Shanghai, 1987), 171–183, Lecture Notes in Math., 1297, Springer, Berlin, 1987.

(6) Yang, Z. H. (1987). Steady problems in thermal ignition. (in Chinese) *Comm. Appl. Math. Comput.* **1**, pp. 8–21.

(7) Yang, Z. H. (1988). An acceleration method in the homotopy Newton's continuation for nonlinear singular problems, *J. Comput. Math.* **6**, pp. 1–6.

(8) Yang, Z. H. (1988). The application of the continuation method in the direct method for computing higher order folds, (in Chinese) *Math. Numer. Sinica* **10**, pp. 6–17.

(9) Yang, Z. H. (1988). Approximation to cusp catastrophe. BAIL V (Shanghai, 1988), 411–416, Boole Press Conf. Ser., 12, Boole, Dún Laoghaire, 1988.

(10) Yang, Z. H. (1988). Global asymptotic behavior of solutions to nonsteady state thermal ignition problems, (in Chinese) *Comm. Appl. Math. Comput.* **2**, pp. 67–73.

(11) Yang, Z. H. (1989). Higher order folds in nonlinear problems with several parameters, *J. Comput. Math.* **7**(3), pp. 262–278.

(12) Yang, Z. H. (1989). Classification of pitchfork bifurcations and their computation, *Sci. China Ser. A* **32**(5), pp. 537–549.

(13) Yang, Z. H and Sleeman, B. D. (1989). Hopf bifurcation in wave solutions of FitzHugh-Nagumo equation, Proceedings of the International Conference on Bifurcation Theory and its Numerical Analysis (Xi'an, 1988), 115–125, Xi'an Jiaotong Univ. Press, Xi'an, 1989.

(14) Yang, Z. H. (1990). An improved scheme for chord methods at singular points, (in Chinese) *Numer. Math. J. Chinese Univ.* **12**(2), pp. 151–157.

(15) Yang, Z. H. (1991). A direct method for pitchfork bifurcation points, *J. Comput. Math.* **9**(2), pp. 149–153.

(16) Yang, Z. H. (1991). Approximation of catastrophe points of cusp form, (in Chinese) *Gaoxiao Yingyong Shuxue Xuebao* **6**(1), pp. 1–12.

(17) Yang, Z. H. and Li, Z. L. (1992). Bifurcation study on the laminar flow in the coiled tube with the triangular cross section. Numerical methods for partial differential equations (Tianjin, 1991), 126–138, World Sci. Publ., River Edge, NJ, 1992.

(18) Yang, Z. H. (1992). Detecting codimension two bifurcations with a pure imaginary and a simple zero eigenvalue, *J. Comput. Math.* **10**, pp. 204–208.

(19) Yang, Z. H. (1992). Symmetry-breaking in two-cell exothermic reaction problems, (in Chinese) *Shanghai Keji Daxue Xuebao* **15**, pp. 44–54.

(20) Ye, R. S. and Yang, Z. H. (1995). The computation of symmetry-breaking bifurcation points in $Z_2 \times Z_2$-symmetric nonlinear problems, *Appl. Math. J. Chinese Univ. Ser. B* **10**, pp. 179–194.

(21) Ye, R. S. and Yang, Z. H, Mahmood, A. (1995). Extended systems for multiple S-breaking turning points in $Z_2 \times Z_2$-symmetric nonlinear problems, *Numer. Math. J. Chinese Univ.* (English Ser.) **4**, pp. 119–132.

(22) Ye, R. S. and Yang, Z. H. (1996). Double S-breaking cubic turning points and their computation, *J. Comput. Math.* **14**, pp. 8–22.
(23) Yang, Z. H and Ye, R. S. (1996). A numerical method for solving nonlinear singular problems and application to bifurcation problems. World Congress of Nonlinear Analysts'92, Vol. I-IV (Tampa, FL, 1992), 1619–1626, de Gruyter, Berlin, 1996.
(24) Yang, Z. H and Ye, R. S. (1996). Double high order S-breaking bifurcation points and their numerical determination, *Appl. Math. Mech.* (English Ed.) **17**, pp. 633–646
(25) Yang, Z. H. and Ye, R. S. (1996). Symmetry-breaking and bifurcation study on the laminar flows through curved pipes with a circular cross section, *J. Comput. Phys.* **127**, pp. 73–87
(26) Yang, Z. H, Mahmood, A. and Ye, R. S. (1997). Fully discrete nonlinear Galerkin methods for Kuramoto-Sivashinsky equation and their error estimates, *J. Shanghai Univ.* **1**, pp. 20–27.
(27) Li, C. P., Yang, Z. H and Wu, Y. J. (1997). Bifurcation and stability of non-trivial solution to Kuramoto-Sivashinsky equation, *J. Shanghai Univ.* **1**, pp. 95–97.
(28) Ye, R. S. and Yang, Z. H. (1997). Classification of simple higher-order symmetry-breaking bifurcations and their computation, *J. Shanghai Univ.* **1**, pp. 175–183.
(29) Li, C. P. and Yang, Z. H. (1998). Bifurcation of two-dimensional Kuramoto-Sivashinsky equation, *Appl. Math. J. Chinese Univ. Ser. B* **13**, pp. 263–270.
(30) Yang, Z. H and Li, C. P. (1998). A numerical approach to Hopf bifurcation points, *J. Shanghai Univ.* **2**, pp. 182–185.
(31) Mahmood, A. and Yang, Z. H. (1998). Numerical results of Galerkin and non-linear Galerkin methods for one-dimensional Kurmoto-Sivashinsky equation, *Proc. Pakistan Acad. Sci.* **35**, pp. 33–37.
(32) Yang, Z. H, Wu, Y. J. and Guo, B. Y. (1999). Computation of nonlinear Galerkin methods with variable modes for 2-D K-S equations. Advances in computational mathematics (Guangzhou, 1997), 545–563, Lecture Notes in Pure and Appl. Math., 202, Dekker, New York, 1999.
(33) Li, C. P. and Yang, Z. H. (2000). A note of nonlinear Galerkin method for steady state Kuramoto-Sivashinsky equation, *Math. Appl.* (Wuhan) **13**, pp. 46–51.
(34) Yang, Z. H and Zhou, W. (2000). A computational method for D6 equivariant nonlinear bifurcation problems, (in Chinese) *Comm. Appl. Math. Comput.* **14**, pp. 1–13.
(35) Li, C. P. and Yang, Z. H. (2001). A nonlinear Galerkin method for K-S equation, *Math. Appl.* (Wuhan) **14**, pp. 22–27.
(36) Li, C. P. and Yang, Z. H. (2001). Remark on Galerkin method for two-dimensional steady state Kuramoto-Sivashinsky equation, *Numer. Math. J.*

Chinese Univ. (English Ser.) **10**, pp. 161–169.

(37) Yang, Z. H and Zhu, Y. (2001). Numerical determination of symmetry-increasing bifurcation of chaotic attractors in a class of planar D_3-equivariant mappings, (in Chinese) *Comm. Appl. Math. Comput.* **15**(2), pp. 1–8.

(38) Li, C. P. and Yang, Z. H. (2002). Error estimates of Galerkin method for high dimensional steady state Kuramoto-Sivashinsky equation, *Numer. Math. J. Chinese Univ.* (English Ser.) **11**, pp. 129–136.

(39) Guo, Q. and Yang, Z. H. (2002). Dynamics of methods for delay differential equations, (in Chinese) *Comm. Appl. Math. Comput.* **16**, pp. 7–14.

(40) Wu, Y. J. and Yang, Z. H. (2002). On the error estimates of the fully discrete nonlinear Galerkin method with variable modes to Kuramoto-Sivashinsky equation. Recent progress in computational and applied PDEs (Zhangjiajie, 2001), 383–397, Kluwer/Plenum, New York, 2002.

(41) Yang, Z. H, Wei, J. Q. and Xiong, B. (2003). Computation of higher-order singular points in nonlinear problems with single parameter. (in Chinese) *Comm. Appl. Math. Comput.* **17**, pp. 1–6.

(42) Li, C. P. and Yang, Z. H. (2004). Symmetry-breaking bifurcation in $O(2) \times O(2)$-symmetric nonlinear large problems and its application to the Kuramoto-Sivashinsky equation in two spatial dimensions, *Chaos Solitons Fractals* **22**, pp. 451–468.

(43) Yang, Z. H. and Guo, Q. (2005). Bifurcation analysis of delayed logistic equation, *Appl. Math. Comput.* **167**, pp. 454–476.

(44) Li, C. P., Yang, Z. H. and Chen, G. R. (2005). On bifurcation from steady-state solutions to rotating waves in the Kuramoto-Sivashinsky equation, *J. Shanghai Univ.* **9**, pp. 286–291.

(45) Yang, Z. H. and Zhou, W. (2005). Bifurcation analysis and computation of double Takens-Bogdanov point in Z_2-equivariable nonlinear equations, *Numer. Math. J. Chinese Univ.* (English Ser.) **14**, pp. 315–324.

(46) Ji, Q. B., Lu, Q. S. and Yang, Z. H. (2007). Computation of D_8-equivariant nonlinear bifurcation problems, *Dyn. Contin. Discrete Impuls. Syst. Ser. B Appl. Algorithms* **14**, suppl. S5, pp. 17–20.

(47) Yang, Z. H, Li, Z. X. and Zhu, H. L. (2008). Bifurcation method for solving multiple positive solutions to Henon equation, *Sci. China Ser. A* **51**, pp. 2330–2342.

(48) Wei, J. Q. and Yang, Z. H. (2009). Approximation to butterfly catastrophe, (in Chinese) *Gongcheng Shuxue Xuebao* **26**, pp. 94–98.

(49) Wei, J. Q. and Yang, Z. H. (2009). Fourier collocation method for a class of reaction-diffusion equations, (in Chinese) *Numer. Math. J. Chinese Univ.* **31**, pp. 232–239.

(50) Li, Z. X. and Yang, Z. H. (2010). Bifurcation method for solving multiple positive solutions to boundary value problem of p-Henon equation on the unit disk, *Appl. Math. Mech.* (English Ed.) **31**, pp. 511–520.

(51) Li, Z. X., Yang, Z. H. and Zhu, H. L. (2010). Computing multiple positive solutions to the p-Henon equation on a square, (in Chinese) *J. Numer. Methods Comput. Appl.* **31**, pp. 161–171.
(52) Li, Z. X., Yang, Z. H. and Zhu, H. L. (2010). Computing the multiple solutions to boundary value problem of p-Henon equation on the disk of plane, *Int. J. Comp. Math. Sci.* **4**, pp. 137-139.
(53) Li, Z. X., Zhu, H. L. and Yang, Z. H. (2011). Bifurcation method for solving multiple positive solutions to Henon equation on the unit cube, *Commun. Nonlinear Sci. Numer. Simul.* **16**, pp. 3673–3683.
(54) Li, Z. X., Yang, Z. H. and Zhu, H. L. (2011). Bifurcation method for solving multiple positive solutions to boundary value problem of Henon equation on unit disk, *Comput. Math. Appl.*, **62**, pp. 3775-3784.

Contents

Chapter 1

Gronwall inequalities[1]

Fanhai Zeng, Jianxiong Cao and Changpin Li*

Department of Mathematics, Shanghai University, Shanghai 200444, PR China

**lcp@shu.edu.cn*

In this chapter, we display the existing continuous and discrete Gronwall type inequalities, including their modifications such as the weakly singular Gronwall inequalities which are very useful to study the fractional integral equations and the fractional differential equations.

Keywords: Gronwall inequality, weakly singular Gronwall inequality

1.1 Introduction

It is well known that Gronwall–Bellman type integral inequalities play important roles in the study of existence, uniqueness, continuation, boundedness, oscillation and stability properties to the solutions of differential and integral equations. In 1919, Gronwall first introduced the famous Gronwall inequality in the study of the solution of the differential equation. Since then, a lot of contributions have been achieved by many researchers. The original Gronwall inequality has been extended to the more general case, including the generalized linear and nonlinear Gronwall type inequalities [Bihari (1956); Willett (1964); Bainov and Simenov (1992); Pachpatte (2002a)], the two and more variables cases [Beckenbach and Bellman (1961); Pachpatte (2002a); Snow (1971); Yeh (1980, 1982b); Bondge and Pachpatte (1979)], and the Gronwall type inequalities for discontinuous functions [Samoilenko and Borysenko (1998); Borysenko and Iovane (2007); Galloa and Piccirillo (2007)]. At the same time, the discrete analogues have also been derived [Yang (1983, 1988); Zhou and Zhang (2010); Salem and Raslan (2004)]. Meanwhile, some useful results of the weakly singular Gronwall inequalities have been established as well [Mckee (1982); Dixo and Mckee (1986); Medveď (1997); Denton and Vatsala (2010)], which are

[1]The present work was supported by the National Natural Science Foundation of China (grant no. 10872119), the Shanghai Leading Academic Discipline Project (grant no. S30104), and the Key Program of Shanghai Municipal Education Commission (grant no. 12ZZ084).

the powerful tools in the theoretical analysis of the integral equations with weakly singular kernels and the fractional differential equations.

In the present chapter, we collect almost all the important existing Gronwall type inequalities, which include the continuous cases, discrete cases, weakly singular cases and their discrete analogues. If some important references happened not to be here, we do apologize for these omissions.

The rest of this chapter is outlined as follows. In Section 1.2, we introduce the continuous Gronwall inequalities. Then we present the discrete Gronwall inequalities in Section 1.3. In Section 1.4, the weakly singular Gronwall integral inequalities and some of their discrete analogues are displayed. And the conclusions are included in the last section.

1.2 The continuous Gronwall inequalities

In this section,we state some continuous integral inequalities of Gronwall type, which can be used in the analysis of various problems in the theory of the nonlinear differential equations and the integral equations.

In 1919, Gronwall first proved the following famous inequality, which is called the Gronwall inequality .

Theorem 1.1 (Gronwall Inequality [Gronwall (1919)]). *Let $u(t)$ be a continuous function defined on the interval $[t_0, t_1]$ and*

$$u(t) \leq a + b \int_{t_0}^{t} u(s)\,\mathrm{d}s, \tag{1.1}$$

where a and b are nonnegative constants. Then we have

$$u(t) \leq a e^{b(t-t_0)}, t \in [t_0, t_1]. \tag{1.2}$$

After more than 20 years, Bellman extended the original Gronwall inequality, which reads in the following theorem.

Theorem 1.2 (Bellman Inequality [Bellman (1943)]). *Let a be a positive constant, $u(t)$ and $b(t), t \in [t_0, t_1]$ be real-valued continuous functions, $b(t) \geq 0$, satisfying*

$$u(t) \leq a + \int_{t_0}^{t} b(s)u(s)\,\mathrm{d}s, \ \forall t \in [t_0, t_1]. \tag{1.3}$$

Then we have

$$u(t) \leq a \exp\left\{ \int_{t_0}^{t} b(s)\,\mathrm{d}s \right\}, \ \forall t \in [t_0, t_1]. \tag{1.4}$$

Bellman also proved that if $u(t)$ and $b(t)$ are continuous functions, $b(t)$ is a nonnegative function, and $u(t) \leq u(t_0) + \int_{t_0}^{t} b(s)u(s)\,\mathrm{d}s, t \in (t_0, t_1)$, then

$$u(t_0) \exp\left\{ -\int_{t_0}^{t} b(s)\,\mathrm{d}s \right\} \leq u(t) \leq u(t_0) \exp\left\{ \int_{t_0}^{t} b(s)\,\mathrm{d}s \right\}, \quad t \in (t_0, t_1).$$

The somewhat more general extensions of the original Gronwall inequality can be found in [Mitrinovic *et al.* (1991); Kuang (2010)].

Theorem 1.3 (Beesack inequality). *Let $a(t), q(t) \in L[t_0, t_1]$, $u(t)$ and $b(t), t \in [t_0, t_1]$, be real valued continuous functions, and $b(t)$ and $q(t)$ be nonnegative functions, satisfying*

$$u(t) \le a(t) + q(t) \int_{t_0}^{t} b(s)u(s)\,\mathrm{d}s, \ \forall t \in [t_0, t_1], \tag{1.5}$$

or

$$u(t) \ge a(t) + q(t) \int_{t_0}^{t} b(s)u(s)\,\mathrm{d}s, \ \forall t \in [t_0, t_1]. \tag{1.6}$$

Then we have

$$u(t) \le a(t) + q(t) \int_{t_0}^{t} a(s)b(s) \exp\Big\{ \int_{s}^{t} q(r)b(r)\,\mathrm{d}r \Big\}\,\mathrm{d}s, \ \forall t \in [t_0, t_1], \tag{1.7}$$

or

$$u(t) \ge a(t) + q(t) \int_{t_0}^{t} a(s)b(s) \exp\Big\{ \int_{s}^{t} q(r)b(r)\,\mathrm{d}r \Big\}\,\mathrm{d}s, \ \forall t \in [t_0, t_1], \tag{1.8}$$

respectively.

In [Pachpatte (2006)], another bound for $u(t)$ was derived: If $u(t)$ satisfies (1.5), $a(t), b(t), q(t)$ and $u(t)$ are all continuous on $[t_0, t_1]$, then

$$u(t) \le a(t) + q(t) \Big\{ \frac{1}{1-p} \int_{t_0}^{t} a(s)b(s)\,\mathrm{d}s \Big\},$$

where $p = \int_{t_0}^{t} q(s)b(s)\,\mathrm{d}s < 1$.

In 1967, Chu and Metcalf [Chu and Metcalf (1967)] got the following generalized result.

Theorem 1.4. *Let the functions $u(t)$ and $f(t)$ be continuous on the interval $[0, 1]$; let the function $K(t, s)$ be continuous and nonnegative on the triangle $0 \le s \le t \le 1$. If*

$$u(t) \le f(t) + \int_{0}^{t} K(t, s)u(s)\,\mathrm{d}s, \quad 0 \le t \le 1,$$

then

$$u(t) \le f(t) + \int_{0}^{t} H(t, s)f(s)\,\mathrm{d}s, \quad 0 \le t \le 1,$$

where $H(t, s) = \sum_{i=1}^{\infty} K_i(t, s), 0 \le s \le t \le 1$ is the resolvent kernel, and the $K_i (i = 1, 2, ...)$ are the iterated kernels of K.

In 1988, Pirapakaran [Pirapakaran (1988)] got the following inequality:

$$u(t) \le c \prod_{t_0<t_k<t} (1+\beta_k) \exp\left\{ \int_{t_0}^{t} g(s)\,\mathrm{d}s \right\}, \quad t_0 \le t,$$

if $u(t)$ and $g(t)$ are continuous functions on $(0,\infty)$, β_k and c are nonnegative constants, $\{t_n\}_{n=1}^{\infty}$ are increasing serous with $t_n \to \infty, t_n > 0$, and

$$u(t) \le c + \int_{t_0}^{t} g(s)u(s)\,\mathrm{d}s + \sum_{t_0<t_k<t} \beta_k u(t_k).$$

Theorem 1.5 (BS Inequality [Bainov and Simenov (1992)]). *Let $u(t), a(t)$ and $b(t)$ be continuous functions in $I = [\alpha, \beta]$, let $a(t)$ and $b(t)$ be nonnegative in I, and suppose*

$$u(t) \le c + \int_{\alpha}^{t} a(s)u(s)\,\mathrm{d}s + \int_{\alpha}^{\beta} b(s)u(s)\,\mathrm{d}s, \tag{1.9}$$

for $t \in I$, where c is a nonnegative constant. If

$$q = \int_{\alpha}^{\beta} b(s) \exp\left\{ \int_{\alpha}^{s} a(\sigma)\,\mathrm{d}\sigma \right\} \mathrm{d}s < 1,$$

then

$$u(t) \le \frac{c}{1-q} \exp\left\{ \int_{t}^{\infty} a(s)\,\mathrm{d}s \right\}, \ \forall t \in I. \tag{1.10}$$

An extension for the above Theorem 1.5 was obtained in [Pachpatte (2002b); Chou and Yang (2005a)]. The functions $a(t)$ and $b(t)$ in the condition (1.9) are replaced by $a(t,s)$ and $b(t,s), \alpha \le s \le t \le \beta$ in [Pachpatte (2002b)], and the similar result as (1.10) was derived. While in [Chou and Yang (2005a)], the constant c in the condition (1.9) was replaced by $c(t)$, and the derived inequality was slightly different form (1.10). In both [Pachpatte (2002b)] and [Chou and Yang (2005a)], the discrete analogues were established.

In 2002, Pachpatte [Pachpatte (2002a)] proved the following inequality and used it in the study of terminal value problems for certain differential equations.

Theorem 1.6 ([Pachpatte (2002a)]). *Let $u(t), a(t)$ and $b(t)$ be real-valued nonnegative continuous functions defined for $t \in (0,\infty)$, and suppose that $a(t)$ is nonincreasing for t. If*

$$u(t) \le a(t) + \int_{t}^{\infty} b(s)u(s)\,\mathrm{d}s, \ \forall t > 0, \tag{1.11}$$

then we have

$$u(t) \le a(t) \exp\left\{ \int_{t}^{\infty} b(s)\,\mathrm{d}s \right\}, \ \forall t > 0. \tag{1.12}$$

Pachpatte also extended Theorem 1.6 in the more general situations, which are given in the following three Theorems.

Theorem 1.7 ([Pachpatte (2002a)]). *Let $u(t), a(t)$ and $b(t)$ be as in Theorem 1.6 and $L : \mathbb{R}_+^2 \to \mathbb{R}_+ (\mathbb{R}_+ = [0, \infty))$ be a continuous function which satisfies the condition*

$$0 \le L(t,u) - L(t,v) \le M(t,u)(u-v),$$

for $u \ge v \ge 0$, where $M(t,v)$ is a real-valued nonnegative continuous function defined for $t, v \in \mathbb{R}_+$. If

$$u(t) \le a(t) + \int_t^\infty b(s)u(s)\,\mathrm{d}s + \int_t^\infty L(s,u(s))\,\mathrm{d}s, \ \forall t \ge 0, \tag{1.13}$$

then we have

$$u(t) \le E(t)\Big\{a(t) + A(t)\exp\Big(\int_t^\infty M(s, E(s)a(s))E(s)\,\mathrm{d}s\Big)\Big\}, \ \forall t \ge 0, \tag{1.14}$$

where

$$E(t) = \exp\Big\{\int_t^\infty b(s)\,\mathrm{d}s\Big\}, \forall t \ge 0,$$

$$A(t) = \int_t^\infty L(s, E(s)a(s))\,\mathrm{d}s, \forall t \ge 0.$$

Theorem 1.8 ([Pachpatte (2002a)]). *Let $u(t), a(t)$ and $b(t)$ be real-valued nonnegative continuous functions defined for $t \in \mathbb{R}_+$ and L, M be as in Theorem 1.7. If*

$$u(t) \le a(t) + b(t)\int_t^\infty L(s,u(s))\,\mathrm{d}s, \ \forall t > 0, \tag{1.15}$$

then we have

$$u(t) \le a(t) + \Big(\int_t^\infty L(s,a(s))\,\mathrm{d}s\Big)\exp\Big(\int_t^\infty M(s,a(s))b(s)\,\mathrm{d}s\Big), \ \forall t \ge 0, \tag{1.16}$$

Theorem 1.9 ([Pachpatte (2002a)]). *Let $u(t), a(t)$ and $b(t)$ be as in Theorem 1.8 and $L : \mathbb{R}_+^2 \to \mathbb{R}_+$ be a continuous function which satisfies the condition*

$$0 \le L(t,u) - L(t,v) \le M(t,u)\phi^{-1}(u-v),$$

for $u \ge v \ge 0$, where $M(t,u)$ is defined in Theorem 1.7, $\phi : \mathbb{R}_+ \to \mathbb{R}_+$ be a continuous and strictly increasing function with $\phi(0) = 0$, ϕ^{-1} is the inverse function of ϕ and

$$\phi^{-1}(uv) \le \phi^{-1}(u)\phi^{-1}(v), \quad u, v \in \mathbb{R}_+.$$

If

$$u(t) \le a(t) + b(t)\phi\Big(\int_t^\infty L(s,u(s))\,\mathrm{d}s\Big), \ \forall t \ge 0, \tag{1.17}$$

then we have

$$u(t) \le a(t) + b(t)\phi\Big(\Big\{\int_t^\infty L(s,a(s))\,\mathrm{d}s\Big\}\exp\Big\{\int_t^\infty M(s,a(s))\phi^{-1}(b(s))\,\mathrm{d}s\Big\}\Big), \ \forall t \ge 0. \tag{1.18}$$

For the two independent variable versions of the Theorems 1.6–1.9, see [Pachpatte (2002a)] for more details.

Theorem 1.10 (B–L Inequality [Bihari (1956)]). *Let $u(t)$ and $f(t)$ be positive continuous functions in $[t_0, t_1]$ and $a, b \geq 0$, further $w(u)$ be a non-negative non-decreasing continuous function for $u \geq 0$. Then the inequality*

$$u(t) \leq a + b \int_{t_0}^{t} f(s)w(u(s))\,\mathrm{d}s, \quad t_0 \leq t \leq t_1 \tag{1.19}$$

implies the inequality

$$u(t) \leq G^{-1}\Big(G(a) + b\int_{t_0}^{t} f(s)\,\mathrm{d}s\Big), \quad t_0 \leq t \leq t_1, \tag{1.20}$$

where

$$G(u) = \int_{u_0}^{u} \frac{\mathrm{d}s}{w(s)}, \quad u_0 > 0, u \geq 0.$$

Lipovan [Lipovan (2000)] extended the above B–L inequality, in which the inequality (1.19) is replaced by $u(t) \leq a + \int_{\alpha(t_0)}^{\alpha(t)} f(s)w(u(s))\,\mathrm{d}s$, where $\alpha(t) \in C^1([t_0, t_1))$ is a nondecreasing function, $\alpha(t) \leq t$ on $[t_0, t_1)$, and the similar result as (1.20) was obtained. In [Györi (1971)], Györi extended the B–L inequality by replacing the condition (1.19) with $u(t) \leq a(t) + b(t)\int_{t_0}^{t} f(s)w(u(s))\,\mathrm{d}s$, and the slight different result was also derived. Another simple generalization of the Gronwall's inequality by using the Viswanatham's Theorem can be found in [Viswanatham (1963)]. For more information, see [Györi (1971); Mitrinovic *et al.* (1991); Viswanatham (1963); Kuang (2010)] and the references cited therein.

There are other extensions of the Gronwall type inequalities, such as in [Pachpatte (1995a)]: Let u, f, g be nonnegative continuous functions on $\mathbb{R}^1_+$, $c_1, c_2 \geq 0$. If

$$u(t) \leq \Big(c_1 + \int_0^t f(s)u(s)\,\mathrm{d}s\Big)\Big(c_2 + \int_0^t g(s)u(s)\,\mathrm{d}s\Big), \quad \forall t \geq 0,$$

and

$$c_1c_2\int_0^t F(s)G(s)\,\mathrm{d}s < 1, \quad \forall t \geq 0,$$

then

$$u(t) \leq \frac{c_1c_2G(t)}{1 - c_1c_2\int_0^t F(s)G(s)\,\mathrm{d}s},$$

where

$$F(t) = g(t)\int_0^t f(s)\,\mathrm{d}s + f(t)\int_0^t g(s)\,\mathrm{d}s, \quad G(t) = \exp\Big(\int_0^t (c_1g(s) + c_2f(s))\,\mathrm{d}s\Big).$$

In [Oguntuase (1998)], Oguntuase proved that: If

$$u(t) \leq a(t) + \Big(\int_{t_0}^{t} K(t,s)[u(s)]^p\,\mathrm{d}s\Big)^{1/p},$$

then

$$u(t) \leq a(t) + \frac{\left(\int_{t_0}^{t} v(s)[a(s)]^p b(s)\,\mathrm{d}s\right)^{1/p}}{1-(1-v(t))^{1/p}},$$

where $v(t) = \exp\left(-\int_{t_0}^{t} K(t,s)\,\mathrm{d}s\right), b(t) = K(t,t) + \int_{t_0}^{t} K(t,s)\,\mathrm{d}s$ for $t_0 \leq s \leq t \leq t_1, p \geq 1$.

In [Abdeldaim and Yakout (2011)], Abdeldaim and Yakout proved the following theorem.

Theorem 1.11 ([Abdeldaim and Yakout (2011)]). *We assume that $x(t), f(t)$ and $h(t)$ are nonnegative real-valued continuous functions defined on I, and satisfy the inequality*

$$x^p(t) \leq \int_0^t f(s)x^p(s)\,\mathrm{d}s + \int_0^t h(s)x^q(s)\,\mathrm{d}s, \quad t > 0,$$

where $x_0 > 0$, and $p > q \geq 0$, are constants. Then

$$x(t) \leq \exp\left(\frac{1}{p}\int_0^t f(s)\,\mathrm{d}s\right)\left[x_0^{p_1} + p_1\int_0^t h(s)\exp\left(-p_1\int_0^s f(\lambda)\,\mathrm{d}\lambda\right)\mathrm{d}s\right]^{\left[\frac{1}{p-q}\right]}, \quad (1.21)$$

for all $t > 0$, where $p_1 = \left[\frac{p-q}{p}\right]$.

In Theorem 1.11, if $h = 0$ and $p = 1$, the inequality in Theorem 1.11 reduces to the well-known Gronwall inequality; if $p = 1$, it reduces to Willett and Wong inequality [Pachpatte (1998a)]; if $q = 1$ and $f = 0$, it reduces to the El-Owaidy, Ragab and Abdeldaim inequality [El-Owaidy *et al.* (1999)]; if $p = 2, q = 1$ and $f = 0$, it reduces to the well known Ou-Inag inequality [Pachpatte (1998a)]. Abdeldaim and Yakout also established other new nonlinear integral inequalities of Gronwall–Bellman–Pachpatte type in [Abdeldaim and Yakout (2011)], which are useful to study the qualitative and the quantitative properties of solutions of some nonlinear ordinary differential and integral equations, see [Abdeldaim and Yakout (2011)] and the references therein for more information.

Theorem 1.12 ([Dafermos (1979)]). *Assume that the nonnegative functions $u(t) \in L^\infty[0,b]$ and $g(t) \in L^1[0,b]$ satisfy the inequality*

$$u^2(t) \leq M^2u^2(0) + \int_0^t \left(2\alpha u^2(s) + 2Ng(s)u(s)\right)\mathrm{d}s, \quad t \in [0,b], \quad (1.22)$$

where α, M, N are nonnegative constants. Then

$$u(t) \leq M\exp(\alpha t)u(0) + N\exp(\alpha t)\int_0^b g(s)\,\mathrm{d}s. \quad (1.23)$$

If $\alpha = 0$, the inequality in Theorem 1.12 reduces to the result in [Ou-Yang (1957)]. In [Pachpatte (1995b); Chou and Yang (2005b)], the inequalities related to Theorem 1.12, which can be seen extensions of Theorem 1.12, were provided, and the corresponding discrete analogues were presented in [Pachpatte (1995b)], while in [Chou and Yang (2005b)], two independent variable versions were established. We do not list these results, and readers can refer to [Pachpatte (1995b); Chou and Yang (2005b)] and the references therein.

Next, we introduce the Gronwall type inequality with two independent variables. We just list the following two inequalities, which can be found in [Beckenbach and Bellman (1961); Pachpatte (2002a)].

Theorem 1.13 (Wendroff Inequality [Beckenbach and Bellman (1961)]). *Let $a(x), b(y) > 0$, $a'(x), b'(y) \geq 0$, $u(x,t)$, $v(x,y) \geq 0$. If*

$$u(x,y) \leq a(x) + b(y) + \int_0^x \int_0^y v(t,s)u(t,s)\,\mathrm{d}t\,\mathrm{d}s, \tag{1.24}$$

then

$$u(x,t) \leq \frac{(a(0)+b(y))(a(x)+b(0))}{a(0)+b(0)} \exp\Big(\int_0^x \int_0^y v(t,s)\,\mathrm{d}t\,\mathrm{d}s\Big). \tag{1.25}$$

Theorem 1.14 ([Pachpatte (2002a)]). *Let $u(x,y), a(x,y), b(x,y)$ be real-valued nonnegative continuous functions defined for $x, y > 0$ and suppose that $a(x,y)$ is nonincreasing in $x > 0$ and $y > 0$. If*

$$u(x,y) \leq a(x,y) + \int_x^\infty \int_y^\infty b(s,t)u(s,t)\,\mathrm{d}t\,\mathrm{d}s, \quad x, y > 0, \tag{1.26}$$

then

$$u(x,t) \leq a(x,y) \exp\Big(\int_x^\infty \int_y^\infty b(s,t)\,\mathrm{d}t\,\mathrm{d}s\Big), \quad x, y > 0. \tag{1.27}$$

In 1971, Nurimov extended Theorem 1.13 to a more general case, see [Kuang (2010)] for details and for other corresponding versions. In [Pachpatte (2002a)], Theorem 1.14 is also extended to a more general form as Theorems 1.7–1.9 by S. B. Pachpatte and B. G. Pachpatte. For more types of Wendroff type inequalities, readers can refer to [Abdeldaim and Yakout (2010); András and Mészáros (2011); Shastri and Kasture (1978); Yeh (1980)] and the references referred therein. For more Gronwall types inequalities with two or more variables, see [Agarwal (1982); Pachpatte (1979); Borysenko (1989); Bondge and Pachpatte (1979); Snow (1971, 1972); Thandapani and Agarwall (1982); Yeh (1982a,b); Yeh and Shih (1982); Young (1973)] and the reference therein, here we omit these results. Some new nonlinear integral inequalities for discontinuous functions with two independent variables (Wendroff type) by including also inequalities with delay were derived in [Borysenko and Iovane (2007)], and some new integral Gronwall-Bellman-Bihari type inequalities for discontinuous functions (integro-sum inequalities) were presented in [Galloa and Piccirillo (2007)]. See [Borysenko and Iovane (2007); Samoilenko and Borysenko (1998)] and the references therein for more details.

1.3 The discrete Gronwall inequalities

In this section, we introduce the discrete Gronwall inequalities The discrete Gronwall inequalities are useful tools in the numerical analysis of differential equations. We first give the following classical discrete Gronwall inequality, which can be found in [Li (1996)].

Theorem 1.15 (Gronwall Inequality [Li (1996)]). *Let $a, b \geq 0$ and η_i satisfy*

$$|\eta_n| \leq b + ah \sum_{i=0}^{n-1} |\eta_i|, n = k, k+1, ..., nh \leq T, \tag{1.28}$$

then

$$|\eta_n| \leq e^{aT}(b + akhM_0), n \geq k, nh \leq T, \tag{1.29}$$

where $M_0 = \max(|\eta_0|, |\eta_1|, ..., |\eta_{k-1}|)$.

A simple generalization of the Theorem 1.15 can be found in [Bohner (2001)].

Theorem 1.16 ([Bohner (2001)]). *Let $a, v \in R$ be given. If y and f are functions defined on N_{v+a}, and $r > 0$ is a constant such that*

$$y(t) \leq f(t) + r \sum_{\tau=v-1}^{t-1} y(\tau) \tag{1.30}$$

for all $t \in N_{v+a}$, then

$$y(t) \leq f(t) + r \sum_{\tau=v-1}^{t-1} f(\tau)(1+\tau)^{t-\tau-1}, \tag{1.31}$$

where $N_{v+a} = \{a+v, a+v+1, ...\}$, $v \in R$.

In 1969, Sugiyama [Sugiyama (1969)] established the following Gronwall Inequality, which also can be found in [Pachpatte (1977)].

Theorem 1.17 ([Sugiyama (1969)]). *Let $x(n)$ and $f(n)$ be real-valued functions defined for $n \in N$,and suppose that $f(n) \geq 0$ for every $n \in N_{n_0}$. If*

$$x(n) \leq x_0 + \sum_{s=n_0}^{n-1} f(s)x(s), \quad n \in N, \tag{1.32}$$

where N_{n_0} is the set of points $n_0 + k(k = 0, 1, 2, ...), n_0 \geq 0$ is a given integer and x_0 is a nonnegative constant, then

$$x(n) \leq x_0 \prod_{s=n_0}^{n-1} \Big(1 + f(s)\Big), \quad n \in N_{n_0}. \tag{1.33}$$

Pachpatte [Pachpatte (1977)] extended the above inequality to a more general form, interesting readers can refer to [Pachpatte (1977)]. Next we give the more general types of discrete Gronwall inequalities in the following two theorems, which can be found in [Heywood and Rannacher (1990)] and [Quarteroni and Valli (1994)], respectively.

Theorem 1.18 ([Heywood and Rannacher (1990)]). *Let* $\Delta t, H$ *and* a_n, b_n, c_n, r_n, *for integers* $n \geq 0$, *be nonnegative numbers such that*

$$a_m + \Delta t \sum_{n=0}^{m} b_n \leq \Delta t \sum_{n=0}^{m} r_n a_n + \Delta t \sum_{n=0}^{m} c_n + H, \quad m \geq 0. \tag{1.34}$$

Suppose that $r_n \Delta t < 1$ *for all* n, *set* $\sigma_n = (1 - r_n \Delta t)^{-1}$. *Then*

$$a_m + \Delta t \sum_{n=0}^{m} b_n \leq \exp\Big(\Delta t \sum_{n=0}^{m} \sigma_n r_n\Big)\Big(\Delta t \sum_{n=0}^{m} c_n + H\Big), \quad m \geq 0. \tag{1.35}$$

Theorem 1.19 ([Quarteroni and Valli (1994)]). *Assume that* k_n *is a nonnegative sequence, and the sequence* ϕ_n *such that*

$$\begin{cases} \phi_0 \leq g_0, \\ \phi_n \leq g_0 + \sum_{j=0}^{n-1} p_j + \sum_{j=0}^{n-1} k_j \phi_j, \quad n \geq 1. \end{cases} \tag{1.36}$$

Then ϕ_n *satisfies*

$$\begin{cases} \phi_1 \leq g_0(1 + k_0) + p_0, \\ \phi_n \leq g_0 \prod_{j=0}^{n-1}(1 + k_j) + \sum_{j=0}^{n-2} p_j \prod_{r=j+1}^{n-1} (1 + k_r) + p_{n-1}, \quad n \geq 2. \end{cases} \tag{1.37}$$

Moreover, if $g_0 \geq 0$ *and* $p_n \geq 0$ *for* $n \geq 0$, *it follows*

$$\phi_n \leq \Big(g_0 + \sum_{j=0}^{n-1} p_j\Big) \exp\Big(\sum_{j=0}^{n-1} k_j\Big), \quad n \geq 1. \tag{1.38}$$

In 2010, Zhou and Zhang [Zhou and Zhang (2010)] generalized a projected discrete Gronwall's inequality given in [Matsunaga and Murakami (2004)] to a general one, which may include both terms of sub-exponential growth inside the summation and non-monotonic terms outside the summation, the main result is displayed in the following theorem.

Theorem 1.20 ([Zhou and Zhang (2010)]). *Suppose that*
(1) $a : \mathbb{Z}_+ \to \mathbb{R}_+$ *is bounded and* $a_* = \inf_{n \geq s_0} a(n)$, s_0 *is a nonnegative integer,*
(2) *the functions* $b(n, s)$ *and* $c(n, s)$ *are both defined for all integers* $0 \leq s \leq n < \infty$, *and both are nonnegative and*

(3) $\eta(n) = \sum_{s=s_0}^{n-1} b(n,s) + \sum_{s=n}^{\infty} c(n,s)$ *is well defined for all* $s_0 \le n$ *and* $\eta_0 = \sup_{n \ge s_0} \eta(n) < 1$. *If* $u(n)$ *satisfies the following inequality*

$$u(n) \le a(n) + \sum_{s=s_0}^{n-1} b(n,s)u(s) + \sum_{s=n}^{\infty} c(n,s)u(s), \quad n \ge s_0, \tag{1.39}$$

then

$$u(n) \le \frac{\tilde{a}(n)}{1-\eta_0} + \frac{1}{(1-\eta_0)^2} \sum_{s=s_0}^{n-1} \Big(\prod_{\tau=s+1}^{n-1} \xi(\tau) \Big) (\tilde{a}(s) - a_*) \tilde{b}(s+1,s), \quad n \ge s_0, \tag{1.40}$$

where $\tilde{a}(n) = \sup_{\tau \ge a(\tau)} a(\tau)$, $\tilde{b}(n,s) = \sup_{\tau \ge n} \tilde{b}(\tau,s)$, $\xi(n) = r(n) + \tilde{b}(n+1,n)/(1-\eta_0)$, $r(n) = \max_{s_0 \le s \le n} B(n,s)$, and

$$B(n,s) = \begin{cases} \tilde{b}(n+1,s)/\tilde{b}(n,s), & \tilde{b}(n,s) \ne 0, \\ 0, & \tilde{b}(n,s) = 0. \end{cases}$$

Now, we introduce several nonlinear generalizations of the Gronwall inequality. In 1965, Willett and Wong [Willett and Wong (1965)] proved the following inequality

Theorem 1.21. *Let* a_n, b_n, c_n *and* $x_n (n = 1, 2, ...)$ *be non-negative real numbers, and let* $r \ge 1$ *be a real number. If*

$$x_n \le a_n + b_n \Big(\sum_{i=1}^{n-1} c_i x_i^r \Big)^{1/r} \quad (n = 1, 2, ...),$$

then

$$\sum_{i=1}^{n} c_i x_i^r \le \Big[1 - (1-p_n)^{1/r} \Big]^r \sum_{i=1}^{n} a_i^r c_i p_i \quad (n = 1, 2, ...),$$

where

$$p_k = \prod_{i=1}^{k} (1 + b_i^r c_i)^{-1} \quad (k = 1, 2, ...).$$

The continuous form of the above inequality was established by Willett [Willett (1964)] in early 1964. A variant of the above theorem for the case that $r \in (0,1]$ was derived in [Alzer (1996)] by Alzer.

In 2011, Roshdy and Mousa [Roshdy and Mousa (2011)] derived the following result.

Theorem 1.22 ([Roshdy and Mousa (2011)]). *Let* $u(n)$,$f(n)$ *and* $g(n)$ *be real-valued nonnegative functions defined on* $\mathbb{N}$, *for which the inequality*

$$u(n) \le u_0 + \sum_{s=n_0}^{n-1} f(s)u(s) + \sum_{s=n_0}^{n-1} f(s) \Big(\sum_{t=n_0}^{s-1} g(t) u^{\alpha}(t) \Big), \quad n \in \mathbb{N} \tag{1.41}$$

holds, where u_0 *is a nonnegative constant and* $0 \le \alpha \le 1$. *Then*

$$u(n) \le u_0 + \sum_{s=n_0}^{n-1} f(t) \prod_{t=n_0}^{s-1} \Big[1 + f(\tau)\Big] \Big[u_0^{1-\alpha} + (1-\alpha) \sum_{\tau=n_0}^{s-1} g(\tau) \prod_{t=n_0}^{\tau} [1 + f(t)]^{\alpha-1}\Big]^{\frac{1}{t-\alpha}} \tag{1.42}$$

for all $n \in \mathbb{N}$.

Roshdy and Mousa also gave some other general forms, see [Roshdy and Mousa (2011)] and the references therein for details.

In 1998, Pachpatte[Pachpatte (1998b)] proved the following inequality and use it in the study of finite difference equations.

Theorem 1.23 ([Pachpatte (1998b)]). *Let* $u(t), a(t), b(t), h(t)$ *be real-valued nonnegative functions and let* c *be a nonnegative constant,* t *is also a nonnegative constant. If*

$$u^2(t) \le c^2 + \sum_{s=0}^{t-1} [u(s+1) + u(s)][a(s)u(s) + h(s)], \tag{1.43}$$

then

$$u(t) \le p(t) \prod_{s=0}^{t-1} \Big[1 + a(s)\Big], \tag{1.44}$$

where

$$p(t) = c + \sum_{s=0}^{t-1} h(s). \tag{1.45}$$

In [Pachpatte (1998b)], variable variants of the inequality in the above Theorem 1.23 were discussed, see [Pachpatte (1998b)] for more information.

Next,we introduce the Gronwall's inequality of discrete type in two and more than two independent variables. We just list the following two inequality

Theorem 1.24 ([Salem and Raslan (2004)]). *Let* u, a, b *be nonnegative functions and* a *nondecreasing.If*

$$u(m,n) \le a(m,n) + \sum_{s=0}^{m-1} \sum_{t=0}^{n-1} b(s,t) u(s,t), \tag{1.46}$$

then

$$u(m,n) \le a(m,n) \prod_{t=0}^{n-1} \Big[1 + \sum_{s=0}^{m-1} b(s,t)\Big]. \tag{1.47}$$

Some other nonlinear discrete inequalities in two independent variables were also established in [Salem and Raslan (2004)], and we do note list the results.

Theorem 1.25 ([Cheung and Ren (2006)]). *Suppose u, b are positive functions on Ω, $c \geq 0, \alpha > 0$, and $\varphi \in C(\mathbb{R}_+, \mathbb{R}_+)$. If φ is non-decreasing with $\varphi(r)$ for $r > 0$ for any $(m,n) \in \Omega$,*

$$u^\alpha(m,n) \leq c + \sum_{s=m_0}^{m-1} \sum_{t=n_0}^{n-1} b(s,t)\varphi(u(s,t)), \tag{1.48}$$

then

$$u(m,n) \leq \left\{\phi_\alpha^{-1}[\phi_\alpha(c) + B(m,n)]\right\}^{\frac{1}{\alpha}}, (m,n)\in[m_0, m_1] \times [n_0, n_1] \cap \Omega, \tag{1.49}$$

where

$$B(m,n) = \sum_{s=m_0}^{m-1} \sum_{t=n_0}^{n-1} b(s,t),$$

ϕ_α^{-1} is the inverse of ϕ_α, and $(m_1, n_1) \in \Omega$ is chosen such that $\phi_\alpha(c) + B(m,n) \in \mathrm{Dom}(\phi_\alpha^{-1})$ for all $(m,n) \in [m_0, m_1] \times [n_0, n_1] \cap \Omega$.

In 2008, Ma [Ma (2008)] established a class of new nonlinear Volterra-Fredholm-type discrete inequalities to generalize Ou-Iang's inequality. Readers can see [Ou-Yang (1957); Pachpatte (1995b); Cheung (2004); Cheung and Ren (2006); Ma (2008)] and the references therein for more information about Gronwall-Bellman-Ou-Iang-type inequalities. For the other discrete analogues of Gronwall type in two and more variables, see [Pachpatte and Singare (1979); Popenda and Agarwal (1999); Feng *et al.* (2011); Yeh (1985a,b)] and the references therein, we omit them here.

1.4 The weakly singular Gronwall inequalities

In this section, we introduce the weakly singular Gronwall type integral inequalities and their discrete analogues. The following inequality can be found in [Dixo and Mckee (1986)].

Theorem 1.26 ([Dixo and Mckee (1986)]). *Let $u(t)$ be continuous and non-negative on $[0,T]$. If*

$$u(t) \leq \varphi(t) + M \int_0^t \frac{u(s)}{(t-s)^\alpha} ds, \quad 0 \leq t \leq T, \tag{1.50}$$

where $0 \leq \alpha < 1$, $\varphi(t)$ is nonnegative monotonic increasing continuous function on $[0,T]$, and M is a positive constant, then

$$u(t) \leq \varphi(t) E_{1-\alpha}(M\Gamma(1-\alpha)t^{1-\alpha}), \quad 0 \leq t \leq T, \tag{1.51}$$

where $E_{1-\alpha}(z)$ is the Mittag-Leffler function defined for all $0 \leq \alpha < 1$ by

$$E_{1-\alpha}(z) := \sum_{n=0}^{\infty} \frac{z^n}{\Gamma(n(1-\alpha)+1)}$$

Medveď derived the nonlinear integral inequalities of Henry type and also their Bihari nonlinear version, which can be also seen as a generalization of inequality in Theorem 1.26.

Theorem 1.27 ([Medveď (1997)]). *Let $a(t)$ be a nondecreasing, nonnegative C^1-function on $[0,T)$ for $T>0$, $F(t)$ be a continuous, nonnegative function on $[0,T)$, $w:\mathbb{R}_+\to\mathbb{R}\,(\mathbb{R}_+=[0,\infty))$ be a continuous, nondecreasing function, $w(0)=0, w(u)>0$ on $(0,T)$, and $u(t)$ be a continuous, nonnegative function on $[0,T)$ with*

$$u(t)\leq a(t)+\int_0^t (t-s)^{\beta-1}F(s)w(u(s))\,\mathrm{d}s,\quad t\in[0,T), \tag{1.52}$$

where $\beta>0$. Then the following assertions hold:

(i) *Suppose $\beta>1/2$ and w satisfies the condition (q) (see the end of this theorem) with $q=2$. Then*

$$u(t)\leq e^t\Big\{\Omega^{-1}\Big[\Omega(2a(t)^2)+g_1(t)\Big]\Big\}^{1/2},\quad t\in[0,T_1], \tag{1.53}$$

where

$$g_1(t)=\frac{\Gamma(2\beta-1)}{4^{\beta-1}}\int_0^t R(s)F(s)^2\,\mathrm{d}s,$$

where Γ is the gamma function, $\Omega(v)=\int_{v_0}^{v}(\mathrm{d}y/w(y)), v_0>0$, Ω^{-1} is the inverse of Ω and $T_1\in\mathbb{R}_+$ is such that $\Omega(2a(t)^2)+g_1(t)\in\mathrm{Dom}(\Omega^{-1})$ for all $t\in[0,T_1]$.

(ii) *Let $\beta\in(0,1/2]$ and w satisfies the condition (q) with $q=z+2$, where $z=(1-\beta)/\beta\,(i.e., \beta=1/(z+1))$. Then*

$$u(t)\leq e^t\Big\{\Omega^{-1}\Big[\Omega(2^{q-1}a(t)^q)+g_2(t)\Big]\Big\}^{1/q},\quad t\in[0,T_1], \tag{1.54}$$

where

$$g_2(t)=2^{q-1}K_z^q\int_0^t F(s)^qR(s)\,\mathrm{d}s,$$

$$K_z=\left[\frac{\Gamma(1-\alpha p)}{p^{1-\alpha p}}\right]^{1/p},\quad \alpha=\frac{z}{z+1}, p=\frac{z+2}{z+1},$$

$T_1\in\mathbb{R}_+$ is such that $\Omega(2^{q-1}a(t)^q)+g_2(t)\in\mathrm{Dom}(\Omega^{-1})$ for all $t\in[0,T_1]$.

Condition (q): Let $q>0$ be a real number and $0<T\leq\infty$. We say that a function $w:\mathbb{R}_+\to\mathbb{R}$ satisfies a condition (q), if

$$e^{qt}[w(u)]^q\leq R(t)w(e^{-qt}u^q)\quad \text{for all } u\in\mathbb{R}_+, t\in[0,T),$$

where $R(t)$ is a continuous, nonnegative function.

In the special case with $w(u)=u$ in (1.52), then the bound for $u(t)$ can be as

$$\begin{cases} u(t)\leq\sqrt{2}a(t)\exp\left[\dfrac{2\Gamma(2\beta-1)}{4^\beta}\displaystyle\int_0^t F(s)^2\,\mathrm{d}s+t\right], & t\in[0,T), \beta>1/2,\\ u(t)\leq 2^{\frac{q-1}{q}}a(t)\exp\left[\dfrac{2^{q-1}}{q}K_z^q\displaystyle\int_0^t F(s)^q\,\mathrm{d}s+t\right], & t\in[0,T), \beta=\dfrac{1}{z+1}, z\geq1, \end{cases}$$

where K_z is defined as in Theorem 1.27.

If the condition in (1.52) is replaced by

$$u(t) \le a(t) + b(t) \int_0^t (t-s)^{\beta-1} F(s) w(u(s))\,\mathrm{d}s,$$

$$u(t) \le a(t) + \int_0^t (t-s)^{\beta-1} s^{\gamma-1} F(s) u(s)\,\mathrm{d}s,$$

or

$$u(t)^2 \le a(t) + \int_0^t (t-s)^{\beta-1} s^{\gamma-1} F(s) w(u(s))\,\mathrm{d}s,$$

where $b(t)$ is nonnegative, integrable functions on $[0,T)$, $\gamma > 0$, then the bound for $u(t)$ was also derived, see [Medveď (1997); Ma and Pecari (2008)] for more details, where some other generalizations of the weakly singular integral inequalities of Gronwall–Bellman type are established [Ma and Pecari (2008)]. Extensions of Theorem 1.27 with two variables in linear and nonlinear cases were derived in [Cheung *et al.* (2008); Wang and Zheng (2010)]. See also [Ma and Yang (2008); Ma and Debnath (2008); Wang and Zheng (2010)] for more similar inequalities.

Ding *et al.* [Ye *et al.* (2007)] gave another bound for $u(t)$ if $u(t)$ satisfies (1.52), see the following theorem.

Theorem 1.28 ([Ye *et al.* (2007)]). *Suppose $\alpha > 0$, $a(t)$ is a nonnegative function locally integrable on $0 \le t < T$ (some $T \le \infty$) and $g(t)$ is a nonnegative, nondecreasing continuous function defined on $0 \le t < T$, $g(t) \le M$ (constant), and suppose $u(t)$ is nonnegative and locally integrable on $0 \le t < T$ with*

$$u(t) \le a(t) + g(t) \int_0^t (t-s)^{\alpha-1} u(s)\,\mathrm{d}s \tag{1.55}$$

on this interval. Then

$$u(t) \le a(t) + \int_0^t \Big[\sum_{n=1}^{\infty} \frac{(g(t)\Gamma(\alpha))^n}{\Gamma(n\alpha)} (t-s)^{n\alpha-1} a(s) \Big] \mathrm{d}s, \quad 0 \le t < T. \tag{1.56}$$

Theorem 1.29 ([Denton and Vatsala (2010)]). *Let $0 < q < 1, p = 1-q$, $J = [a,b]$, $v \in C_p(J,\mathbb{R}_+)$ and $y \in C(J,\mathbb{R}_+)$, such that*

$$v(t) \le \frac{v_a}{\Gamma(q)} (t-a)^{q-1} + \frac{1}{\Gamma(q)} \int_0^t (t-s)^{q-1} v(s) y(s)\,\mathrm{d}s, \tag{1.57}$$

then

$$v(t) \le \frac{v_a}{\Gamma(q)} \sum_{k=0}^{\infty} T_y^k (t-a)^{q-1}, \tag{1.58}$$

where $C_p(J,\mathbb{R}) = \{u \in C((a,b],\mathbb{R}) | u(t)(t-a)^p \in C(J,\mathbb{R})\}$ and T_y is an operator defined by

$$T_y \phi = \frac{1}{\Gamma(q)} \int_a^t (t-s)^{q-1} \phi(s) y(s)\,\mathrm{d}s.$$

Next, we give the discrete analogue for Theorem 1.26, which was proved by Mckee in 1982.

Theorem 1.30 ([Mckee (1982)]). *If $u_j, j = 0, 1, ..., N$ is a sequence of real numbers with*

$$|u_0|\leq\delta, \quad |u_i|\leq h^{1-\alpha} M \sum_{j=0}^{i-1} (i-j)^{-\alpha}|u_j| + \delta, \quad i = 1, 2, ..., N,$$

where $M > 0$ and independent of $h, \delta > 0$ and $0\leq\alpha < 1$ (independent of h), then

$$\max_{0\leq j\leq N} |u_j|\leq(\delta' + hM'T^{n-1-n\alpha}\delta)\exp(M'T^{n-n\alpha}),$$

where

$$M' = \begin{cases} M, & n = 1 \\ M^n \prod_{k=1}^{n-1} B(k(1-\alpha), 1-\alpha), & n\geq 2, \end{cases}$$

and

$$\delta' = \delta\Big\{(1 + h^{1-\alpha}M)\sum_{j=0}^{n-2} \gamma^j + \gamma^{n-1}\Big\}, \quad n\geq 1,$$

where $\gamma = \frac{MT^{1-\alpha}}{1-\alpha}$ and n is the smallest positive integer such that $\alpha\leq(n-1)/n$, $B(\cdot,\cdot)$ is the Beta function.

In Theorem 1.30, if δ is replaced by a monotonic increasing sequence of non-negative real numbers $\psi_i (i = 0, 1, 2, ...)$, then another simple bound for u_i was derived as [Dixo and Mckee (1986)]

$$|u_i| \leq \psi_i E_{1-\alpha}(M\Gamma(1-\alpha)(ih)^{1-\alpha}), \quad 0 \leq i \leq N.$$

Mckee also gave another Abel's type inequality with singularity in [Mckee (1982)].

Theorem 1.31 ([Mckee (1982)]). *If $u_j, j = 0, 1, ..., N$ is a sequence of real numbers with*

$$|u_0|\leq\delta, \quad |u_i|\leq hM \sum_{j=0}^{i-1} \left(i^{\frac{n+1}{n}} - j^{\frac{n+1}{n}}\right)^{\frac{n}{n+1}} |u_j| + \delta, \quad i = 1, 2, ..., N,$$

where $M > 0$ and independent of $h, \delta > 0$ and $n\in\mathbb{N}$ (independent of h), then

$$\begin{aligned} \max_{0\leq j\leq N} |u_j| \leq \delta\Big\{ \sum_{j=0}^{n} (nMT)^j + h\Big(\frac{n}{n+1}\Big)^n T^n M^{n+1} \prod_{j=1}^{n} B\Big(\frac{1}{n+1}, \frac{j}{n+1}\Big)\Big\} \\ \times \exp\Big\{\Big(\frac{n}{n+1}\Big)^n M^{n+1}T^{n+1} \prod_{j=1}^{n} B\Big(\frac{1}{n+1}, \frac{j}{n+1}\Big)\Big\}. \end{aligned} \tag{1.59}$$

An extension of Theorem 1.30 was derived by Huang and Lü in [Huang and Lü (2003)], which reads as

Theorem 1.32 ([Huang and Lü (2003)]). *Let N be a positive integer, A, L and T are positive constants. If $\{e_j\}$ satisfies*

$$\begin{cases} |e_0| = 0, \\ |e_i| \leq A + \sum_{j=1}^{i-1} B_{ij}|e_j|, \quad 1 \leq i \leq N, \end{cases}$$

where $B_{ij} = Lh(x_i - x_j)^{\alpha}(\ln|x_i - x_j|)^{\beta}$ for $i > j$, $-1 < \alpha < 0$ and $\beta = 0$ or 1. $x_i = ih$ and $hN \leq T$. Then,

$$|e_i| \leq CA(\ln h)^{\beta}, \quad 0 \leq i \leq N,$$

where C is a positive constant in dependent of h and i.

There are some further results in the development. Some systems of two discrete inequalities of Gronwall type are discussed in [Salem (1997)]. Wang *et al.* [Wang at al. (2008)] got a new generalized Gronwall inequality with impulse, mixed-type integral operator, and B-norm that is much different from classical Gronwall inequality, which is used in the discussion on integro-differential equation of mixed type, see [Wang at al. (2008)] for more information. The Gronwall-Bellman type integral inequalities and the corresponding integral equations for scalar functions of several variables involving abstract Lebesque integrals are considered in [Győri and Horváth (1997)], see [Győri and Horváth (1997)] and the references therein for more details and the related problems. Popenda [Popenda (1995)] provided an algebraic version of Gronwall inequalities from which many of the familiar Gronwall inequalities are shown to be derivable.

1.5 Conclusion

In this paper, we collect the main results of the Gronwall type inequalities, including the more generalized linear extensions and the nonlinear cases. The extensions of the Gronwall inequality to the multi-dimensional case are also mentioned, meanwhile, the weakly singular Gronwall type inequalities and the Gronwall inequality corresponding to the discontinuous functions are further surveyed.

Bibliography

Abdeldaim, A. and Yakout, M. (2010). On Wendroff's inequality and applications, *Int. Journal of Math. Analysis* **4**, pp. 607–616.

Abdeldaim, A. and Yakout, M. (2011). On some new integral inequalities of Gronwall–Bellman–Pachpatte type, *Appl. Math. Comput.* **217**, pp. 7887–7899.

Alzer, H.(1996). Discrete analogues of a Gronwall-type inequality, *Acta Math. Hungar.* **73**, pp. 209–213.

Agarwal, R. P. (1982). On an integral inequality in n-independent variables, *J. Math. Anal. Appl.* **85**, pp. 192–196.

András, S. and Mészáros, A. (2011). Remarks on some wendroff type inequalities, *J. Math. Inequal.* **5**, pp. 401–411.

Bainov, D. and Simenov, P. (1992). *Integral Inequalities and Applications* (Kluwer Academic Publishers, Dordrecht).

Beckenbach, E. F. and Bellman, R. (1961). *Inequalities* (Springer-Verlag, Berlin).

Bellman, R. (1943). The stability of solutions of linear differential equations, *Duke Math. J.* **10**, pp. 643–647.

Bihari, I. A. (1956). generalization of a lemma of Bellman and its application to uniqueness problems of differential equations, *Acta Math. Acad. Sci. Hungar.* **7**, pp. 81–94.

Bohner, M. and Peterson, A. (2001). *Dynamic Equations on Time Scales: An Introduction with Application* (Birkhauser, Boston, MA).

Borysenko, S. D. (1989). Integro-sum inequalities for functions of many independent variables, *Differ. Equ.* **25**, pp. 1638–1641.

Borysenko, S. and Iovane, G. (2007). About some new integral inequalities of Wendroff type for discontinuous functions, *Nonlinear Anal.* **66**, pp. 2190–2203.

Bondge, B. K. and Pachpatte, B. G. (1979). On some fundamental integral inequalities in two independent variables, *J. Math. Anal. Appl.* **72**, pp. 533–544.

Cheung, W. S., Ma, Q. H. and Tseng, S. (2008). Some new nonlinear weakly singular integral inequalities of Wendroff type with applications, *J. Inequal. Appl.* **2008**, Article ID 909156, 13 pages.

Cheung, W. S. (2004). Some discrete nonlinear inequalities and applications to boundary value problems for difference equations, *J. Differential Equation Appl.* **10**, pp. 213–223.

Cheung, W. S. and Ren, J. L. (2006). Discrete non-linear inequalities and applications to boundary value problems, *J. Math. Anal. Appl.* **319**, pp. 708–724.

Chou, C. N. and Yang, G. S. (2005a). A variant of the inequality given by Bainov and Simeonov, *Tamkang J. Math.* **36**, pp. 359–363.

Chou, C. N. and Yang, G. S. (2005b). Some integral inequalities of the Volterra type,

Tamkang J. Math. **36**, pp. 167–178.

Chu, S. C. and Metcalf, F. T. (1967). On Gronwall's inequality, *Proc. Amer. Math. Soc.* **18**, pp. 439–440.

Dafermos, C. M. (1979). The second law of thermodynamics and stability, *Arch. Rational Mech. Anal.* **70**, pp. 167–179.

Denton, Z. and Vatsala, A. S. (2010). Fractional integral inequalities and applications, *Comput. Math. Appl.* **59**, pp. 1087–1094.

Dixo, J. and Mckee, S. (1986). Weakly singular discrete gronwall inequalities, *Z. angew. Math. Mech.* **66**, pp. 535–544.

El-Owaidy, H., Ragab, A. A. and Abdeldaim, A. (1999). On some new integral inequalities of Gronwall–Bellman type, *Appl. Math. Comput.* **106**, pp. 289–303.

Feng, Q. H, Meng, F. W. and Zhang, Y. M. (2011). Generalized Gronwall-Bellman-type discrete inequalities and their applications, *J. Inequal. Appl.* **2011**, 2011:47, 12 pages.

Galloa, A. and Piccirillo, A. M. (2007). About new analogies of Gronwall-Bellman-Bihari type inequalities for discontinuous functions and estimated solutions for impulsive differential systems, *Nonlinear Anal.* **67**, pp. 1550–1559.

Gronwall, T. H. (1919). Note on the derivative with respect to a parameter of the solutions of a system of differential equations, *Ann. of Math.* **20**, pp. 292–296.

Györi, I. (1971). A generalization of Bellman's inequality for Stieltjes integrals and a uniqueness theorem, *Studia Sci. Math. Hungar.* **6**, pp. 137–145.

Győri, I. and Horváth, L. (1997). Gronwall-Bellman type integral inequalities for abstract Lebesgue integral, *Math. Comput. Modelling* **25**, pp. 51–58.

Huang, Y. and Lü, T. (2003). A generalization of discrete Gronwall inequality and its application to weakly singular Volterra integral equation of the second kind, *J. Math. Anal. Appl.* **282**, pp. 56–62.

Kuang, J. C. (2010). *Applied Inequalities* (Shandong Science and Technology Press, Jinan).

Heywood, J. G. and Rannacher, R. (1990). Finite element approximation of the nonstationary Navier–Stokes problem, *SIAM J. Numer. Anal.* **2**, pp. 353–384.

Li, R. (1996). *Numerical Solution of Differential Equation* (High Education Press, Beijing, China).

Lipovan, O. (2000). A retarded Gronwall-like inequality and its applications, *J. Math. Anal. Appl.* **252**, pp. 389–401

Ma, Q. H. and Debnath, L. (2008). On some new nonlinear delay and weakly singular integral inequalities, *J. Appl. Math. & Informatics* **26**, pp. 877– 888.

Ma, Q. H. and Pecari, J. (2008). Some new explicit bounds for weakly singular integral inequalities with applications to fractional differential and integral equations, *J. Math. Anal. Appl.* **341**, pp. 894–905.

Ma, Q. H. (2008). Some new nonlinear Volterra-Fredholm-type discrete inequalities and their applications, *J. Comput. Appl. Math.* **216**, pp. 451–466.

Ma, Q. H. and Yang, E. H. (2002). Estimates on solutions of some weakly singular Volterra integral inequalities, *Acta Math. Appl. Sin.* **25**, pp. 505–515.

Matsunaga, H. and Murakami, S. (2004). Some invariant manifolds for functional difference equations with infinite delay, *J. Differ. Equ. Appl.* **10**, pp. 661–689.

Mckee, S. (1982). Generalized discrete Gronwall Lemmas, *ZAMM.* **62**, pp. 420–434.

Medveď, M. (1997). A new approach to an analysis of henry type integral inequalities and their Bihari type versions, *J. Math. Anal. Appl.* **214**, pp. 349–366.

Mitrinovic, D. S., Pecaric, J. and Fink, A. M. (1991). *Inequalities Involving Functions And Their Integrals And Derivatives* (Kluwer Acdemic Publishers, Dordrencht, The Netherlands).

Oguntuase, J. A. (1998). A note on Gronwall–Bellman integral inequality, *Zb. Rad. (Kragu-*

jevac) **20**, pp. 77–81.

Ou-Yang, L. (1957). The boundedness of solutions of linear differential equations $y + A(t)y = 0$, *Shuxue Jinzhan* **3**, pp. 409–415.

Pachpatte, B. G. (1977). On discrete inequalities related to Gronwall's inequality, *Proc. Indian Acad. Sci.* **85**, pp. 26–40.

Pachpatte, B. G. (1995a). On a new inequality suggested by the study of certain epidemic models, *J. Math. Anal. Appl.* **195**, pp. 638–644.

Pachpatte, B. G. (1995b). On some new inequalityies related to certain inequalityies in the theory of differential equations, *J. Math. Anal. Appl.* **189**, pp. 128–144.

Pachpatte, B. G. (1998a). *Inequalities for Differential and Integral Equations* (Academic Press, New York and London).

Pachpatte, B. G. (1998b). Inequalities applicable in the theory of finite difference equations, *J. Math. Anal. Appl.* **222**, pp. 438–459

Pachpatte, B. G. (2002a). Inequalities for terminal value problems for differential equations, *Tamkang J. Math.* **33**, pp. 199–208.

Pachpatte, B. G. (2002b). A note on certain integral inequality, *Tamkang J. Math.* **33**, pp. 353–358.

Pachpatte, B. G. (2006). *Integral and finite difference inequalities and applications* (Elsevier Science B. V., Amsterdam).

Pachpatte, B. G. (1979). On some new integral and integrodifferential inequalities in two independent variables and their applications, *J. Differential Equations* **33**, pp. 249–272.

Pachpatte, B.G. and Singare, S. M. (1979). Discrete generalized Gronwall inequalities in three independent variables, *Pacific J. Math.* **82**, pp. 197–210.

Pirapakaran, R. (1988). Impulsive integral inequalities of Gronwall-Bihari type, *Differential equations and applications* **I-II**, pp. 296–308.

Popenda, J. (1995). Gronwall type inequalities, *Z. angew. Math. Mech* **75**, pp. 669–677.

Popenda, J. and Agarwal, R. P. (1999) Discrete Gronwall inequalities in many variables, *Comput. Math. Appl.* **38**, pp. 63–70.

Quarteroni, A. and Valli, A. (1994). *Numerical Approximation of Partial Differential Equations* (Springer Series in Computational Mathematics, 23, Springer-Verlag).

Roshdy, E. M. and Mousa, M. S. (2011). Discrete inequality of Gronwall-Bellman type, *IJRRAS* **4**, pp. 416–418.

Salem, S. (1997). On Some systems of two discrete inequalities of Gronwall type, *J. Math. Anal. Appl.* **208**, pp. 553–566.

Salem, Sh. and Raslan, K. R. (2004). Some new discrete inequalities and their applications, *Inequal. Pure Appl. Math.* **5**, Article 2, 9 pages.

Samoilenko, A. M. and Borysenko, S. D. (1998). On functional inequalities of Bihari type for discontinuous functions, *Uspekhi Mat. Nauk.* **53**, pp. 147–148.

Shastri, R. P. and Kasture, D. Y. (1978). Wendroff type inequalities, *Proc. Ameri. Math. Soc.* **72**, pp. 248–250.

Snow, D. R. (1971). A two independent variables Gronwall type inequality, inequalities III, in: *Proceedings Symposium on Inequalities* (Academic Press, New York), pp. 330–340.

Snow, D. R. (1972). Gronwall's inequality for systems of partial differential equations in two independent variables, *Proc. Amer. Math. Soc.* **33**, pp. 46–54.

Sugiyama, S. (1969). On the stability problems of difference equations, *Sci. Engrg. Res. Lab. Waseda Univ.* **45**, pp. 140–144.

Thandapani, E. and Agarwall, R. P. (1982). On some new inequalities in n-independent variables, *J. Math. Anal. Appl.* **86**, pp. 542–561.

Viswanatham, B. (1963). A generalisation of Bellman's Lemma, *Proc. Amer. Math. Soc.* **14**, pp. 15–18.

Wang, J. R. Xiang, X. Wei, W. and Chen, Q. (2008). The Generalized Gronwall inequality and its application to periodic solutions of integrodifferential impulsive periodic system on Banach space, *J. Inequal. Appl.* **2008**, Article ID 430521, 22 pages.

Wang, H. and Zheng, K. L. (2010). Some nonlinear weakly singular integral inequalities with two variables and applications, *J. Inequal. Appl.* **2010**, Article ID 345701, 12 pages.

Willett, D. (1964). Nonlinear vector integral equations as contraction mappings, *Arch. Rat. Mech. Anal.* **15**, pp. 79–86.

Willett, D. and Wong, J. S. W. (1965). On the discrete analogues of some generalizations of Gronwall's inequality, *Monatsh. Math.* **69**, pp. 362–367.

Yang, E. H. (1983). On some new discrete inequalities of the Bellman-Bihari type, *Nonlinear Anal.* **7**, pp. 1237–1246.

Yang, E. H. (1988). On some new discrete generalizations of Gronwall's inequality, *J. Math. Anal. Appl.* **129**, pp. 505–516.

Ye, H. Gao, J. and Ding, Y. (2007). A generalized Gronwall inequality and its application to a fractional differential equation, *J. Math. Anal. Appl.* **328**, pp. 1075–1081.

Yeh, C. C. (1980). Wendroff's inequality in n-independent variables, *J. Math. Anal. Appl.* **78**, pp. 78–87.

Yeh, C. C. [1982a). On some integral inequalities in n-independent variables and their applications, *J. Math. Anal. Appl.* **86**, pp. 387–410.

Yeh, C. C. (1982b). Bellman-Bihari inequalities in several variables, *J. Math. Anal. Appl.* **87**, pp. 311–321.

Yeh, C. C. and Shih, M. H. (1982). The Gronwall-Bellman inequality in several variables, *J. Math. Anal. Appl.* **86**, pp. 157–167.

Yeh, C. C. (1985a). Discrete inequalities of Gronwall-Bellman type in n independent variables, *J. Math. Anal. Appl.* **105**, pp. 322–332.

Yeh, C. C. (1985b). Discrete inequalities of Gronwall-Bellman type in n independent variables, II, *J. Math. Anal. Appl.* **106**, pp. 282–285.

Young, E. C. (1973). Gronwall's inequality in n-independent variables, *Proc. Amer. Math. Soc.* **41**, pp. 241–244.

Zhou, L. F. and Zhang, W. N. (2010). A projected discrete Gronwall's inequality with sub-exponential growth, *J. Differ. Equ. Appl.* **16**, pp. 931–943.

Chapter 2

Existence and uniqueness of the solutions to the fractional differential equations[1]

Yutian Ma[a,b], Fengrong Zhang[a,c] and Changpin Li[a,*]

[a] *Department of Mathematics, Shanghai University, Shanghai 200444, PR China*
** lcp@shu.edu.cn*

[b] *Department of Mathematics, Fuyang Normal college, Fuyang 236029, Anhui, PR China*

[c] *College of Sciences, China University of Petroleum (East China), Qingdao 266555, Shandong, PR China*

Recently, fractional differential equations have drawn more and more attention due to their wide applications in different research areas and engineering, which leads to intensive development of the theory of fractional calculus. More and more results about the existence and uniqueness of solutions appear. In this paper, a brief overview on the recent existence and uniqueness results of fractional differential equations are provided. These equations include initial value problem for fractional differential equations, boundary value problem for fractional differential equations, fractional differential equations with time-delay.

Keywords: Existence, uniqueness, fractional differential equation

2.1 Introduction

The fractional calculus has been 300 years history, and the development of fractional calculus theory is mainly focused on the pure mathematical field at the first stage. The earliest more or less systematic studies seem to have been made in the 19th century by Liouville, Riemann, Leibniz, etc. [Oldham and Spanier (1974); Samko *et al.* (1993)]. In the last two decades, fractional differential equations (FDEs) have been used to model various stable physical phenomena with anomalous decay, say, that are not of exponential type [Matignon (1998)]. We can refer to [Machado

[1]The present work was supported by the National Natural Science Foundation of China (grant no. 10872119), the Shanghai Leading Academic Discipline Project (grant no. S30104), and the Key Program of Shanghai Municipal Education Commission (grant no. 12ZZ084).

et al. (2010)] for the very recent studies of fractional calculus. As we all know, many mathematical models of real problems arising in various fields of science and engineering are either linear systems or nonlinear systems. Nevertheless, the most differential systems used to describe physical phenomena are integer-order systems. With the development of fractional calculus, it has been found that the behaviors of many physical systems can be described by using the fractional differential systems [Bagley and Calico (1991); Sun *et al.* (1984); Ichise *et al.* (1971); Laskin (2000); Kusnezov *et al.* (1999); Tavazoei and Haeri (2009); Torvik and Bagley (1984)].

In recent years, FDEs have attracted increasing interests due to their applications in modeling anomalous diffusion, time dependent materials and process with long range dependence, allometric scaling laws, as well as power law in complex systems. That leads to intensive development of the theory of fractional calculus [Kou *et al.* (2012); Li and Ma (2012); Li *et al.* (2010)], and more and more results about the existence and uniqueness of solutions appear. In this paper, we present and discuss some basic results on existence and uniqueness of the solutions of the FDEs. The analysis of the existence and uniqueness of solutions for FDEs is more complex than that for classical differential equations, since fractional derivatives are nonlocal and have weakly singular kernels.

This chapter is organized as follows. In Section 2.2, some definitions and propositions are introduced. In Section 2.3, some results on existence and uniqueness of solutions for FDEs with initial value problems are presented. In Section 2.4, some results on existence and uniqueness of solutions for FDEs with boundary value problems are described. And Section 2.5 deals with the existence and uniqueness of solutions for FDEs with time-delay. The conclusions are finally included in Section 2.6.

2.2 Preliminaries and notations

In this section, we introduce some definitions and results which will be used throughout the paper.

Let us denote by $\mathbb{R}$ the set of real numbers, $\mathbb{R}^n$ the real n-dimensional Euclidean space, $\mathbb{R}^+$ the set of positive real numbers and by $\mathbb{Z}^+$ the set of positive integer numbers, denote by $\mathbb{C}$ the set of complex numbers. We denote the real part of complex number α by $Re(\alpha)$.

In general, two kinds of fractional derivatives, i.e., the Riemann-Liouville derivative and Caputo derivative, have been often used in FDEs. We briefly introduce these two fractional derivatives which will be frequently used hereafter. Firstly, we introduce the definition of fractional integral [Podlubny (1999)].

Definition 2.1. The fractional integral (or, the Riemann-Liouville integral) $D_{t_0,t}^{-\alpha}$

with fractional order $\alpha \in \mathbb{R}^+$ of function $x(t)$ is defined below:

$$D_{t_0,t}^{-\alpha} x(t) = \frac{1}{\Gamma(\alpha)} \int_{t_0}^{t} (t-\tau)^{\alpha-1} x(\tau) d\tau,$$

where $t = t_0$ is the initial time, $\Gamma(\cdot)$ is the Euler's gamma function.

Definition 2.2. The Riemann-Liouville derivative with order α of function $x(t)$ is defined below:

$$_{RL}D_{t_0,t}^{\alpha} x(t) = \frac{1}{\Gamma(m-\alpha)} \frac{d^m}{dt^m} \int_{t_0}^{t} (t-\tau)^{m-\alpha-1} x(\tau) d\tau = \frac{d^m}{dt^m} D_{t_0,t}^{-(m-\alpha)} x(t),$$

where $m - 1 \leq \alpha < m \in \mathbb{Z}^+$.

Definition 2.3. The Caputo derivative with order α of function $x(t)$ is defined below:

$$_{C}D_{t_0,t}^{\alpha} x(t) = \frac{1}{\Gamma(m-\alpha)} \int_{t_0}^{t} (t-\tau)^{m-\alpha-1} x^{(m)}(\tau) d\tau = D_{t_0,t}^{-(m-\alpha)} \frac{d^m}{dt^m} x(t),$$

where $m - 1 < \alpha \leq m \in \mathbb{Z}^+$.

Proposition 2.1. *From Definition 2.2 and integration by parts, we obtain*

$$D_{t_0,t}^{-\alpha}\ {}_{RL}D_{t_0,t}^{\alpha} x(t) = x(t) - \sum_{j=1}^{m} \frac{[D_{t_0,t}^{\alpha-j} x(t)]_{t=t_0}}{\Gamma(\alpha-j+1)} (t-t_0)^{\alpha-j},$$

$$_{RL}D_{t_0,t}^{\alpha}\left({}_{RL}D_{t_0,t}^{\beta} x(t)\right) = {}_{RL}D_{t_0,t}^{\alpha+\beta} x(t) - \sum_{j=1}^{n} [{}_{RL}D_{t_0,t}^{\beta-j} x(t)]_{t=t_0} \frac{(t-t_0)^{-\alpha-j}}{\Gamma(1-\alpha-j)},$$

and

$$_{RL}D_{t_0,t}^{\beta}\left({}_{RL}D_{t_0,t}^{\alpha} x(t)\right) = {}_{RL}D_{t_0,t}^{\alpha+\beta} x(t) - \sum_{j=1}^{m} [{}_{RL}D_{t_0,t}^{\alpha-j} x(t)]_{t=t_0} \frac{(t-t_0)^{-\beta-j}}{\Gamma(1-\beta-j)},$$

in which $m-1 \leq \alpha < m, n-1 \leq \beta < n, m, n \in \mathbb{Z}^+$, *so*

$$_{RL}D_{t_0,t}^{p}({}_{RL}D_{t_0,t}^{q} x(t)),\ {}_{RL}D_{t_0,t}^{q}({}_{RL}D_{t_0,t}^{p} x(t)) \text{ and } {}_{RL}D_{t_0,t}^{p+q} x(t)$$

are not generally equal to each other.

Proposition 2.2. *Suppose that $x(t)$ satisfies the definitions of Riemann-Liouville derivative and Caputo derivative with order $\alpha, m-1 < \alpha < m, m \in \mathbb{Z}^+$, then they have the following connection:*

$$_{C}D_{t_0,t}^{\alpha} x(t) = {}_{RL}D_{t_0,t}^{\alpha} x(t) - \sum_{j=0}^{m-1} \frac{x^{(j)}(t_0)}{\Gamma(j-\alpha+1)} (t-t_0)^{j-\alpha}.$$

$$_{C}D_{t_0,t}^{\alpha} x(t) = {}_{RL}D_{t_0,t}^{\alpha} x(t),$$

if and only if

$$x'(t_0) = x''(t_0) = \ldots = x^{(m-1)}(t_0) = 0.$$

2.3 Existence and uniqueness of initial value problems for fractional differential equations

2.3.1 *Initial value problems with Riemann-Liouville derivative*

In 1938, Pitcher and Sewell [Pitcher and Sewell (1938)] first considered the nonlinear fractional differential equation:

$$ {}_{RL}D^{\alpha}_{t_0,t}x(t) = f(t,x), \quad 0 < \alpha < 1, \tag{2.1}$$

with initial condition

$$ {}_{RL}D^{\alpha-1}_{t_0,t}x(t)|_{t=t_0} = x_0. \tag{2.2}$$

Provided that $f(t,x)$ is bounded in a special region G lying in $\mathbb{R}\times\mathbb{R}$ and satisfies the Lipschitz condition with respect to x:

$$ |f(t,x_1) - f(t,x_2)| \le L|x_1 - x_2|, \tag{2.3}$$

where the constant $L > 0$ does not depend on x. They proved the existence of the continuous solution $x(t)$ with $0 < \alpha < 1, x_0 = 0$. But the main result given by Pither and Sewell ([Pitcher and Sewell (1938)], Theorem 4.2 for the fractional differential equation (2.1)) with $0 < \alpha < 1$ was not correct because they used the compositional relation $D^{-\alpha}_{t_0,t}{}_{RL}D^{\alpha}_{t_0,t}x(t) = x(t)$, which is generally not correct.

Al-Bassam discussed the fractional differential equation (2.1) with $0 < \alpha \le 1$ in the space of continuous function $C[a,b]$, provided that $f(t,x)$ is a real-valued continuous function in a domain $G \subset \mathbb{R}\times\mathbb{R}$ satisfying $\sup\limits_{(t,x)\in G} \mid f(t,x) \mid \le \infty$ and satisfying the Lipschitz condition (2.3). Applying the operator ${}_{RL}D^{-\alpha}_{t_0,t}$ to both sides of (2.1), he reduced the problem (2.1)-(2.2) to a nonlinear Volterra integral equation:

$$ x(t) = \frac{x_0(t-t_0)^{\alpha-1}}{\Gamma(\alpha)} + \frac{1}{\Gamma(\alpha)}\int_{t_0}^{t}(t-\tau)^{\alpha-1}f(\tau,x(\tau))d\tau. \tag{2.4}$$

by using the method of successive approximation, he established the existence result of the continuous solution $x(t)$ of equation (2.4) [Al-Bassam (1965)].

Delbosco and Rodino [Delbosco and Rodino (1996)] dealt with the following nonlinear fractional differential equation

$$ \begin{cases} {}_{RL}D^{\alpha}_{0,t}x(t) = f(x) \quad (0 < \alpha < 1), \\ x(a) = b \quad (a > 0, b \in \mathbb{R}). \end{cases} \tag{2.5}$$

Let $C_{1-\alpha}([0,h]) = \{x(t) : t^{1-\alpha}x(t) \in C([0,h])\}$. They got the following theorem:

Theorem 2.1. *Assume that $f : \mathbb{R} \to \mathbb{R}$ satisfies $f(0) = 0$ and Lipschitz condition (2.3). Then the initial value problem (2.5) has a unique solution $x(t) \in C_{1-\alpha}([0,h])$ for all $h > a$, provided $a < a_0$, where a_0 is a suitable positive constant depending on α and L.*

Moreover, they also proved the unique solution for the weighted Cauchy type problem:

$$\begin{cases} {}_{RL}D_{0,t}^{\alpha}x(t) = f(x), \quad (0 < \alpha < 1), \\ t^{1-\alpha}x(t)|_{t=0} = x_0. \end{cases} \tag{2.6}$$

Theorem 2.2. *Assume that $f : \mathbb{R} \to \mathbb{R}$ satisfies $f(0) = 0$ and Lipschitz condition (2.3). Then for all $x_0 \in \mathbb{R}$, the initial value problem (2.6) has a unique solution $x(t) \in C_{1-\alpha}([0,h])$ for all $h > 0$.*

Theorems 2.1 and 2.2 require $f(x)$ to satisfy Lipschitz condition, In [Yu and Gao (2005)], Yu and Gao modified the existence results of equation:

$$ {}_{RL}D_{0,t}^{\alpha}x(t) = f(t,x), \quad (0 < \alpha < 1). \tag{2.7}$$

Theorem 2.3. *If $f(t,x)$ is a continuous function on $[0,T] \times \mathbb{R}$ and suppose the following holds:*

$$|f(t,x_1) - f(t,x_2)| \le \lambda(t)h(r),$$

where $h(r)$ is continuous on $[0,\infty)$, and $h(0) = 0, r = |x_1 - x_2|, |D_{0,t}^{-\alpha}\lambda(t)| < M$ for $t \in [0,T]$, then there exists a continuous solution $x(t)$ of equation (2.7) with initial condition (2.2) which is defined on $[0,\overline{T}]$ for a suitable $\overline{T} < T$.

For equation (2.6), Zhang [Zhang (2009)] also obtained the existence and uniqueness results of the following equation.

$$\begin{cases} {}_{RL}D_{0,t}^{\alpha}x(t) = f(t,x), \quad t \in (0,T], 0 < T < \infty, (0 < \alpha < 1), \\ t^{1-\alpha}x(t)|_{t=0} = x_0. \end{cases} \tag{2.8}$$

Before giving the existence and uniqueness theorem of (2.8), we introduce the definition of upper and lower solutions.

Definition 2.4. A function $\widetilde{x}(t) \in C_{1-\alpha}([0,T])$ is called an upper solution of problem (2.8), if it satisfies:

$$\begin{cases} {}_{RL}D_{0,t}^{\alpha}\widetilde{x}(t) \ge f(t,\widetilde{x}(t)), \quad t \in (0,T], \\ t^{1-\alpha}\widetilde{x}(t)|_{t=0} \ge x_0. \end{cases}$$

Similarly, function $\widehat{x}(t) \in C_{1-\alpha}([0,T])$ is called a lower solution of problem (2.8), if it satisfies:

$$\begin{cases} {}_{RL}D_{0,t}^{\alpha}\widehat{x}(t) \le f(t,\widehat{x}(t)), \quad t \in (0,T], \\ t^{1-\alpha}\widehat{x}(t)|_{t=0} \le x_0. \end{cases}$$

In what follows, we assume that:

$$\widetilde{x}(t) \geq \widehat{x}(t), t \in (0,T]; t^{1-\alpha}\widetilde{x}(t)|_{t=0} \geq t^{1-\alpha}\widehat{x}(t)|_{t=0}, \tag{2.9}$$

and define that sector

$$\langle \widehat{x}, \widetilde{x} \rangle = \langle x \in C_{1-\alpha}([0,T]) : \widehat{x}(t) \leq x(t) \leq \widetilde{x}(t), t \in (0,T],$$
$$t^{1-\alpha}\widehat{x}(t)|_{t=0} \leq t^{1-\alpha}x(t)|_{t=0} \leq t^{1-\alpha}\widetilde{x}(t)|_{t=0} \rangle.$$

Next, assume that $f(t,x)$ satisfies the following condition:

$$f(t,x_1) - f(t,x_2) \geq -\underline{d}(x_1 - x_2), \quad \widehat{x} \leq x_2 \leq x_1 \leq \widetilde{x}, \tag{2.10}$$

where $\underline{d} > -\frac{\Gamma(1+\alpha)}{T^\alpha}$ is a constant and $\widehat{x}, \widetilde{x} \in C_{1-\alpha}([0,T])$ are lower and upper solutions of problem (2.8). We also suppose that there exists a constant $\bar{d}$ satisfying $\frac{\Gamma(1+\alpha)}{T^\alpha} > \bar{d} \geq -\underline{d}$, such that:

$$f(t,x_1) - f(t,x_2) \leq \bar{d}(x_1 - x_2), \quad \widehat{x} \leq x_2 \leq x_1 \leq \widetilde{x}, \tag{2.11}$$

where $\widehat{x}, \widetilde{x} \in C_{1-\alpha}([0,T])$ are lower and upper solutions of problem (2.8). The following is the existence and uniqueness theorem for problem (2.8).

Theorem 2.4. *Assume that $\widehat{x}, \widetilde{x} \in C_{1-\alpha}([0,T])$ are lower and upper solutions of problem (2.8), such that (2.9) holds, $f \in C([0,T] \times \mathbb{R})$ and satisfies (2.10). Then, problem (2.8) has one solution in the sector $\langle \widehat{x}, \widetilde{x} \rangle$. Also, if condition (2.11)) holds, then problem (2.8) has one unique solution in the sector $\langle \widehat{x}, \widetilde{x} \rangle$.*

Remark 2.1. (i) Li et al. studied the problem (2.8) when $1 < \alpha < 2$ [Li *et al.* (2010)]:

$$\begin{cases} {}_{RL}D_{0,t}^{\alpha}x(t) = f(t,x), \quad t \in (0,T], 0 < T < \infty, (1 < \alpha < 2), \\ t^{2-\alpha}x(t)|_{t=0} = x_0, t^{2-\alpha}x'(t)|_{t=0} = x_1, \end{cases} \tag{2.12}$$

and obtained a similar result.

(ii) Furati and Tatar [Furati and Tatar (2004)] investigated the nonlocal problem of equation (2.6):

$$\begin{cases} {}_{RL}D_{0,t}^{\alpha}x(t) = f(t,x) + \int_0^t g(t,\tau,x(\tau))d\tau, \quad t > 0, (0 < \alpha < 1), \\ t^{1-\alpha}x(t)|_{t=0} = x_0, \end{cases} \tag{2.13}$$

and gave an existence result.

By applying the well known fixed point theorem, Hayek et al. extended the fractional differential equation in $\mathbb{R}$ to $\mathbb{R}^n$, namely:

Theorem 2.5 ([Hayek *et al.* (1999)]). *Let $U \in \mathbb{R}^n$ be an open connected set, assume $(0,T) \subset \mathbb{R}^+$, and define $\mathbb{D} = (0,T) \times U$. Suppose that $f(t,x)$ is a real valued vectorial function on $\overline{\mathbb{D}}$. Consider the following system of fractional differential equations of order α, $(0 < \alpha \leq 1)$:*

$${}_{RL}D_{0,t}^{\alpha}x(t) = f(t,x), \tag{2.14}$$

under the following conditions,

(i) $f(t,x)$ *is continuous in* $\overline{\mathbb{D}}$,

(ii) $f(t,x)$ *satisfies a Lipschitz condition with respect to* x *in* $\overline{\mathbb{D}}$, *i.e., there exists a constant* $L \in \mathbb{R}^+$, *such that for* $\forall(t,x_1),(t,x_2) \in \overline{\mathbb{D}}$,

$$\|f(t,x_1) - f(t,x_2)\| \le L\|x_1 - x_2\|,$$

where $\|\cdot\|$ *is the supremum norm.*

Then, $\forall(t_0,x_0) \in \mathbb{D}$, *a positive number* β *can be found such that the closed interval* $J = [t_0-\beta, t_0+\beta]$ *is contained in* $(0,T)$, *and there exists a unique continuous function* $x : J \to U$, *such that:*

$$_{RL}D_{0,t}^{\alpha} x(t) = f(t,x), \quad \forall t \in J, \quad x(t_0) = x_0.$$

2.3.2 *Initial value problems with Caputo derivative*

For fractional differential equation (2.1) with $n-1 < \alpha < n$, according to the standard mathematical theory [Samko *et al.* (1993)], the initial conditions corresponding to (2.1) must be of the form:

$$_{RL}D_{0,t}^{\alpha-k} x(t) = x_k, \quad k = 1,\cdots,n,\ n-1 < \alpha < n$$

with given values x_k. Thus we are forced to specify some fractional derivatives of the function $x(t)$. In practical applications, these values are not easily available, and it may not even be clear what their physical meaning are. So Diethelm analyzed the following system [Diethelm and Jord (2002)]:

$$\begin{cases} _{RL}D_{0,t}^{\alpha}(x - T_{n-1}(x)) = f(t,x), & n-1 < \alpha < n, \\ x^{(k)}(0) = x_0^k, & k = 0,\cdots,n-1, \end{cases} \tag{2.15}$$

where $T_{n-1}(x)$ is the Taylor polynomial of order $n-1$ for x, centered at 0.

Theorem 2.6. *Assume that* $D = [0,T^*] \times [x_0^{(0)} - a, x_0^{(0)} + a]$ *with some* $T^*, a > 0, T = \min\{T^*, (a\Gamma(\alpha+1)/\|f\|_\infty)^{1/\alpha}\}$.

(i) If $f : D \to \mathbb{R}$ *is continuous, then there exists a function* $x(t) : [0,T] \to \mathbb{R}$ *solving the initial value problem* (2.15).

(ii) If $f : D \to \mathbb{R}$ *is bounded on* D *and satisfies Lipschitz condition* (2.3), *then, there exists at most a function* $x(t) : [0,T] \to \mathbb{R}$ *solving the initial value problem* (2.15).

Actually, in equation (2.15), $_{RL}D_{0,t}^{\alpha}(x - T_{n-1}(x))$ is Caputo derivative (If $x(t) \in AC^n[0,T]$, where $AC^n[0,T] = \{x(t) : x^{(n-1)}(t) \in AC[0,T]\}, AC[0,T]$ is the space of functions $x(t)$ which are absolutely continuous on $[0,T]$ [Kilbas *et al.* (2006)]). In (2.15), Kilbas and Marzan [Kilbas and Marzan (2005)] considered equation with complex order $\alpha, Re(\alpha) > 0$, and they obtained the uniqueness results under some conditions.

Remark 2.2. Daftardar-Gejji and Babakhani considered a system of equation (2.15) with order $0 < \alpha < 1$, and got similar result [Daftardar and Babakhani (2004)].

Yu and Gao weakened the condition and got the following theorem [Yu and Gao (2005)]:

Theorem 2.7. *Assume that* $D = [0,T] \times [x_0^{(0)} - a, x_0^{(0)} + a], f(t,x) \in C(D)$ *with some* $a > 0$, *where* $C(D)$ *is the class of all continuous function defined on* D.

(i) If $f(t,x)$ *satisfies* $|f(t,x_1) - f(t,x_2)| \leq \lambda(t)|x_1 - x_2|$ *and* ${}_{RL}D_{0,t}^{-\alpha}\lambda(t) \leq \omega < 1$, *then equation* (2.15) *has a unique solution* $x(t) \in C([0,T])$.
(ii) If $f(t,x)$ *satisfies* $|f(t,x_1) - f(t,x_2)| \leq Lt^{\gamma}|x_1 - x_2|$ *for any* $\alpha + \gamma \geq 0, \gamma > -1$, *with constant* $L > 0, Lh^{\alpha+\gamma} < 1$, *then there exists a unique solution of* (2.15) *on* $[0,h]$, *where* $h = \min\{T, (a\Gamma(\alpha+1)M^{-1})^{1/\alpha}\}, M = \max_{(t,x)\in D} |f(t,x)|$.

Momani et al. investigated the following equations [Momani *et al.* (2007)]:

$$\begin{cases} {}_CD_{0,t}^{\alpha}x(t) = f(t,x) + \int_0^t K(t,\tau,x(\tau))d\tau, & 0 < \alpha \leq 1, \\ x(0) = x_0 > 0, \end{cases} \tag{2.16}$$

where $f \in C[I \times \mathbb{R}^n, \mathbb{R}^n], K \in C[I \times I \times \mathbb{R}^n, \mathbb{R}^n], I = [0,T]$.

Theorem 2.8. *Assume that* f, K *satisfy the following:*

$$|f(t,x) - f(t,y)| \leq \phi(|x-y|),$$

$$\int_s^t |K(\sigma,s,x(s)) - K(\sigma,s,y(s))d\sigma| \leq M\phi(|x-y|),$$

M *is a a positive constant and* $\phi(x)$ *is a continuous non-decreasing function on* $(0,A]$ *with* $\phi(0) = 0$ *and*

$$\int_0^A \frac{dx}{\phi(x)} = +\infty,$$

then initial value problem (2.16) *has a unique solution on the interval* $(0,T]$.

Remark 2.3. Using Gronwall's inequality, the authors also gave a global uniqueness result for the initial value problem (2.16).

Recently, Balachandran and Trujillo dealt with the fractional quasilinear integrodifferential equation in Banach space [Balachandran and Trujillo (2010)]:

$$\begin{cases} {}_CD_{0,t}^{\alpha}x(t) = A(t,x)x(t) + f(t,x,\int_0^t h(t,\tau,x(\tau))d\tau), & 0 \leq t \leq T, 0 < \alpha < 1, \\ x(0) = x_0, \end{cases} \tag{2.17}$$

where $A(t,x)$ is a bounded linear operator on X and $f : I \times X \times X \to X, h : \Delta \times X \to X$ are continuous, X is a Banach space, $I = [0,T], \Delta = \{(t,s) : 0 \leq s \leq t \leq T\}$.

Theorem 2.9. *Let the following conditions be satisfied:*

(i) *$A : I \times X \to B(X)$ is a continuous bounded linear operator and there exists a constant $M > 0$ such that,*

$$\|A(t,x) - A(t,y)\|_{B(X)} \leq M\|x - y\|, \text{ for all } x,\ y \in X.$$

(ii) *$f : I \times X \times X \to$ is continuous and there exists a constant $L > 0$ such that,*

$$\|f(t,x,u) - f(t,y,v)\| \leq L[\|x-y\| + \|u-v\|], \text{ for all } x,\ y,\ u,\ v \in X.$$

(iii) *$h : \Delta \times X \to X$ is continuous and there exists a constant $L^* > 0$ such that,*

$$\|h(t,s,x) - h(t,s,y)\| \leq L^*\|x-y\|, \text{ for all } x,\ y \in X.$$

(iv) *$\|x_0\| + (Mr+K)r\gamma + \gamma M_0 \leq r, p = \gamma(2Mr + K + L + LL^*T)$ such that $0 \leq p < 1$, where $r > 0, \gamma = \frac{T^\alpha}{\Gamma(\alpha+1)}, K = \sup_{t \in I}\|A(t,0)\|, N = \max_{t \in I}\|f(t,0,0)\|, N^* = \max_{(t,s)\in\Delta}\|h(t,s,0)\|, M_0 = Lr + LL^*Tr + LN^*T + N$.*

Then, the equation (2.17) has a unique solution continuous in I.

For equation (2.17), the unique result is discussed when replacing initial condition $x(0) = x_0$ with nonlocal condition $x(0) + g(x) = x_0$, where $g : C(I,X) \to X$ is a given function.

2.3.3 *The positive solution to fractional differential equation*

Zhang first defined the positive solution to equation (2.7) with $0 < t < 1, f : [0,1] \times [0,\infty) \to [0,\infty)$ [Zhang (2000)].

Theorem 2.10. *Let $f : [0,1] \times [0,\infty) \to [0,\infty)$ be continuous, $f(t,\cdot)$ be nondecreasing for each $t \in [0,1]$. Assume that there exists a positive constant a, such that $f(t,\cdot)$ is strictly increasing on $[0,a]$ for each $t \in [0,1]$ and one of the following conditions is true.*

(i) *v_0, w_0 are lower and upper solutions of (2.7), satisfying $v_0 \leq w_0, 0 \leq t \leq 1$.*
(ii) *$0 < \lim_{x\to\infty} f(t,x(t)) < +\infty$ for each $t \in [0,1]$.*
(iii) *$0 \leq \lim_{x\to\infty} \max_{0\leq t\leq 1} \frac{f(t,x(t))}{x(t)} < +\infty$.*

Then equation (2.7) has one positive solution $x(t)$, namely $x(t)$ satisfies $x(0) = 0, x(t) > 0, t \in (0,1], x(t) \in C[0,1]$.

Theorem 2.11. *Assume that $f(t,\cdot)$ is decreasing convex for each $t \in [0,1]$; $0 < f(t,0) < \Gamma(1+\alpha), \forall t \in [0,1]$; $f(t,1) \geq cf(t,0), c \geq \frac{1}{2}, \forall t \in [0,1]$. Then, equation (2.7) has a unique positive solution $x(t)$.*

Furthermore, Daftardar-Gejji studied a system of non-autonomous fractional differential equations [Daftardar-Gejji (2005)]:

$$_{RL}D_{0,t}^{\alpha_i} x_i(t) = f_i(t, \overline{x}), \tag{2.18}$$

where $0 \le t \le T, \overline{x} = (x_1, \cdots, x_n)$, and $0 < \alpha_i < 1, x_i(0) = 0, f_i : [0,T] \times [0,\infty)^n \to [0,\infty), i = 1, \cdots, n$, $[0,\infty)^n$ denotes $\underbrace{[0,\infty) \times \cdots \times [0,\infty)}_{n-times}$.

Theorem 2.12. *i) Let $f_i : [0,T] \times [0,\infty)^n \to [0,\infty), i = 1, \cdots, n$ be continuous. Assume that there exist two distinct positive constants m, M such that $m \le f_i(t, \overline{x}) \le M, i = 1, \cdots, n$, then the system of equations* (2.18) *has at least one positive solution.*

ii) Let $f_i : [0,T] \times [0,\infty)^n \to [0,\infty), i = 1, \cdots, n$ be continuous and $f_i(t, \cdot)$ increasing for each $t \in [0,T], i = 1, \cdots, n$. Suppose there exist $\overline{x}_0, \overline{y}_0$ satisfying $A\overline{x}_0 \le \overline{x}_0, A\overline{y}_0 \ge \overline{y}_0$ and $0 \le \overline{x}_0 \le \overline{y}_0, 0 \le t \le T$, then the system of equations (2.18) *has a positive solution. Where A is defined: $A\overline{x}(t) = (A_1\overline{x}_1(t), \cdots, A_n\overline{x}_n(t))$,*

$$A_i\overline{x}_i(t) = \frac{1}{\Gamma(\alpha_i)} \int_0^t (t-\tau)^{\alpha_i - 1} f(\tau, \overline{x}(\tau)) d\tau, \quad i = 1, \cdots, n.$$

iii) Let $f_i : [0,T] \times [0,\infty)^n \to [0,\infty), i = 1, \cdots, n$ be continuous and $f_i(t, \cdot)$ increasing for each $t \in [0,T], i = 1, \cdots, n$. Further, if $0 < \lim\limits_{\|\overline{x}\| \to \infty} f_i(t, \overline{x}(t)) < \infty, i = 1, \cdots, n$, and for each $t \in [0,T]$. Then the system of equations (2.18) *has a positive solution.*

iv) Let $f_i : [0,T] \times [0,\infty)^n \to [0,\infty), i = 1, \cdots, n$ be continuous. Further assume $\|f(t, \overline{x}(t)) - f(t, \overline{y}(t))\| \le L\|\overline{x}(t) - \overline{y}(t)\|, \overline{x}(t), \overline{y}(t) \in [0,\infty)^n, t \in [0,t]$ and $\frac{LT^\alpha}{\gamma} < 1$, where $\gamma = min\Gamma(\alpha_1 + 1), \cdots, \Gamma(\alpha_n + 1)$, then the system of equations (2.18) *has unique solution which is positive.*

Remark 2.4. For equation (2.18), Bonilla et al. extended $0 < \alpha_i < 1$ to $n_i - 1 < Re(\alpha_i) \le n_i, n_i = -[-Re(\alpha_i)], \alpha_i \in \mathbb{C}, i = 1, \cdots, m$, with initial conditions ${}_{RL}D_{0,t}^{\alpha_i - j_i} x_i(t)|_{t=0} = b_{j_i} \in \mathbb{C}, i = 1, \cdots, m$, $j_i = 1, \cdots, n_i$, and they proved the existence and uniqueness results [Bonilla *et al.* (2000)].

In [Babakhani and Daftardar-Gejji (2005)], the authors extended the analysis for multiple-order fractional differential equations:

$$\mathfrak{L}(D)x(t) = f(t,x), \quad (0 < t < 1), \tag{2.19}$$

where $\mathfrak{L}(D) = {}_{RL}D_{0,t}^{\alpha} - \sum\limits_{j=1}^{n-1} a_j {}_{RL}D_{0,t}^{\alpha - j}, a_j > 0$ with initial conditions $x(0) = 0$, $[D^{\alpha-n+1}x(t)]|_{t=0} = x_{n-1} \ge 0, [D^{\alpha-n+j}x(t)]|_{t=0} = x_{n-j}, x_{n-j} \ge \sum\limits_{k=1}^{j-1} a_k x_{k+n-j}, j = 2, 3, \ldots, n-1, n-1 < \alpha < n, n \in \mathbb{N}, f : [0,1] \times [0,\infty) \to [0,\infty)$ is a continuous function.

Theorem 2.13. *i) Let $f : [0,1] \times [0,\infty) \to [0,\infty)$ be continuous. $f(t, \cdot)$ is nondecreasing for each $t \in [0,1]$, and there exists a positive constant a, such that $f(t, \cdot)$ is strictly increasing on $[0,a]$ for each $t \in [0,1]$.*

ii) Suppose there exist v_0, w_0 *satisfying* $\mathfrak{L}(D)v_0 \leq f(t, v_0), \mathfrak{L}(D)w_0 \geq f(t, w_0)$, $0 \leq v_0 \leq w_0, 0 \leq t \leq 1$, *or* $0 \leq \lim\limits_{x\to\infty} \max\limits_{0\leq t\leq 1} \frac{f(t,x(t))}{x(t)} < +\infty$, *or* $0 < \lim\limits_{x\to\infty} f(t, x(t)) < +\infty$ *for each* $t \in [0, 1]$.

Then equation (2.19) *has one positive solution* $x(t)$.

Theorem 2.14. *Let* $f : [0, 1] \times [0, \infty) \to [0, \infty)$ *be continuous and Lipschitz continuous with respect to second variable with Lipschitz constant* L.

(i) a_j *satisfy the following conditions:* $a_j > 0, j = 1, 2, \ldots, n-1$, *and* $0 < \frac{L}{\Gamma(\alpha+1)} + \sum\limits_{j=1}^{n-1} \frac{a_j}{\Gamma(j+1)} < 1$, *then equation* (2.19) *has unique solution, which is also positive.*

(ii) a_j *satisfy* $0 < \frac{L}{\Gamma(\alpha+1)} + \sum\limits_{j=1}^{n-1} \frac{|a_j|}{\Gamma(j+1)} < 1$, *then equation* (2.19) *has unique solution, which necessarily not be positive.*

Remark 2.5. For the following equation:

$$L(D)x(t) = f(t, x), \quad (0 < t < 1), \tag{2.20}$$

where $L(D) = {}_{RL}D_{0,t}^{\alpha_n} - \sum\limits_{j=1}^{n-1} a_j {}_{RL}D_{0,t}^{\alpha_j}, 0 < \alpha_1 < \cdots < \alpha_n < 1, a_j > 0, j = 1, 2, \cdots, n-1, x(0) = 0, f : [0, 1] \times [0, \infty) \to [0, \infty)$ is a continuous function. Theorems 2.13 and 2.14 are also correct except replacing $\Gamma(\alpha + 1), \Gamma(j + 1)$ with $\Gamma(\alpha_n + 1), \Gamma(\alpha_n - \alpha_j + 1)$, respectively [Babakhani and Daftardar-Gejji (2003)].

2.4 Existence and uniqueness of the boundary value problems

In this section, we give some existence and uniqueness results for various classes of boundary value problems.

2.4.1 *Boundary value problems with Riemann-Liouville derivative*

Belmekki et al. studied the nonlinear fractional differential equation [Belmekki *et al.* (2009)]:

$$\begin{cases} {}_{RL}D_{0,t}^{\alpha}x(t) - \lambda x(t) = f(t, x), & t \in I = (0, 1], 0 < \alpha < 1, \\ \lim\limits_{t\to 0+} t^{1-\alpha}x(t) = x(1), \end{cases} \tag{2.21}$$

where f is continuous and $\lambda \in \mathbb{R}$.

Theorem 2.15. *Assume that* f *satisfies Lipschitz condition* (2.3).

(i) *If there exists a constant* $M > 0$ *such that* $|f(t, x)| \leq M, \quad \forall t \in [0, 1], x \in \mathbb{R}$, *then the problem* (2.21) *has at least one solution in* $C_{1-\alpha}[0, 1]$.

(ii) The problem (2.21) *has a unique solution in* $C_{1-\alpha}[0,1]$, *if*

$$L\frac{(\Gamma(\alpha))^2}{\Gamma(2\alpha)}E_{\alpha,\alpha}(|\lambda|)(\frac{\Gamma(\alpha)E_{\alpha,\alpha}(|\lambda|)}{|1-\Gamma(\alpha)E_{\alpha,\alpha}(|\lambda|)|}+1)<1.$$

In [Bai and Lü (2005)], the authors considered the following problem:

$$_{RL}D_{0,t}^{\alpha}x(t)+f(t,x(t))=0,\quad 0<t<1, \tag{2.22}$$

$$x(0)=x(1)=0, \tag{2.23}$$

where $1<\alpha\leq 2$, $f:[0,1]\times[0,\infty)\to[0,\infty)$ is continuous.

Before introducing the theorem, we need give some notations:

$$M=\left(\int_0^1 G(s,s)ds\right)^{-1},\quad N=\left(\int_{1/4}^{3/4}\gamma(s)G(s,s)ds\right)^{-1},$$

in which

$$G(s,s)=\frac{1}{\Gamma(\alpha)}[s(1-s)]^{\alpha-1},\quad s\in(0,1),\quad \gamma(s)=\begin{cases}\frac{[\frac{3}{4}(1-s)]^{\alpha-1}-(\frac{3}{4}-s)^{\alpha-1}}{[s(1-s)]^{\alpha-1}}, & s\in(0,r],\\ \frac{1}{(4s)^{\alpha-1}}, & s\in[r,1),\end{cases}$$

where $\frac{1}{4}<r<\frac{3}{4}$ solves the equation below

$$[\frac{3}{4}(1-s)]^{\alpha-1}-(\frac{3}{4}-s)^{\alpha-1}=\frac{1}{4^{\alpha-1}}(1-s)^{\alpha-1}.$$

Theorem 2.16. *Assume that* $f:[0,1]\times[0,\infty)\to[0,\infty)$ *is continuous.*

(i) If there exist two constants $r_2>r_1>0$ *such that* $f(t,x)\leq Mr_2$ *for* $(t,x)\in[0,1]\times[0,r_2]$, *and* $f(t,x)\geq Nr_1$ *for* $(t,x)\in[0,1]\times[0,r_1]$, *then problem* (2.22)-(2.23) *has at least one positive solution.*

(ii) If there exist constants $0<a<b<c$ *such that* $f(t,x)<Ma$ *for* $(t,x)\in[0,1]\times[0,a]$, $f(t,x)\geq Nb$ *for* $(t,x)\in[1/4,3/4]\times[b,c]$, *and* $f(t,x)\leq Mc$ *for* $(t,x)\in[0,1]\times[0,c]$, *then the boundary value problem* (2.22)-(2.23) *has at least three positive solutions.*

Remark 2.6. Xu et al. discussed (2.22) ($3<\alpha\leq 4$) with boundary condition: $x(0)=x(1)=x'(0)=x'(1)=0$ and also got the existence results [Xu *et al.* (2009)].

El-Sayed et al. proved some local and global existence theorems for a nonlocal boundary value problem [El-Sayed and Abd El-Salam (2009)]:

$$\begin{cases} _{RL}D_{0,t}^{\alpha}x(t)+f(t,x(\phi(t)))=0,\quad 0<t<1, 1<\alpha<2,\\ _{RL}D_{0,t}^{-\beta}x(t)|_{t=0}=0, \lambda x(\eta)=x(1),\quad 0<\beta\leq 1, 0<\eta<1, 0<\lambda\eta^{\alpha-1}<1.\end{cases} \tag{2.24}$$

To facilitate the discussion, let us first state the following assumptions:

(I) $f:(0,1)\times\mathbb{R}\to\mathbb{R}$, $f(t,\cdot)$ is continuous for each $t\in(0,1)$; $f(\cdot,x)$ is measurable for each $x\in\mathbb{R}$; there exist two real functions $a(t),b(t)$ such that

$$f(t,x)\le a(t)+b(t)|x(t)|,\quad \text{for each } t\in(0,1),\quad x\in\mathbb{R}.$$

where $a(\cdot)\in L_1(0,1)$ and $b(\cdot)$ are measurable and bounded, and $L_1(0,1)$ is the Lebesgue integrable function space defined on (0,1).

(II) $\phi:(0,1)\to(0,1)$ is nondecreasing and there exists a constant $M>0$ such that $\phi'\ge M$ a. e. on (0,1).

(III) Assume that every solution $x(\cdot)\in L_1(0,1)$ to the equation

$$x(t)=l(-{}_{RL}D_{0,t}^{-\alpha}f(t,x(\phi(t))))-\frac{\lambda t^{\alpha-1}}{1-\lambda\eta^{\alpha-1}}\int_0^\eta\frac{(\eta-s)^{\alpha-1}}{\Gamma(\alpha)}f(s,x(\phi(s)))ds$$
$$+\frac{t^{\alpha-1}}{1-\lambda\eta^{\alpha-1}}\int_0^1\frac{(1-s)^{\alpha-1}}{\Gamma(\alpha)}f(s,x(\phi(s)))ds)\ \ a.\ e.\ on\ (0,1),$$

holds for $\|x\|\ne r$, (r is arbitrary but fixed, $l\in(0,1)$).

Theorem 2.17. *i) Let the assumptions (I) and (II) be satisfied. If* $\sup|b(t)|<MK$, *where* $K=(1-\lambda\eta^{\alpha-1})\Gamma(\alpha+1)$, *then the nonlocal boundary value problem* (2.24) *has a solution* $x\in B_r$, *where* $B_r=\{x\in L_1(0,1),\|x\|<r,r>0\}$, $r\le\frac{\frac{1}{K}\|a\|}{1-\frac{1}{MK}\sup|b(t)|}$.

ii) Let the conditions (I)-(III) be satisfied, then the nonlocal boundary value problem (2.24) *has at least one solution* $x\in L_1(0,1)$.

Hu and Wang studied a nonlinear fractional differential equation with integral boundary condition of the following form [Hu and Wang (2011)]:

$$\begin{cases}{}_{RL}D_{0,t}^{\alpha}x(t)=f(t,x,{}_{RL}D_{0,t}^{\beta}x),\ \ 0<t<1,\ 1<\alpha\le2,\ 0<\beta<1,\\ x(0)=0,\ \ x(1)=\int_0^1 g(s)x(s)ds,\end{cases}\tag{2.25}$$

where $f:[0,1]\times\mathbb{R}\times\mathbb{R}\to\mathbb{R}$ is continuous, $g\in L^1[0,1]$ satisfies $1-\int_0^1 g(s)s^{\alpha-1}ds>0$.

Theorem 2.18. *Assume that* $f:[0,1]\times\mathbb{R}\times\mathbb{R}\to\mathbb{R}$ *is continuous.*

(i) If there exists a nonnegative function $\phi\in L([0,1])$ *such that* $|f(t,x,y)|\le\phi(t)+c_1|x|^{\sigma_1}+c_2|y|^{\sigma_2}$, $c_1,c_2\in\mathbb{R}$ *are nonnegative constants,* $0<\sigma_1,\sigma_2<1$, *or* $\sigma_1,\sigma_2>1$. *Then equation* (2.25) *has a solution.*

(ii) If there exists a constant $k>0$ *such that* $|f(t,x,y)-f(t,\overline{x},\overline{y})|\le k(|x-\overline{x}|+|y-\overline{y}|)$ *for each* $t\in[0,1]$ *and all* $x,y,\overline{x},\overline{y}\in\mathbb{R}$ *and* $k\rho<1$, *where*

$$\rho=\frac{1+\Theta}{\Gamma(\alpha+1)}+\frac{\Theta}{\alpha\Gamma(\alpha-\beta)}+\frac{1}{\Gamma(\alpha-\beta+1)}+$$
$$\left(\frac{\Theta}{\Gamma(\alpha-\beta)}+\frac{\Theta}{\Gamma(\alpha)}\right)\int_0^1\int_s^1|g(r)|(r-s)^{\alpha-1}drds,$$
$$\Theta=\left(1-\int_0^1 g(s)s^{\alpha-1}ds\right)^{-1}.$$

Then equation (2.25) *has a unique solution.*

For the boundary value problem of the coupled system

$$ {}_{RL}D^{\alpha}_{0,t}x(t) = f(t, y(t), {}_{RL}D^{\mu}_{0,t}y(t)), \quad 0 < t < 1, \tag{2.26}$$

$$ {}_{RL}D^{\beta}_{0,t}y(t) = f(t, x(t), {}_{RL}D^{\nu}_{0,t}x(t)), \quad 0 < t < 1, \tag{2.27}$$

$$ x(0) = x(1) = y(0) = y(1) = 0, \tag{2.28}$$

where $1 < \alpha, \beta < 2, 0 < \mu, \nu, 1 \leq \alpha - \nu, 1 \leq \beta - \mu, f, g : [0,1] \times \mathbb{R} \times \mathbb{R} \to \mathbb{R}$ are continuous. Su gave the following theorem [Su (2009)]:

Theorem 2.19. *Suppose that one of the following conditions is satisfied*

(i) If there exist two nonnegative functions $a(t), b(t) \in L[0,1]$ *such that* $|f(t,x,y)| \leq a(t) + c_1|x|^{\rho_1} + c_2|y|^{\rho_2}$ *and* $|g(t,x,y)| \leq b(t) + d_1|x|^{\theta_1} + d_2|y|^{\theta_2}$ *where* $c_i, d_i \geq 0, 0 < \rho_i, \theta_i < 1$ *for* $i = 1, 2$.

(ii) $|f(t,x,y)| \leq c_1|x|^{\rho_1} + c_2|y|^{\rho_2}$ *and* $|g(t,x,y)| \leq d_1|x|^{\theta_1} + d_2|y|^{\theta_2}$ *where* $c_i, d_i > 0, \rho_i, \theta_i > 1$ *for* $i = 1, 2$.

Then problem (2.25)-(2.28) has a solution.

Remark 2.7. Ahmad and Nieto discussed equations (2.25)-(2.27) with boundary conditions [Ahmad and Nieto (2009)]:

$$ x(0) = 0, \;\; x(1) = \gamma x(\eta), \;\; y(0) = 0, \;\; y(1) = \gamma x(\eta), \;\; \gamma\eta^{\alpha-1} < 1, \;\; \gamma\eta^{\beta-1} < 1, \tag{2.29}$$

and they also gave the existence condition.

2.4.2 *Boundary value problems with Caputo derivative*

Benchohra et al. dealt with equation:

$$ {}_{C}D^{\alpha}_{0,t}x(t) = f(t, x), \quad t \in I = [0, T], \tag{2.30}$$

with the boundary condition:

$$ ax(0) + bx(T) = c, \tag{2.31}$$

where $0 < \alpha < 1, f : [0,T] \times \mathbb{R} \to \mathbb{R}$ is a continuous function, a, b, c are real constants with $a + b \neq 0$, they obtained the following theorem [Benchohra *et al.* (2008b)]:

Theorem 2.20. *i) Assume that* $f : [0,T] \times \mathbb{R} \to \mathbb{R}$ *is continuous, and there exists a constant* $M > 0$ *such that* $f(t,x) \leq M$ *for each* $t \in [0,T]$ *and all* $x \in \mathbb{R}$, *then the boundary value problem (2.30)-(2.31) has at least one solution on* $[0,T]$.

ii) Assume that $f(t,x)$ *satisfies Lipschitz condition (2.3) and* $\frac{LT^{\alpha}\left(1+\frac{|b|}{|a+b|}\right)}{\Gamma(\alpha+1)} < 1$, *then the boundary value problem (2.30)-(2.31) has a unique solution on* $[0,T]$.

Remark 2.8. i) In [Benchohra *et al.* (2008b)], they also discussed existence and uniqueness results for nonlocal problems, namely, replacing (2.31) with

$$x(0) + g(x) = x_0, \tag{2.32}$$

$g : C([0,T], \mathbb{R}) \to \mathbb{R}$ is a continuous function. They gave the conditions when (2.30) with nonlocal condition (2.32) has a unique solution and when (2.30) with nonlocal condition (2.32) has at least one solution on $[0,T]$.

ii) In [Benchohra and Slimani (2009)], the authors considered equation (2.30) with order $1 < \alpha \leq 2$, and nonlocal boundary condition:

$$x(0) = g(x), x(T) = x_T, \tag{2.33}$$

$g : C([0,T], \mathbb{R}) \to \mathbb{R}$ is a continuous function, $x_T \in \mathbb{R}$. They got the conditions when (2.30) ($1 < \alpha \leq 2$) with nonlocal condition (2.33) has a unique solution and when (2.30) ($1 < \alpha \leq 2$) with nonlocal condition (2.33) has at least one solution on $[0,T]$.

iii) In [Benchohra *et al.* (2008a)], the authors considered equation (2.30) with order $1 < \alpha \leq 2$, and non-linear integral conditions of the form:

$$x(0) - x'(0) = \int_0^T g(\tau, x(\tau))d\tau, x(T) + x'(T) = \int_0^T h(\tau, x(\tau))d\tau, \tag{2.34}$$

where $f, g, h : [0,T] \times \mathbb{R} \to \mathbb{R}$ are continuous functions. They got the conditions when (2.30) ($1 < \alpha \leq 2$) with integral condition (2.34) has a unique solution and when (2.30) ($1 < \alpha \leq 2$) with integral condition (2.34) has at least one solution on $[0,T]$.

The author considered the existence of positive solution for the following nonlinear boundary-value problem [Zhang (2006)]:

$$_CD_{0,t}^{\alpha} x(t) = f(t,x), \quad 0 < t < 1, \tag{2.35}$$

with boundary condition:

$$x(0) + x'(0) = 0, x(1) + x'(1) = 0, \tag{2.36}$$

where $1 < \alpha \leq 2, f : [0,1] \times [0,+\infty) \to [0,+\infty)$ is continuous. By means of the fixed-point theorem on a cone, the existence result of positive solutions is obtained.

Theorem 2.21. *Assume that $f(t,x)$ is continuous on $[0,1] \times [0,\infty)$ and satisfies one of the following conditions.*

(i) There exist $0 < \mu_1, \nu_1 < 1$ such that $\lim\limits_{x\to\infty} \frac{f(t,x)}{x^{\mu_1}} = 0; \lim\limits_{x\to 0} \frac{f(t,x)}{x^{\nu_1}} = \infty, t \in [0,1]$.

(ii) There exist $\mu_2, \nu_2 \geq 1$ such that $\lim\limits_{x\to\infty} \frac{f(t,x)}{x^{\mu_2}} = \infty; \lim\limits_{x\to 0} \frac{f(t,x)}{x^{\nu_2}} = 0, t \in [0,1]$.

Then problem (2.35) with boundary condition (2.36) has one positive solution.

Remark 2.9. When boundary condition (2.36) is replaced by:

$$x(0) = x_0 \neq 0, \quad x(1) = x_1 \neq 0, \tag{2.37}$$

the existence of solution for equation (2.35) with order $\alpha(1 < \alpha < 2)$ was also obtained too [Zhang (2006)].

Integral boundary problems have various applications in applied fields such as blood flow problems, chemical engineering, thermoelasticity, underground water flow, population dynamics, and so forth. In [Ahmad and Nieto (2009)], Ahmad and Nieto considered the following boundary value problem with integral boundary conditions

$$\begin{cases} {}_C D_{0,t}^{\alpha} x(t) = f(t, x(t), (\chi x)(t)), \quad 0 < t < 1, 1 < \alpha \leq 2, \\ ax(0) + bx'(0) = \int_0^1 q_1(x(\tau))d\tau, ax(1) + bx'(1) = \int_0^1 q_2(x(\tau))d\tau, \end{cases} \tag{2.38}$$

where $f : [0,1] \times X \times X \to X$, for $\gamma : [0,1] \times [0,1] \to [0,\infty), (\chi x)(t) = \int_0^t \gamma(t,\tau)x(\tau)d\tau, q_1, q_2 : X \to X$ and $a > 0, b \geq 0$ are real constants. Here, $(X, \|\cdot\|)$ is a Banach space.

Theorem 2.22. *Assume that $f : [0,1] \times X \times X \to X$ is jointly continuous and maps bounded subsets of $[0,1] \times X \times X$ into relatively compact subsets of $X, \gamma : [0,1] \times [0,1] \to [0,\infty)$ is continuous with $\gamma_0 = \max\{\gamma(t,s) : (t,s) \in [0,1] \times [0,1]\}$ and $q_1, q_2 : X \to X$ are continuous functions. Further, there exist positive constants $L_1, \overline{L}_1, L_2, L_3, M_2, M_3$ such that*

(I) $\|f(t, x(t), (\chi x)(t)) - f(t, y(t), (\chi y)(t))\| \leq L_1\|x - y\| + \overline{L}_1\|\chi x - \chi y\|$ for all $t \in [0,1], x, y \in X$.
(II) $\|q_1(x) - q_1(y)\| \leq L_2\|x - y\|, \|q_2(x) - q_2(y)\| \leq L_3\|x - y\|$ with $\|q_1(x)\| \leq M_2, \|q_2(x)\| \leq M_3$.

Then, we have the following conclusions:

(i) the boundary value problem (2.38) has a unique solution provided that

$$(L_1 + \gamma_0\overline{L}_1)[\frac{2a+b}{a\Gamma(\alpha+1)} + \frac{ab+b^2}{a^2\Gamma(\alpha)}] + \frac{a+b}{a^2}(L_2 + L_3) < 1,$$

with

$$L_1 + \gamma_0\overline{L}_1 \leq \frac{1}{2}(\frac{2a+b}{a\Gamma(\alpha+1)} + \frac{ab+b^2}{a^2\Gamma(\alpha)})^{-1}.$$

(ii) Assume that (I)-(II) hold with $\|f(t, x(t), (\chi x)(t))\| \leq \mu(t)$ for all

$$(t, x(t), (\chi x)(t)) \in [0,1] \times X \times X,$$

where $\mu \in L^1([0,1], \mathbb{R}^+)$ and

$$(L_1 + \gamma_0\overline{L}_1)[\frac{a+b}{a\Gamma(\alpha+1)} + \frac{ab+b^2}{a^2\Gamma(\alpha)}] + \frac{a+b}{a^2}(L_2 + L_3) < 1.$$

Then the boundary value problem (2.38) has at least one solution on $[0,1]$

It is worthwhile mentioning that the nonlinear term f in many papers is independent of fractional derivative of unknown function $x(t)$. But the opposite case is more difficult and complicated. In [Su and Liu (2007)], the authors discussed the following problem

$$ {}_CD_{0,t}^{\alpha}x(t) = f(t, x, {}_CD_{0,t}^{\beta}x), \tag{2.39} $$

$$ x(0) = x(1) = 0, \quad or \quad x'(0) = x(1) = 0, \quad or \quad x(0) = x'(1) = 0, \tag{2.40} $$

where $0 < t < 1, 1 < \alpha \leq 2, 0 < \beta \leq 1$, f is continuous on $[0, 1] \times \mathbb{R} \times \mathbb{R}$.

Theorem 2.23. *Suppose that f satisfies the following conditions,*

(i) there exist nonnegative functions $a(t), h(x, y)$ such that $|f(t, x, y)| \leq a(t) + h(x, y)$, where $a(t) \in L(0, 1)$ and $h(x,\ y)$ is continuous on $\mathbb{R} \times \mathbb{R}$.

(ii) $\lim\limits_{|x|+|y|\to\infty} \frac{h(x,y)}{|x|+|y|} < \frac{\Gamma(\alpha+1)\Gamma(2-\beta)}{(3\alpha+1)\Gamma(2-\beta)+2\alpha}$ for $0 < \beta < 1$, $\lim\limits_{|x|+|y|\to\infty} \frac{h(x,y)}{|x|+|y|} < \frac{\Gamma(\alpha+1)}{3\alpha+1}$ for $\beta = 1$.

Then, there exists a solution $x(t)$ for problem (2.39)-(2.40).

Remark 2.10. i) Recently, Khan et al. considered equation (2.39) ($0 \leq t \leq T, 1 < \alpha < 2, 0 < \beta < 1$), with integral boundary conditions,

$$ ax(0) - bx'(0) = \int_0^T g(\tau, x(\tau))d\tau, cx(T) + dx'(T) = \int_0^T h(\tau, x(\tau))d\tau, \tag{2.41} $$

where $f : [0, T] \times \mathbb{R} \times \mathbb{R} \to \mathbb{R}, g, h : [0, T] \times \mathbb{R} \to \mathbb{R}$ are continuous, $a, d \geq 0, b, c > 0$ (or $a, d > 0, b, c \geq 0$). They gave conditions when equation (2.39) with integral boundary conditions (2.41) has a unique solution and when equation (2.39) with integral boundary conditions (2.41) has at least one solution on $[0, T]$ [Khan *et al.* (2011)].

ii) In [Liu *et al.* (2009)], for equation (2.39), the authors extended $1 < \alpha \leq 2$ to $1 < \alpha$ ($\beta = 1, 0 \leq t \leq 1$), with the boundary conditions:

$$ x(0) = \int_0^1 g(\tau, x(\tau))d\tau, x(1) = \int_0^1 h(\tau, x(\tau))d\tau, x^{(k)}(0) = 0, k = 2, 3, \cdots, [\alpha] - 1, \tag{2.42} $$

and obtained the existence and uniqueness results.

In [Qiu and Bai (2008)], Qiu et al. discussed the nonlinear fractional differential equation:

$$ \begin{cases} {}_CD_{0,t}^{\alpha}x(t) + f(t, x(t)) = 0, & 0 < t < 1, \\ x(0) = x'(1) = x''(0) = 0, \end{cases} \tag{2.43} $$

where $2 < \alpha \leq 3, f : (0, 1] \times [0, \infty) \to [0, \infty)$ is continuous with $\lim\limits_{t\to 0+} f(t, \cdot) = \infty$ (namely, f is singular at $t = 0$).

Theorem 2.24. *Let $0 < \sigma < 1, t^{\sigma}f(t, x)$ is continuous on $[0, 1] \times [0, \infty]$.*

(i) Assume that there exist two distinct positive constants $\rho, \mu, (\rho > \mu)$ *such that*

$$t^\sigma f(t,x) \leq \rho \frac{\Gamma(\alpha-\sigma)}{\Gamma(1-\sigma)}, \quad for \quad (t,x) \in [0,1] \times [0,\rho],$$

$$t^\sigma f(t,x) \geq \mu \frac{\Gamma(\alpha-\sigma)}{\Gamma(1-\sigma)}, \quad for \quad (t,x) \in [0,1] \times [0,\mu].$$

Then (2.43) *has at least one positive solution.*

(ii) There exists a continuous, nondecreasing function $\varphi : [0,\infty) \to [0,\infty)$ *with* $t^\sigma f(t,x) \leq \varphi(x)$*, for* $(t,x) \in [0,1] \times [0,\infty)$*, and* $r > 0$ *with* $\frac{r}{\varphi(r)} > \frac{\Gamma(\alpha-\sigma)}{\Gamma(1-\sigma)}$. *Then* (2.43) *has one positive solution.*

Remark 2.11. El-Shahed discussed the nonlinear fractional differential equation:

$$_C D_{0,t}^\alpha x(t) + \lambda h(t) f(x(t)) = 0, \quad 0 < t < 1, 2 < \alpha \leq 3, \tag{2.44}$$

and obtained the existence and nonexistence of positive solutions for different boundary conditions:

$$x(0) = x'(0) = x'(1) = 0, \tag{2.45}$$

or

$$x(0) + x''(0) = 0, ax'(1) + bx''(1) = 0, \tag{2.46}$$

where λ is a positive parameter, $h : (0,1) \to [0,\infty)$ is continuous with $\int_0^1 h(t)dt > 0$ and $f : [0,\infty) \to [0,\infty)$ is continuous, a, b are constants [El-Shahed (2007a,b)].

2.4.3 *Fractional differential equations with impulsive boundary conditions*

Impulsive differential equations have become important in recent years. Benchohra and Slimani first considered fractional differential equation with impulsive boundary conditions [Benchohra and Slimani (2009)]:

$$_C D_{0,t}^\alpha x(t) = f(t,x), \quad t \in I = [0,T], t \neq t_k, \tag{2.47}$$

$$\Delta x|_{t=t_k} = H_k(x(t_k^-)), \tag{2.48}$$

$$x(0) = x_0, \tag{2.49}$$

where $k = 1, \cdots, m, 0 < \alpha \leq 1, f : I \times \mathbb{R} \to \mathbb{R}, H_k : \mathbb{R} \to \mathbb{R}, x_0 \in \mathbb{R}, 0 = t_0 < t_1 < \cdots < t_m < t_{m+1} = T, \Delta x|_{t=t_k} = x(t_k^+) - x(t_k^-), x(t_k^+) = \lim_{h\to 0^+} x(t_k + h)$ and $x(t_k^-) = \lim_{h \to 0^-} x(t_k + h)$ represent the right and left limits of $x(t)$ at $t = t_k$.

Theorem 2.25. *Assume that:*

(i) *There exists a constant* $L > 0$ *such that* $|f(t,x) - f(t,y)| \leq L|x-y|$, *for each* $t \in I$, $x, y \in \mathbb{R}$, *and a constant* $L^* > 0$ *such that* $|H_k(x) - H_k(y)| \leq L^*|x-y|$, *for each* $x, y \in \mathbb{R}$ *and* $k = 1, \cdots, m$, *if* $\frac{T^\alpha L(m+1)}{\Gamma(\alpha+1)} + mL^* < 1$, *then* (2.47)-(2.49) *has a unique solution on* I.

(ii) $f : I \times \mathbb{R} \to \mathbb{R}$ *is continuous, there exists a constant* $M > 0$ *such that* $|f(t,x) \leq M$ *for each* $t \in I$, $x \in \mathbb{R}$. $H_k : \mathbb{R} \to \mathbb{R}$ *are continuous, there exists a constant* $M^* > 0$ *such that* $|H_k(x) \leq M^*$ *for each* $x \in \mathbb{R}, k = 1, \cdots, m$, *then* (2.47)-(2.49) *has at least one solution on* I.

Remark 2.12. i) In [Benchohra and Slimani (2009)], the nonlocal condition is also discussed, namely replace (2.49) by

$$x(0) + g(x) = x_0, \tag{2.50}$$

where $g : PC(I,\mathbb{R}) \to \mathbb{R}$ is continuous. $PC(I,\mathbb{R}) = \{x : I \to \mathbb{R} : x \in C((t_k, t_{k+1}],\mathbb{R}), k = 0, \cdots, m$ and there exist $x(t_k^-)$ and $x(t_k^+)$, $k = 1, \cdots, m$, with $x(t_k^-) = x(t_k^+)\}$.

ii) Agarwal et al. extended $0 < \alpha \leq 1$ to $1 < \alpha \leq 2$ [Agarwal *et al.* (2008)],

$$_C D_{0,t}^{\alpha} x(t) = f(t,x), \quad t \in I = [0,T], t \neq t_k, \tag{2.51}$$

$$\Delta x|_{t=t_k} = H_k(x(t_k^-)), \Delta x'|_{t=t_k} = \overline{H}_k(x(t_k^-)), \tag{2.52}$$

$$x(0) = x_0, x'(0) = x_1, \tag{2.53}$$

for nonlocal condition, (2.53) is replaced by

$$x(0) + g(x) = x_0, x'(0) = x_1, \tag{2.54}$$

similar conclusions to [Benchohra and Slimani (2009)] are given too.

iii) Ahmad and Sivasundaram in [Ahmad and Sivasundaram (2009)] considered (2.51)–(2.52) ($T = 1$) with boundary condition

$$x(0) + x'(0) = 0, x(1) + x'(1) = 0, \tag{2.55}$$

and got the existence and uniqueness result.

iv) In [Ahmad and Sivasundaram (2010)], the authors considered (2.51)-(2.52) ($T = 1$) with integral boundary condition given by

$$ax(0) + bx'(0) = \int_0^1 h_1(x(\tau))d\tau, \quad ax(1) + bx'(1) = \int_0^1 h_2(x(\tau))d\tau, \tag{2.56}$$

where $h_1, h_2 : \mathbb{R} \to \mathbb{R}, a > 0, b \geq 0$, and got existence and uniqueness result too.

Recently, Zhang and Wang [Zhang and Wang (2011)] considered fractional differential equations with impulses and anti-periodic boundary condition:

$$\begin{cases} _C D_{0,t}^{\alpha} x(t) = f(t,x), & t \in I = [0,T], t \neq t_k, 1 < \alpha \leq 2, \\ \Delta x|_{t=t_k} = H_k(x(t_k^-)), & \Delta x'|_{t=t_k} = \overline{H}_k(x(t_k^-)), \\ x(0) = -x(T), & x'(0) = -x'(T), \end{cases} \tag{2.57}$$

Theorem 2.26. *Assume one of the following conditions is satisfied,* *(i)* $\lim_{x\to 0} \frac{f(t,x)}{x} = 0, \lim_{x\to 0} \frac{H_k(x)}{x} = 0, \lim_{x\to 0} \frac{\overline{H}_k(x)}{x} = 0.$

(ii) *There exist positive constants* $L_i(i = 1, 2, 3)$ *such that* $|f(t,x) \leq |L_1, |H_k(x) \leq |L_2, |\overline{H}_k(x) \leq |L_3$ *for each* $t \in I$, $x \in \mathbb{R}$ *and* $k = 1, \cdots, m$.

Then, problem (2.57) *has at least one solution.*

2.5 Existence and uniqueness of the fractional differential equations with time-delay

It is well known that fractional differential equations with time-delay provide mathematical models for real-world problems in which the fractional rate of change depends on the influence of their hereditary effects. Lakshmikantham discussed the following equation [Lakshmikantham (2008)]

$$\begin{cases} {}_C D^{\alpha}_{t_0,t} x(t) = f(t, x(t+s)), & -\tau \leq s \leq 0,\ t_0 \leq t \leq T, \\ x(t_0 + s) = \phi_0(s) \in \mathfrak{J}, \end{cases} \tag{2.58}$$

where $\mathfrak{J} = C([-\tau, 0], \mathbb{R})$, for any $\phi \in \mathfrak{J}$, $|\phi|_0 = \max_{-\tau \leq s \leq 0} |\phi_s|$.

Theorem 2.27. *Let* $\mathfrak{J}_\rho = \{\phi \in \mathfrak{J} : |\phi|_0 < \rho\}, f \in C([t_0, T] \times \mathfrak{J}_\rho, \mathbb{R})$. *Then given an initial function* $\phi_0 \in \mathfrak{J}$ *at* $t = t_0$, *there exists a constant* $a > 0$ *such that there is a solution* $x(t)$ *of* (2.58) *existing on* $[t_0 - \tau, t_0 + \alpha]$.

El-Sayed dealt with the following problem [El-Sayed (1998)]:

$$\begin{cases} {}_C D^{\alpha}_{0,t} x(t) = f(t, x, {}_C D^{\alpha_1}_{0,t} x(t-r), \cdots, {}_C D^{\alpha_n}_{0,t} x(t-nr)), & t \in I = (0,1), \\ x^{(j)}(0) = 0, \quad for\ t \leq 0,\ j = 0, 1, \cdots, n, \end{cases} \tag{2.59}$$

where $n < \alpha \leq n+1, k-1 < \alpha_k \leq k, k = 1, 2, \cdots, n$, and $\alpha_0 = 0, r$ is a positive constant.

Theorem 2.28. *Assume that* f *satisfies the following conditions:*

(i) $f(t,U) : I \times \mathbb{R}^{n+1} \to \mathbb{R}$ *satisfies the Carathéodory conditions, i.e. it is measurable in* $t \in I$ *for any* $U \in \mathbb{R}^{n+1}$ *and continuous in* $U \in \mathbb{R}^{n+1}$ *for almost all* $t \in I$.

(ii) $f(t,U) = f(t, u_1(t), u_2(t), \cdots, u_n(t))$ *is nondecreasing for all variables, i.e. if* $u_k(t_1) \leq u_k(t_2))$ *for* $t_1 \leq t_2$, *then,* $f(t_1, U_1) \leq f(t_2, U_2)$, *where* $U_1 = U(t_l)$, *and* $U_2 = U(t_2)$.

(iii) *There exists a function* $a(t) \in L_1$ *and constants* $b_k > 0$, *such that* $|f(t,U)| \leq a(t) + \sum_{k=0}^{n} b_k |u_k(t)|$ *holds for all* $(t,U) \in I \times \mathbb{R}^{n+1}$ *and* $\frac{1}{\Gamma(1+\alpha-n)} \sum_{k=0}^{n} \frac{b_k}{\Gamma(1+\alpha-n)} < 1$.

Then problem (2.59) *has at least one (nondecreasing) solution* $x(t) \in C^{n-1}(I)$.

Remark 2.13. For Riemann-Liouville derivative, Bai considered the equation:

$$\begin{cases} {}_{RL}D^{\alpha}_{0,t}x(t) = f(t, x(t+s)), \quad 0 \le t \le 1, \\ x(s) = \phi(s), \quad -\tau \le s \le 0, \ x'(0) = \cdots = x^{n-2}(0) = 0, \ x^{n-2}(1) = 0, \end{cases} \tag{2.60}$$

where $n-1 < \alpha \le n, n \ge 3, f(t, x(t+s)) : [0,1] \times \mathcal{J}$ is continuous, $\mathcal{J} = C([-\tau, 0], \mathbb{R}), \phi$ is an element of the space $\mathcal{J}^+ = \{\varphi \in \mathcal{J} : \varphi(s) \ge 0, s \in [-\tau, 0], \varphi(0) = 0\}$, and gave the existence results of positive solutions [Bai (2010)].

Zhang considered the following problem [Zhang (2008)]:

$$\begin{cases} {}_{RL}D^{\alpha}_{0,t}x(t) = A_0 x(t) + A_1 x(t-r) + f(t), \quad 0 \le t \le T, \\ x(t) = \phi(t), \quad t \in [-r, 0], \end{cases} \tag{2.61}$$

where $0 < a < 1, \phi$ is a given continuous function on $[-r, 0]$, A_0 and A_1 are constants system matrices of appropriate dimensions, and $r > 0$ is a constant, $f(t)$ is a given continuous function on $[0, T]$.

Theorem 2.29. *Let $\phi \in C$ be continuous, fix $T > 0$, and assume $f(t) \in C([0,T], \mathbb{R}^n)$, then the initial value problem* (2.61) *exists a unique continuous solution $x(t)$ defined on $[0, T]$.*

Remark 2.14. Furthermore, replacing $A_0 x(t) + A_1 x(t-r) + f(t)$ with $f(t, x(t), x(t-r))$, Abbas also proved the existence results [Abbas (2011)].

Recently, Guo and Liu discussed the fractional differential equations with infinite delay [Guo and Liu (2011)]:

$$\begin{cases} {}_{RL}D^{\alpha}_{0,t}(x_i(t) - g_i(t, x_{it})) = f_i(t, x_{1t}, x_{2t}), \quad 0 \le t \le b, \\ x_i(t) = \phi(t), \quad t \in (-\infty, 0], \end{cases} \tag{2.62}$$

where $0 < \alpha < 1, x_{it} = x_i(t+s), s \in (-\infty, 0], f_i : [0,b] \times B^2 \to \mathbb{R}, g_i : [0,b] \times B \to \mathbb{R}$ such that $g_i(0, \phi_i) = 0, \phi_i \in B, \phi_i(0) = 0, i = \{1, 2\}$ and B is a phase space. In the theory of functional differential equations, a usual choice of the phase space B is a seminormed space of functions mapping $(-\infty, 0]$ into $\mathbb{R}$ satisfying the following fundamental axioms:

If $x : (-\infty, b] \to \mathbb{R}$ and $x_0 \in B$, then for all $t \in [0, b]$, the following conditions hold:

(i) x_t is in B,
(ii) $\|x_t\|_B \le K(t) \sup\{|x(s)| : 0 \le s \le t\} + M(t)\|x_0\|_B$,
(iii) $|x(t)| \le H\|x_t\|_B$,

where $H > 0$ is a constant, $k : [0, b] \to [0, +\infty)$ is continuous, $M : [0, +\infty) \to [0, +\infty)$ is locally bounded.

Theorem 2.30. *Assume that there exist $l_1, l_2, c_1, c_2 > 0$ such that $|f_i(t, u_1, v_1) - f_i(t, u_2, v_2)| \le l_i \max\{\|u_1 - u_2\|_B + \|v_1 - v_2\|_B\}, |g_i(t, u_1) - g_i(t, u_2)| \le c_i\|u_1 -$*

$u_2\|_B, \forall t \in [0, b], u_i, v_i \in B, i = \{1, 2\}$. *If* $[c\Gamma(\alpha+1)+b^\alpha]\max\{K_{1b}, K_{2b}\} < \Gamma(\alpha+1)$, *where* $c = \max\{c_1, c_2\}, l = \max\{l_1, l_2\}, K_{ib} = \sup\{|k_i(t)| : t \in [0, b]\}$, *then equation* (2.62) *has a unique solution.*

2.6 Conclusions

In this chaper, we present a brief survey on the existence and uniqueness results of fractional differential equations. These equations include initial value problem for fractional differential equations, boundary value problem for fractional differential equations, fractional differential equations with time-delay. These collected results reflect the present recent advances of the various kinds of fractional differential equations.

Bibliography

Abbas, S. (2011). Existence of solutions to fractional order ordinary and delay differential equations, *Electronic Journal of Differential Equations* **09**, 2011, pp. 1–11.

Agarwal, R. P., Benchohra, M. and Slimani, B. A. (2008). Existence results for differential equations with fractional order and impulses, *Memoirs on Differential Equations and Mathematical Physics* **44**, pp. 1–21.

Ahmad, B. and Nieto, J. J. (2009). Existence results for nonlinear boundary value problems of fractional integrodifferential equations with integral boundary conditions, *Boundary Value Problems* **2009**, Article ID 708576, 11 pages.

Ahmada, B. and Nieto, J. J. (2009). Existence results for a coupled system of nonlinear fractional differential equations with three-point boundary conditions, *Comp. Math. Appl.* **58**, pp. 1838–1843.

Ahmad, B. and Sivasundaram, S. (2009). Existence results for nonlinear impulsive hybrid boundary value problems involving fractional differential equations, *Nonlinear Analysis: Hybrid Systems* **3**, pp.251–258.

Ahmad, B. and Sivasundaram, S. (2010). Existence of solutions for impulsive integral boundary value problems of fractional order, *Nonlinear Analysis: Hybrid Systems* **4**, pp. 134–141.

Al-Bassam, M. A. (1965). Some existence theorems on differential equations of generalized order, *J. Reine. Angew. Math.* **1**, 218, pp. 70–78.

Babakhani, A. and Daftardar-Gejji, V. (2003). Existence of positive solutions of nonlinear fractional differential equations, *J. Math. Anal. Appl.* **278**, pp. 434–442.

Babakhani, A. and Daftardar-Gejji, V. (2005). Existence of positive solutions for N-term non-autonomous fractional differential equations, *Positivity* **9**, pp. 193–206.

Bagley, R. L.and Calico, R. A. (1991). Fractional order state equations for the control of viscoelastically damped structures, *J. Guid. Contr. Dyn.* **14**, pp. 304–311.

Bai, C. Z. (2010). Existence of Positive Solutions for a Functional Fractional Boundary Value Problem, *Abstract and Applied Analysis* **2010**, Article ID 127363 13 pages.

Bai, Z. B., and Lü, H. S. (2005). Positive solutions for boundary value problem of nonlinear fractional differential equation, *J. Math. Anal. Appl.* **311**, pp. 495–505.

Balachandran, K. and Trujillo, J. J. (2010). The nonlocal Cauchy problem for nonlinear fractional integrodifferential equations in Banach spaces, *Nonlinear Analysis* **72**, pp. 4587–4593.

Belmekki, M., Nieto, J. J. and Rodríguez-López, R. (2009). Existence of periodic solution for a nonlinear fractional differential equation, *Boundary Value Problems* **2009**, Article ID 324561, 18 pages.

Benchohra, M., Graefb, J. R. and Hamani, S. (2008a) Existence results for boundary value problems with non-linear fractional differential equations, *Applicable Analysis*, **7**, 87, pp. 851–863.

Benchohra, M., Hamani, S. and Ntouyas, S. K. (2008b). Boundary value problems for differential equations with fractional order, *Surveys in Mathematics and its Applications* **3**, pp. 1–12.

Benchohra, M., Hamani, S. and Ntouyas, S. K. (2009). Boundary value problems for differential equations with fractional order and nonlocal conditions, *Nonlinear Analysis* **71**, pp. 2391–2396.

Benchohra, M. and Slimani, B. A. (2009). Existence and uniqueness of solutions to impulsive fractional differential equations, *Electronic Journal of Differential Equations* **10**, 2009, pp. 1–11.

Bonilla, B.,Kilbas, A. A. and Trujillo, J. J. (2000). Systems of nonlinear fractional differential equations in the space of summable functions, *Tr. Inst. Mat. Minsk* **6**, pp. 38–46.

Daftardar-Gejji, V. (2005). Positive solutions of a system of non-autonomous fractional differential equations, *J. Math. Anal. Appl.* **302**, pp. 56–64.

Daftardar-Gejji, V. and Babakhani, A. (2004). Analysis of a system of fractional differential equations, *J. Math. Anal. Appl.* **293**, pp. 511–522.

Delbosco, D. and Rodino, L. (1996). Existence and uniqueness for a nonlinear fractional differential equaiton, *J. Math. Anal. Appl.* **2**, **204**, pp. 609–625.

Diethelm, K. and Jord, N. (2002). Analsysis of fractional differential equations, *J. Math. Anal. Appl.* **265**, pp. 229–248.

El-Sayed, A. M. A. (1998). Nonlinear functional differential equations of arbitrary orders, *Nonlinear Analysis* **4**, 33, pp. 181–186.

El-Sayed, A. M. A. and Abd El-Salam, Sh. A. (2009). Nonlocal Boundary Value Problem of a Fractional-Order Functional Differential Equation, *International Journal of Nonlinear Science* **4**, 7, pp. 436–442.

El-Shahed, M. (2007a). Positive Solutions for Boundary Value Problem of Nonlinear Fractional Differential Equation, *Abstract and Applied Analysis* Article ID 10368, pp. 1–8.

El-Shahed, M. (2007b). On the existence of positive solutions for a boundary valued problem of fractional Order, *Thai Journal of Mathematics* **1**, 5, pp. 143–150.

Furati, K. M. and Tatar, N. (2004). An existence result for a nonlocal fractional differential problem, *J. Frac. Cal.* **26**, pp. 43–51.

Guo, Z. Y. and Liu, M. (2011). Unique solutions for systems of fractional order differential equations with infinite delay, *Bulletin of Mathematical Analysis and Applications* **1**, 3, pp. 142–147.

Hayek, N., Trujillo, J., Rivero, M., Bonilla, B. and Moleno, J. C. (1999) An extension of Picard-Lindelöff theorem to fractional differential equations, *Appl. Anal* **3-4**, 70, pp.347–361.

Hu, M. and Wang, L. L. (2011). Existence of solutions for a nonlinear fractional differential equation with integral boundary condition, *International Journal of Mathematical and Computer Sciences* 7:1, **2011**, 4 pages.

Ichise, M., Nagayanagi, Y. and Kojima, T. (1971). n analog simulation of noninteger order transfer functions for analysis of electrode process, *J. Electroanal. Chem* **33**, pp. 253–265.

Khan, R. A., Rehman, M. U. and Henderson, J. (2011). Existence and uniqueness of solutions for nonlinear fractional differential differential with integral boundary conditions, *Fractional Differential Equaitons* **22**, 1, pp. 29–43.

Kilbas, A. A. and Marzan, S. A. (2005). Nonlinear differential equations with the Caputo fractional derivative in the space of continuously differentiable functions, *Differential Equations* **1**, 41, pp. 84–89.

Kilbas, A. A., Srivastava, H. M. and Trujillo, J. J. (2006). *Theory and Applications of Fractional Differential Equationss* (Elsevier, New York).

Kou, C. H., Zhou, H. C. and Li, C. P. (2012). Existence and continuation theorems of R-L type fractional differential equations, *Int. J. Bifurcation Chaos* **22(4)**, 1250077 (12 pages).

Kusnezov, D., Bulgac, A. and Dang, G. D. (1999). Quantum levy processes and fractional kinetics, *Phys. Rev. Lett* **82**, pp. 1136–1139.

Lakshmikantham, V. (2008). Theory of fractional functional differential equations, *Nonl. Anal.* **69**, pp. 3337–3343.

Laskin, N. (2000). Fractional market dynamics, *Phys. A: Statistical Mechanics and its Applications* **287**, pp. 482–492.

Li, C. P. and Ma, Y. T. (2012). Fractional dynamical system and its linearization theorem, *Nonlinear Dynamics*, DOI: 10.1007/s11071-012-0601-1, in press.

Li, C. P., Gong, Z. Q., Qian, D. L. and Chen, Y. Q. (2010). On the bound of the Lyapunov exponents for the fractional differential systems, *Chaos* **20(1)**, 013127.

Li, Q. P., Sun, S. R., Han, Z. L. and Zhao, Y. G. (2010) On the existence and uniqueness of solutions for initial value problem of nonlinear fractional differential equations, (2010) *2010 Sixth IEEE/ASME International Conference on Mechatronic and Embedded Systems and Applications* (Qingdao, China), pp. 452–457.

Liu, X. P., Jia. M., and Wu, B. F. (2009). Existence and uniqueness of solution for fractional differential equations with integral boundary conditions, *Electronic Journal of Qualitative Theory of Differential Equations* **69**, pp. 1–10.

Machado, J. T., Kiryakova, V. and Mainardi, F. (2010). Recent history of fractional calculus, *Commun. Nonlinear Sci. Numer. Simulat.* `doi:10.1016/j.cnsns.2010.05.027`.

Matignon, D. (1998). Stability properties for generalized fractional differential systems, *Proceedings of the Colloquium FDS'98: Fractional Differential Systems: Models, Methods and Applications, Paris, 1998*, edited by D. Matignon, G. Montseny, Vol. 5 (SMAI, Paris, 1998), pp. 145–158.

Momani, S., Jameel, A. and Al-azawi, S. (2007). Local and global uniqueness theorems on fractional integro-differential equations via Bihari's and Gronwall's inequalities, *Soochow Journal of Mathematics* **4**, 33, pp. 619–627.

Oldham, K. B. and Spanier, J. (1974). *The Fractional Calculus* (Academic Press, New York).

Pither, E. and Sewell, W. E. (1938). Existence theorems for solutions of differential equations of non-integral order, *Bull. Amer. Math. Soc.* **55**, 152, pp. 100–107.

Podlubny, I. (1999) *Fractional Differential Equations* (Academic Press, New York).

Qiu, T. T. and Bai, Z. B. (2008). Existence of positve solutions for singular fractional differential equations, *Electronic Journal of Differential Equations* **146**, pp. 1–9.

Samko, S. G., Kilbas, A. A. and Marichev, O. I. (1993). *Fractional Integrals and Derivatives: Theory and Applications* (Gordon and Breach, New York).

Su, X. (2009). Boundary value problem for a coupled system of nonlinear fractional differential equations, *Appl. Math. Lett.* **22**, pp. 64–69.

Su, X. W., and Liu, L. D. (2007). Existence of solution for boundart value problem of nonlinear fractional differentianl equation, *Appl. Math. J. Chinese Univ. Ser. B* **22**, pp. 291–298.

Sun, H. H., Abdelwahab, A. A. and Onaral, B. (1984). Linear approximation of transfer function with a pole of fractional power, *IEEE Trans. Auto. Contr.* **29**, 5, pp. 441–444.

Tavazoei, M. S. and Haeri, M. (2009). A note on the stability of fractional order systems. Mathematics and Computers in Simulation, *Mathematics and Computers in Simulation* **79**, 5, pp. 1566–1576.

Torvik, P. J. and Bagley, R. L. (1984). On the appearance of the fractional derivatives in the behavior of real materials, *J. Appl. Mech.* **51**, pp. 294–298.

Xu, X. J., Jiang, D. Q. and Yuan, C. J. (2009). Multiple positive solutions for the boundary value problem of a nonlinear fractional differential equation, *Nonlinear Analysis* **71**, pp. 4676–4688.

Yu, C. and Gao, G. Z. (2005). Existence of fractional differential equations, *J. Math. Anal. Appl.* **310**, pp. 26–29.

Yu, C., and Gao, G. Z. (2005). On the solution of nonlinear fractional order differential equation, *Nonlinear Analysis* **63**, pp. 971–976.

Zhang, S. Q. (2000). The existence of positive solution for a nonlinear fractional differential equation, *J. Math. Anal. Appl.* **252**, pp. 804–812.

Zhang, S. Q. (2006). Positive solutions for boundary-value problems of nonlinear fractional differential equations, *Electronic Journal of Differential Equations*, **36**, pp. 1–12.

Zhang, S. Q. (2006). Existence of solution for a boundary value problem of fractional order, *Acta Mathematica Scientia* **2**, pp. 220–228.

Zhang, S. Q. (2009). Monotone interative method for initial value problem involving Riemann-Liouville fractional derivative, *Nonlinear Analysis* **71**, pp. 2087–2093.

Zhang, L. H. and Wang, G. T. (2011). Existence of solutions for nonlinear fractional differential equations with impulses and anti-periodic boundary conditions, *Electronic Journal of Qualitative Theory of Differential Equations* **7**, pp. 1–11.

Zhang, X. Y. (2008). Some results of linear fractional order time-delay system, *Appl. Math. Comput.* **197**, pp. 407–411.

Chapter 3

Finite element methods for fractional differential equations[1]

Changpin Li* and Fanhai Zeng

Department of Mathematics, Shanghai University, Shanghai 200444, PR China

**lcp@shu.edu.cn*

In this chapter, the existing finite element methods (FEMs) for the fractional differential equations (FDEs) are displayed. Recently, the FEMs for FDEs mainly focus on the fractional kinetic equations of diffusion or dispersion with time, space and time-space derivatives. We introduce the theoretical framework for the variational solutions of the considered equations. The existence and uniqueness results are displayed, and the error estimates are also given.

keywords: fractional differential equations, finite element method, Caputo derivative, Riemman–Liouville derivative

3.1 Introduction

Fractional derivatives are as old as their familiar counterparts of the classical one, which can be dated back to 1695 in a letter to L'Hospital by Leibniz. In the past few decades, fractional calculus has attracted much attention and interest due to their applications in science and engineering [Agrawal and Baleanu (2007); Kilbas *et al.* (2006); Li and Deng (2007); Podlubny (1999); Oldham and Spanier (1974); Schmidt and Gaul (2002); Sorrentino and Fasana (2007)]. Recently, fractional derivatives have provided excellent tools to describe various materials and processes with memory and hereditary properties, and model anomalous diffusion. There are several ways to solve the FDEs, such as the analytical methods of the Fourier transform method, the Laplace transform method, the Mellin transform method and the Green function method, which are used to seek the closed-form analytical solutions of FDEs, while there are few simple cases in which the analytical

[1]The present work was supported by the National Natural Science Foundation of China (grant no. 10872119), the Shanghai Leading Academic Discipline Project (grant no. S30104), and the Key Program of Shanghai Municipal Education Commission (grant no. 12ZZ084).

solutions are available [Podlubny (1999)]. Hence, developing efficient and reliable numerical methods for solving FDEs is of special usefulness in applications.

There are various numerical methods for solving the classical differential equations (DEs), while numerical methods for FDEs are relatively limited. Furthermore, the numerical methods for FDEs are more complicated than those for the classical DEs, since the fractional differential operators are not local operators and the adjoint of a fractional differential operator is not the negative of itself [Ervin and Roop (2006, 2007)]. Meanwhile, the coefficient matrix derived from these methods are often not sparse [Roop (2006)], and at each time level n $(n > 0)$, all the numerical values of the previous time levels (generally, from time level 0 to $n-1$) are often needed to get the numerical solution of the present time level n, which contributes to the complexity and expensiveness in real applications. Up to now, there are some works in this respect, such as the short memory principle, see [Deng (2007); Ford and Simpson (2001); Li *et al.* (2011); Podlubny (1999)] and the references therein.

At present, there are several numerical methods to solve FDEs, such as the finite difference methods [Langlands and Henry (2005); Li *et al.* (2011); Sun and Wu (2006); Tadjeran and Meerschaert (2007); Yuste (2006); Li and Zeng (2012a,b); Zhuang *et al.* (2008)], the spectral methods [Hanert (2010); Li and Xu (2009); Li *et al.* (2012)], the homotopy perturbation method (HPM), the variational iteration method (VIM) and the decomposition method, and so on [Momani and Odibat (2007); Li and Wang (2009); Vanani and Aminataei (2011)]. In this chapter, we just review almost all the existing FEMs for the FDEs of different types. If some important references happened not to be here, we do apologize for these omissions. In recent publications, FEMs for the FDEs almost concentrate on the kinetic equations of diffusion (subdiffusion and superdiffusion), diffusion-advection, and Fokker–planck type, and so on. We will outline the existing framework of FEMs for some typical FDEs and their main results (including consistency, stability and convergence), which can be seen as the extension of the classical DEs. Meanwhile, we also introduce some related problems such as the least squares methods for FDEs and the variational principles developed for fractional variational problems.

This chapter is organized as follows. In Section 3.2, we introduce several definitions for different types of fractional derivatives, the spaces and related notations used in the numerical methods for FDEs. The existing framework of FEMs and some related problems are surveyed in Section 3.3. Conclusions and remarks are included in the last section.

3.2 Preliminaries and notations

In this section, we will introduce the definitions of different fractional derivatives and notations used to describe the numerical methods prescribed in this chapter. There are several different ways to define the fractional derivatives, and the most common used fractional derivatives are the Grünwald–Letnikov derivative, the Riemann–

Liouville derivative and the Caputo derivative [Li *et al.* (2009); Li and Deng (2007); Li *et al.* (2011); Li and Zhao (2011); Podlubny (1999); Samko *et al.* (1993)]. We just introduce these definitions as follows.

Definition 3.1. *The fractional integral (or the Riemann–Liouville integral) with order* $\alpha > 0$ *of the given function* $f(t)$ *is defined as*

$$D_{a,t}^{-\alpha} f(t) = \frac{1}{\Gamma(\alpha)} \int_a^t (t-s)^{\alpha-1} f(s)\,\mathrm{d}s, \tag{3.1}$$

where $\Gamma(\cdot)$ *is the Euler's gamma function.*

Definition 3.2. *The left and right Grünwald–Letnikov derivatives with order* $\alpha > 0$ *of the given function* $f(t)$ *are defined as*

$${}_{GL}D_{a,t}^{\alpha} f(t) = \lim_{\substack{h\to 0\\ Nh=t-a}} h^{-\alpha} \sum_{j=0}^{N} (-1)^j \binom{\alpha}{j} f(t-jh), \tag{3.2}$$

and

$${}_{GL}D_{t,b}^{\alpha} f(t) = \lim_{\substack{h\to 0\\ Nh=b-t}} h^{-\alpha} \sum_{j=0}^{N} (-1)^j \binom{\alpha}{j} f(t+jh), \tag{3.3}$$

respectively.

Definition 3.3. *The left and right Riemann–Liouville derivatives with order* $\alpha > 0$ *of the given function* $f(t)$ *are defined as*

$$\begin{aligned} {}_{RL}D_{a,t}^{\alpha} f(t) &= \frac{\mathrm{d}^n}{\mathrm{d}t^n}\left[D_{a,t}^{-(n-\alpha)} f(t)\right] \\ &= \frac{1}{\Gamma(n-\alpha)} \frac{\mathrm{d}^n}{\mathrm{d}t^n} \int_a^t (t-s)^{n-\alpha-1} f(s)\,\mathrm{d}s, \end{aligned} \tag{3.4}$$

and

$${}_{RL}D_{t,b}^{\alpha} f(t) = \frac{(-1)^n}{\Gamma(n-\alpha)} \frac{\mathrm{d}^n}{\mathrm{d}t^n} \int_t^b (s-t)^{n-\alpha-1} f(s)\,\mathrm{d}s, \tag{3.5}$$

respectively, where n *is a nonnegative integer and* $n-1 \le \alpha < n$.

Definition 3.4. *The left and right Caputo derivatives with order* $\alpha > 0$ *of the given function* $f(t)$ *are defined as*

$$\begin{aligned} {}_{C}D_{a,t}^{\alpha} f(t) &= D_{a,t}^{-(n-\alpha)}\left[f^{(n)}(t)\right] \\ &= \frac{1}{\Gamma(n-\alpha)} \int_a^t (t-s)^{n-\alpha-1} f^{(n)}(s)\,\mathrm{d}s, \end{aligned} \tag{3.6}$$

and

$${}_{C}D_{t,b}^{\alpha} f(t) = \frac{(-1)^n}{\Gamma(n-\alpha)} \int_t^b (s-t)^{n-\alpha-1} f^{(n)}(s)\,\mathrm{d}s, \tag{3.7}$$

respectively, where n *is a nonnegative integer and* $n-1 < \alpha \le n$.

Another type of the fractional derivative is the Riesz derivative [Samko *et al.* (1993)], which has several kinds of forms, and we just introduce one case of definitions that will be used in the FDEs [Zhang *et al.* (2010)].

Definition 3.5. *The Riesz derivative with order* $\alpha > 0$ *of the given function* $f(t)$, $t \in (a,b)$ *is defined as*

$$ {}_{RZ}D_t^{\alpha} f(t) = c_{\alpha}\Big({}_{RL}D_{a,t}^{\alpha} f(t) + {}_{RL}D_{t,b}^{\alpha} f(t)\Big). \tag{3.8}$$

In the above Definition 3.5, the parameter c_α is chosen to be $-\frac{1}{2\cos(\frac{\pi\alpha}{2})}$ in [Zhang *et al.* (2010)], detailed information refers to [Agrawal (2007); Samko *et al.* (1993); Zhang *et al.* (2010)] and the references cited therein.

Remark 3.1. Generally speaking, the above definitions of fractional derivatives are not equivalent, the properties and relations of these definitions are discussed in details in [Kilbas *et al.* (2006); Li and Deng (2007); Li *et al.* (2009); Samko *et al.* (1993); Podlubny (1999)], and we just list one case as follows

$$ {}_{RL}D_{a,t}^{\alpha} f(t) = {}_{C}D_{a,t}^{\alpha} f(t) + \sum_{k=0}^{n-1} \frac{f^{(k)}(a)(t-a)^{k-\alpha}}{\Gamma(k+1-\alpha)}, \tag{3.9}$$

where $f \in C^{n-1}[a,t]$ and $f^{(n)}$ is integrable on $[a,t]$. Furthermore, if $f \in C^{n}[a,t]$, then

$$ {}_{GL}D_{a,t}^{\alpha} f(t) = {}_{RL}D_{a,t}^{\alpha} f(t).$$

Definition 3.6 (Directional Integral [Ervin and Roop (2006)]). *Let* $\alpha > 0$, $\theta \in [0, 2\pi)$ *be given. The* α*th order fractional integral in the direction* θ *is given by*

$$ D_\theta^{-\alpha} u(x,y) = \frac{1}{\Gamma(\alpha)} \int_0^{\infty} \xi^{\alpha-1} u(x-\xi\cos\theta, y-\xi\sin\theta)\, \mathrm{d}\xi. \tag{3.10}$$

Definition 3.7 (Directional Derivative [Ervin and Roop (2006)]). *Let* $\alpha > 0$, $\theta \in [0, 2\pi)$ *be given. Let* n *be the smallest integer greater than* α, $n-1 \le \alpha < n$, *and define* $\sigma = n - \alpha$. *Then the* α*th order fractional derivative in the direction* θ *is defined by*

$$ D_\theta^{\alpha} u(x,y) = D_\theta^{n} D_\theta^{-\sigma}, \tag{3.11}$$

where

$$ D_\theta^{n} u(x,y) = \Big(\cos\theta \frac{\partial}{\partial x} + \sin\theta \frac{\partial}{\partial y}\Big)^n u(x,y) = ([\cos\theta, \sin\theta]^T \cdot \nabla)^n u(x,y).$$

In order to better illustrate the FEMs for the FDEs, we introduce a series of spaces and norms. Let Ω be a finite domain, denote by $H^s(\Omega)$ and $H_0^s(\Omega)$ the fractional Sobolev spaces with s being a nonnegative number. Define $H^r(\Omega)$ and $H_0^r(\Omega) (r \in \mathbb{N})$ as the Sobolev spaces. Denote $(\cdot,\cdot)$ and $\|\cdot\|$ to be the inner product and norm in $L^2(\Omega)$, respectively.

Definition 3.8 (Left Fractional Derivative Space [Roop (2004)]). *Let $\alpha > 0$, $u(x)$ is defined on $\mathbb{R}$. Define the semi-norm*

$$|u|_{J_L^\alpha(\mathbb{R})} = \|_{RL}D^\alpha_{-\infty,x}u\|_{L^2(\mathbb{R})}, \tag{3.12}$$

and norm

$$\|u\|_{J_L^\alpha(\mathbb{R})} = (\|u\|^2_{L^2(\mathbb{R})} + |u|^2_{J_L^\alpha(\mathbb{R})})^{\frac{1}{2}}, \tag{3.13}$$

and let $J_L^\alpha(\mathbb{R})$ denote the closure of $C^\infty(\mathbb{R})$ with respect to $\|\cdot\|_{J_L^\alpha(\mathbb{R})}$.

Similarly, the right fractional derivative space can be defined as follows.

Definition 3.9 (Right Fractional Derivative Space [Roop (2004)]). *Let $\alpha > 0$, $u(x)$ is defined on $\mathbb{R}$. Define the semi-norm*

$$|u|_{J_R^\alpha(\mathbb{R})} = \|_{RL}D^\alpha_{x,\infty}u\|_{L^2(\mathbb{R})}, \tag{3.14}$$

and norm

$$\|u\|_{J_R^\alpha(\mathbb{R})} = (\|u\|^2_{L^2(\mathbb{R})} + |u|^2_{J_R^\alpha(\mathbb{R})})^{\frac{1}{2}}, \tag{3.15}$$

and let $J_R^\alpha(\mathbb{R})$ denote the closure of $C^\infty(\mathbb{R})$ with respect to $\|\cdot\|_{J_R^\alpha(\mathbb{R})}$.

Definition 3.10 (Symmetric Fractional Derivative Space [Roop (2004)]). *Let $\alpha > 0$, $\alpha \neq n - 1/2, n \in \mathbb{N}$, $u(x)$ is defined on $\mathbb{R}$. Define the semi-norm*

$$|u|_{J_S^\alpha(\mathbb{R})} = |({}_{RL}D^\alpha_{-\infty,x}u, {}_{RL}D^\alpha_{x,\infty}u)|^{1/2}, \tag{3.16}$$

and norm

$$\|u\|_{J_S^\alpha(\mathbb{R})} = (\|u\|^2_{L^2(\mathbb{R})} + |u|^2_{J_S^\alpha(\mathbb{R})})^{\frac{1}{2}}, \tag{3.17}$$

and let $J_S^\alpha(\mathbb{R})$ denote the closure of $C^\infty(\mathbb{R})$ with respect to $\|\cdot\|_{J_S^\alpha(\mathbb{R})}$.

Definition 3.11 ([Fix and Roop (2004)]). *For $s = m + \alpha$, where m is a non-negative integer and $0 \leq \alpha < 1$, define*

$$J^s(\Omega) = \{u \in H^m(\Omega) | {}_{RL}D^\alpha_{a,x}u, {}_{RL}D^\alpha_{x,b}u \in H^m(\Omega)\} \tag{3.18}$$

with norm $\|\cdot\|_s$ and seminorm $|\cdot|_s$ given by

$$\|u\|_s = (\|u\|^2_m + \|_{RL}D^s_{a,x}u\|^2)^{\frac{1}{2}}, \qquad |u|_s = \|_{RL}D^s_{a,x}u\|, \tag{3.19}$$

where $\Omega = (a, b)$.

For $s \geq 0$, the spaces $J^s(\Omega)$ and $H^s(\Omega)$ are equivalent.

Definition 3.12. Let $\alpha > 0$, $u(x)$ is defined on $\mathbb{R}$. Define the semi-norm

$$|u|_{H^\alpha(\mathbb{R})} = \||\omega|^\alpha \hat{u}\|_{L^2(\mathbb{R})}, \tag{3.20}$$

and norm

$$\|u\|_{H^\alpha(\mathbb{R})} = (\|u\|^2_{L^2(\mathbb{R})} + |u|^2_{H^\alpha(\mathbb{R})})^{\frac{1}{2}}, \tag{3.21}$$

and let $H^\alpha(\mathbb{R})$ denote the closure of $C^\infty(\mathbb{R})$ with respect to $\|\cdot\|_{H^\alpha(\mathbb{R})}$, where $\hat{u} = \hat{u}(w)$ denotes the Fourier transform of u.

The spaces $J_L^\alpha(\Omega)$, $J_R^\alpha(\Omega)$ and $J_S^\alpha(\Omega)$ are defined as the closures of $C^\infty(\Omega)$ under their respective norms. Similarly, the spaces $J_{L,0}^\alpha(\Omega)$, $J_{R,0}^\alpha(\Omega)$ and $J_{S,0}^\alpha(\Omega)$ are defined as the closures of $C_0^\infty(\Omega)$ under their respective norms.

Remark 3.2. If $\alpha > 0$, then the spaces $J_L^\alpha(\mathbb{R})$ and $J_R^\alpha(\mathbb{R})$ are equivalent in the sense of equivalent semi-norms and norms. For $\alpha > 0$, $\alpha \neq n - 1/2, n \in \mathbb{N}$, the spaces $J_L^\alpha(\mathbb{R})$ and $J_S^\alpha(\mathbb{R})$ are equivalent in the sense of equivalent semi-norms and norms.

If $\alpha > 0$, then the spaces $J_{L,0}^\alpha(\Omega)$, $J_{R,0}^\alpha(\Omega)$ and $H_0^\alpha(\Omega)$ are equivalent. And if $\alpha \neq n - 1/2, n \in \mathbb{N}$, the spaces $J_{L,0}^\alpha(\Omega)$, $J_{R,0}^\alpha(\Omega)$, $J_{S,0}^\alpha(\Omega)$ and $H_0^\alpha(\Omega)$ are equivalent. See [Fix and Roop (2004); Roop (2004)] for more details.

3.3 Finite element methods for fractional differential equations

In this section, we introduce the existing theoretical framework established for the Galerkin finite element approximation to the FDEs of fractional diffusion, advection-dispersion equations, and the fractional Fokker–Planck equations, and so on.

We first introduce some useful notations. Let S_h denote a partition of Ω such that $\bar{\Omega} = \{\cup K : K \in S_h\}$. Suppose that there exist positive constants c_1, c_2 such that $c_1 h \leq h_K \leq c_2 h$, where h_K is the width of the subinterval K, and $h = \max_{K \in S_h} h_K$. For $m \in \mathbb{N}$, denote $P_m(K)$ to be the space of polynomials of degree at most m on $K \in S_h$. Associated with S_h, define the finite element space $X_h^{\alpha,m}$ as

$$X_h^{\alpha,m} = \{v \in H^\alpha(\Omega) \cap C^0(\bar{\Omega}) : v|_K \in P_m(K), \forall K \in S_h\}.$$

If $\alpha = r \in \mathbb{N}$ is a positive integer, then we denote by X_h^m as

$$X_h^m = \{v \in H^r(\Omega) \cap C^0(\bar{\Omega}) : v|_K \in P_m(K), \forall K \in S_h\}.$$

Denote by $X_{h,0}^{\alpha,m} = X_h^{\alpha,m} \cap H_0^\alpha(\Omega)$ and $X_{h,0}^m = X_h^m \cap H_0^r(\Omega)$, respectively. Let τ denote the step size for t so that $t_n = n\tau, n = 0, 1, ..., n_T \in \mathbb{N}$. For convenience, denote $v^n = v(\cdot, t_n)$ and

$$d_t v^n = \frac{v^n - v^{n-1}}{\tau}.$$

Now, we introduce the theoretical framework for the fractional differential equations. The early work in this field dated back to [Fix and Roop (2004)], where the least squares finite-element approximation was established for the fractional order differential equation of the form [Fix and Roop (2004)]

$$\begin{cases} {}_{RL}D_{a,x}^\alpha\left(k\dfrac{\mathrm{d}\phi}{\mathrm{d}x}\right) = f(x), \quad x \in [a,b], 0 < \alpha \leq 1, \\ \phi(a) = \phi_0, \quad \phi(b) = \phi_1. \end{cases} \tag{3.22}$$

In order to formulate the least squares finite-element method, the following model problem is considered in [Fix and Roop (2004)]:

Problem 1. For $0 < \alpha < 1, f \in L^2[0,1]$, find (ϕ, u) satisfying

$$\begin{cases} D\phi - u = 0, & x \in [0,1], \\ {}_{RL}D^{\alpha}_{a,x}u = f(x), & x \in [0,1], \\ \phi(0) = \phi_0, & \phi(1) = \phi_1, \end{cases} \tag{3.23}$$

where $Du = \frac{\mathrm{d}u}{\mathrm{d}x}$.

In [Fix and Roop (2004)], Problem 1 is decomposed into two problems as:

Problem 1A. For $0 < \alpha < 1, f \in L^2[0,1]$, find (ϕ_a, u_a) satisfying

$$\begin{cases} D\phi_a - u_a = 0, & x \in [0,1], \\ {}_{RL}D^{\alpha}_{0,x}u_a = f(x), & x \in [0,1], \\ \phi_a(0) = 0, & \phi_a(1) = 0; \end{cases} \tag{3.24}$$

and Problem 1B. For $0 < \alpha < 1, f \in L^2[0,1]$, find (ϕ_b, u_b) satisfying

$$\begin{cases} D\phi_b - u_b = 0, & x \in [0,1], \\ {}_{RL}D^{\alpha}_{0,x}u_b = 0, & x \in [0,1], \\ \phi_b(0) = \phi_0, & \phi_b(1) = \phi_1. \end{cases} \tag{3.25}$$

The solution of Problem 1 is given by $\phi = \phi_a + \phi_b$. Therefore, the attention is restricted to the solution of Problem 1A. Define the functional

$$L(\psi, v) = \frac{1}{2}(\|D\psi - v\|^2 + \|{}_{RL}D^{\alpha}_{0,x}v - f\|^2), \psi \in H^1_0(\Omega), v \in J^{\alpha}(\Omega). \tag{3.26}$$

For $\phi, \psi \in H^1_0(\Omega)$ and $u, v \in J^{\alpha}(\Omega)$, define the following bilinear form

$$B([\phi, u], [\psi, v]) = (D\phi - u, D\psi - v) + ({}_{RL}D^{\alpha}_{0,x}u, {}_{RL}D^{\alpha}_{0,x}v), \tag{3.27}$$

and the linear functional

$$F(\psi, v) = ({}_{RL}D^{\alpha}_{0,x}v, f), \quad \forall \psi \in H^1_0(\Omega), \forall v \in J^{\alpha}(\Omega).$$

Taking the first variation of $L(\psi, v)$, the least squares solution (ϕ, u) of (3.26) satisfies the following variational form

$$B([\phi, u], [\psi, v]) = F(\psi, v), \quad \forall \psi \in H^1_0(\Omega), \forall v \in J^{\alpha}(\Omega). \tag{3.28}$$

Fix and Roop [Fix and Roop (2004)] proved that there exists a unique solution $(\phi, u) \in H^1_0(\Omega) \times J^{\alpha}(\Omega)$ by using the Lax–Milgram theorem. The least squares finite-element solution of (3.28) is given by: Find $\phi_h \in X^1_{h,0}$ and $u_h \in X^{\alpha,1}_h$, such that

$$B([\phi_h, u_h], [\psi, v]) = F(\psi, v), \quad \forall \psi \in X^1_{h,0}, \forall v \in X^{\alpha,1}_h. \tag{3.29}$$

The scheme (3.29) has the unique solutions $(\phi_h, u_h) \in X^1_{h,0} \times X^{\alpha,1}_h$, which are convergent of order $O(h^2)$ and $O(h)$ both in L^2 norms, see [Fix and Roop (2004)] for more details.

In fact, the least squares method has already been used to solve the integral equations. In [Bedivan and Fix (1998b)], the least squares method was applied to solve the Volterra equations of the form

$$\lambda u - Vu = f(x), \quad x \in \Omega, \tag{3.30}$$

where λ is a real number, $f \in X$ is a given function and $V \in L(X)$ is an integral operator defined by

$$(Vu)(x) = \int_\Omega k(x,y)u(y)\,\mathrm{d}y.$$

The least squares method was derived by minimizing the following functional

$$\text{Minimize}\|(\lambda I - V)u - f\|^2 \ \text{ over all } u \in X, \tag{3.31}$$

which leads to the equivalent form as

$$((\lambda I - V)u, (\lambda I - V)v) = (f, (\lambda I - V)v), \quad \forall v \in X. \tag{3.32}$$

Therefore, the least squares finite-element method for (3.30) is given by: Find $u_h \in X_h$, such that

$$((\lambda I - V)u_h, (\lambda I - V)v) = (f, (\lambda I - V)v), \quad \forall v \in X_h, \tag{3.33}$$

where X_h is a finite dimensional subspace of X. The optimal error estimates for the method (3.33) were proved both in the L^2-norm and H^1-norm senses. For the composite operator $V = V_1V_2$, the least square methods for (3.30) was also established, which is similar to that of (3.33). In [Bedivan and Fix (1998a)], the Galerkin finite element method was established for (3.30) of the noncoercive case, and the optimal error estimates were also obtained in L^2-norm and H^1-norm, respectively.

In [Ervin and Roop (2006)], the theoretical framework of the Galerkin finite element approximation to the steady state fractional advection dispersion equation (FADE) was presented. The FADE reads as

$$\begin{cases} -Da(p\,{}_{RL}D_{0,x}^{-\beta} + q\,{}_{RL}D_{x,1}^{-\beta})Du + b(x)Du + c(x)u = f(x), & x \in \Omega, \\ u = 0, \quad x \in \partial\Omega, \end{cases} \tag{3.34}$$

where D represents a single spatial derivative, $0 \le \beta < 1, a > 0, b(x) \in C^1(\bar{\Omega})$, $c(x) \in C(\bar{\Omega})$ with $c - Db/2 \ge 0$, and $0 \le p, q \le 1$, satisfying $p + q = 1$.

Denote by $\alpha = (2-\beta)/2$, then the Galerkin variational solution for (3.34) is given by: Find $u \in H_0^\alpha$, such that

$$B(u,v) = F(v), \quad \forall v \in H_0^\alpha(\Omega), \tag{3.35}$$

where

$$B(u,v) = a\,p\langle {}_{RL}D_{0,x}^{-\beta}Du, Dv\rangle + a\,q\langle {}_{RL}D_{x,1}^{-\beta}Du, Dv\rangle + \langle b\,Du, v\rangle + (c\,u, v),$$

and $F(v) = \langle f, v\rangle$, $\langle\cdot,\cdot\rangle$ denotes the duality pairing of $H^{-\mu}(\Omega)$ and $H_0^\mu(\Omega), \mu \ge 0$.

The finite element approximation to (3.35) is given by: Find $u_h \in X_{h,0}^{\alpha,m}$, such that

$$B(u_h, v) = F(v), \quad \forall v \in X_{h,0}^{\alpha,m}. \tag{3.36}$$

Ervin and Roop [Ervin and Roop (2006)] proved that there exists unique solutions of (3.35) and (3.36) in the spaces H_0^α and $X_{h,0}^{\alpha,m}$ by applying the Lax–Milgram theorem.

They also obtained the error estimates for the method (3.36) in L^2-norm and H^α-norm, respectively, see Corollary 4.3 and Theorem 4.4 in [Ervin and Roop (2006)] for more details. The finite element method for the FADE as (3.34) with boundary layer was presented by Roop in [Roop (2008)]. In his method, the boundary layer was incorporated into the finite element basis, and the numerical experiments were conducted to verify the accuracy and efficiency of the derived method.

In [Roop (2004)], the semi-discrete Galerkin approximation was established for the following time dependent FADE of the form

$$\begin{cases} \partial_t u - a\partial_x (p\,{}_{RL}D_{0,x}^{-\beta} + q\,{}_{RL}D_{x,1}^{-\beta})\partial_x u \\ \qquad + b(x,t)\partial_x u + c(x,t)u = f, (x,t) \in \Omega \times (0,T], \\ u = 0, \quad (x,t) \in \partial\Omega \times (0,T], \\ u(x,0) = g, \quad x \in \Omega. \end{cases} \tag{3.37}$$

The derived semi-discrete Galerkin scheme is similar to (3.36). The uniqueness and error estimate were also proved. Choosing appropriate base functions, the established method was transformed into a system of ordinary differential equations, which was solved by backward Euler method.

In [Ervin *et al.* (2007)], the finite element approximation to the time dependent fractional order diffusion equations containing a nonlocal quadratic nonlinearity was developed. The equation reads as

$$\begin{cases} \partial_t u - \mathcal{D}^{2\alpha} u - \nabla \cdot (uB(u)) = f(x), & (x,t) \in \Omega \times (0,T], \\ u = 0, & (x,t) \in \partial\Omega \times (0,T], \\ u(x,0) = u^0(x), & x \in \Omega, \end{cases} \tag{3.38}$$

where $\mathcal{D}^{2\alpha}$ denotes a general fractional order diffusion operator of order $2\alpha, 1/2 < \alpha \leq 1$. The term $\nabla \cdot (uB(u))$ models particle interactions.

Ervin *et al.* [Ervin *et al.* (2007)] established a fully discrete scheme for (3.38) as follows: For $n = 1, 2, ..., n_T$, find $u_h^n \in X_{h,0}^{\alpha,m}$ such that

$$(d_t u_h^n, v) + \langle \mathcal{D}^{2\alpha} u_h^n, v\rangle + (u_h^n B(u_h^{n-1}), \nabla v) = (f^n, v), \quad \forall v \in X_{h,0}^{\alpha,m}. \tag{3.39}$$

They proved that the derived scheme (3.39) was convergent with order $O(\tau + h^{m+1-\alpha})$ under suitable conditions.

Deng [Deng (2008)] considered the following time-space fractional Fokker–Planck equation

$${}_C D_{a,t}^{\alpha} u(x,t) = \left[\frac{\partial}{\partial x}\frac{U'(x)}{\eta_\alpha} + \kappa_\alpha \nabla^\mu\right] u(x,t), \quad (x,t) \in (a,b) \times (0,T], \tag{3.40}$$

with suitable initial and boundary conditions. $0 < \alpha < 1$, $1 < \mu < 2$, and ∇^μ is defined by

$$\nabla^\mu = \frac{1}{2}({}_{RL}D_{a,x}^{\mu} + {}_{RL}D_{x,b}^{\mu}). \tag{3.41}$$

The time direction is discretitized by the L1 method [Langlands and Henry (2005); Oldham and Spanier (1974)], the semi-discrete finite element scheme and the fully discrete finite element scheme are established as follows:

Semi-discrete scheme: Find $u^n \in H_0^{\frac{\mu}{2}}(\Omega)$ such that

$$(\widetilde{{}_C D^{\alpha}_{0,t_n}} u^n, v) = \left(\left[\frac{\partial}{\partial x}\frac{U'(x)}{\eta_\alpha} + \kappa_\alpha \nabla^\mu\right] u^n, v\right), \quad \forall v \in H_0^{\frac{\mu}{2}}(\Omega). \tag{3.42}$$

Fully discrete scheme: Find $u_h^n \in X_{h,0}^m$ such that

$$(\widetilde{{}_C D^{\alpha}_{0,t_n}} u_h^n, v) = \left(\left[\frac{\partial}{\partial x}\frac{U'(x)}{\eta_\alpha} + \kappa_\alpha \nabla^\mu\right] u_h^n, v\right), \quad \forall v \in X_{h,0}^m, \tag{3.43}$$

where $\widetilde{{}_C D^{\alpha}_{a,t_n}}$ is the time discrete operator defined by L1 method [Langlands and Henry (2005)]. The error estimates were proved for both the semi-discrete scheme and the fully discrete scheme with convergence order of $O(\tau^{2-\alpha} + h^\mu)$ for the fully discrete scheme under certain conditions, see [Deng (2008)] for more details.

Li *et al.* [Li *et al.* (2011)] developed a numerical scheme for the following nonlinear time-space diffusion equation

$${}_C D^{\alpha}_{0,t} u - \lambda_1 \, {}_{RL}D^{1+\beta}_{a,x} u + \lambda_2 B(u)\partial_x u = f(x,t), \quad (x,t) \in (a,b) \times (0,T], \tag{3.44}$$

with suitable initial and boundary conditions, $0 < \beta < 1$, λ_1 and λ_2 are constants satisfying some suitable conditions for error estimates. If $0 < \alpha < 1$, the equation (3.44) is called the subdiffusion equation. If $1 < \alpha < 2$, the equation (3.44) is called the superdiffusion equation.

For both cases of the subdiffusion and superdiffusion, Li *et al.* [Li *et al.* (2011)] established the semi-discrete and fully discrete schemes for (3.44), which can be written as:

Semi-discrete scheme: For each time step n, find $u^n \in H_0^{\frac{1+\beta}{2}}(\Omega)$ such that

$$(\widetilde{{}_C D^{\alpha}_{0,t_n}} u^n, v) = \lambda_1({}_{RL}D^{\beta}_{a,x} u^n, \nabla v) + \lambda_2(B(u^n)\partial_x u^{n-1}, v) + (f^n, v), \quad \forall v \in H_0^{\frac{1+\beta}{2}}(\Omega). \tag{3.45}$$

Fully discrete scheme: For each time step n, find $u_h^n \in X_{h,0}^m$ such that

$$(\widetilde{{}_C D^{\alpha}_{0,t_n}} u_h^n, v) = \lambda_1({}_{RL}D^{\beta}_{a,x} u_h^n, \nabla v) + \lambda_2(B(u_h^n)\partial_x u_h^{n-1}, v) + (f^n, v), \quad \forall v \in X_{h,0}^m. \tag{3.46}$$

The operator $\widetilde{{}_C D^{\alpha}_{0,t_n}}$ is defined by the L1 method [Oldham and Spanier (1974); Langlands and Henry (2005)] for $\alpha \in (0,1)$, and L2 method [Oldham and Spanier (1974); Lynch *et al.* (2003)] for $\alpha \in (1,2)$. For both cases of the schemes (3.45) and (3.46), the existence and uniqueness of the weak solutions were proved, the stability and error estimates were investigated carefully. The convergence orders are $O(\tau^{2-\alpha} + \tau^{-\alpha}h^2)$ and $O(\tau + \tau^{-\alpha}h^2)$ for the subdiffusion linear and nonlinear problems, respectively. For the supdiffusion problem, the convergence order is $O(\tau^2 + \tau^{2-\alpha}h^2)$. Meanwhile, numerical examples are provided to verify the theoretical analysis. Recently, Zhao and Li [Zhao and Li (2012a,b)] have developed the finite element methods for the generalized nonlinear fractional Fokker-Planck equation and the time-space fractional telegraph equation.

Remark 3.3. The existence and uniqueness of the solutions of the variational formulae (3.35), (3.36), (3.39), (3.42), (3.43), (3.45) and (3.46) are all proved by the Lax–Milgram theorem. For instance, the bilinear $B(\cdot,\cdot)$ in (3.35) satisfies the following inequalities:

$$\begin{aligned} B(u,u) &\geq C\|u\|^2_{H^\alpha(\Omega)}, \quad \forall u \in H_0^\alpha(\Omega), \\ B(u,v) &\leq C\|u\|_{H^\alpha(\Omega)}\|u\|_{H^\alpha(\Omega)}, \quad \forall u,v \in H_0^\alpha(\Omega), \end{aligned} \tag{3.47}$$

where C is a positive constant. Therefore, Eq. (3.35) has a unique solution in $H_0^\alpha(\Omega)$ for the continuous $F(\cdot)$.

In [Zheng *et al.* (2010a)], Zheng *et al.* studied the finite element solution to the FPDE of the form

$$\begin{cases} \partial_t u - a\partial_x u + b\,{}_C D^\alpha_{a,x} u + cu = f(x,t), & (x,t) \in \Omega \times (0,T], \\ u = \varphi(x), & (x,t) \in \partial\Omega \times (0,T], \\ u(x,0) = u_0(x), & x \in \Omega, \end{cases} \tag{3.48}$$

where a, b, c and f are functions of x and t, $1 < \alpha \leq 2$.

Zheng *et al.* [Zheng *et al.* (2010a)] first transformed the original equation (3.48) into the following equation with a homogeneous boundary condition

$$\begin{cases} \partial_t v - a\partial_x v + b\,{}_C D^\alpha_{a,x} v + cv = g, & (x,t) \in \Omega \times (0,T], \\ v = 0, & (x,t) \in \partial\Omega \times (0,T], \\ v(x,0) = v_0(x), & x \in \Omega, \end{cases} \tag{3.49}$$

where $u = v + u_\varphi$ satisfying $u_\varphi = \varphi$ on the boundary $\partial\Omega \times [0,T]$.

The fully discrete finite element solution to (3.49) is given by: For each time step $n, n = 1, 2, ..., n_T$, find $v_h^n \in X^m_{h,0}$, such that

$$\begin{aligned} &(d_t v_h^n, w) - (v_h^n, \partial_x(aw)) - ({}_C D^{\alpha-1}_{a,x} v_h^n, \partial_x(bw)) + (cv_h^n, w) \\ =&(g^n, w), \quad \forall w \in X^m_{h,0}. \end{aligned} \tag{3.50}$$

Zheng *et al.* [Zheng *et al.* (2010a)] proved the convergence of the method (3.50) with the convergence order of $O(\tau + h^{m+1-\frac{\alpha}{2}})$. Meanwhile, numerical experiments were conducted to verify the theoretical results.

In [Zheng and Zhao (2011)], Zheng and Zhao proposed a fully discrete Galerkin scheme for a type of nonlinear space-fractional anomalous advection-diffusion equation, and the error estimate was also derived. In [Zheng *et al.* (2010b)], Zheng *et al.* proposed a discontinuous Galerkin finite element method to solve the nonlinear space fractional Fokker–Planck equation in the Caputo sense. The existence, uniqueness and the error estimate were also proved, and numerical experiments were conduced to verify the effectiveness of the method.

Zhang *et al.* [Zhang *et al.* (2010)] considered the following symmetric space FPDE with the symmetric Riesz fractional derivative

$$\begin{cases} \partial_t u = {}_{RZ}D_x^{2\alpha}(Au) + {}_{RZ}D_x^{2\beta}(Bu) + f(x,t), & (x,t) \in \Omega \times (0,T], \\ u = 0, & (x,t) \in \partial\Omega \times (0,T], \\ u(x,0) = u_0(x), & x \in \Omega, \end{cases} \tag{3.51}$$

where $1 < 2\alpha \le 2, 0 < 2\beta < 1, A > 0, B \ge 0, \Omega = (a,b)$, ${}_{RZ}D_x^{2\mu}$ is the Riesz fractional derivative defined by (3.8) with $c_\mu = -\frac{1}{2\cos(\frac{\mu\pi}{2})}$, $\mu = 2\alpha, 2\beta$.

Time direction is discretized by the backward difference method, Zhang *et al.* established an implicit Galerkin finite element method for (3.51) as: For each time step n, find $u_h^n \in X_{h,0}^m$ such that

$$(d_t u_h^n, v) + a(u_h^n, v) = (f^n, v), \quad \forall v \in X_{h,0}^m, \tag{3.52}$$

where

$$\begin{aligned} a(u,v) = & \frac{A}{2\cos(\alpha\pi)}\Big[({}_{RL}D_{a,x}^{\alpha}u, {}_{RL}D_{x,b}^{\alpha}v) + ({}_{RL}D_{x,b}^{\alpha}u, {}_{RL}D_{a,x}^{\alpha}v)\Big] \\ & + \frac{B}{2\cos(\beta\pi)}\Big[({}_{RL}D_{a,x}^{\beta}u, {}_{RL}D_{x,b}^{\beta}v) + ({}_{RL}D_{x,b}^{\beta}u, {}_{RL}D_{a,x}^{\beta}v)\Big]. \end{aligned} \tag{3.53}$$

The existence and uniqueness were investigated, the stability and convergence were proved with convergence order of $O(\tau + h^r)$ in L^2-norm, where $\alpha \le r \le m+1$. The numerical experiments were also provided to illustrate the theoretical analysis.

In [Huang *et al.* (2008)], Huang *et al.* developed a finite element solution to the FADE of the form

$$\partial_t u = -\nu\partial_x u + B\partial_x\Big(\frac{1+\beta}{2}{}_{C}D_{0,x}^{\alpha-1}u + \frac{1-\beta}{2}{}_{C}D_{x,L}^{\alpha-1}u\Big), \quad (x,t) \in (0,L) \times (0,T], \tag{3.54}$$

with the suitable initial and boundary conditions, $1 < \alpha < 2$, $-1 \le \beta \le 1$, ν and B are constants. Their derived method is given by: Find $u_h \in X_h^m$ such that

$$(\partial_t u_h, v) = -\nu(\partial_x u_h, v) + B(\partial_x(\frac{1+\beta}{2}{}_{C}D_{0,x}^{\alpha-1}u_h + \frac{1-\beta}{2}{}_{C}D_{x,L}^{\alpha-1}u_h), v), \quad v \in X_{h,0}^m. \tag{3.55}$$

Huang *et al.* discussed different boundary conditions and considered the singularity at the boundary for the established scheme. In real computation, the linear basis functions were used to solve (3.55) and the time derivative was approximated by the back Euler method. They proved that the derived scheme was unconditionally stable. The method was also used to simulate a laboratory experiment of atrazine movement in a saturated soil column.

For the time fractional equations, Liu *et al.* [Liu *et al.* (2011)] presented a finite element approximation for a modified anomalous subdiffusion equation of the form

$$\partial_t u = (\mu\, {}_{RL}D_{0,t}^{1-\alpha} + \nu\, {}_{RL}D_{a,t}^{1-\beta})\partial_x^2 u + f(x,t), \quad (x,t) \in (a,b) \times (0,T], \tag{3.56}$$

with the suitable initial and boundary conditions. $0 < \beta \le \alpha < 1$, μ and ν are positive constants. They established a semi-discrete scheme and a fully-discrete scheme which read as:

Semi-discrete scheme: Find $u^n \in H_0^1(\Omega)$ such that

$$(d_t u^n, v) + \left((\mu \widetilde{{}_{RL}D_{0,t_n}^{1-\alpha}} + \nu \widetilde{{}_{RL}D_{0,t_n}^{1-\beta}}) \partial_x u^n, \partial_x v \right) = (f^n, v), \quad \forall v \in H_0^1(\Omega). \quad (3.57)$$

Fully discrete scheme: Find $u_h^n \in X_{h,0}^m$ such that

$$(d_t u_h^n, v) + \left((\mu \widetilde{{}_{RL}D_{0,t_n}^{1-\alpha}} + \nu \widetilde{{}_{RL}D_{0,t_n}^{1-\beta}}) \partial_x u_h^n, \partial_x v \right) = (f^n, v), \quad \forall v \in X_{h,0}^m, \quad (3.58)$$

where $\widetilde{{}_{RL}D_{0,t_n}^{\gamma}}$ $(\gamma = 1-\alpha, 1-\beta)$ is the operator for time discretization defined by the new method proposed in [Zhuang *et al.* (2008)]. The stability for both schemes were investigated and the convergence for the two schemes were proved, which was convergent of order $O(\tau + h^{m+1})$ in L^2 norm and $O(\tau + h^m)$ in H^1 norm for the fully discrete scheme.

Jiang and Ma [Jiang and Ma (2011)] presented a high-order finite element method for the time-fractional partial differential equation of the form

$$\begin{cases} {}_C D_{0,t}^{\alpha} u = \partial_x^2 u + f(x,t), & (x,t) \in (a,b) \times (0,T], \\ u = 0, & (x,t) \in \partial\Omega \times [0,T], \\ u(x,0) = u_0(x), & x \in \Omega, \end{cases} \quad (3.59)$$

where $0 < \alpha < 1$. In their method, the time fractional derivative was discretized by the L1 method and the spacial direction was approximated by the finite element method . They got the optimal convergence error rate with order $O(\tau^{2-\alpha} + h^{m+1})$. In the numerical simulation, they adopted the basis functions of the cubic Lagrange interpolation for computation, which verified the theoretical analysis.

In [Schmidt and Gaul (2002)], finite element formulation for three dimensional time fractional viscoelastic constitutive equations was developed, in which the time fractional derivative was approximated by the Grünwald–Letnikov approach [Podlubny (1999)]. Finite element methods were also presented in [Galucio *et al.* (2004); Cortés and Elejabarrieta (2007a,b)] to simulate the numerical solutions of the time fractional equations describing the viscoelastic materials with the time fractional derivative being approximated by the Grünwald-Letnikov approach. For more information and applications, see [Galucio *et al.* (2004); Cortés and Elejabarrieta (2007a,b); Sorrentino and Fasana (2007)] and the references therein.

For the high dimensional problems, Ervin and Roop [Ervin and Roop (2007)] investigated the variational solution to the steady state FADE in $\mathbb{R}^2$ of the form

$$\begin{cases} -\int_0^{2\pi} (D_\theta \, a \, D_\theta^\beta u) M(d\theta) + \mathbf{b}(x,y) \cdot \nabla u + c(x,y) u = f(x,y), & (x,y) \in \Omega, \\ u = 0, \quad (x,y) \in \partial\Omega, \end{cases} \quad (3.60)$$

where $0 < \beta \leq 1, \alpha = (1+\beta)/2, a > 0$, $\mathbf{b}(x,y) = [b_1(x,y), b_2(x,y)]^T \in (C^1(\bar{\Omega}))^2, c(x,y) \in C^1(\bar{\Omega})$ with $c - \frac{1}{2}\nabla \cdot \mathbf{b} \geq 0$, $D_\theta^{-\beta}$ is the β order fractional integral defined by (3.10), and $M(d\theta)$ is a probability density function on the unit sphere in $\mathbb{R}^2$.

They established the variational solution for (3.60), which is given by: Find $u \in H_0^\alpha(\Omega)$ such that

$$B(u, v) = F(v), \quad \forall v \in H_0^\alpha(\Omega), \tag{3.61}$$

where

$$B(u, v) = -\int_0^{2\pi} a(D_\theta^\alpha u, D_{\theta+\pi}^\alpha v) M(d\theta)$$
$$+ (D_0^\alpha u, D_\pi^{1-\alpha}(b_1 v)) + (D_{\pi/2}^\alpha u, D_{3\pi/2}^{1-\alpha}(b_2 v)) + (cu, v),$$

and

$$F(v) = (f, v).$$

The finite element solution for (3.60) is given by: Find $u_h \in X_{h,0}^{\alpha,m-1}$ such that

$$B(u_h, v) = F(v), \quad \forall v \in X_{h,0}^{\alpha,m-1}.$$

The existence and uniqueness of the variational solution was proved by the Lax-Milgram theorem, and the error estimate was proved with convergence of order $O(h^{r-\alpha})(\alpha \le r \le m)$ and $O(h^r)(0 \le r \le m)$ in the H^α-norm and L^2-norm, respectively, where r is related to the smoothness of the solution of Eq. (3.60). In [Roop (2006)], the computational aspects of the finite element approximation of the variational solution to (3.60), and the corresponding time dependent FADE on the bounded domains in $\mathbb{R}^2$ were carefully investigated by using the continuous piecewise polynomial basis functions on a regular triangulation of the domain.

For the fractional variational problems, some advances have been achieved. The earlier works in this field can be found in [Riewe (1996, 1997)], where the Euler–Lagrange equation of motion with the nonconservative force was investigated. Agrawal [Agrawal (2002)] presented the Euler–Lagrange equations for the unconstrained and constrained fractional variational problems in the Riemann–Liouville sense, which can be seen as the generalization of the classical calculus of variations. In [Agrawal (2007)], the generalized Euler–Lagrange equations and the generalized boundary conditions for fractional variational problems defined in terms of Riesz and Riesz–Caputo fractional derivatives were presented and applied to several different problems. The fractional variational principles within Riemann–Liouville and Caputo derivatives in the presence of delay were investigated in [Baleanu (2008b)] and [Baleanu *et al.* (2008); Jarad *et al.* (2010)], respectively, and the corresponding Euler–Lagrange equations were obtained. Agrawal [Agrawal (2008)] presented a general finite element formulation for a class of fractional variational problems in the Riemann–Liouville sense, and the Euler–Lagrange equation and the transversality/natural conditions were presented for a fractional variational problem. Other related works about the fractional variational problems and applications refer to [Agrawal (2004); Agrawal and Baleanu (2007); Almeida and Torres (2009); Atanacković *et al.* (2008); Baleanu (2008a,b, 2009); Bastos *et al.* (2011); Malinowskaa and Torres (2010); Tarasov and Zaslavsky (2005, 2006)] and the references therein.

3.4 Conclusion

In this chapter, we present a survey on the FEMs for the FDEs which characterizes anomalous diffusion or dispersion. In these papers, the FEMs and their numerical analysis (convergence and stability) for FDEs mainly focus on the linear cases, little work has been done for nonlinear ones. The main reason is possibly due to the non-locality of the fractional derivative themselves and nonlinearity of the considered equations, which contribute to the complexity and expensiveness in the computational cost, and very limited work has been done in this regards. Some numerical methods can be seen as a generalization of the numerical methods for classical equations and the technique for analysis of the classical DEs can be extended to the case of FDEs. From this review chapter, there exist very limited numerical studies for the nonlinear FPDEs. We hope more studies for the nonlinear problems will appear somewhere in the future.

Bibliography

Agrawal, O. P. (2002). Formulation of Euler–Lagrange equations for fractional variational problems, *J. Math. Anal. Appl.* **272**, pp. 368–379.

Agrawal, O. P. (2004). A general formulation and solution scheme for fractional optimal control problems, *J. Nonlinear Dyn.* **38**, pp. 323–337.

Agrawal, O. P. (2007). Fractional variational calculus in terms of Riesz fractional derivatives, *J. Phys. A: Math. Theor.* **40**, pp. 6287–6303.

Agrawal, O. P. (2008). A general finite element formulation for fractional variational problems, *J. Math. Anal. Appl.* **337**, pp. 1–12.

Agrawal, O. P. and Baleanu, D. (2007). A Hamiltonian formulation and a direct numerical scheme for fractional optimal control problems, *J. Vib. Control* **13**, pp. 1269–1281.

Almeida, R. and Torres, D. F. M. (2009) Calculus of variations with fractional derivatives and fractional integrals, *Appl. Math. Lett.* **22**, pp. 1816–1820.

Atanacković, T. M., Konjik, S. and Pilipović, S. (2008). Variational problems with fractional derivatives: Euler–Lagrange equations, *J. Phys. A: Math. Theor.* **41**, 095201 (12pp).

Baleanu, D. (2008). Fractional variational principles and their applications, *PAMM Proc. Appl. Math. Mech.* **7**, pp. 1030201–1030202.

Baleanu, D. (2008). New applications of fractional variational principles, *Rep. Math. Phys.* **61**, pp. 199–206.

Baleanu, D. (2009). About fractional quantization and fractional variational principles, *Commun. Nonlinear Sci. Numer. Simulat.* **14**, pp. 2520–2523.

Baleanu, D., Maaraba, T. and Jarad, F. (2008). Fractional variational principles with delay, *J. Phys. A: Math. Theor.* **41**, 315403 (8pp).

Bastos, N. R. O., Ferreira, R. C. and Torres, D. F. M. (2011). Discrete-time fractional variational problems, *Signal Processing* **91**, pp. 513–524.

Bedivan, D. M. and Fix, G. J. (1998). Analysis of finite element approximation and quadrature of Volterra integral equations, *Numer. Meth. Part. D. E.* **13**, pp. 663–672.

Bedivan, D. M. and Fix, G. J. (1998). Leat square methods for Volterra integral equations and generaliztions, *Numer. Meth. Part. D. E.* **14**, pp. 679–693.

Cortés, F. and Elejabarrieta, M. J. (2007a). Finite element formulations for transient dynamic analysis in structural systems with viscoelastic treatments containing fractional derivative models, *Int. J. Numer. Meth. Engng.* **69**, pp. 2173–2195.

Cortés, F. and Elejabarrieta, M. J. (2007b). Homogenised finite element for transient dynamic analysis of unconstrained layer damping beams involving fractional derivative models, *IComput. Mech.* **40**, pp. 313–324.

Deng, W. H. (2007). Short memory principle and a predictor-corrector approach for frac-

tional differential equations, *J. Comput. Appl. Math.* **206**, pp. 174–188.
Deng, W. H. (2008). Finite element method for the space and time fractional Fokker–Planck equation, *SIAM J. Math. Anal.* **47**, pp. 204–226.
Ervin, V. J., Heuer, N. and Roop, J. P. (2007). Numerical approximation of a time dependent, nonlinear, space-fractional diffusion equation, *SIAM J. Math. Anal.* **45**, pp. 572–591.
Ervin, V. J. and Roop, J. P. (2006). Variational formulation for the stationary fractional advection dispersion equation, *Numer. Meth. Part. D. E.* **22**, pp. 558–576.
Ervin, V. J. and Roop, J. P. (2007). Variational Solution of Fractional Advection Dispersion Equations on Bounded Domains in R^d, *Numer. Meth. Part. D. E.* **23**, pp. 256–281.
Fix, G. J. and Roop, J. P. (2004). Least squares finite-element solution of a fractional order two-point boundary value problem, *Comput. Math. Appl.* **48**, pp. 1017–1033.
Ford, N. J. and Simpson, A. C. (2001). The numerical solution of fractional differential equations:Speed versus accuracy, *Numer. Algor.* **26**, pp. 333–346.
Galucio, A. C., Deü, J. F. and Ohayon, R. (2004). Finite element formulation of viscoelastic sandwich beams using fractional derivative operators, *Comput. Mech.* **33**, pp. 282–291.
Hanert, E. (2010). On the numerical solution of space-time fractional diffusion models, *Computers & Fluids* **46**, 33–39.
Huang, Q. Z., Huang, G. H. and Zhan, H. B. (2008). A finite element solution for the fractional advection-dispersion equation, *Adv. Water Resour.* **31**, pp. 1578–1589.
Jiang, Y. J. and Ma, J. T. (2011). High-order finite element methods for time-fractional partial differential equations, *J. Comput. Appl. Math.* **235**, pp. 3285–3290.
Jarad, F., Abdeljawad, T. and Baleanui, D. (2010). Fractional variational principles with delay within caputo derivatives, *Rep. Math. Phys.* **65**, pp. 17–28.
Kilbas, A. A., Srivastava, H. M. and Trujillo, J. J. (2006). *Theory and Applications of Fractional Differential Equations* (Netherlands, Elsevier).
Langlands, T. A. M. and Henry, B. I. (2005). The accuracy and stability of an implicit solution method for the fractional diffusion equation, *J. Comput. Phys.* **205**, pp. 719–736.
Li, C. P., Chen, A. and Ye, J. J. (2011). Numerical approach to fractional calculus and fractional ordinary differential equations, *J. Comput. Phys.* **230**, pp. 3352–3368.
Li, C. P., Dao, X. H., and Guo, P. (2009). Fractional derivatives in complex planes, *Nonlinear Analysis: TMA* **71**, pp. 1857–1869.
Li, C. P. and Deng, W. H. (2007). Remarks on fractional derivatives, *Appl. Math. Comput.* **187**, pp. 777–784.
Li, C. P., Qian, D. L. and Chen, Y. Q. (2011). On Riemann–Liouville and Caputo derivatives, *Discrete Dynamics in Nature and Society* **2011** Article ID 562494, 15 pages.
Li, C. P. and Wang, Y. H. (2009). Numerical algorithm based on Adomian decomposition for fractional differential equations, *Comput. Math. Appl.* **57**, pp. 1672–1681.
Li, C. P. and Zhao, Z. G. (2011). Introduction to fractional integrability and differentiability, *Eur. Phys. J. Special Topics* **193**, pp. 5–26.
Li, C. P., Zhao, Z. G. and Chen, Y. Q. (2011). Numerical approximation of nonlinear fractional differential equations with subdiffusion and superdiffusion, *Comput. Math. Appl.* **62**, pp. 855–875.
Li, X. J. and Xu, C. J. (2009). A space-time spectral method for the time fractional diffusion equation, *SIAM J. Numer. Anal.* **47**, pp. 2108–2131.
Li, C. P. and Zeng, F. H. (2012a). Finite difference methods for fractional differential equations, *Int. J. Bifurcation Chaos* **22(4)**, 1230014 (28 pages).
Li, C. P. and Zeng, F. H. (2012b). Finite difference methods for fractional ordinary differ-

ential equations, *Numer. Func. Anal. Opt.*, **34**(1), 1230014 (2013), in press.

Li, C. P., Zeng, F. H. and Liu, F. (2012b). Spectral approximations to the fractional integral and derivative, *Frac. Cal. Appl. Anal.* **15(3)**, pp. 383–406.

Liu, Q., Liu, F., Turner, I. and Anh, V. (2011). Finite element approximation for a modified anomalous subdiffusion equation, *Appl. Math. Model.* **35** pp. 4103–4116.

Lynch, V. E., Carreras, B. A., del-Castillo-Negrete, D., Ferreira-Mejias, K. M. and Hicks, H. R. (2003). Numerical methods for the solution of partial differential equations of fractional order, *J. Comput. Phys.* **192**, pp. 406–42.

Malinowskaa, A. B. and Torres, D. F. M. (2010). Generalized natural boundary conditions for fractional variational problems in terms of the Caputo derivative, *Comput. Math. Appl.* **59**, pp. 3110–3116.

Momani, S. and Odibat, Z. (2007). Comparison between the homotopy perturbation method and the variational iteration method for linear fractional partial differential equations, *Comput. Math. Appl.* **54**, pp. 910–919.

Oldham, K. and Spanier, J. (1974). *The Fractional Calculus* (Acdemic Press, New York).

Podlubny, I. (1999). *Fractional Differential Euations* (Acdemic Press, San Dieg).

Roop, J. P. (2004). *Variational Solution of the Fractional Advection Dispersion Equation,* Ph.D. thesis, Clemson University, Clemson, SC,.

Roop, J. P. (2006). Computational aspects of FEM approximation of fractional advection dispersion equations on bounded domains in R^2, *J. Comput. Appl. Math.* **193**, pp. 243–268.

Roop, J. P. (2008). Numerical approximation of a one-dimensional space fractional advection-dispersion equation with boundary layer, *Comput. Math. Appl.* **56**, pp. 1808–1819.

Schmidt, A. and Gaul, L. (2002). Finite element formulation of viscoelastic constitutive equations using fractional time derivatives, *Nonlinear Dynamics* **17**, pp. 37–55.

Sorrentino, S. and Fasana, A. (2007). Finite element analysis of vibrating linear systems with fractional derivative viscoelastic models, *J. Sound Vib.* **299**, pp. 839–853.

Riewe, F. (1996). Nonconservative Lagrangian and Hamiltonian mechanics, *Phys. Rev. E* **53**, pp. 1890–1899.

Riewe, F. (1997). Mechanics with fractional derivatives, *Phys. Rev. E* **55**, pp. 3581–3592.

Samko, S. G., Kilbas, A. A. and Marichev, O. I. (1993). *Fractional Integrals and Derivatives: Theory and Applications* (Published 1993 by Gordon and Breach Science Publishers in Switzerland, Philadelphia, Pa., USA).

Sun, Z. Z. and Wu, X. N. (2006). A fully discrete difference scheme for a diffusion-wave system, *Appl. Numer. Math.* **56**, pp. 193–209.

Tadjeran, C. and Meerschaert, M. M. (2007). A second-order accurate numerical method for the two-dimensional fractional diffusion equation, *J. Comput. Phys.* **220**, pp. 813–823.

Tarasov, V. E and Zaslavsky, G. M. (2005). Fractional Ginzburg–Landau equation for fractal media, *Physica A* **354**, pp. 249–261.

Tarasov, V. E and Zaslavsky, G. M. (2006). Nonholonomic constraints with fractional derivatives, *J. Phys. A: Math. Gen.* **39**, pp. 9797–9815.

Vanani, S. K. and Aminataei, A. (2011). Tau approximate solution of fractional partial differential equations, *Comput. Math. Appl.* **62**, pp. 1075–1083.

Yuste, S. B. (2006). Weighted average finite difference methods for fractional diffusion equations, *J. Comput. Phys.* **216** pp. 264–274.

Zhang, H., Liu, F. and Anh, V. (2010). Galerkin finite element approximation of symmetric space-fractional partial differential equations, *Appl. Math. Comput.* **217**, pp. 2534–2545.

Zhao, Z. G. and Li, C. P. (2012a). A numerical approach to the generalized nonlinear fractional Fokker-Planck equation, *Comput. Math. Appl.*, DOI: 10.1016/j.camwa.2012.01.067, in press.

Zhao, Z. G. and Li, C. P. (2012b). Fractional difference/finite element approximations for the time-space fractional telegraph equation, *Comput. Math. Appl.*, DOI: 10.1016/j.amc.2012.09.022, in press.

Zheng, Y. Y., Li, C. P. and Zhao, Z. G. (2010a). A note on the finite element method for the space-fractional advection diffusion equation, *Comput. Math. Appl.* **59**, pp. 1718–1726.

Zheng, Y. Y., Li, C. P. and Zhao, Z. G. (2010b). A fully discrete discontinuous Galerkin method for nonlinear fractional Fokker–Planck equation, *Mathematical Problems in Engineering* **2010**, Article ID 279038, 26 pages.

Zheng, Y. Y. and Zhao, Z. G. (2011). A fully discrete Galerkin method for a nonlinear space-fractional diffusion equation, *Mathematical Problems in Engineering* **2011**, Article ID 171620, 20 pages.

Zhuang, P., Liu, F., Anh, V. and Turner, I. (2008). New solution and analytical techniques of the implicit numerical method for the anomalous subdiffusion equation, *SIAM J. Numer. Anal.* **46**, pp. 1079–1095.

Chapter 4

Fractional step method for the nonlinear conservation laws with fractional dissipation[1]

Can Li and Weihua Deng*

School of Mathematics and Statistics, Lanzhou University, Lanzhou 730000, Gansu, PR China

**dengwh@lzu.edu.cn*

A fractional step method is proposed for solving the one-dimensional nonlinear conservation laws with fractional dissipation

$$u_t(x,t) + (f(u))_x = \varepsilon \nabla^{\mu} u(x,t), \qquad \mu \in (0,1) \cup (1,2].$$

To numerically solve the above equation, we split it into two steps: one is for the classical hyperbolic conservation laws by using Lax-Friedrichs scheme; another one is for the fractional dissipative operator by applying the pseudo-spectral approximation. The computational cost of this numerical method is $O(NM \log(M))$ (for the usual method to solve this kind of nonlocal problem, the cost is $O(NM^2)$), where the N and M are respectively the number of points used in the time and space directions. The convergent order in the time direction is simply verified, and more attentions we focus on are to use the provided scheme to solve the above equation with discontinuous solutions. A striking phenomena we observe is that the dissipation decreases correspondingly when μ decreases from 2 to 0, and ε plays the similar role as it works for the classical problem.

Keywords: Fractional conservation laws, fractional step method

4.1 Introduction

Differential equations involving fractional calculus are widely accepted and used in depicting various natural phenomena, e.g., anomalous diffusion, which covers

[1]This research was partially supported by the Program for New Century Excellent Talents in University (grant no. NCET-09-0438), the National Natural Science Foundation of China (grant no. 10801067), the Fundamental Research Funds for the Central Universities (grant no. lzujbky-2010-63 and no. lzujbky-2012-k26).

almost all the subjects, such as, physics, chemistry, biology, etc. [Metzler and Klafter (2004); Zaslavsky (2002)]. In this work, we consider the numerical method for the one-dimensional fractional conservation laws

$$u_t(x,t) + (f(u))_x = \varepsilon \nabla^\mu u(x,t), \qquad \mu \in (0,1) \cup (1,2], \tag{4.1}$$

with the periodic boundary conditions, where $u(x,t)$ is the probability distribution of the Brownian particle (or concentration of some chemical compounds) at position x and time t, $\nabla^\mu u(x,t)$ is the fractional Riesz derivative defined as

$$\nabla^\mu u(x,t) = \frac{-1}{2\cos(\pi\mu/2)} \left[{}_{-\infty}D_x^\mu + {}_xD_\infty^\mu \right] u(x,t), \tag{4.2}$$

and ${}_{-\infty}D_x^\mu$ and ${}_xD_\infty^\mu$ are the left and right Rieman-Liouville fractional derivatives of order μ, respectively. The one dimensional fractional derivative operator (4.2) is equivalent to the fractional Laplacian $-(-\Delta)^{\frac{\mu}{2}}$, i.e., [Samko *et al.* (1993); Brockmann and Sokolov (2002); Yang *et al.* (2010)]

$$-(-\Delta)^{\frac{\mu}{2}} u(x,t) = \frac{-1}{2\cos(\pi\mu/2)} \left[{}_{-\infty}D_x^\mu + {}_xD_\infty^\mu \right] u(x,t), \tag{4.3}$$

where the fractional Laplacian $-(-\Delta)^{\frac{\mu}{2}}$ is usually defined through the Fourier transform

$$-(-\Delta)^{\frac{\mu}{2}} u(\cdot,t) := -\mathcal{F}^{-1}(|\xi|^\mu \mathcal{F}(u(\cdot,t))), \ \mu \in (0,2].$$

For the well-posedness, regularity, and asymptotic behaviors of the solutions of (4.1), one can see [Biler *et al.* (1998); Droniou and Imbert (2006); Karch *et al.* (2008); Alibaud (2007); Alibaud *et al.* (2010)]. There are already some well developments for the numerical methods of anomalous diffusion equations, e.g., [Podlubny (1999); Gorenflo *et al.* (1999); Meerschaert and Tadjeran (2004); Ilic *et al.* (2005); Ervin *et al.* (2007); Deng (2008); Diethelm (2010); Li *et al.* (2011)]. There are also a limited number of works for the numerical methods of nonlinear conservation laws with the fractional dissipation. Stanescu et al. [Stanescu *et al.* (2005)] develops a numerical method based on the interacting particles approximation for the solution of a large class of evolution problems involving the fractional Laplacian operator and a non-local quadratic-type non-linearity. Droniou in [Droniou (2010)] constructs a class of finite difference schemes for (4.1). It is proved that the numerical solutions converge toward Alibaud's entropy solution. By means of the Fourier spectral methods, Ohkitani in [Ohkitani and Sakajo (2010)] studies the long-time evolution of the surface quasi-geostrophic equation with fractional Laplacian. Constantin et al. in [Constantin *et al.* (2011)] discusses the finite-time singularities of surface quiasi-geostrophic equation with fractional Laplacian. The discontinuous Galerkin approximation of fractional conservation law is discussed in [Cifani *et al.* (2011)] and a Kuznetsov type of theory is established and applied to obtain the error estimates.

Usually, it is not easy or convenient to solve (4.1) by using direct numerical approach since the fractional derivative operator is no longer a differential but an integro-differential operator. Fractional step (or operator splitting) methods are popularly used for the complex unsteady problems containing a range of multi-variables, convection dominated problems [Yanenko (1971); Holden *et al.* (2010)]. The splitting shares some advantages, after splitting the individual operators can be approximated by methods that are well suited for their own characteristic. On the other hand, the weak coupled subsystems produced after splitting can be solved independently and in parallel. So for the hyperbolic-conservation-laws part of (4.1), any high order resolution, essentially non-oscillatory shock-capturing methods [Shu (2009)] can be used. The smooth diffusion part can be approximated by some higher order methods which are well suited for the smooth diffusion operator. In this chapter, the classical Lax-Friedrichs scheme is used to approximate hyperbolic conservation laws, and a pseudo-spectral approximation is presented for the fractional dissipative operator. The computational cost for this numerical method is $O(NM\log(M))$, where N and M are respectively the number of points used in the time and space directions. In fact, for the usual method to solve this kind of nonlocal problem, its cost is $O(NM^2)$. In other words, by using the pseudo-spectral method to solve the fractional operator, the computational cost is reduced besides it is high accurate. We simply verify the convergent order in the time direction, and then turn our main attentions to simulate the discontinuous solutions of (4.1). We observe that the dissipation decreases correspondingly when μ decreases from 2 to 0, and ε plays the similar role as it works for the classical problem.

This chapter is organized as follows. In Section 4.2, we briefly describe the fractional step method. The original equation is split into two subsystems, namely, the hyperbolic conservation law and fractional diffusion equation. And the numerical approximations for each subsystems are presented. In Section 4.3, the convergence rate is simply verified and the extensive numerical experiments are performed to simulate the discontinuous solutions of (4.1). Concluding remarks and a plan for future works are included in the last section.

4.2 Fractional step algorithm

We first split (4.1) into the conservation law

$$u_t(x,t) + (f(u))_x = 0, \tag{4.4}$$

and the linear fractional diffusion equation

$$u_t(x,t) = \varepsilon \nabla^{\mu} u(x,t). \tag{4.5}$$

Let $t_n = n\Delta t$, where Δt is the time stepsize. Combining with the above two sub-equations, the first order Godunov splitting strategy [Godunov (1959)] is used for designing the numerical scheme. The procedure is as follows

$$\text{Conservation law} \quad \begin{cases} \partial_t v + f(v)_x = 0, & \text{for } t \in (t^n, t^{n+1}] \\ v(t^n) = u^n, \end{cases} \tag{4.6}$$

and

$$\text{Diffusion term} \quad \begin{cases} \partial_t u - \varepsilon \nabla^\mu u = 0, & \text{for } t \in (t^n, t^{n+1}] \\ u(t^n) = v^{n+1}. \end{cases} \tag{4.7}$$

4.2.1 *Discretization of the fractional calculus*

There are several different definitions of the fractional calculus, for example the Riemann-Liouville, Grünwald-Letnikov, and Caputo [Podlubny (1999); Samko *et al.* (1993)]. Among these definitions, it seems that the Grünwald-Letnikov fractional derivative is more suitable for numerical computation, which is defined as [Lubich (1986); Podlubny (1999)]

$$_{GL}D^{\mu}_{a,x}u(x,t) = \lim_{h\to 0} \frac{1}{h^\mu} \sum_{m=0}^{[\frac{x-a}{h}]} \omega_m^{(\mu)} u(x - mh, t), \tag{4.8}$$

where $w_m^{(\mu)} := \frac{\Gamma(m-\mu)}{\Gamma(-\mu)\Gamma(m+1)} = (-1)^m \binom{\mu}{m}$ is the normalized Grünwald weights. Usually, in order to ensure the stability of the numerical schemes, the Grünwald-Letnikov fractional derivative need to be modified as (shifted Grünwald-Letnikov fractional derivative) [Meerschaert and Tadjeran (2004)]

$$_{GL}D^{\mu}_{a,x}u(x,t) = \lim_{h\to 0} \frac{1}{h^\mu} \sum_{m=0}^{[\frac{x-a}{h}]} \omega_m^{(\mu)} u(x - (m-1)h, t). \tag{4.9}$$

There are also some more than one order accuracy approximations, such as the G1, G2, L1, L2, R2-algorithms [Oldham and Spanier (1974)] and Lubich approximations [Lubich (1986)]. Another idea is to discretize the (Caputo's, Grünwald-Letnikov's and Riemann-Liouville's) fractional operators in the Fourier space. Then we can straightforwardly calculate the fractional calculus in the Fourier space. The pseudo-spectral method [Guo (1998)] may be the most suitable one for the fractional dissipation sub-equation (4.5), which has the spectral accuracy.

The Fourier transforms of $_{-\infty}D^{\mu}_{x}$ and $_{x}D^{\mu}_{\infty}$ are given, respectively, by [Podlubny (1999); Samko *et al.* (1993)]

$$\mathcal{F}\{_{-\infty}D^{\mu}_{x}u(x,t)\} = (ik)^\mu \widehat{u}(k,t), \quad \mathcal{F}\{_{x}D^{\mu}_{\infty}u(x,t)\} = (-ik)^\mu \widehat{u}(k,t), \quad k \in \mathbf{R}.$$

Further using the fact

$$(\pm ik)^\mu = |k|^\mu \exp\big(\pm i\mu\pi \mathrm{sgn}(k)/2\big) = |k|^\mu \Big(\cos(\frac{\mu\pi}{2}) \pm i\sin(\frac{\mu\pi}{2})\mathrm{sgn}(k)\Big),$$

for (4.7), we get an ODE system in frequency domain, i.e,

$$\begin{cases} \frac{d}{dt}\widehat{u}(k,t) = -\varepsilon|k|^\mu \widehat{u}(k,t), & t \in (t_n, t_{n+1}]; \\ \widehat{u}(k,t_n) = \widehat{v^{n+1}}(k), \end{cases} \tag{4.10}$$

where $k = 0, 1, \cdots, M-1$. Hence,

$$\widehat{u}(k,t_{n+1}) = \exp\big(-\varepsilon|k|^\mu\big)\widehat{v^{n+1}}(k), \qquad k = 0, 1, \cdots, M-1. \tag{4.11}$$

In the above two equations, $\{\widehat{u}(1,t), \widehat{u}(2,t), \cdots, \widehat{u}(M-1,t)\}$ is the discrete Fourier transform of $\{u(x_1,t), u(x_2,t), \cdots, u(x_{M-1},t)\}$, where x_j, $j = 1, 2, \cdots, M-1$ are the M points used in the space domain. Performing the inverse Fourier transform to (4.11), then we get $\{u(x_1,t_{n+1}), u(x_2,t_{n+1}), \cdots, u(x_{M-1},t_{n+1})\}$. By using the technique of fast Fourier transform, the computational costs for both Fourier transform and inverse Fourier transform are $O(M \log M)$ [Hesthaven *et al.* (2007)].

4.2.2 *Discretization of the conservation law*

The conservation law is computed by the classical three-point Lax-Friedrichs scheme

$$u_j^{n+1} = \frac{1}{2}(u_{j-1}^n + u_{j+1}^n) - \frac{\Delta t}{2\Delta x}[f(u_{j+1}^n) - f(u_{j-1}^n)], \tag{4.12}$$

where $u_j^n = u(x_j, t_n)$. The numerical scheme is stable under the CFL restriction

$$\frac{\Delta t}{\Delta x} \max_u |f'(u)| \leq 1. \tag{4.13}$$

The fractional step method is of first order accuracy in time for smooth solution. This is confirmed by numerical examples shown in the next section.

4.3 Numerical results

To confirm the effectiveness of our numerical schemes, the first example with smooth initial values is given to show the convergence rate. All the numerical examples are computed with the periodic boundary conditions

$$u(x + 2\pi, t) = u(x, t). \tag{4.14}$$

Our programs are implemented by MATLAB, and the MATLAB functions `fft` and `ifft` are used in computing the fractional diffusion equation.

Example 4.1 (Accuracy test). *Consider the problem* (4.1) *with the Burgers flux* $f(u) = u^2/2$ *on* $(-\pi, \pi)$, *and the initial condition*

$$u(x, 0) = -\sin(x), \quad x \in (-\pi, \pi),$$

and the boundary condition (4.14).

In Fig. 4.1, we present the maximum norm errors and the convergence rates in time steps for different μ. The error is calculated relative to a fine grid solution ($\Delta t = 2^{-9}$). These numerical results were computed using $\Delta t = 2^{-2}, \cdots, 2^{-8}$ for fixed $\Delta x = 2\pi/512$.

Example 4.2 (Nonlinear conservation laws with fractional Laplacian). *We further consider the following fractal Burgers equation*

$$u_t(x, t) + uu_x = -\varepsilon(-\Delta)^{\frac{\mu}{2}} u(x, t), \quad x \in (-\pi, \pi), \tag{4.15}$$

with discontinuous initial conditions.

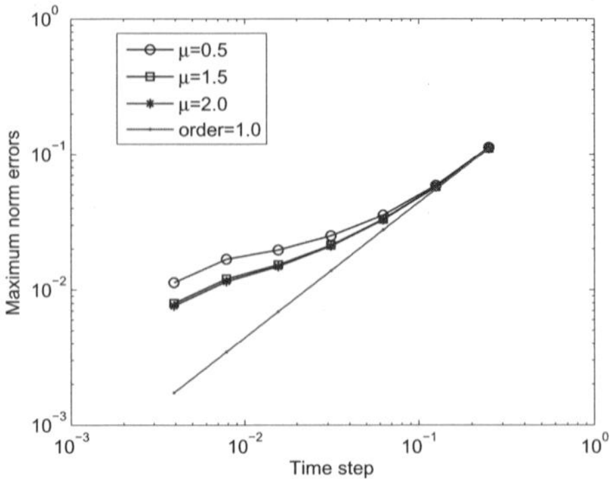

Fig. 4.1 Numerical convergence rate in time with $\varepsilon = 1$ at time $T = 1.0$.

For $\mu \in (1, 2)$, the initial value problem (4.15) admits an unique, smooth, global solution for any bounded initial data [Biler *et al.* (1998); Droniou and Imbert (2006); Droniou (2010)] and the large time asymptotic behavior of solutions is described by rarefaction waves [Karch *et al.* (2008)]. In the case $\mu \in (0, 1)$, the shock solution will be produced even if the initial value is smooth enough [Alibaud (2007)] and the nonlinearity of the equation is negligible in the large time asymptotic expansion of solutions [Alibaud *et al.* (2010)]. First, in order to observe the phenomenon of shock creation in finite time for $0 < \mu < 1$, we plot the numerical solution of (4.15) with the following initial condition

$$u(x, 0) = \begin{cases} 1, & \text{if} \quad x \in (-\pi, 0), \\ -1, & \text{if} \quad x \in (0, \pi). \end{cases}$$

Fig. 4.2 clearly shows that the dissipative behavior decreases when μ decreases from 2 to 0.

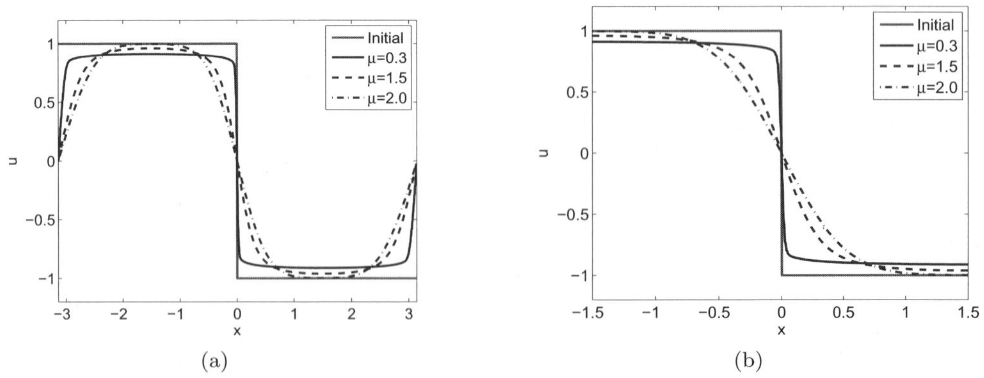

Fig. 4.2 The smoothness effect of fractional dissipation for different μ with $T = 0.1, \varepsilon = 1$.

On the other hand, in Figs. 4.3-4.5, we illustrate the evolution of the smooth profile for different balances between the convective and diffusive forces, where ε and μ are two important parameters.

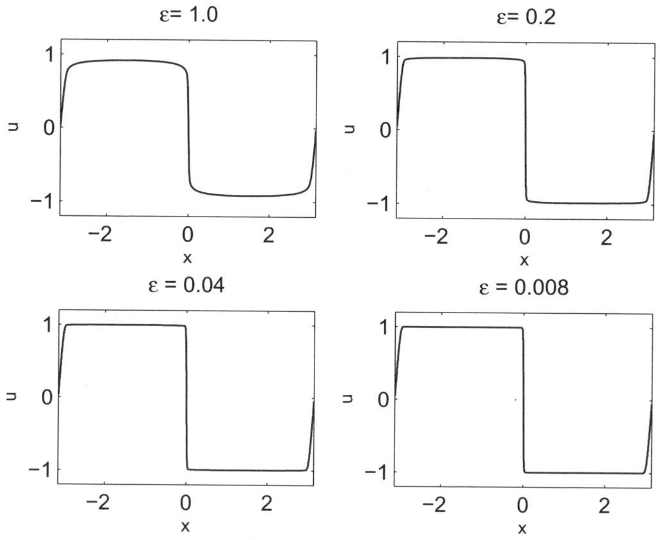

Fig. 4.3 The smoothness effect of fractional dissipation for different ε with fixed $\mu = 0.5$ at time $T = 0.1$.

Secondly, in order to observe the phenomenon of shock creation in finite time for $0 < \mu < 1$, we plot the numerical solution of (4.15), Fig. 4.6, with $\varepsilon = 1$ and the following initial condition

$$u(x, 0) = \min(1, \max(-1.5x, -1)).$$

Finally, we consider (4.15) with the smooth initial value (P.22, Example 2.5 [Holden *et al.* (2010)])

$$u(x, 0) = \exp(-4\sin((x + 2)/2)^2).$$

In Figs. 4.7-4.9, we show that the shock creation for different μ and ε at time $T = 6$. The appearances presented in Figs. 4.7-4.9 are similar to the results reported in [Holden *et al.* (2010)] for each μ. As we can see that, for large values of ε, the diffusive forces dominate and the solution profile evolves towards a flat surface. As ε decreases, the convective forces take over and the solution develops into a viscous shock that travels to the right for both $\mu \in (0, 1)$ and $\mu \in (1, 2)$ with small ε. The hyperbolic behaviors of (4.15) is observed.

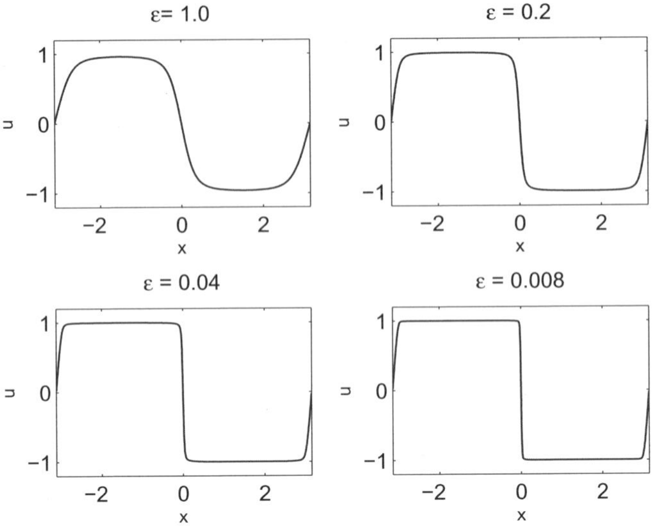

Fig. 4.4 The smoothness effect of fractional dissipation for different ε with fixed $\mu = 1.5$ at time $T = 0.1$.

4.4 Concluding remarks

In this work, we have developed fractional step method to solve a one-dimensional nonlinear conservation laws with fractional dissipation. The behaviors of the solutiond are discussed using the algorithm. The numerical examples show that the proposed splitting method is effectively for both smooth and discontinuous initial values. In the hyperbolic conservation laws, any high order resolution, essentially non-oscillatory shock-capturing method (such as ENO/WENO schemes [Shu (2009)]) can be used. The smooth diffusion (dispersion) step are approximated by some higher order methods which are well suited for the smooth diffusion operator. Higher order fractional step method will be reported in our further works.

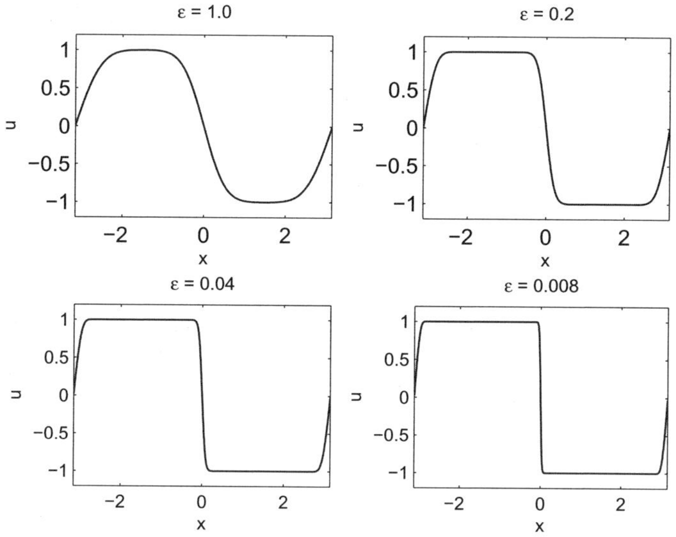

Fig. 4.5 The smoothness effect of fractional dissipation for different ε with fixed $\mu = 2.0$ at time $T = 0.1$.

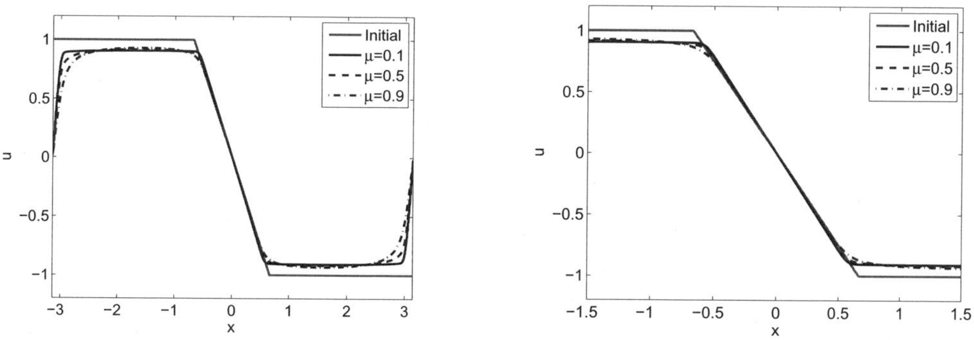

(a) (b)

Fig. 4.6 Shock creation for different $\mu < 1$ with $T = 0.1, \varepsilon = 1$.

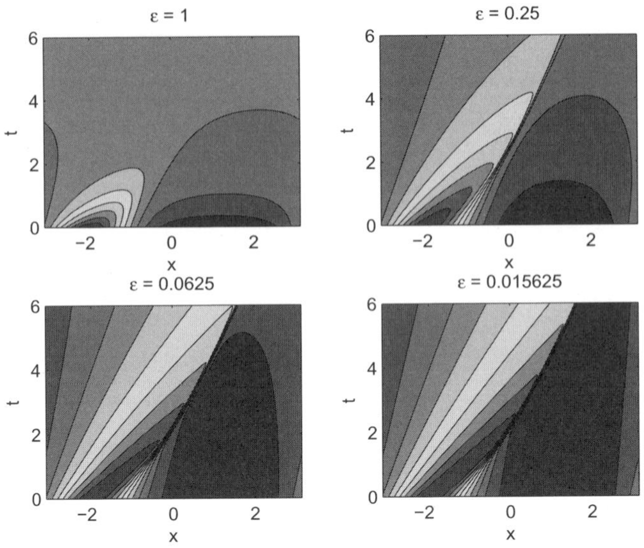

Fig. 4.7 The phenomenon of shock creation for different ε with $\mu = 0.2$.

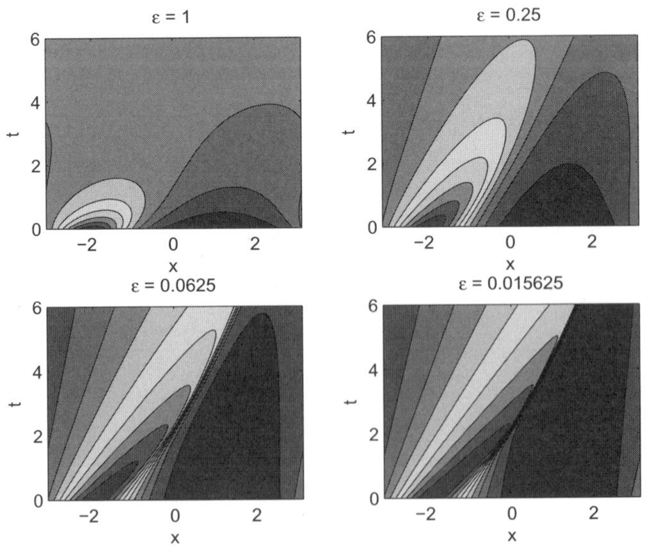

Fig. 4.8 The phenomenon of shock creation for different ε with $\mu = 1.2$.

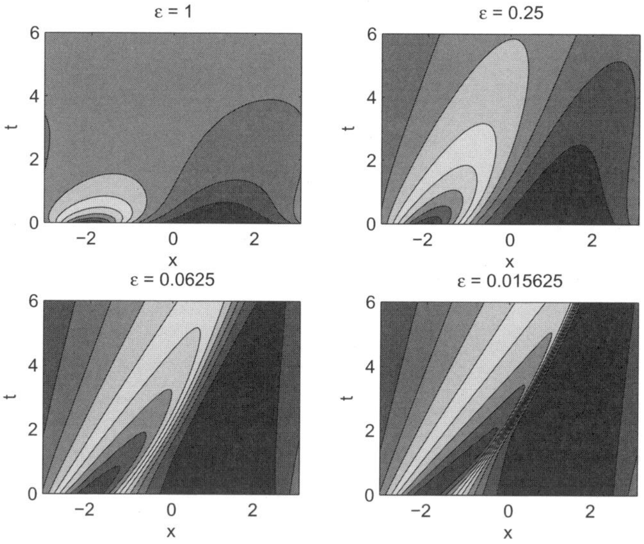

Fig. 4.9 The phenomenon of shock creation for different ε with $\mu = 2.0$.

Bibliography

Alibaud, N. (2007). Entropy formulation for fractal conservation laws, *J. Evol. Equ.* **7**, pp. 145–175.

Alibaud, N., Imbert, C. and Karch, G. (2010). Asymptotic properties of entropy solutions to fractal burgers equation, *SIAM J. Math. Anal.* **42**, pp. 354–376.

Biler, P., Funaki, T. and Woyczyński, W. A. (1998). Fractal burgers equations, *J. Differential Equations* **148**, pp. 9–46.

Brockmann, D. and Sokolov, I. M. (2002). Lévy flights in external force fields: from models to equations, *Chem. Phys.* **284**, pp. 409–421.

Cifani, S. Jakobsen, E. R. and Karlsen, K. H. (2011). The discontinuous Galerkin method for fractal conservation laws, *IMA J. Numer. Anal.* **31**, pp. 1090–1122.

Constantin, P., Lai, M. C., Sharma, R., Tseng, Y. H. and Wu, J. H. (2011). New numerical results for the surface quasi-geostrophic equation, *J. Sci. Comput.* `DOI10.1007/s10915-011-9471-9`.

Deng, W. H. (2008). Finite element method for the space and time fractional fokker-planck equations, *SIAM J. Numer. Anal.* **47**, pp. 204–226.

Diethelm, K. (2010). *The analysis of fractional differential equations:an application-oriented exposition using differential operators of Caputo type* (Springer-Verlag, New York).

Droniou, J. (2010). A numerical method for fractal conservation laws, *Math. Comp.* **79**, pp. 95–124.

Droniou, J. and Imbert, C. (2006). Fractal first-order partial differential equations, *Arch. Ration. Mech. Anal.* **182**, pp. 299–331.

Ervin, V. J., Heuer, N. and Roop, J. P. (2007). Numerical approximation of a time dependent, nonlinear, space-fractional diffusion equation, *SIAM J. Numer. Anal.* **45**, pp. 572–591.

Godunov, S. K. (1959). A difference scheme for numerical computation of discontinuous solutions of equations in fluid dynamics, *Mat. Sab.* **47**, pp. 271–306.

Gorenflo, R., Fabritiis, G. D. and Mainardi, F. (1999). Discrete random walk models for symmetric lévy-feller diffusion processes, *Phys. A* **269**, pp. 79–89.

Guo, B. Y. (1998). *Spectral methods and their applications* (World Scientific, Singapore).

Hesthaven, J. S., Gottlieb, S. and Gottlieb, D. (2007). *Spectral methods for time-dependent problems* (Cambridge University Press, Cambridge).

Holden, H., Karlsen, K. H., Lie, K.-A. and Risebro, N. H. (2010). *Splitting methods for partial differential equations with rough solutions, analysis and Matlab programs* (European Math. Soc. Publishing House, Zurich).

Ilic, M., Liu, F., Turner, I. and Anh, V. (2005). Numerical approximation of a fractional-in-space diffusion equation, *Fract. Calc. Appl. Anal.* **8**, pp. 323–341.

Karch, G., Miao, C. and Xu, X. (2008). On convergence of solutions of fractal burgers equation toward rarefaction waves, *SIAM J. Math. Anal.* **39**, pp. 1536–1549.

Li, C. P., Zhao, Z. G. and Chen, Y. Q. (2011). Numerical approximation of nonlinear fractional differential equations with subdiffusion and superdiffusion, *Comput. Math. Appl.* **62**, pp. 855–875.

Lubich, C. (1986). Discretized fractional calculus, *SIAM J. Appl. Math.* **17**, pp. 704–719.

Meerschaert, M. M. and Tadjeran, C. (2004). Finite difference approximations for fractional advection-dispersion flow equations, *J. Comput. Appl. Math.* **172**, pp. 65–77.

Metzler, R. and Klafter, J. (2004). The restaurant at the end of the random walk: recent developments in the description of anomalous transport by fractional dynamics, *J. Phys. A: Math. Gen.* **37**, pp. 161–208.

Ohkitani, K. and Sakajo, T. (2010). Oscillatory damping in long-time evolution of the surface quasi-geostrophic equations with generalized viscosity: a numerical study, *Nonlinearity* **23**, pp. 3029–3051.

Oldham, K. B. and Spanier, J. (1974). *The fractional calculus: theory and applications of differentiation and integration to arbiteary* (Academic Press, New York).

Podlubny, I. (1999). *Fractional differential equations: An introduction to fractional derivatives, fractional differential equations, to the methods of their solution and some of their applications* (Academic Press, San Diego).

Samko, S., Kilbas, A. and Marichev, O. (1993). *Fractional Integrals and Derivatives: Theory and Applications* (Gordon and Breach, London).

Shu, C.-W. (2009). High order weighted essentially non-oscillatory schemes for convection dominated problems, *SIAM Rev.* **51**, pp. 82–126.

Stanescu, D., Kim, D. and Woyczynski, W. A. (2005). Numerical study of interacting particles approximation for integro-differential equations, *J. Comput. Phys.* **206**, pp. 706–726.

Yanenko, N. N. (1971). *The methods of fractional steps. The solution of problems of mathmatical physics in several variables* (Springer-Verlag, Berlin).

Yang, Q. Q., Liu, F. and Turner, I. (2010). Numerical methods for fractional partial differential equations with riesz space fractional derivatives, *Appl. Math. Modell.* **34**, pp. 200–218.

Zaslavsky, G. M. (2002). Chaos, fractional kinetic, and anomaloustransport, *Phys. Rep.* **371**, pp. 461–580.

Chapter 5

Error analysis of spectral method for the space and time fractional Fokker–Planck equation[1]

Tinggang Zhao[a,*] and Haiyan Xuan[b]

[a]*School of Mathematics, Lanzhou City University, Lanzhou 730070, Gansu, PR China*

**zhaotingg@gmail.com*

[b]*School of Science, Lanzhou University of Technology, Lanzhou 730070, Gansu, PR China*

In this paper, error analysis was carried out for the numerical approximation to the space and time fractional Fokker-Planck equation by spectral method. The time fractional derivative is the Caputo type which is used to characterize the traps while the space fractional derivative is the Riemann-Liouville type which depicts the flights. The finite difference method is used to discretize the time fractional derivative and polynomial approximation for the space fractional derivative. Convergence and stability are established rigorously. The convergence rate is $O(N^{3\mu/4-r} + k^{2-\alpha})$ for exact solution on H^r, where N is the degree of truncated polynomials, k time step, μ and α are orders of derivative with respect to space and time variables respectively.

Keywords: Spectral method, fractional Fokker–Planck equation, Caputo derivative, Riemann–Liouville derivative

5.1 Introduction

While the fractional calculus is extensively used in dealing with engineering and physical problems such as material and mechanics, signal processing and systems identification, anomalous diffusion, control and robotics, wave propagation, turbulence, seepage in fractal media, friction modelling, etc. [Agrawal, Machado and Sabatier (2004)], more and more fractional differential equations appeared in publications. The advantage of the fractional derivatives basically lies in the straightforward way of including external force terms and of calculating boundary value prob-

[1]This research was supported by the National Natural Science Foundation of China (grant no. 11161026).

lems. And the complexity of these equations comes of involving pseudo-differential operators that are nonlocal and having the character of history dependence and universe mutuality.

Normal diffusion operator may not adequately describe single particle trajectories (Brownian motion or random walk process) because of derivating from the long time dynamics of system with anomalous scale [Metzler *et al.* (2009)]. Anomalous diffusion (subdiffusion) is concluded. The fractional diffusion operator describes anomalous diffusion, where the underlying stochastic process is a Lévy flight. Anomalous diffusion is also observed in various fields of physics, for instance, transport of fluid in porous media, surface growth, diffusion of plasma, diffusion at liquid surfaces, two-dimensional rotating flow [Bouchaud and Georges (1990); Goychuk *et al.* (2006); Lenzi *et al.* (2003); Solomon *et al.* (1993)].

In this paper we study the numerical approximation to the time and space fractional Fokker–Planck equation which describes a process with both traps and flights. The time fractional derivative of the equation is used to characterize the traps while the space fractional derivative of the equation to depict the flights.

For the numerical algorithm of time fractional Fokker-Planck type equation and space fractional Fokker-Plank type equation, there are already some progresses covering finite difference method, finite element method, and some random approaches [Deng (2007b); Ervin and Roop (2005); Ervin, Heuer and Roop (2007); Gorenflo *et al.* (2002); Heinsalu *et al.* (2006); Lin and Xu (2007); Liu *et al.* (2004); Meerschaert *et al.* (2006); Yuste and Acedo (2005)]. However, published papers on the numerical algorithm of space and time fractional Fokker-Planck equation are relatively sparse. We only know that the finite element method to approximate the space and time fractional Fokker-Planck equation are analyzed in [Deng (2008)] . Using the Monte Carlo approach for this equation, Magdziarz and Weron [Magdziarz and Weron (2007)] positively answer a question raised by Metzler and Klafter [Metzler and Klafter (2000)]: Can one see a competition between subdiffusion and Lévy flights in the framework of the fractional Fokker-Planck dynamics? Although spectral method is an important numerical algorithm for ordinary and partial differential equations, there seems few works on fractional differential equations.

More recently, it is worthy to mention Li and Xu's work in [Li and Xu (2009, 2010)], in which treatment of the time-space fractional derivative equations is based on the frame of Sobolev spaces defined as

$$B^{s,\sigma}(Q) := H^s(I; L^2(\Omega)) \cap L^2(I; H_0^\sigma(\Omega)),$$

where $I = (0,T)$ is the range of time variable t, $\Omega = (-1,1)^d$ is the domain of space variable x and $Q = I \times \Omega$. This maybe mean that time and space are in the same viewpoint. In this paper, one of the most distinguished aspect is that the time discretization is different from the space discretization, specifically, we firstly employ Galerkin spectral method to obtain a semi-discretization scheme and then finite difference for time discretization. In the spectral method context to calculate fractional derivative, the most cost is to evaluate the integration appeared in the weak formulation. Although the fractional derivative of a polynomial results in a non-polynomial function, it is just a polynomial except a factor like $(x-a)^\mu$ or $(b-x)^\mu$ due to the following well-known facts

$$ {}_aD_x^\mu (x-a)^\nu = \frac{\Gamma(1+\nu)}{\Gamma(1+\nu-\mu)}(x-a)^{\nu-\mu}, $$

and

$$ {}_xD_b^\mu (b-x)^\nu = \frac{\Gamma(1+\nu)}{\Gamma(1+\nu-\mu)}(b-x)^{\nu-\mu}, $$

where the fractional derivative operators ${}_aD_x^\mu$ and ${}_xD_b^\mu$ defined as in later section. Moreover, we have the links between Jacobi polynomials $P_n^{\alpha,\beta}(x)$ and their fractional integral [Askey (1975)] as follows:

$$ (1+x)^{\beta+\mu}\frac{P_n^{\alpha-\mu,\beta+\mu}(x)}{P_n^{\beta+\mu,\alpha-\mu}(1)} = \frac{\Gamma(\beta+\mu+1)}{\Gamma(\beta+1)\Gamma(\mu)}\int_{-1}^{x}(1+y)^\beta\frac{P_n^{\alpha,\beta}(y)}{P_n^{\beta,\alpha}(1)}(x-y)^{\mu-1}dy, \quad (5.1) $$

and

$$ (1-x)^{\alpha+\mu}\frac{P_n^{\alpha+\mu,\beta-\mu}(x)}{P_n^{\alpha+\mu,\beta-\mu}(1)} = \frac{\Gamma(\alpha+\mu+1)}{\Gamma(\alpha+1)\Gamma(\mu)}\int_{x}^{1}(1-y)^\alpha\frac{P_n^{\alpha,\beta}(y)}{P_n^{\alpha,\beta}(1)}(y-x)^{\mu-1}dy. \quad (5.2) $$

These relations tells that it is possible to develop certain approach by using relationships as above and properties of Jacobi polynomials such as orthogonality.

There are several ways to discrete the time fractional derivative, and speed its computation [Deng (2007a); Lin and Xu (2007); Lubich (1986); Podlubny (1999)]. Here we use the one provided by Lin and Xu in [Lin and Xu (2007)] to discrete the left hand side time Caputo derivative and exploit the spectral method to approximate the right hand side space fractional derivative. This approach based on temporal backward differentiation and spatial spectral method obtains estimates of $(2-\alpha)-$order convergence in time and $(3\mu/4-r)$-order convergence in space for exact solution on H^r. And the time-stepping scheme is shown to be unconditionally stable.

5.2 Preliminaries

The space and time fractional Fokker-Planck equation [Magdziarz and Weron (2007); Metzler and Klafter (2000); Metzler and Nonnenmacher (2002); Zaslavsky (2002)], describing the competition between Lévy flights and traps under the influence of an external potential $U(x)$, has the form

$$ \frac{\partial}{\partial t}p(x,t) = {}_0D_t^{1-\alpha}\left[\frac{\partial}{\partial x}\frac{U'(x)}{\eta_\alpha} + \kappa_\alpha\nabla^\mu\right]p(x,t), \quad (5.3) $$

where $p(x,t)$ is the probability density, a prime stands for the derivative with respect to the space coordinate, κ_α denotes the anomalous diffusion coefficient with physical dimension $[m^\mu s^{-\alpha}]$, η_α represents the generalized friction coefficient possessing the dimension $[kgs^{\alpha-2}]$, $\alpha \in (0,1)$ and $\mu \in (1,2)$ throughout this paper. Here, the operators

$$ {}_0D_t^{1-\alpha}p(x,t) = \frac{1}{\Gamma(\alpha)}\frac{\partial}{\partial x}\int_0^t (t-\tau)^{\alpha-1}p(x,\tau)d\tau $$

and $\nabla^\mu = \frac{1}{2}\,{}_aD_x^\mu + \frac{1}{2}\,{}_xD_b^\mu$, ${}_aD_x^\mu$ and ${}_xD_b^\mu$ are left and right Riemann-Liouville space fractional derivative of order μ, respectively, described by

$$ {}_aD_x^\mu p(x,t) = D^2\,{}_aD_x^{-(2-\mu)}p(x,t) = \frac{1}{\Gamma(2-\mu)}\frac{d^2}{dx^2}\int_a^x (x-\xi)^{1-\mu}p(\xi,t)d\xi, $$

and

$$ {}_xD_b^\mu p(x,t) = (-D)^2\,{}_xD_b^{-(2-\mu)}p(x,t) = \frac{1}{\Gamma(2-\mu)}\frac{d^2}{dx^2}\int_x^b (\xi-x)^{1-\mu}p(\xi,t)d\xi. $$

Letting ${}_0D_t^{\alpha-1}$ perform on both sides of (5.3) and according to the attributes of Riemann-Liouville derivative and Caputo derivative [Deng (2007b)], we obtain the equivalent form (5.3) as

$$ D_*^\alpha p(x,t) = \left[\frac{\partial}{\partial x}\frac{U'(x)}{\eta_\alpha} + \kappa_\alpha \nabla^\mu\right]p(x,t), \tag{5.4} $$

where D_*^α is Caputo derivative and it is defined by

$$ D_*^\alpha p(x,t) = \frac{1}{\Gamma(1-\alpha)}\int_0^t (t-\tau)^{-\alpha}\frac{\partial p(x,\tau)}{\partial \tau}d\tau. $$

For specification and without loss of generality, we take $a = -1$, $b = 1$ in the operator ∇^μ. Letting $T > 0$, $\Omega = (-1,1)$, rewriting (5.4) and making it subject to the given initial and boundary conditions, we have

$$ D_*^\alpha p(x,t) = \left[\frac{\partial}{\partial x}\frac{U'(x)}{\eta_\alpha} + \kappa_\alpha \nabla^\mu\right]p(x,t), \quad 0 < t \le T, \quad x \in \Omega \tag{5.5} $$

with initial and boundary conditions,

$$ p(x,0) = g(x), \quad x \in \Omega, \tag{5.6} $$

and

$$ p(-1,t) = p(1,t) = 0, \quad 0 < t \le T. \tag{5.7} $$

For any nonnegative real number r, $H^r(\Omega)$ and $H_0^r(\Omega)$ are the standard Sobolev spaces with norm $\|\cdot\|_{H^r(\Omega)}$ and seminorm $|\cdot|_{H_0^r(\Omega)}$. The inner product and norm of $L^2(\Omega) = H^0(\Omega)$, denoted by $(\cdot,\cdot)$ and $\|\cdot\|$, takes as

$$ (f,g) = \int_\Omega f(x)g(x)dx, \qquad \|f\| = \sqrt{(f,f)}. $$

Now let us introduce a bilinear form defined by

$$ B(p,q) := -\left(\left[\frac{\partial}{\partial x}\frac{U'(x)}{\eta_\alpha} + \kappa_\alpha \nabla^\mu\right]p, q\right), \qquad \forall p,q \in H_0^{\frac{\mu}{2}}(\Omega). \tag{5.8} $$

A variational formulation of (5.5)-(5.7) is:

$$\begin{cases} \text{Find } p(t) \in H_0^{\frac{\mu}{2}}(\Omega), \text{ such that } \forall q \in H_0^{\frac{\mu}{2}}(\Omega), \\ (D_*^{\alpha} p, q) + B(p, q) = 0, \quad 0 < t \leq T, \\ (p(0), q) = (g, q). \end{cases} \tag{5.9}$$

For $\alpha > 0$, we use the following notation:

$$D_*^{-\alpha} u(t) = \frac{1}{\Gamma(\alpha)} \int_0^t (t-\tau)^{\alpha-1} u(\tau) d\tau,$$

which is the fractional Riemann-Liouville integral of order α.

In the following, we introduce several spaces about the left, right, and symmetric fractional derivatives which are used later.

Definition 5.1 (Left fractional derivative [Deng (2008)]). *Let q be a function defined on R, $\beta > 0$, n be the smallest integer greater than β $(n-1 \leq \beta < n)$, and $\sigma = n - \beta$. Then the left fractional derivative of order β is defined as*

$$\mathcal{D}^{\beta} q := D^n \ {}_{-\infty}D_x^{-\sigma} q(x) = \frac{1}{\Gamma(\sigma)} \frac{d^n}{dx^n} \int_{-\infty}^{x} (x-\xi)^{\sigma-1} q(\xi) d\xi.$$

Definition 5.2 (Right fractional derivative [Deng (2008)]). *Let q be a function defined on R, $\beta > 0$, n be the smallest integer greater than β $(n-1 \leq \beta < n)$, and $\sigma = n - \beta$. Then the right fractional derivative of order β is defined as*

$$\mathcal{D}^{\beta *} q := (-D)^n \ {}_{x}D_{\infty}^{-\sigma} q(x) = \frac{(-1)^n}{\Gamma(\sigma)} \frac{d^n}{dx^n} \int_{x}^{\infty} (\xi - x)^{\sigma-1} q(\xi) d\xi.$$

Remark 5.1. If $\text{supp}(q) \subset (a, b)$, then $\mathcal{D}^{\beta} q = {}_aD_x^{\beta} q$ and $\mathcal{D}^{\beta *} q = {}_xD_b^{\beta} q$ where ${}_aD_x^{\beta}$ and ${}_xD_b^{\beta}$ are the left and right Riemann-Liouville fractional derivatives of order β defined as

$${}_aD_x^{\beta} q = \frac{1}{\Gamma(\sigma)} \frac{d^n}{dx^n} \int_a^x (x-\xi)^{\sigma-1} q(\xi) d\xi,$$

and

$${}_xD_b^{\beta} q = \frac{(-1)^n}{\Gamma(\sigma)} \frac{d^n}{dx^n} \int_x^b (\xi - x)^{\sigma-1} q(\xi) d\xi.$$

Definition 5.3 (Left fractional derivative space [Deng (2008)]). *Let $\beta > 0$. Define the semi-norm*

$$|q|_{J_L^{\beta}(R)} := \|\mathcal{D}^{\beta} q\|_{L^2(R)},$$

and

$$\|q\|_{J_L^{\beta}(R)} := \left(\|q\|_{L^2(R)}^2 + |q|_{J_L^{\beta}(R)}^2 \right)^{1/2},$$

and let $J_L^{\beta}(R)$ denote the closure of $C^{\infty}(R)$ with respect to $\| \cdot \|_{J_L^{\beta}(R)}$.

Definition 5.4 (Right fractional derivative space [Deng (2008)]). *Let $\beta > 0$. Define the semi-norm*

$$|q|_{J_R^\beta(R)} := \|\mathcal{D}^{\beta *} q\|_{L^2(R)},$$

and

$$\|q\|_{J_R^\beta(R)} := \left(\|q\|^2_{L^2(R)} + |q|^2_{J_R^\beta(R)}\right)^{1/2},$$

and let $J_R^\beta(R)$ denote the closure of $C^\infty(R)$ with respect to $\|\cdot\|_{J_R^\beta(R)}$.

Definition 5.5 (Symmetric fractional derivative space [Deng (2008)]). *Let $\beta > 0, \beta \neq n - 1/2, n \in N$. Define the semi-norm*

$$|q|_{J_S^\beta(R)} := \left|\left(\mathcal{D}^\beta q, \mathcal{D}^{\beta *} q\right)\right|^{1/2},$$

and norm

$$\|q\|_{J_S^\beta(R)} := \left(\|q\|^2_{L^2(R)} + |q|^2_{J_S^\beta(R)}\right)^{1/2},$$

and let $J_S^\beta(R)$ denote the closure of $C^\infty(R)$ with respect to $\|\cdot\|_{J_S^\beta(R)}$.

Definition 5.6 ([Deng (2008)]). *Let $\beta > 0$, $\hat{q}$ is the Fourier transform of q. Define the semi-norm*

$$|q|_{H^\beta(R)} := \||\Omega|^\beta \hat{q}\|_{L^2(R)},$$

and

$$\|q\|_{H^\beta(R)} := \left(\|q\|^2_{L^2(R)} + |q|^2_{H^\beta(R)}\right)^{1/2},$$

and let $H^\beta(R)$ denote the closure of $C^\infty(R)$ with respect to $\|\cdot\|_{H^\beta(R)}$.

Lemma 5.1 ([Deng (2008)]). *Let $\beta > 0$ be given. Then*

$$\left(\mathcal{D}^\beta q, \mathcal{D}^{\beta *} q\right) = \cos(\beta\pi)\|\mathcal{D}^\beta q\|^2. \tag{5.10}$$

Definition 5.7 ([Deng (2008)]). *Define the spaces $J_L^\beta(\Omega), J_R^\beta(\Omega), J_S^\beta(\Omega)$ as the closures of $C^\infty(\Omega)$ under their respective norms.*

Definition 5.8 ([Deng (2008)]). *Define the spaces $J_{L,0}^\beta(\Omega), J_{R,0}^\beta(\Omega), J_{S,0}^\beta(\Omega)$ as the closures of $C_0^\infty(\Omega)$ under their respective norms.*

Lemma 5.2. *Let $\beta > 0$. Then the spaces $J_{L,0}^\beta(\Omega), J_{R,0}^\beta(\Omega)$ and $H_0^\beta(\Omega)$ are equivalent. Also, if $\beta \neq n - \frac{1}{2}$, $n \in N$, the semi-norms and norms of $J_{L,0}^\beta(\Omega), J_{R,0}^\beta(\Omega)$ and $H_0^\beta(\Omega)$ are equivalent; and the $J_{S,0}^\beta(\Omega)$ and $J_{L,0}^\beta(\Omega)$ ($J_{R,0}^\beta(\Omega)$ or $H_0^\beta(\Omega)$) are equivalent with equivalent semi-norms and norms.*

Lemma 5.3 ([Ervin, Heuer and Roop (2007)]). *For $q \in H_0^\beta(\Omega)$, we have*

$$\|q\| \leq c|q|_{H_0^\beta(\Omega)}.$$

From Lemma 5.3 we know, for $\beta > 0$,

$$\|q\|^2_{H_0^\beta(\Omega)} = \|q\|^2 + |q|^2_{H_0^\beta(\Omega)} \le c|q|^2_{H_0^\beta(\Omega)}.$$

Lemma 5.4 (Fractional Gronwall-Bellman lemma [Alikhanov (2010)]). *Let a nonnegative absolutely continuous function $y(t)$ satisfy the inequality*

$$D_*^\alpha y(t) \le c_1 y(t) + c_2(t), \forall t \in [0, T],$$

where $c_1 > 0$ and $c_2(t)$ is an integrable nonnegative function on $[0, T]$. Then

$$y(t) \le y(0) E_\alpha(c_1 t^\alpha) + \Gamma(\alpha) E_{\alpha,\alpha}(c_1 t^\alpha) D_*^{-\alpha} c_2(t),$$

where

$$E_\alpha(z) = \sum_{n=0}^\infty \frac{z^n}{\Gamma(\alpha n + 1)}, \qquad E_{\alpha,\mu}(z) = \sum_{n=0}^\infty \frac{z^n}{\Gamma(\alpha n + \mu)},$$

are the Mittag-Leffler functions.

Another useful result is as follow:

Lemma 5.5 ([Deng (2008)]). *The left and right Riemann-Liouville fractional integral operators are adjoint with respect to the $L^2(a, b)$ inner product, i.e.,*

$$\left({}_aD_x^{-\beta} p,\ q\right)_{L^2(a,b)} = \left(p,\ {}_xD_b^{-\beta} q\right)_{L^2(a,b)}, \qquad \forall \beta > 0,$$

where

$${}_aD_x^{-\beta} p = \frac{1}{\Gamma(\beta)} \int_a^x (x - \xi)^{\beta - 1} p(\xi) d\xi,$$

and

$${}_xD_b^{-\beta} q = \frac{1}{\Gamma(\beta)} \int_x^b (\xi - x)^{\beta - 1} q(\xi) d\xi.$$

5.3 Spectral method

Let $\mathbb{P}_N(\Omega)$ be the space of all algebraic polynomials of degree at most N on Ω,

$$V_N = \mathbb{P}_N(\Omega) \cap H_0^{\frac{\mu}{2}}(\Omega).$$

The weak formulation (5.9) leads to the following semidiscrete Galerkin method for (5.5)-(5.7), which reads as

$$\begin{cases} \text{Find } p_N(t) \in V_N, \text{ such that } \forall q \in V_N, \\ (D_*^\alpha p_N(t), q) + B(p_N(t), q) = 0, \quad 0 < t \le T, \\ (p_N(0), q) = (g, q). \end{cases} \tag{5.11}$$

Let $\varphi_j, j = 0, ..., N-2$ be a set of basis of V_N, that is,

$$V_N = \text{span}\{\varphi_0, \varphi_1, \cdots, \varphi_{N-2}\}, \tag{5.12}$$

Let us expand the approximation solution $p_N(t)$ to (5.11) as

$$p_N(t) = \sum_{k=0}^{N-2} a_k(t)\varphi_k(x). \tag{5.13}$$

Inserting (5.13) into (5.11) and taking trial function $q = \varphi_j(x)$ $(j = 0, 1, \cdots, N-2)$, we have

$$\begin{cases} \mathbb{M}D_*^\alpha \mathbf{a}(t) + \mathbb{B}\mathbf{a}(t) = 0, \\ \mathbf{a}(0) = \mathbb{M}^{-1}\mathbf{g}, \end{cases} \tag{5.14}$$

here

$$\begin{aligned} &\mathbf{a}(t) = (a_0(t), a_1(t), \cdots, a_{N-2}(t))^T, \\ &\mathbb{M} = (m_{j,i})_{(N-2)\times(N-2)}, \quad m_{j,i} = (\varphi_i, \varphi_j), \\ &\mathbb{B} = (b_{j,i})_{(N-2)\times(N-2)}, \quad b_{j,i} = B(\varphi_i, \varphi_j), \\ &\mathbf{g} = (g_0, g_1, \cdots, g_{N-2})^T, \quad g_j = (g, \varphi_j). \end{aligned} \tag{5.15}$$

The elements of matrix $\mathbb{M}$ and vector $\mathbf{g}$ are easily obtained. For evaluating b_{ij}, we recall the definition of $B(\cdot, \cdot)$,

$$\begin{aligned} &B(\varphi_i, \varphi_j) \\ &= -\left(\left[\frac{\partial}{\partial x}\frac{U'(x)}{\eta_\alpha} + \kappa_\alpha \nabla^\mu\right]\varphi_i, \varphi_j\right) \\ &= \left(\frac{U'(x)}{\eta_\alpha}\varphi_i, \varphi_j'\right) - \kappa_\alpha\left(\nabla^\mu \varphi_i, \varphi_j\right) \\ &= \left(\frac{U'(x)}{\eta_\alpha}\varphi_i, \varphi_j'\right) - \frac{\kappa_\alpha}{2}\left({}_{-1}D_x^{\frac{\mu}{2}}\varphi_i,\ {}_xD_1^{\frac{\mu}{2}}\varphi_j\right) - \frac{\kappa_\alpha}{2}\left({}_xD_1^{\frac{\mu}{2}}\varphi_i,\ {}_{-1}D_x^{\frac{\mu}{2}}\varphi_j\right). \end{aligned} \tag{5.16}$$

The first integral in the right-hand side of (5.16) cannot be evaluated exactly when $U'(x)$ is a nonlinear function such as $\sin(x), e^x$, etc. For the last two integrals in the right-hand side of (5.16), they can be calculated by using Gauss-Jacobi quadratures which have high accuracy or even exact if right nodes are chosen.

5.4 Stability and convergence

5.4.1 *Semi-discrete of space spectral method*

Let us cite the following result [Alikhanov (2010)]:

Lemma 5.6. *For any function $v(t)$ absolutely continuous on $[0, T]$, it holds*

$$\frac{1}{2}D_*^\alpha v^2(t) \le v(t)D_*^\alpha v(t), \qquad 0 < \alpha < 1. \tag{5.17}$$

Proof. Let us rewrite inequality (5.17) in the form

$$
\begin{aligned}
& v(t)D_*^\alpha v(t) - \frac{1}{2}D_*^\alpha v^2(t) \\
&= \frac{1}{\Gamma(1-\alpha)}v(t)\int_0^t \frac{v_\tau(\tau)d\tau}{(t-\tau)^\alpha} - \frac{1}{2\Gamma(1-\alpha)}\int_0^t \frac{2v(\tau)v_\tau(\tau)d\tau}{(t-\tau)^\alpha} \\
&= \frac{1}{\Gamma(1-\alpha)}\int_0^t \frac{v_\tau(\tau)(v(t)-v(\tau))d\tau}{(t-\tau)^\alpha} = \frac{1}{\Gamma(1-\alpha)}\int_0^t \frac{v_\tau(\tau)d\tau}{(t-\tau)^\alpha}\int_\tau^t v_\eta(\eta)d\eta \\
&= \frac{1}{\Gamma(1-\alpha)}\int_0^t v_\eta(\eta)d\eta\int_0^\eta \frac{v_\tau(\tau)d\tau}{(t-\tau)^\alpha} =: I.
\end{aligned}
$$

Therefore, to prove the lemma, it suffices to show that the integral I is nonnegative. The integral I takes nonnegative values, since

$$
\begin{aligned}
I &= \frac{1}{\Gamma(1-\alpha)}\int_0^t (t-\eta)^\alpha \frac{v_\eta(\eta)d\eta}{(t-\eta)^\alpha}\int_0^\eta \frac{v_\tau(\tau)d\tau}{(t-\tau)^\alpha} \\
&= \frac{1}{2\Gamma(1-\alpha)}\int_0^t (t-\eta)^\alpha \frac{\partial}{\partial x}\left(\int_0^\eta \frac{v_\tau(\tau)d\tau}{(t-\tau)^\alpha}\right) d\eta \\
&= \frac{\alpha}{2\Gamma(1-\alpha)}\int_0^t (t-\eta)^{\alpha-1}\left(\int_0^\eta \frac{v_\tau(\tau)d\tau}{(t-\tau)^\alpha}\right)^2 d\eta \geq 0.
\end{aligned}
$$

The proof of the lemma is complete. □

We need the following property of the bilinear form $B(p,q)$ defined by (5.8).

Lemma 5.7. *Let $U''(x) < 0$ for any $x \in \Omega$. Then the bilinear form $B(\cdot,\cdot)$ is coercive on $H_0^{\frac{\mu}{2}}(\Omega)$, that is,*

$$
B(q,q) \geq \kappa_\alpha \left|\cos\left(\frac{\mu}{2}\pi\right)\right| |q|^2_{H_0^{\frac{\mu}{2}}(\Omega)}, \qquad \forall q \in H_0^{\frac{\mu}{2}}(\Omega). \tag{5.18}
$$

For the proof, please see [Deng (2008)] (Lemma 3.1 in [Deng (2008)]).

Now taking $q = p_N(t)$ in the second equation of (5.11) and using the results of Lemmas 5.6 and 5.7, we have

$$
D_*^\alpha \|p_N\|^2 + 2\kappa_\alpha \left|\cos\left(\frac{\mu}{2}\pi\right)\right| |p_N|^2_{H_0^{\frac{\mu}{2}}(\Omega)} \leq 0.
$$

Then by performing operator $D_*^{-\alpha}$ on the equation above, we come to the following stability result.

Theorem 5.1. *Let $U''(x) < 0$ for any $x \in \Omega$, $p_N(t)$ be the solution to (5.11), then it holds*

$$
\|p_N(t)\|^2 + cD_*^{-\alpha}|p_N|^2_{H_0^{\frac{\mu}{2}}(\Omega)} \leq \|g\|^2, \qquad \forall t \in (0,T], \tag{5.19}
$$

where $c = 2\kappa_\alpha \left|\cos\left(\frac{\mu}{2}\pi\right)\right|$.

The following lemma is significant for the convergence of the spectral Galerkin method.

Lemma 5.8. *The bilinear form* $B(\cdot,\cdot)$ *is continuous on* $H_0^{\frac{\mu}{2}}(\Omega)\times H_0^{\frac{\mu}{2}}(\Omega)$*, that is,*

$$B(p,q)\le c_B\|p\|_{H_0^{\frac{\mu}{2}}(\Omega)}\|q\|_{H_0^{\frac{\mu}{2}}(\Omega)},\qquad \forall p,q\in H_0^{\frac{\mu}{2}}(\Omega). \tag{5.20}$$

Proof. From the definition (5.8) we have

$$\begin{aligned}|B(p,q)| &\le \left|\left(\frac{U''(x)}{\eta_\alpha}p,q\right)\right|+\left|\left(\frac{U'(x)}{\eta_\alpha}\frac{\partial p}{\partial x},q\right)\right|+\kappa_\alpha\left|(\nabla^\mu p,q)\right|\\ &=: I_1+I_2+I_3.\end{aligned}$$

For the first term in the right-hand side of the equation above, we have

$$I_1\le\left\|\frac{U''(x)}{\eta_\alpha}\right\|_\infty\|p\|\|q\|.$$

We bound the second term in the right-hand side of the equation above as follows:

$$\begin{aligned}I_2 &= \left(\mathfrak{D}^{\frac{\mu}{2}}p,\mathfrak{D}^{(1-\frac{\mu}{2})*}\left(\frac{U'(x)}{\eta_\alpha}q\right)\right)\le|p|_{J_{L,0}^{\frac{\mu}{2}}(\Omega)}\left|\frac{U'(x)}{\eta_\alpha}q\right|_{J_{R,0}^{1-\frac{\mu}{2}}(\Omega)}\\ &\le c\|p\|_{H_0^{\frac{\mu}{2}}(\Omega)}\|q\|_{H_0^{\frac{\mu}{2}}(\Omega)},\end{aligned}$$

and the last term can be bounded as

$$\begin{aligned}I_3 &= \frac{\kappa_\alpha}{2}\left[({}_aD_x^\mu p,q)+({}_xD_b^\mu p,q)\right]\\ &= \frac{\kappa_\alpha}{2}\left[(\mathfrak{D}^{\frac{\mu}{2}}p,\mathfrak{D}^{\frac{\mu}{2}*}q)+(\mathfrak{D}^{\frac{\mu}{2}*}p,\mathfrak{D}^{\frac{\mu}{2}}q)\right]\\ &\le \frac{\kappa_\alpha}{2}\left[|p|_{J_{L,0}^{\frac{\mu}{2}}(\Omega)}|q|_{J_{R,0}^{\frac{\mu}{2}}(\Omega)}+|p|_{J_{R,0}^{\frac{\mu}{2}}(\Omega)}|q|_{J_{L,0}^{\frac{\mu}{2}}(\Omega)}\right]\\ &\le c\frac{\kappa_\alpha}{2}\|p\|_{H_0^{\frac{\mu}{2}}(\Omega)}\|q\|_{H_0^{\frac{\mu}{2}}(\Omega)}.\end{aligned}$$

The proof is completed by combining the estimates above. □

We introduce a projection operator which is $P_N: L^2(\Omega)\to\mathbb{P}_N$ such that

$$(P_Nv,\phi)=(v,\phi),\qquad\forall\phi\in\mathbb{P}_N. \tag{5.21}$$

its approximation property (Theorem 6.1 and (6.8) in [Bernardi and Maday (1997)], page 258,261) is:

Lemma 5.9. *For any* $u\in H^r(\Omega)(r>1)$*, there exist a positive constant* C *independent on* N *such that*

$$\|u-P_Nu\|\le CN^{-r}\|u\|_r, \tag{5.22}$$

and

$$\|u-P_Nu\|_s\le\begin{cases}CN^{3s/2-r}\|u\|_r, & s\le 1,\\ CN^{2s-1/2-r}\|u\|_r, & s\ge 1.\end{cases} \tag{5.23}$$

Next, we come to the convergence of the (5.11). Let $p(t)$ and $p_N(t)$ be the solutions to (5.9) and (5.11) respectively. We denote $p^* = P_N p$, $e_N(t) = p_N(t) - p^*(t)$. Then we have the following result.

Theorem 5.2. *Let $U''(x) < 0$ for any $x \in \Omega$, and assume that $p \in C(0,T; H^r(\Omega))$ and $r > max\{\frac{\mu}{2}, 1\}$. Then it holds estimate*

$$\|e_N(t)\|^2 + c_1 D_*^{-\alpha} |e_N(t)|^2_{H_0^{\frac{\mu}{2}}(\Omega)} \leq cN^{3\mu/4-r} \|p\|_{H^r(\Omega)}. \qquad \forall t \in (0,T]. \tag{5.24}$$

Proof. From (5.9),(5.11) and (5.21) we get

$$(D_*^\alpha e_N(t), q) + B(e_N, q) = B(p - p^*, q), \qquad \forall q \in V_N. \tag{5.25}$$

By taking $q = e_N$ and using Lemma 5.6, Lemma 5.7 and Lemma 5.8, we have

$$\frac{1}{2} D_*^\alpha \|e_N(t)\|^2 + c_1 |e_N(t)|^2_{H_0^{\frac{\mu}{2}}(\Omega)} \leq c\varepsilon \|p - p^*\|^2_{H_0^{\frac{\mu}{2}}(\Omega)} + c\frac{1}{4\varepsilon} \|e_N(t)\|^2_{H_0^{\frac{\mu}{2}}(\Omega)}. \tag{5.26}$$

Making use of the equivalence of semi-norm and norm in $H_0^\beta(\Omega)$ and choosing appropriate small ε, we get

$$D_*^\alpha \|e_N(t)\|^2 + c_1 |e_N|^2_{H_0^{\frac{\mu}{2}}(\Omega)} \leq c_2 \|p - p^*\|^2_{H_0^{\frac{\mu}{2}}(\Omega)}. \tag{5.27}$$

Thus, it follows from the fractional Gronwall-Bellman lemma

$$\begin{aligned} &\|e_N(t)\|^2 + c_1 D_*^{-\alpha} |e_N(t)|^2_{H_0^{\frac{\mu}{2}}(\Omega)} \\ &\leq \left(\|e_N(0)\|^2 + c_1 D_*^{-\alpha} |e_N(0)|^2_{H_0^{\frac{\mu}{2}}(\Omega)} \right) + c_2 D_*^{-\alpha} \|p - p^*\|^2_{H_0^{\frac{\mu}{2}}(\Omega)}. \end{aligned} \tag{5.28}$$

Thanks to the approximation property of P_N, we get

$$\|e_N(t)\|^2 + c_1 D_*^{-\alpha} |e_N(t)|^2_{H_0^{\frac{\mu}{2}}(\Omega)} \leq cN^{3\mu/4-r} \|p\|_{H^r(\Omega)}.$$

The proof is completed. □

Remark 5.2. We obtain the convergence rate $N^{3\mu/4-r}$ in Theorem 5.2 which the comparison function $p* = P_N p$ is a L^2−orthogonal projection of p. The convergence rate may be improved by considering $H_0^{\frac{\mu}{2}}$−orthogonal projection of p as instead.

5.4.2 *The time discretization of Caputo derivative*

For the time discretization of the Caputo derivative in (5.9), letting $t_m := mk$, $m = 0, 1, \cdots, M$, and $k := \frac{T}{M}$ is the time step size, so we have the following formulation of Caputo derivative:

$$
\begin{aligned}
&D_*^{\alpha} p(t_{m+1}) \\
&= \frac{1}{\Gamma(1-\alpha)} \sum_{j=0}^{m} \int_{t_j}^{t_{j+1}} (t_{m+1}-\tau)^{-\alpha} p'(\tau) d\tau \\
&= \frac{1}{\Gamma(1-\alpha)} \sum_{j=0}^{m} \frac{p(t_{j+1})-p(t_j)}{k} \int_{t_j}^{t_{j+1}} (t_{m+1}-\tau)^{-\alpha} d\tau + r_k^{m+1} \\
&= \frac{1}{\Gamma(1-\alpha)} \sum_{j=0}^{m} \frac{p(t_{j+1})-p(t_j)}{k} \frac{k^{1-\alpha}}{1-\alpha} ((m+1-j)^{1-\alpha} - (m-j)^{1-\alpha}) + r_k^{m+1} \\
&= \frac{1}{\Gamma(2-\alpha)} \sum_{j=0}^{m} \frac{p(t_{m+1-j})-p(t_{m-j})}{k^{\alpha}} d_j + r_k^{m+1}.
\end{aligned}
\tag{5.29}
$$

where $d_j = (j+1)^{1-\alpha} - j^{1-\alpha}$, and r_k^{m+1} is the truncated error.

The following equality which is very useful in giving a good bound of the truncation error can be proved easily by using integration by parts.

$$
f(t) = f(s) + f'(s)(t-s) + \int_s^t f''(\tau)(t-\tau) d\tau. \tag{5.30}
$$

From (5.30), we have

$$
\frac{p(t_{j+1})-p(t_j)}{k} - p'(\tau) = \frac{1}{k}\left[\int_{\tau}^{t_{j+1}} p''(s)(t_{j+1}-s) ds - \int_{t_j}^{\tau} p''(s)(s-t_j) ds\right].
$$

Then we have

$$
\begin{aligned}
&\left|r_k^{m+1}\right| \\
&= \frac{1}{\Gamma(1-\alpha)k} \sum_{j=0}^{m} \left| \int_{t_j}^{t_{j+1}} \int_{\tau}^{t_{j+1}} \frac{p''(s)(t_{j+1}-s) ds d\tau}{(t_{m+1}-\tau)^{\alpha}} - \int_{t_j}^{t_{j+1}} \int_{t_j}^{\tau} \frac{p''(s)(s-t_j) ds d\tau}{(t_{m+1}-\tau)^{\alpha}} \right| \\
&= \frac{1}{\Gamma(1-\alpha)k} \sum_{j=0}^{m} \left| \int_{t_j}^{t_{j+1}} \int_{t_j}^{s} \frac{p''(s)(t_{j+1}-s) d\tau ds}{(t_{m+1}-\tau)^{\alpha}} - \int_{t_j}^{t_{j+1}} \int_{s}^{t_{j+1}} \frac{p''(s)(s-t_j) d\tau ds}{(t_{m+1}-\tau)^{\alpha}} \right| \\
&= \frac{1}{\Gamma(1-\alpha)k} \sum_{j=0}^{m} \left| \int_{t_j}^{t_{j+1}} p''(s)(t_{j+1}-s) \left(\int_{t_j}^{s} \frac{d\tau}{(t_{m+1}-\tau)^{\alpha}} \right) ds \right. \\
&\left. - \int_{t_j}^{t_{j+1}} p''(s)(s-t_j) \left(\int_{s}^{t_{j+1}} \frac{d\tau}{(t_{m+1}-\tau)^{\alpha}} \right) ds \right| \\
&= \frac{1}{\Gamma(2-\alpha)k} \sum_{j=0}^{m} \left| \int_{t_j}^{t_{j+1}} p''(s) R(s) ds \right|,
\end{aligned}
$$

where

$$
\begin{aligned}
R(s) =& (t_{j+1}-s)[(t_{m+1}-t_j)^{1-\alpha} - (t_{m+1}-s)^{1-\alpha}] \\
& - (s-t_j)[(t_{m+1}-s)^{1-\alpha} - (t_{m+1}-t_{j+1})^{1-\alpha}].
\end{aligned}
$$

Because of $R(t_j) = R(t_{j+1}) = 0$ and $R''(s) = k\alpha(1-\alpha)(t_{m+1}-s)^{-\alpha-1} \geq 0, \forall s \in [t_j, t_{j+1}](j = 0, 1, ..., m)$, we have $R(s) \leq 0$, $\forall s \in [t_j, t_{j+1}](j = 0, 1, ..., m)$. Now assuming that

$$\max_{t\in[0,T]} |p''(t)| \leq C, \tag{5.31}$$

then we have

$$\left|r_k^{m+1}\right| \leq \frac{C}{\Gamma(2-\alpha)k} \left| \sum_{j=0}^{m} \int_{t_j}^{t_{j+1}} R(s)ds \right|. \tag{5.32}$$

The estimation of the error term r_k^{m+1} is then reduced to computing the integrals in the right hand side:

$$\begin{aligned}
&\sum_{j=0}^{m} \int_{t_j}^{t_{j+1}} R(s)ds \\
&= \sum_{j=0}^{m} \int_{t_j}^{t_{j+1}} \Big[(t_{j+1}-s)(t_{m+1}-t_j)^{1-\alpha} + (s-t_j)(t_{m+1}-t_{j+1})^{1-\alpha} \\
&\quad -k(t_{m+1}-s)^{1-\alpha}\Big]ds \\
&= \sum_{j=0}^{m} \left[\frac{k^2}{2}\left((t_{m+1}-t_j)^{1-\alpha} + (t_{m+1}-t_{j+1})^{1-\alpha}\right)\right. \\
&\quad \left. -\frac{k}{2-\alpha}\left((t_{m+1}-t_j)^{2-\alpha} - (t_{m+1}-t_{j+1})^{2-\alpha}\right)\right] \\
&= k^{3-\alpha} \sum_{j=0}^{m} \left\{ \frac{(m+1-j)^{1-\alpha} + (m-j)^{1-\alpha}}{2} - \frac{(m+1-j)^{2-\alpha} - (m-j)^{2-\alpha}}{2-\alpha} \right\} \\
&= \frac{k^{3-\alpha}}{2} \sum_{i=0}^{m} \left\{ \left[(i+1)^{1-\alpha} + i^{1-\alpha}\right] - \frac{2}{2-\alpha}\left[(i+1)^{2-\alpha} - i^{2-\alpha}\right] \right\}.
\end{aligned}$$

Let

$$s(m) = \sum_{i=0}^{m} \left\{ \left[(i+1)^{1-\alpha} + i^{1-\alpha}\right] - \frac{2}{2-\alpha}\left[(i+1)^{2-\alpha} - i^{2-\alpha}\right] \right\}. \tag{5.33}$$

In Fig. 5.1, we plot $s(m)$ as a function of m for different α. It is clear that $s(m)$

is bounded. Next we rigorously prove it.

$$\begin{aligned}
&\left|\left[(i+1)^{1-\alpha}+i^{1-\alpha}\right]-\frac{2}{2-\alpha}\left[(i+1)^{2-\alpha}-i^{2-\alpha}\right]\right| \\
&= i^{1-\alpha}\left|\frac{2i}{2-\alpha}\left(\left(1+\frac{1}{i}\right)^{2-\alpha}-1\right)-\left(1+\frac{1}{i}\right)^{1-\alpha}-1\right| \\
&= i^{1-\alpha}\left|\frac{2i}{2-\alpha}\left(1+(2-\alpha)\frac{1}{i}+\frac{(2-\alpha)(1-\alpha)}{2!}\frac{1}{i^2}+\frac{(2-\alpha)(1-\alpha)(-\alpha)}{3!}\frac{1}{i^3}\right.\right. \\
&\quad \left.+\frac{(2-\alpha)(1-\alpha)(-\alpha)(-\alpha-1)}{4!}\frac{1}{i^4}+\cdots-1\right) \\
&\quad \left.-\left(1+(1-\alpha)\frac{1}{i}+\frac{(1-\alpha)(-\alpha)}{2!}\frac{1}{i^2}+\frac{(1-\alpha)(-\alpha)(-\alpha-1)}{3!}\frac{1}{i^3}+\cdots\right)-1\right| \\
&= i^{1-\alpha}\left|\left(\frac{1}{2!}-\frac{2}{3!}\right)(1-\alpha)\alpha\frac{1}{i^2}+\left(\frac{1}{3!}-\frac{2}{4!}\right)(1-\alpha)\alpha(-\alpha-1)\frac{1}{i^3}+\cdots\right| \\
&= \frac{(1-\alpha)\alpha}{6}\frac{1}{i^{1+\alpha}}\left|1+\frac{(-\alpha-1)}{2}\frac{1}{i}+\frac{3(-\alpha-1)(-\alpha-2)}{20}\frac{1}{i^2}+\cdots\right| \\
&\le \frac{(1-\alpha)\alpha}{6}\frac{1}{i^{1+\alpha}}\left|1+\frac{1}{i}+\frac{1}{i^2}+\cdots\right| \le C\frac{1}{i^{1+\alpha}}.
\end{aligned}$$

Therefore the series

$$\sum_{i=0}^{\infty}\left|\left[(i+1)^{1-\alpha}+i^{1-\alpha}\right]-\frac{2}{2-\alpha}\left[(i+1)^{2-\alpha}-i^{2-\alpha}\right]\right|$$

converges for all $\alpha \in (0,1)$. Consequently, $s(m)$ is bounded.

As a result of the above estimates, we have the following theorem.

Theorem 5.3. *For the truncated error, if $p''(t) \in C([0,T])$, we have*

$$|r_k^{m+1}| \le \frac{C_1}{\Gamma(2-\alpha)}k^{2-\alpha}, \tag{5.34}$$

where C_1 depends on $p''(t)$ but independent of k and m.

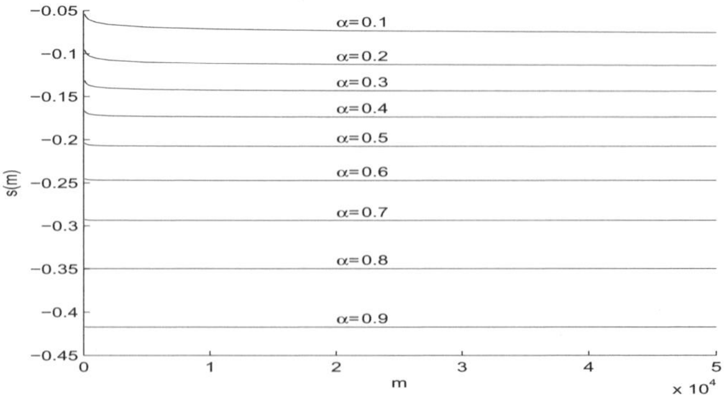

Fig. 5.1 $s(m)$ as a function of m for different α.

Table 5.1 Convergence order for $p(t) = t^3$ at $t = 1$.

	$\alpha = 0.1$		$\alpha = 0.5$		$\alpha = 0.9$	
k	Error	Order	Error	Order	Error	Order
1.00e-01	2.7224e-03	-	3.8321e-02	-	2.0036e-01	-
2.00e-02	1.6521e-04	1.74	3.7454e-03	1.44	3.5263e-02	1.10
1.00e-02	4.8033e-05	1.78	1.3493e-03	1.47	1.6523e-02	1.10
2.00e-03	2.6212e-06	1.81	1.2360e-04	1.49	2.8239e-03	1.10
1.00e-03	7.3969e-07	1.83	4.3938e-05	1.49	1.3181e-03	1.10
2.00e-04	3.8387e-08	1.84	3.9582e-06	1.50	2.2452e-04	1.10
1.00e-04	1.0659e-08	1.85	1.4018e-06	1.50	1.0475e-04	1.10
2.00e-05	5.3711e-10	1.86	1.2566e-07	1.50	1.7837e-05	1.10
1.00e-05	1.4765e-10	1.87	4.4451e-08	1.50	8.3211e-06	1.10
2.00e-06	7.2995e-12	1.87	3.9785e-09	1.50	1.4168e-06	1.10

The estimate (5.34) is also confirmed by sufficient numerical tests. We report some in Table 5.1.

Let us define the discrete fractional differential operator L_t^α by

$$L_t^\alpha p(x, t_{m+1}) = \frac{1}{\Gamma(2-\alpha)} \sum_{j=0}^{m} d_j \frac{p(x, t_{m+1-j}) - p(x, t_{m-j})}{k^\alpha}.$$

So we have

$$D_*^\alpha p(x, t_{m+1}) = L_t^\alpha p(x, t_{m+1}) + r_k^{m+1}. \tag{5.35}$$

By replacing D_*^α with L_t^α in (5.9), we obtain the following semi-discrete scheme

$$\begin{cases} \text{Find } p(t) \in H_0^{\frac{\mu}{2}}(\Omega), \text{ such that } \forall q \in H_0^{\frac{\mu}{2}}(\Omega), \\ (L_t^\alpha p, q) + B(p, q) = 0, \quad 0 < t \leq T, \\ (p(0), q) = (g, q). \end{cases} \tag{5.36}$$

We denote $v^j(x) := v(x, t_j)$ by v^j. The scheme (5.36) can be rewritten as

$$\begin{cases} \text{Find } p^{m+1} \in H_0^{\frac{\mu}{2}}(\Omega), \text{ such that } \forall q \in H_0^{\frac{\mu}{2}}(\Omega), \\ (p^{m+1}, q) + \alpha_0 B(p^{m+1}, q) = (1 - d_1)(p^m, q) \\ \qquad + \sum\limits_{j=1}^{m-1} (d_j - d_{j+1})(p^{m-j}, q) + d_m(p^0, q), \quad m = 1, \cdots, M-1, \\ (p^1, q) + \alpha_0 B(p^1, q) = (p^0, q), \\ (p^0, q) = (g, q). \end{cases} \tag{5.37}$$

The following theorem gives the stability results for the problem above.

Theorem 5.4. *Let $U''(x) \leq 0$ for any $x \in \Omega$. The weak semi-discrete scheme (5.36) or (5.37) is unconditionally stable in the sense that for any time-step length $k > 0$, it holds*

$$\|p^{m+1}\|_{H_0^{\frac{\mu}{2}}(\Omega)} \leq \|p^0\|_{L^2}, \quad m = 0, 1, 2, \cdots, M-1.$$

Proof. It is proved by Deng in [Deng (2008)]. Here we copy it for completeness. For convenience, we introduce an equivalent norm defined as:

$$\|p\|_{H_0^{\frac{\mu}{2}}(\Omega)} = \left[\|p\|^2_{L^2(\Omega)} + \alpha_0\kappa\left|\cos\left(\frac{\mu}{2}\pi\right)\right| |p|^2_{H_0^{\frac{\mu}{2}}(\Omega)}\right]^{\frac{1}{2}}. \tag{5.38}$$

When $m = 0$, we have

$$(p^1, q) + \alpha_0 B(p^1, q) = (p^0, q), \quad \forall q \in H_0^{\frac{\mu}{2}}(\Omega).$$

Taking $q = p^1$ and from Lemma 5.8 and the fact $\|p\|_{L^2(\Omega)} \le \|p\|_{H_0^{\frac{\mu}{2}}(\Omega)}$ we obtain

$$\|p^1\|_{H_0^{\frac{\mu}{2}}(\Omega)} \le \|p^0\|_{L^2(\Omega)}.$$

Now we assume that for $j = 1, 2, \cdots, m$

$$\|p^j\|_{H_0^{\frac{\mu}{2}}(\Omega)} \le \|p^0\|_{L^2(\Omega)}. \tag{5.39}$$

From the first equation in scheme (5.36), we have

$$\begin{aligned}
&(p^{m+1}, p^{m+1}) + \alpha_0 B(p^{m+1}, p^{m+1}) \\
&= (1-d_1)(p^m, p^{m+1}) + \sum_{j=1}^{m-1}(d_j - d_{j+1})(p^{m-j}, p^{m+1}) + d_m(p^0, p^{m+1}).
\end{aligned}$$

By Lemma 5.8, (5.38) and the assumption (5.39), it holds

$$\begin{aligned}
&\|p^{m+1}\|^2_{H_0^{\frac{\mu}{2}}(\Omega)} \\
&\le (1-d_1)\|p^m\|_{L^2(\Omega)}\|p^{m+1}\|_{L^2(\Omega)} \\
&\quad + \sum_{j=1}^{m-1}(d_j - d_{j+1})\|p^{m-j}\|_{L^2(\Omega)}\|p^{m+1}\|_{L^2(\Omega)} + d_m\|p^0\|_{L^2(\Omega)}\|p^{m+1}\|_{L^2(\Omega)} \\
&\le \left[(1-d_1) + \sum_{j=1}^{m-1}(d_j - d_{j+1}) + d_m\right]\|p^0\|_{L^2(\Omega)}\|p^{m+1}\|_{H_0^{\frac{\mu}{2}}(\Omega)} \\
&= \|p^0\|_{L^2(\Omega)}\|p^{m+1}\|_{H_0^{\frac{\mu}{2}}(\Omega)}.
\end{aligned}$$

Hence,

$$\|p^{m+1}\|_{H_0^{\frac{\mu}{2}}(\Omega)} \le \|p^0\|_{L^2(\Omega)}.$$

The induction is completed. □

The following theorem shows that the scheme (5.36) is of $(2-\alpha)$-order accuracy.

Theorem 5.5. *Let $U''(x) \le 0$ for any $x \in \Omega$. p and $\{p^m\}_{m=0}^M$ be the solution to (5.9) and (5.36) respectively , then it holds*

$$\|p(t_m) - p^m\|_{H_0^{\frac{\mu}{2}}(\Omega)} \le c_{p,\alpha}T^\alpha k^{2-\alpha}, \quad m = 1, 2, \cdots, M.$$

Proof. It is also proved by Deng in [Deng (2008)]. Here we copy it for the same reason. We introduce the denotation

$$r^{m+1} := \alpha_0 r_k^{m+1}. \tag{5.40}$$

Then we have

$$|r^{m+1}| = \Gamma(2-\alpha)k^{\alpha}|r_k^{m+1}| \le \Gamma(2-\alpha)\tilde{c}_p k^2 := c_p k^2. \tag{5.41}$$

Next, we would like to prove the estimate

$$\|p(t_j) - p^j\|_{H_0^{\frac{\mu}{2}}(\Omega)} \le c_p d_{j-1}^{-1} k^2, \quad j = 1, 2, \cdots, M. \tag{5.42}$$

Denote $e^m = p(t_m) - p^m$. For $j = 1$, the error equation is

$$(e^1, q) + \alpha_0 B(e^1, q) = (r^1, q), \quad \forall q \in H_0^{\frac{\mu}{2}}(\Omega).$$

Taking $q = p^1$ we have

$$\|e^1\|^2_{H_0^{\frac{\mu}{2}}(\Omega)} \le \|r^1\|_{L^2}\|e^1\|_{L^2} \le c_p d_0^{-1} k^2.$$

Suppose that (5.42) holds for all $j = 1, 2, \cdots, m$, we would prove it holds for $j = m+1$. Note that the error equation is

$$(e^{m+1}, q) + \alpha_0 B(e^{m+1}, q) = (1-d_1)(e^m, q) + \sum_{j=1}^{m-1}(d_j - d_{j+1})(e^{m-j}, q) + (r^{m+1}, q).$$

Taking $q = e^{m+1}$ yields

$$\begin{aligned}\|e^{m+1}\|^2_{H_0^{\frac{\mu}{2}}(\Omega)} \le\ & (1-d_1)\|e^m\|_{L^2}\|e^{m+1}\|_{L^2} + \sum_{j=1}^{m-1}(d_j - d_{j+1})\|e^{m-j}\|_{L^2}\|e^{m+1}\|_{L^2} \\ & + \|r^{m+1}\|_{L^2}\|e^{m+1}\|_{L^2}.\end{aligned}$$

According to the induction assumption, we have

$$\begin{aligned}\|e^{m+1}\|_{H_0^{\frac{\mu}{2}}(\Omega)} &\le \left((1-d_1)d_{m-1}^{-1} + \sum_{j=1}^{m-1}(d_j - d_{j+1})d_{m-j-1}^{-1}\right) c_p k^2 + c_p k^2 \\ &\le \left((1-d_1) + \sum_{j=1}^{m-1}(d_j - d_{j+1}) + d_m\right) c_p d_m^{-1} k^2 = c_p d_m^{-1} k^2.\end{aligned}$$

Therefore, the estimate (5.42) is proved.

A direct computation shows that $m^{-\alpha} d_{m-1}^{-1} = 1$ when $m = 1$, and $m^{-\alpha} d_{m-1}^{-1}$ increasingly tends to $\frac{1}{1-\alpha}$ as $m \to +\infty$. Hence

$$m^{-\alpha} d_{m-1}^{-1} \le \frac{1}{1-\alpha}, \quad m = 1, 2, \cdots, M.$$

For all m such that $mk \le T$, we have

$$\begin{aligned}\|p(t_m) - p^m\|_{H_0^{\frac{\mu}{2}}(\Omega)} &\le c_p d_{m-1}^{-1} k^2 = c_p m^{-\alpha} d_{m-1}^{-1} m^{\alpha} k^2 \\ &\le c_p \frac{1}{1-\alpha}(mk)^{\alpha} k^{2-\alpha} = c_{p,\alpha} T^{\alpha} k^{2-\alpha}.\end{aligned}$$

This is just the desired result. □

5.5 Fully discretization and its error analysis

Using $L_t^\alpha p(x,t_{m+1})$ as an approximation of $D_*^\alpha p(x,t_{m+1})$ leads to the fully discretization of (5.9) as

$$\begin{cases} \text{Find } p_N^{m+1} \in V_N, \text{ such that } \forall q \in V_N, \\ (L_t^\alpha p_N^{m+1}, q) + B(p_N^{m+1}, q) = 0, \quad m = 0, 1, \cdots, M-1, \\ (p_N^0, q) = (g, q). \end{cases} \tag{5.43}$$

For computation, we can rewrite the scheme (5.43) as:

$$\begin{cases} \text{Find } p_N^{m+1} \in V_N, \text{ such that } \forall q \in V_N, \\ (p_N^{m+1}, q) + \alpha_0 B(p_N^{m+1}, q) = (1-d_1)(p_N^m, q) \\ \qquad + \sum\limits_{j=1}^{m-1} (d_j - d_{j+1})(p_N^{m-j}, q) + d_m(p_N^0, q), \quad m = 1, \cdots, M-1, \\ (p_N^1, q) + \alpha_0 B(p_N^1, q) = (p_N^0, q), \\ (p_N^0, q) = (g, q), \end{cases} \tag{5.44}$$

where $\alpha_0 = \Gamma(2-\alpha)k^\alpha$.

Theorem 5.6. *Let $U''(x) \le 0$ for any $x \in \Omega$. $p_N^{m+1} (m = 0, 1, ..., M-1)$ are computed by scheme (5.44). Then we have the following estimate for any time-step length $k > 0$*

$$\|p_N^{m+1}\|_{H_0^{\frac{\mu}{2}}(\Omega)} \le \|p_N^0\|_{L^2(\Omega)}, \qquad m = 0, 1, \cdots, M-1. \tag{5.45}$$

This means scheme (5.44) is unconditionally stable.

Proof. The result will be concluded by induction. From (5.44), we have $p_N^0 = P_N g$ and

$$\|p_N^1\|^2 + \alpha_0 B(p_N^1, p_N^1) = (p_N^0, p_N^1).$$

Then from Lemma 5.8 and the fact $\|p\|_{L^2(\Omega)} \le \|p\|_{H_0^{\frac{\mu}{2}}(\Omega)}$ we obtain

$$\|p_N^1\|_{H_0^{\frac{\mu}{2}}(\Omega)} \le \|p_N^0\|_{L^2(\Omega)}.$$

Now we assume that for $j = 1, 2, \cdots, m$

$$\|p_N^j\|_{H_0^{\frac{\mu}{2}}(\Omega)} \le \|p_N^0\|_{L^2(\Omega)}. \tag{5.46}$$

From the first equation in scheme (5.44), we have

$$\begin{aligned} &(p_N^{m+1}, p_N^{m+1}) + \alpha_0 B(p_N^{m+1}, p_N^{m+1}) \\ &= (1-d_1)(p_N^m, p_N^{m+1}) + \sum_{j=1}^{m-1} (d_j - d_{j+1})(p_N^{m-j}, p_N^{m+1}) + d_m(p_N^0, p_N^{m+1}). \end{aligned}$$

By Lemma 5.8, (5.38) and the assumption (5.46), it holds

$$\begin{aligned}\|p_N^{m+1}\|^2_{H_0^{\frac{\mu}{2}}(\Omega)} &\le (1-d_1)\|p_N^m\|_{L^2(\Omega)}\|p_N^{m+1}\|_{L^2(\Omega)} \\ &\quad + \sum_{j=1}^{m-1}(d_j-d_{j+1})\|p_N^{m-j}\|_{L^2(\Omega)}\|p_N^{m+1}\|_{L^2(\Omega)} \\ &\quad + d_m\|p_N^0\|_{L^2(\Omega)}\|p_N^{m+1}\|_{L^2(\Omega)} \\ &\le \left[(1-d_1)+\sum_{j=1}^{m-1}(d_j-d_{j+1})+d_m\right]\|p_N^0\|_{L^2(\Omega)}\|p_N^{m+1}\|_{H_0^{\frac{\mu}{2}}(\Omega)} \\ &= \|p_N^0\|_{L^2(\Omega)}\|p_N^{m+1}\|_{H_0^{\frac{\mu}{2}}(\Omega)}.\end{aligned}$$

Hence,

$$\|p_N^{m+1}\|_{H_0^{\frac{\mu}{2}}(\Omega)} \le \|p_N^0\|_{L^2(\Omega)}.$$

The induction is completed. □

Let $p(x,t)$ and p_N^{m+1} be solutions to (5.9) and (5.43), respectively. Denote $p_*(t) = P_N p$ (projection with respect to space variable x) and $e_N^{m+1} = p_N^{m+1} - p_*^{m+1}$.

Theorem 5.7. *Let $U''(x) \le 0$ for any $x \in \Omega$ and $p(x,t) \in \left(H_0^{\frac{\mu}{2}}(\Omega) \cap H^r(\Omega)\right) \times C^2([0,T])$. Then it holds*

$$\|e_N^{m+1}\|_{H_0^{\frac{\mu}{2}}(\Omega)} \le C(N^{3\mu/4-r} + k^{2-\alpha}), \quad m = 0, 1, \cdots, M-1. \tag{5.47}$$

Proof. It is obviously that $e_N^0 = 0$ and they satisfy

$$\left(L_t^\alpha e_N^{m+1}, q\right) + B(e_N^{m+1}, q) = B(p_*^{m+1} - p^{m+1}, q) + \left((L_t^\alpha - D_*^\alpha)p_*^{m+1}, q\right). \tag{5.48}$$

We rewrite (5.48) as

$$\begin{cases} (e_N^{m+1}, q) + \alpha_0 B(e_N^{m+1}, q) = (1-d_1)(e_N^m, q) + \sum\limits_{j=1}^{m-1}(d_j - d_{j+1})(e_N^{m-j}, q) \\ \quad + d_m(e_N^0, q) + B(p_*^{m+1} - p^{m+1}, q) + \left((L_t^\alpha - D_*^\alpha)p_*^{m+1}, q\right), m = 1, \cdots, M-1, \\ (e_N^1, q) + \alpha_0 B(e_N^1, q) = (e_N^0, q) + B(p_*^1 - p^1, q) + \left((L_t^\alpha - D_*^\alpha)p_*^1, q\right). \end{cases} \tag{5.49}$$

The second equation in (5.49) by taking $q = e_N^1$ tells us

$$\|e_N^1\|_{H_0^{\frac{\mu}{2}}(\Omega)} \le c_B\|p_*^1 - p^1\|_{H_0^{\frac{\mu}{2}}(\Omega)} + \|(L_t^\alpha - D_*^\alpha)p_*^1\|_{L^2(\Omega)}.$$

For $m > 1$, by taking $q = e_N^{m+1}$ in the first equation of (5.49), we obtain

$$\begin{aligned}\|e_N^{m+1}\|_{H_0^{\frac{\mu}{2}}(\Omega)} &\le (1-d_1)\|e_N^m\|_{L^2(\Omega)} + \sum_{j=1}^{m-1}(d_j - d_{j+1})\|e_N^{m-j}\|_{L^2(\Omega)} \\ &\quad + c_B\|p_*^{m+1} - p^{m+1}\|_{H_0^{\frac{\mu}{2}}(\Omega)} + \|(L_t^\alpha - D_*^\alpha)p_*^{m+1}\|_{L^2(\Omega)}.\end{aligned}$$

Using the approximation results in Lemma 5.10 and Theorem 5.3, by the same argument as in Theorem 5.5, we can conclude the desired result. The details are omitted. □

5.6 Conclusion remarks

In this paper, we analyze the numerical error of finite difference/spectral method for time and space fractional Fokker-Planck equation which the time fractional derivative is in sense of Caputo and the space fractional derivative is in Riemann-Liouville sense. The discrete of time Caputo derivative is by finite difference method whereas we discrete the space Riemann-Liouville derivative by spectral method. The stability and convergence rate are set up rigorously both for semi and full-discrete scheme. We come to a theoretical result as $O(N^{3\mu/4-r} + k^{2-\alpha})$.

Bibliography

Agrawal P. Om, Machado J. A. Tenreiro and Sabatier J.(2004). Introduction, *Nonlinear Dynam.* **38**, pp. 1–2.

Alikhanov, A. A. (2010). A priori estimates for solutions of boundary value problems for fractional-order equations, *Diff. Eqns.* **46**, 5, pp. 660–666.

Askey, R. (1975). Inequalities via fractional integration, *Lect. Notes in Math.* **457**, pp. 106–115.

Bernardi, C. and Maday, Y. (1997). *Spectral methods*, in *Handbook of Numerical Analysis*, Vol. V, Techniques of Scientific Computing (Part 2), P. G. Ciarlet and J. L. Lions, eds., (North-Holland, Amsterdam), pp. 209–486.

Bouchaud, J. and Georges, A. (1990). Anomalous diffusion in disordered media: Statistical mechanisms, models and physical applications, *Phys. Rep.* **195**, pp. 127–293.

Deng, W. H. (2007a). Short memory principle and a predictor-corrector approach for fractional differential equations, *J. Comput. Appl. Math.* **206**, pp. 174–188.

Deng, W. H. (2007b). Numerical algorithm for the time fractional Fokker-Planck equation, *J. Comput. Phys.* **227**, 2, pp. 1510–1522.

Deng, W. H. (2008). Finite element method for the space and time fractional Fokker–Planck equation, *SIAM J. Numer. Anal.* **47**, 1, pp. 204–226.

Ervin, V. J. and Roop, J. P. (2005). Variational formulation for the stationary fractional advection dispersion equation, *Numer. Methods P. D. E.* **22**, 3, pp. 558–576.

Ervin, V. J., Heuer, N. and Roop, J. P. (2007). Numerical approximation of a time dependent, nonliear, space-fractional diffusion equation, *SIAM J. Numer. Anal.* **45**, 2, pp. 572–591.

Gorenflo, R., Mainardi, F., Moretti, D. and Paradisi, P. (2002). Time fractional diffusion: A discrete random walk approach, *Nonlinear Dyn.* **29**, 1-4, pp. 129–143.

Goychuk, I., Heinsalu, E., Patriarca, M., Schmid, G. and Hänggi, P. (2006). Current and universal scaling in anomalous transport, *Phys. Rev. E* **73**, 2, article 020101.

Heinsalu, E., Patriarca, M., Schmid, G. and Hänggi, P. (2006). Fractional Fokker–Planck dynamics: Numerical algorithm and simulations, *Phys. Rev. E* **73**, 4, article 046133.

Lenzi, E. K., Mendes, R. S., Fa, K. S. and Malacarne, L. C. (2003). Anomalous diffusion: Fractional Fokker–Planck equation and its solutions, *J. Math. Phys.* **44**, 5, pp. 2179–2185.

Li, X. J. and Xu, C. J. (2009). A space-time spectral method for the time fractional diffusion equation, *SIAM J. Numer. Anal.* **47**, 3, pp. 2108–2131.

Li, X. J. and Xu, C. J. (2010). Existence and uniqueness of the weak solution of the space-time fractional diffusion equation and a spectral method approximation, *Commun. Comput. Phys.* **8**, 5, pp. 1016–1051.

Lin, Y. M. and Xu, C. J. (2007). Finite difference/spectral approximations for the time-fractional diffusion equation, *J. Comput. Phys.* **225**, 2, pp. 1533–1552.

Liu, F. W, Ann, V. and Xu, C. J. (2004). Numerical solution of the space fractional Fokker-Planck equation, *J. Comput. Appl. Math.* **166**, 1, pp. 209–219.

Lubich, C. (1986). Discretized fractional calculus, *SIAM J. Math. Anal.* **17**, 3, pp. 704–719.

Magdziarz, M. and Weron, A. (2007). Competition between subdiffusion and lévy flights: A Monte Carlo approach, *Phys. Rev. E* **75**, 5, article:056702.

Meerschaert, M. M. , Scheffler, H. P. and Tadjeran, C. (2006). Finite difference methods for two-dimensional fractional dispersion equation, *J. Comput. Phys.* **211**, 1, pp. 249–261.

Metzler, R., Tejedor, V., Jeon, J. H., He, Y., Deng ,W. H, Burov, S. and Barkai, E. (2009). Analysis of single particle trajectories: from normal to anomalous diffusion, *ACTA Physica Polonica B* **40**, 5, pp. 1533–1552.

Metzler, R. and Klafter, J. (2000). The random walk's guide to anomalous diffusion: A fractional dynamics approach, *Phys. Rep.* **339**, 1, pp. 1–77.

Metzler, R. and Nonnenmacher, T. F. (2002). Space- and time-fractional diffusion and wave equations, fractional Fokker-Planck equations, and physical motivation, *Chem. Phys.* **284**, 1-2, pp. 67–90.

Podlubny, I. (1999). *Fractional Differential Equations* (Academic Press, New York.)

Solomon, T. H., Weeks, E. R. and Swinney, H. L. (1993). Observations of anomalous diffusion and Lévy flights in a 2-dimensional rotating flow, *Phys. Rev. Lett.* **71**, 24, pp. 3975–3979.

Sugimoto, N. (1991). Burgers equation with a fractional derivative: Hereditary effects on nonlinear acoustic waves, *J. Fluid Mech.* **225**, pp. 631–653.

Yuste, S. B. and Acedo, L. (2005). An explicit finite difference method and a new von Neumann-type stability analysis for fractional diffusion equations, *SIAM J. Numer. Anal.* **42**, 5, pp. 1862–1874.

Zaslavsky, G. M. (2002). Chaos, fractional kinetics, and anomalous transport, *Phys. Rep.* **371**, 6, pp. 461–580.

Chapter 6

A discontinuous finite element method for a type of fractional Cauchy problem[1]

Yunying Zheng

School of Mathematical Sciences, Huaibei Normal University, Huaibei 235000, Anhui, PR China

zhengyunying1973@163.com

In this chapter, a new discontinuous Galerkin numerical method for a type of the nonlinear fractional Cauchy problem is formulated. By introducing an auxiliary function, we change the fractional differential equation into a system of equations. With the help of modified discontinuous Galerkin method, a new numerical scheme is derived. The existence and uniqueness of the solution to the discrete system are proved. The error estimates are also derived in details. Numerical results are included which are in line with the theoretical results.

Keywords: Fractional Cauchy problem, discontinuous Galerkin method, Caputo derivative

6.1 Introduction

In this chapter we study the fractional Cauchy problem in the following form

$$a(t)\frac{du}{dt} + b(u)_C D_{0,t}^{\alpha} u = f(t,u), t > 0, \alpha \in (0,1),$$

where $\frac{du}{dt}$ is the typical derivative, and ${}_C D_{0,t}^{\alpha}$ denotes the left fractional derivative with fractional order α in the sense of Caputo.

The study of the numerical methods for fractional Cauchy problems have received considerable attention. And there are several methods to approximate the solution of the differential equation involving the time fractional derivative. One of them is to change the differential equation into a Volterra-type integral equation, then approximate the unknown function with the quadratic polynomials [Kumar and Agrawal (2006)], or the Adams-type predictor-corrector method [Diethelm *et al.*

[1]This work is partially supported by the National Natural Science Foundation of China (grant no. 10872119) and Shanghai Leading Academic Discipline Project (grant no. S30104).

(2004)], or the finite element method [Agrawal (2008); Adolfsson *et al.* (2004)]. The second is to introduce a fractional-order linear multiple step method to approximate the fractional differential equation [Lin and Liu (2007)]. The third is to apply the finite difference scheme to approximate the derivative with noninteger order in the definition of Caputo derivative [Liu *et al.* (2007)].

It has been found that there are some advantages to apply discontinuous finite element method: it allows for arbitrariness in the time step as well as in the approximation order, it is suitable to seek high-order accurate solutions both on structured and unstructured grids. The first analysis of a Galerkin method for ODEs, seemed to be contained in [Lesaint and Raviart (1974)]. Generalizations of this method were proposed [Delfour *et al.* (1981)]. DG methods have also been applied successfully to ODEs and PDEs [Thomée (2006); Tsallis and Lenzia (2002); Wang *et al.* (2009); Schötzau and Schwab (2000); Brunner and Schötzau (2006); Shaw and Whiteman (1996); Larsson *et al.* (1998)] and FDEs [Zheng *et al.* (2010a,b); Zheng and Zhao (2011)].

In the present work, we numerically approximate the fractional nonlinear ordinary differential equations by the DG finite element methods, and derive a priori error bound that is completely explicit in the time step. The outline of the rest of this chapter is as follows. In Section 6.2 we introduce the fractional derivative and its properties. In Section 6.3 we describe the DG method and prove existence and uniqueness of discrete DG solution. In Section 6.4 we analyze the convergence in detail. In the last section, we perform the numerical experiments to verify our theoretical results.

6.2 Fractional derivative space

Here we firstly introduce the fractional integral, the Caputo derivative and some of their properties which will be used later on.

Definition 6.1. The α-th order left and right Riemann-Liouville integrals of function $u(x)$ are defined as follows,

$$D_{a,x}^{-\alpha}u(x) = \frac{1}{\Gamma(\alpha)}\int_a^x \frac{u(s)}{(x-s)^{1-\alpha}}\mathrm{d}s,$$

$$D_{x,b}^{-\alpha}u(x) = \frac{1}{\Gamma(\alpha)}\int_x^b \frac{u(s)}{(s-x)^{1-\alpha}}\mathrm{d}s,$$

where $\alpha > 0$ and $a < x < b$.

Definition 6.2. The α-th order Caputo derivatives of function $u(x)$ are defined as follows,

$$_C D_{a,x}^{\alpha}u(x) = D_{a,x}^{\alpha-n}\, D^n u(x),$$

$$_C D_{x,b}^{\alpha}u(x) = D_{x,b}^{\alpha-n}\, (-D)^n u(x),$$

where $n-1 < \alpha < n \in Z^+$ and $a < x < b$.

$W^{\alpha,2}(\Omega)$ is rewritten as $H^\alpha(\Omega)$ which is equipped with norm $\|\cdot\|_{H^\alpha}$ or $\|\cdot\|_\alpha$ and semi-norm $|\cdot|_{H^\alpha}$ or $|\cdot|_\alpha$. The following are some useful results.

Lemma 6.1 ([Roop (2006)]). *For $u \in J^{-\alpha}([a,b])$, $0<\beta<\alpha$, then*

$$D_{a,x}^{-\alpha}u(x) = D_{a,x}^{-\alpha-\beta}\, D_{a,x}^{-\beta}u.$$

Lemma 6.2. *The left and the right Riemann-Liouville fractional integral operators satisfy the following relation. If $u, v \in L_{loc}(a,b)$, then*

$$(D_{a,x}^{-\alpha}u, v) = (u,\ D_{x,b}^{-\alpha}v), \forall \alpha > 0.$$

Lemma 6.3 ([Ervin and Roop (2006)]). *For $u \in L^2(R)$, one has*

$$(D_{-\infty,t}^{-\alpha}u(t), D_{t,\infty}^{-\alpha}u(t)) = \cos(\alpha\pi)\int_{-\infty}^{+\infty}(D_{-\infty,x}^{-\alpha}u(t))^2 dt.$$

Lemma 6.4. *Let $\mu > 0$. The following mapping properties hold*

$${}_C D_{a,x}^{\mu} : J^\mu(\Omega) \to L^2(\Omega) \quad \textit{is a bounded linear operator},$$
$${}_C D_{x,b}^{\mu} : J^\mu(\Omega) \to L^2(\Omega) \quad \textit{is a bounded linear operator}.$$

The proof is similar to [Roop (2006)].

6.3 The discontinuous Galerkin finite element approximation

In this section we will formulate a discontinuous Galerkin finite element method for a type of the nonlinear fractional Cauchy problem.

Problem 6.1. (Fractional Cauchy problem) For some $T > 0$ and $0 < \alpha < 1$, we consider the following problem.

$$\begin{cases} a(t)u' + b(t)\,{}_C D_{0,t}^{\alpha}u = f(t,u), & t \in (0,T], \\ u(0) = u_0, \end{cases}$$

where the coefficients a and b satisfy

$$0 < M_1 < a(t), b(t) < M_2.$$

Throughout the paper, we always assume that the following Lipschitz continuity conditions on f holds: there exists a positive constant L for $t \in (0,T]$ and $s, r \in R$, such that

$$|f(t,s) - f(t,r)| \leq L|s-r|. \tag{6.1}$$

In order to derive a variational form of Problem 6.1, we suppose that u is a sufficiently smooth solution. Following the standard approach of local discontinuous

Galerkin methods, we introduce the auxiliary variable v, and rewrite Problem 6.1 as

$$\begin{cases} u'(t) = v(t), & \text{in } (0,T), \\ a(t)v(t) + b(t)D_{0,t}^{\alpha-1}v(t) = f(u), & \text{in } (0,T). \\ u(0) = u_0. \end{cases} \tag{6.2}$$

Using standard notation, let the nodes $0 = t_0 < t_1 < \cdots < t_N = T$, and the subdivision $I_n = (t_{n-1}, t_n]$. Let k_n be the length of I_n, i.e. $k_n = t_n - t_{n-1}$, $n = 1, \cdots N$. Set $k = \max\limits_{1\le n\le N} k_n$. Define the mesh $\mathcal{M} = \{I_n\}_{n=1}^N$, and associated with the mesh $\mathcal{M}$, we define the broken Sobolev spaces

$$L^2(I, \mathcal{M}) = \{u : I \to \mathbb{R}, u|_{I_n} \in L^2(I_n), n = 1, 2, \cdots, N\},$$

and

$$H^1(I, \mathcal{M}) = \{u : I \to \mathbb{R}, u|_{I_n} \in H^1(I_n), n = 1, 2, \cdots, N\}.$$

For a function $u \in H^1(I, \mathcal{M})$, we denote the superscripts '-' and '+' as left and right limits respectively,

$$u^{\pm}(t_n) = u(t_n^{\pm}) = \lim_{t\to t_n^{\pm}} u(t)$$

and the jump is described by

$$[u]_n = \lim_{\varepsilon\to 0}[u(t_n + |\varepsilon|) - u(t_n - |\varepsilon|)] = u(t_n^+) - u(t_n^-).$$

So, the exact solution (u, v) of (6.2) belongs to

$$H^1(I; H^1(I, \mathcal{M})) \times L^2(I; L^2(I, \mathcal{M})).$$

Under this assumption, we can give the equations, where (u, v) satisfies

$$\begin{cases} \int_I u'\varphi dt + \sum\limits_{n=2}^{N-1} [u]_{n-1}\varphi_{n-1}^+ + u_0^+\varphi_0^+ = \int_I v\varphi dt + u_0\varphi_0^+, \\ \\ \int_I av\psi dt + \int_I bD_{0,t}^{\alpha-1}v\psi dt = \int_I f(u)\psi dt, \end{cases} \tag{6.3}$$

for all $(\varphi, \psi) \in H^1(I, \mathcal{M}) \times L^2(I, \mathcal{M})$.

In each time slab, it gives

$$\begin{cases} \int_{I_n} u'\varphi dt + [u]_{n-1}\varphi_{n-1}^+ = \int_{I_n} v\varphi dt, \\ \\ \int_{I_n} av\psi dt + \int_{I_n} bD_{0,t}^{\alpha-1}v\psi dt = \int_{I_n} f(u)\psi dt, \end{cases} \tag{6.4}$$

for all $(\varphi, \psi) \in H^1(I, \mathcal{M}) \times L^2(I, \mathcal{M})$.

In the following discussion, we restrict the trial and test functions φ, ψ to be in the finite dimensional subspaces $\Im_k \subset H^1(I, \mathcal{M})$. On each time slab I_n, we choose $\Im_k$ to be the discontinuous, piecewise polynomial function space

$$\Im_k = \{v : I_n \to R;\ v|_{t\in I_n} \in P_q(I_n)\}.$$

The functions of the space $\Im_k$ is discontinuous at the time node t_n, but is left-continuous and right-continuous. And the functions of the space $\Im_k$ are polynomials on time slab I_n, whose degree is no more than q.

Furthermore, we can give the numerical schemes of the discontinuous finite element. We define U, V as the approximations of u, v respectively in the space $\Im_k$, then the time discontinuous Galerkin scheme for Problem 6.1 can be now formulated as follows: find $(U, V) \in H^1(I; \Im_k) \times L^2(I; \Im_k)$, such that for all $\varphi, \psi \in \Im_k$ the following relations are satisfied

$$\begin{cases} \sum\limits_{n=1}^{N} \int_{I_n} U'\varphi dt + \sum\limits_{n=2}^{N-1} [U]_{n-1}\varphi_{n-1}^{+} + U_0^{+}\varphi_0^{+} = \sum\limits_{n=1}^{N} \int_{I_n} V\varphi dt + U_0\varphi_0^{+}, \\ \\ \sum\limits_{n=1}^{N} \int_{I_n} aV\psi dt + \sum\limits_{n=1}^{N} \int_{I_n} bD_{0,t}^{\alpha-1} V\psi dt = \sum\limits_{n=1}^{N} \int_{I_n} f(U)\psi dt. \end{cases} \tag{6.5}$$

As a time stepping scheme, we give the scheme at each time step. If the value of (U, V) is given on the time slab $I_n, 1 \le k \le n-1$, we can find the value (U, V) at node t_n by solving the following equations.

$$\begin{cases} \int_{I_n} U'\varphi dt + [U]_{n-1}\varphi_{n-1}^{+} = \int_{I_n} V\varphi dt, \\ \\ \int_{I_n} aV\psi dt + \int_{I_n} bD_{0,t}^{\alpha-1} V\psi dt = \int_{I_n} f(U)\psi dt, \end{cases} \tag{6.6}$$

for all $\varphi, \psi \in \Im_k$, where $U_0^- = u_0$.

When we select the shape function as a constant function, we find that

$$\int_{I_n} U'\varphi dt = 0.$$

Furthermore

$$\|U\|_{I_n}^2 = \int_{I_n} \|U\|^2 dt = k_n \|U_{n-1}^{+}\|^2 = k_n \|U_n^{-}\|^2 = 0.$$

The scheme (6.5) can be changed into

$$\begin{cases} (U_{n-1}^{+} - U_{n-1}^{-})\varphi_{n-1}^{+} = \int_{I_n} V\varphi dt, \\ \\ \int_{I_n} aV\psi dt + \int_{I_n} bD_{0,t}^{\alpha-1} V\psi dt = \int_{I_n} f(U)\psi dt. \end{cases} \tag{6.7}$$

If we select the shape function V as a constant shape function, then Scheme (6.6) is rewritten as the following form.

$$a\frac{U_n^- - U_{n-1}^-}{k_n} + bD_{t_{n-1},t_n}^{\alpha-1} \frac{U_n^- - U_{n-1}^-}{k_n} + bD_{0,t_{n-1}}^{\alpha-1} du = f(U).$$

If we approximate $f(U)$ on the time slab with $f(U_{n-1})$, this scheme is similar to the Euler forward difference scheme. If we take $\frac{dU(t_i)}{dt} \approx \frac{U_i - U_{i-1}}{k_i}$ at every time slab $I_i, i = 1, \cdots, n-1$, then the scheme is

$$a_n\frac{U_n^- - U_{n-1}^-}{k_n} + b_n D_{t_{n-1},t_n}^{\alpha-1} \frac{U_n^- - U_{n-1}^-}{k_n} + \sum_{i=1}^{n-1} b_i D_{I_i}^{\alpha-1} \frac{U_i^- - U_{i-1}^-}{k_i} = f(U_i).$$

It can be considered as a non-local difference scheme. If we adapt $U' = 0$ at time slab $(0, t_{n-1})$, then the scheme is changed as

$$a_n \frac{U_n^- - U_{n-1}^-}{k_n} + b_n D_{t_{n-1},t_n}^{\alpha-1} \frac{U_n^- - U_{n-1}^-}{k_n} = f(U_{i-1}).$$

It is a standard Euler forward scheme.

Next, we investigate the uniqueness and existence about the discontinuous numerical scheme. First, we give the discrete Gronwall inequality [Larsson *et al.* (1998)].

Lemma 6.5. *Let $\mathcal{M}$ be a partition of $(0, T)$ with $k = max_{n=1}^N\{k_n\}$. Let $\{a_n\}_{n=1}^N$ and $\{b_n\}_{n=1}^N$ be sequences of numbers with $0 \leq b_1 \leq b_2 \leq \cdots \leq b_N$. Assume that there is a constant $K \geq 0$ such that*

$$a_1 \leq b_1, \quad a_n \leq b_n + K \sum_{m=1}^{n} \omega_{n,m}(\alpha) a_n, \quad n = 2, \cdots, N,$$

where $\omega_{n,m}(\alpha) = \int_{I_m} (t_n - t)^{-\alpha} dt$. Assume further that $\delta = \frac{Kk^{1-\alpha}}{1-\alpha} \leq 1$. Then we have

$$a_n \leq b_n, n = 1, 2, \cdots, N,$$

with a constant $C > 0$ that solely depends on δ, K, α, and T.

Lemma 6.6 ([Schötzau and Schwab (2000)]). *Let $I = (a, b)$, and $k = b - a > 0$. Then*

$$\int_a^b \|\varphi\|^2 dt \leq \frac{1}{k} \sum_{i=1}^{d} (\int_a^b \varphi_i(t) dt)^2 + \frac{1}{2} \int_a^b (b - t)(t - a) \|\varphi'(t)\|^2 dt$$

for all $\varphi(t) = (\varphi_1(t), \cdots, \varphi_1(t)) \int P^r((a, b); \mathbb{R}^d), r \in \mathbb{N}_0$. (This estimate is, in particular, independent of r.)

Lemma 6.7 ([Brunner and Schötzau (2006)]). *For $f \in L^2(0, \tau)$, and $\alpha \in (0, 1)$ there holds*

$$\int_0^\tau \Big(\int_0^t (t - s)^{-\alpha} f(s) ds \Big)^2 dt \leq \frac{\tau^{1-\alpha}}{(1 - \alpha)} \int_0^\tau (\tau - t)^{-\alpha} \Big(\int_0^t f^2(s) ds \Big) dt.$$

Now we address the existence and uniqueness of discrete solutions.

Theorem 6.1. *Assume that $(\mathcal{M}, r)$ is an hp-discretization of $(0, T)$ with $\dfrac{M_2 T^{1-\alpha}}{2M_1 \Gamma(2 - \alpha)} \dfrac{k^{1-\alpha}}{1 - \alpha} < 1$ and $\max\{\dfrac{\sqrt{5}kL}{\sqrt{6}m}, kL\} \leq 1$. Then there exists a unique solution $(U, V) \in \Im k \times \Im k$.*

Proof. Given the initial value U_{n-1}^{-} ($U_0^{-} = U_0$) it is sufficient to show that the problem 6.1 on time slab I_n has a unique solution $U \in P^r(I_n; R^d)$. Consider first the case where $r = 0$. The computation scheme is that

$$\begin{cases} U_{n-1}^{+}\varphi_{n-1}^{+} = U_{n-1}^{-}\varphi_n^{+} + \int_{I_n} V\varphi dt, \\ \int_{I_n} aV\psi dt + \int_{I_n} bD_{0,t}^{\alpha-1}V\psi dt = \int_{I_n} f(U)\psi dt. \end{cases}$$

To prove the uniqueness, set (U, V) and $(\widetilde{U}, \widetilde{U})$ are two solutions of above equations. Then $(U - \widetilde{U}, V - \widetilde{V})$ is the solution of the following equations.

$$\begin{cases} (U - \widetilde{U})_{n-1}^{+}\varphi_{n-1}^{+} = \int_{I_n}(V - \widetilde{V})\varphi dt, \\ \int_{I_n} a(V - \widetilde{V})\psi dt + \int_{I_n} bD_{0,t}^{\alpha-1}(V - \widetilde{V})\psi dt = \int_{I_n}(f(U) - f(\widetilde{U}))\psi dt. \end{cases} \tag{6.8}$$

We set $\varphi = U - \widetilde{U}$. Since $\|W\|_{I_n}^2 = k_n\|W_{n-1}^{+}\|^2$, then

$$\begin{aligned} &\|U - \widetilde{U}\|_{I_n}^2 = k_n\|(U - \widetilde{U})_{n-1}^{+}\|^2 \\ &= k_n \int_{I_n}(V - \widetilde{V})(U - \widetilde{U})dt \\ &\le k_n\|V - \widetilde{V}\|_{I_n}\|U - \widetilde{U}\|_{I_n}. \end{aligned} \tag{6.9}$$

So we can conclude that

$$\|U - \widetilde{U}\|_{I_n} \le k_n\|V - \widetilde{V}\|_{I_n}. \tag{6.10}$$

We set $\psi = V - \widetilde{V}$, then one has

$$\begin{aligned} &\int_{I_n} a(V - \widetilde{V})^2 dt + \int_{I_n} bD_{0,t}^{\alpha-1}(V - \widetilde{V})(V - \widetilde{V})dt \\ &= \int_{I_n}(f(U) - f(\widetilde{U}))(V - \widetilde{V})dt \\ &\le L\int_{I_n} |U - \widetilde{U}| \cdot (V - \widetilde{V})dt \\ &\le L\|U - \widetilde{U}\|_{I_n}\|V - \widetilde{V}\|_{I_n}. \end{aligned} \tag{6.11}$$

Take (6.10) into (6.11), one has

$$\begin{aligned} &\int_{I_n} a(V - \widetilde{V})^2 dt + \int_{I_n} bD_{0,t}^{\alpha-1}(V - \widetilde{V})(V - \widetilde{V})dt \\ &\le L\|U - \widetilde{U}\|_{I_n}\|V - \widetilde{V}\|_{I_n} \\ &\le Lk_n\|V - \widetilde{V}\|_{I_n}^2. \end{aligned} \tag{6.12}$$

Furthermore, one has

$$\begin{aligned} \int_{I_n} a(V - \widetilde{V})^2 dt &\le Lk_n\|V - \widetilde{V}\|_{I_n}^2 - \int_{I_n} bD_{0,t}^{\alpha-1}(V - \widetilde{V})(V - \widetilde{V})dt \\ &\le Lk_n\|V - \widetilde{V}\|_{I_n}^2 + \int_{I_n} |bD_{0,t}^{\alpha-1}(V - \widetilde{V})(V - \widetilde{V})|dt. \end{aligned} \tag{6.13}$$

With the boundedness of $a(t)$ and $b(t)$, we can give the following estimation.

$$\begin{aligned} &M_1 \int_{I_n}(V - \widetilde{V})^2 dt \\ &\le Lk_n\|V - \widetilde{V}\|_{I_n}^2 + M_2 \int_{I_n} |D_{0,t}^{\alpha-1}(V - \widetilde{V})(V - \widetilde{V})|dt \\ &\le Lk\|V - \widetilde{V}\|_{I_n}^2 + M_2 \int_{I_n} |D_{0,t}^{\alpha-1}(V - \widetilde{V})(V - \widetilde{V})|dt. \end{aligned} \tag{6.14}$$

Summing n from 1 to m, with the help of Lemma 6.7, the above estimate yields

$$\begin{aligned}
&\textstyle\int_0^{t_m}(V-\widetilde{V})^2dt\\
&\le \frac{Lk}{M_1}\|V-\widetilde{V}\|^2_{(0,t_m)}+\frac{M_2}{M_1}\textstyle\int_0^{t_m}|D^{\alpha-1}_{0,t}(V-\widetilde{V})(V-\widetilde{V})|dt\\
&\le \frac{Lk}{M_1}\|V-\widetilde{V}\|^2_{(0,t_m)}+\frac{M_2}{M_1\Gamma(1-\alpha)}\left(\textstyle\int_0^{t_m}(\int_0^t(t_m-t)^{-\alpha}|V-\widetilde{V}|ds)^2dt\right)^{\frac{1}{2}}\\
&\quad\times\left(\textstyle\int_0^{t_m}(V-\widetilde{V})^2dt\right)^{\frac{1}{2}}\\
&\le \frac{Lk}{M_1}\|V-\widetilde{V}\|^2_{(0,t_m)}+\frac{M_2t_m^{1-\alpha}}{2M_1\Gamma(2-\alpha)}(\textstyle\int_0^{t_m}(t_m-t)^{-\alpha}\int_0^t|V-\widetilde{V}|^2ds)dt)\\
&\quad+\frac{1}{2}\textstyle\int_0^{t_n}(V-\widetilde{V})^2dt.
\end{aligned}\tag{6.15}$$

For the sufficiently small k, one gets

$$\int_0^{t_m}(V-\widetilde{V})^2dt\le\frac{M_2t_m^{1-\alpha}}{M_1\Gamma(2-\alpha)}\sum_{n=1}^{m}(\int_0^{t_m}\int_{I_n}(t_m-t)^{-\alpha}dt)(\int_0^{t_n}(V-\widetilde{V})^2dt).\tag{6.16}$$

Setting $a_m=\int_0^{t_m}(V-\widetilde{V})^2dt$ and $b_m=0$, the Gronwall inequality gives

$$\int_0^{t_m}(V-\widetilde{V})^2dt=0,\qquad m=1,\cdots,N,\tag{6.17}$$

provided that $\dfrac{M_2t_m^{1-\alpha}}{M_1\Gamma(2-\alpha)}\dfrac{k^{1-\alpha}}{1-\alpha}<1$ is satisfied. It follows that $V-\widetilde{V}=0$ and $V=\widetilde{V}$. That means that V is uniqueness. Afterwards we sum up the two equations and set $V=0$, then

$$U^+_{n-1}\psi^+_{n-1}=U^-_{n-1}\psi^+_n+\int_{I_n}f(U)\psi dt.$$

Set $\widetilde{U}$ is the other solution to above equation, one gets

$$\begin{aligned}
&((U-\widetilde{U})^+_{n-1})^2=+\textstyle\int_{I_n}(f(U)-f(\widetilde{U}))dt\\
&\le L\|U-\widetilde{U}\|^2_{I_n}.
\end{aligned}$$

So

$$\|U-\widetilde{U}\|^2\le k_nL\|U-\widetilde{U}\|^2_{I_n}\le kL\|U-\widetilde{U}\|^2_{I_n}.$$

If $kL<1$, then $\|U-\widetilde{U}\|\le 0$. This means that U is unique. Combing with the proof of the uniqueness of v, we end the proof for the case of $r=0$.

Next we consider the case of $r\ge 1$.

We also set (U,V) and $(\widetilde{U},\widetilde{V})$ are two solutions for (6.6). Then $(U-\widetilde{U},V-\widetilde{V})$ is the solution of the following equations.

$$\begin{cases}\int_{I_n}(U-\widetilde{U})'\varphi dt+[U-\widetilde{U}]^+_{n-1}\varphi^+_{n-1}=\int_{I_n}(V-\widetilde{V})\varphi dt,\\ \int_{I_n}a(V-\widetilde{V})\psi dt+\int_{I_n}bD^{\alpha-1}_{0,t}(V-\widetilde{V})\psi dt=\int_{I_n}(f(U)-f(\widetilde{U}))\psi dt.\end{cases}\tag{6.18}$$

Taking $\varphi=(t-t_{n-1})(U-\widetilde{U})'$ into the first equation of (6.18), one gets

$$\begin{aligned}
&\textstyle\int_{I_n}(t-t_n)((U-\widetilde{U})')^2dt+(U-\widetilde{U})^+_{n-1}(U-\widetilde{U})^+_{n-1}(t_{n-1}-t_{n-1})\\
&=\textstyle\int_{I_n}(V-\widetilde{V})(t-t_{n-1})(U-\widetilde{U})'dt.
\end{aligned}$$

That means

$$\begin{aligned}
&\int_{I_n}(t-t_n)((U-\widetilde{U})')^2dt\\
&=\int_{I_n}(V-\widetilde{V})(t-t_{n-1})(U-\widetilde{U})'dt\\
&\le(\int_{I_n}(V-\widetilde{V})^2(t-t_{n-1})dt)^{\frac{1}{2}}(\int_{I_n}((U-\widetilde{U})')^2(t-t_{n-1})dt)^{\frac{1}{2}}.
\end{aligned}$$

So

$$\int_{I_n}(t-t_n)((U-\widetilde{U})')^2dt\le\int_{I_n}(V-\widetilde{V})^2(t-t_{n-1})dt\le k_n\int_{I_n}(V-\widetilde{V})^2dt. \tag{6.19}$$

By integration by parts, the first equation of (6.18) is equivalent to

$$-\int_{I_n}(U-\widetilde{U})\varphi dt+(U-\widetilde{U})_n^-\varphi_n^-=\int I_n(V-\widetilde{V})\varphi dt. \tag{6.20}$$

Set $\varphi=t_n-t$, and take it into (6.20), we have

$$\begin{aligned}
&\int_{I_n}(U-\widetilde{U})dt=\int_{I_n}(V-\widetilde{V})(t_n-t)dt\\
&\le\|V-\widetilde{V}\|_{I_n}(\int_{I_n}(t_n-t)^2dt)^{\frac{1}{2}}=\frac{k_n^{1.5}}{\sqrt{3}}\|V-\widetilde{V}\|_{I_n}.
\end{aligned}$$

So

$$\Big(\int_{I_n}U-\widetilde{U}dt\Big)^2\le\frac{k_n^3}{3}\|V-\widetilde{V}\|_{I_n}^2. \tag{6.21}$$

According to Lemma 6.6, and combing with (6.19) and (6.21), one gets

$$\begin{aligned}
\|U-\widetilde{U}\|^2&\le\frac{1}{k_n}(\int_{I_n}U-\widetilde{U}dt)^2+\frac{1}{2}\int_{I_n}(t-t_{n-1})(t_n-t)\|(U-\widetilde{U})'\|^2dt\\
&\le\frac{k_n^2}{3}\|V-\widetilde{V}\|_{I_n}^2+\frac{k_n}{2}\int_{I_n}(t-t_{n-1}))((U-\widetilde{U})')^2dt\\
&\le\frac{k_n^2}{3}\|V-\widetilde{V}\|_{I_n}^2+\frac{k_n^2}{2}\int_{I_n}(V-\widetilde{V})^2dt\\
&\le\frac{5k_n^2}{6}\|V-\widetilde{V}\|_{I_n}^2.
\end{aligned} \tag{6.22}$$

Set $\psi=V-\widetilde{V}$, and take it into the second equation of (6.18). With the similar proof, we have

$$\begin{aligned}
&\int_0^{t_m}(V-\widetilde{V})^2dt\\
\le&\frac{\sqrt{5}Lk}{\sqrt{6}M_1}\|V-\widetilde{V}\|_{(0,t_m)}^2+\frac{M_2}{M_1}\int_0^{t_m}|D_{0,t}^{\alpha-1}(V-\widetilde{V})(V-\widetilde{V})|dt\\
\le&\frac{\sqrt{5}Lk}{\sqrt{6}M_1}\|V-\widetilde{V}\|_{(0,t_m)}^2+\frac{M_2}{M_1\Gamma(1-\alpha)}\Big(\int_0^{t_m}(\int_0^t(t_m-t)^{-\alpha}|V-\widetilde{V}|ds)^2dt\Big)^{\frac{1}{2}}\\
&\times\Big(\int_0^{t_m}(V-\widetilde{V})^2dt\Big)^{\frac{1}{2}}\\
\le&\frac{\sqrt{5}Lk}{\sqrt{6}M_1}\|V-\widetilde{V}\|_{(0,t_m)}^2+\frac{M_2t_m^{1-\alpha}}{2M_1\Gamma(2-\alpha)}(\int_0^{t_m}\int_0^t(t_m-t)^{-\alpha}\int_0^t|V-\widetilde{V}|^2ds)dt)\\
&+\frac{1}{2}\int_0^{t_m}(V-\widetilde{V})^2dt.
\end{aligned}$$

So

$$(\frac{1}{2}-\frac{\sqrt{5}Lk}{\sqrt{6}M_1})\int_0^{t_m}(V-\widetilde{V})^2dt \le \frac{M_2 t_m^{1-\alpha}}{2M_1\Gamma(2-\alpha)}(\int_0^{t_m}\int_0^t(t_m-t)^{-\alpha}\int_0^t|V-\widetilde{V}|^2ds)dt)$$

If k is sufficiently small, then $\frac{\sqrt{5}Lk}{\sqrt{6}M_1}\int_0^{t_m}(V-\widetilde{V})^2dt$ is far less than $\frac{1}{2}\int_0^{t_m}(V-\widetilde{V})^2dt$. So one has

$$\frac{1}{2}\int_0^{t_m}(V-\widetilde{V})^2dt \le \frac{M_2 t_m^{1-\alpha}}{2M_1\Gamma(2-\alpha)}\sum_{n=1}^{m}(\int_0^{t_m}\int_{I_n}(t_m-t)^{-\alpha}dt)(\int_0^{t_n}(V-\widetilde{V})^2dt). \tag{6.23}$$

Setting $a_m = \int_0^{t_m}(V-\widetilde{V})^2dt$ and $b_m = 0$, the Gronwall inequality in Lemma 6.6 gives

$$\int_0^{t_m}(V-\widetilde{V})^2dt = 0, \qquad m = 1, \cdots, N, \tag{6.24}$$

provided that $\frac{M_2 t_m^{1-\alpha}}{M_1\Gamma(2-\alpha)}\frac{k^{1-\alpha}}{1-\alpha} < 1$ is satisfied. This means that V is uniqueness.

As to the uniqueness of U, set $V-\widetilde{V}=0$ in the first equation in (6.18), then we have

$$\int_{I_n}(U-\widetilde{U})'\varphi dt + [U-\widetilde{U}]_{n-1}^{+}\varphi_{n-1}^{+} = \int_{I_n}(f(U)-f(\widetilde{U}))\varphi dt.$$

And the following inequality is hold by almost the same reasoning

$$\|U-\widetilde{U}\|_{I_n} \le \frac{\sqrt{5}kL}{\sqrt{6}M_1}\|U-\widetilde{U}\|_{I_n}.$$

Since $\frac{\sqrt{5}kL}{\sqrt{6}M_1} < 1$, it immediately follows that

$$\|U-\widetilde{U}\|_{I_n} = 0.$$

This means U is unique. As problem (6.6) is finite dimensional, the existence of solutions follows from their uniqueness. This completes the proof. □

6.4 Error estimation

Now we turn to analyzing the error estimate of the DG scheme.

Let U be the approximate solution on time slab I_n. Denote $P_h u$ as a Rits-Galerkin Projection operator defined as follows

$$(u-P_h u, \varphi) = 0, \forall\varphi \in P^r(I_n), k \ge 0,$$

and for all the interval $I_n = (t_{n-1}^{+}, t_n^{-}), n = 1, \cdots, N$, $P_h^{\pm}$ are defined to satisfy that

$$P_h u_n^{-} = u(x_n^{-}) = u_n^{-},\ P_h u_n^{+} = u(x_n^{+}) = u_n^{+}.$$

Let $\theta_u = U - P_h u, \rho_u = P_h u - u$, then one has

$$U - u = U - P_h u + P_h u - u = \theta_u + \rho_u.$$

Let $\theta_v = V - P_h v, \rho_v = P_h v - v$, then one gets

$$V - v = V - P_h v + P_h v - v = \theta_v + \rho_v.$$

Theorem 6.2. *Let U and u are the solutions of equations (6.6) and (6.4) respectively. Then we have the following estimates.*

$$\max_{(0,t_n]} \|u - U\| \le ck^r(\|v\|_{(0,t_n)} + \|D_{0,t}^{\alpha-1} v\|_{(0,t_n)} + \|u\|_{(0,t_n)} + \|u'\|_{(0,t_n)}),$$
$$\|u - U\|_{(0,t_n]} \le ck^{r+\frac{1}{2}}(\|v\|_{(0,t_n)} + \|D_{0,t}^{\alpha-1} v\|_{(0,t_n)} + \|u\|_{(0,t_n)} + \|u'\|_{(0,t_n)}).$$

Proof. Combining (6.6) and (6.4), and using the definitions of $\rho_u, \theta_u, \rho_v, \theta_v$ into it, we get the error equation below.

$$\begin{cases} \int_{I_n} \theta_u' \varphi dt + \theta_{u,n-1}^+ \varphi_{n-1}^+ = \theta_{u,n-1}^- \varphi_{n-1}^+ + \int_{I_n} \theta_v \varphi dt - \int_{I_n} \rho_u' \varphi dt \\ \qquad -\rho_{u,n-1} \varphi_{n-1}^+ + \rho_{u,n-1}^+ \varphi_{u,n-1}^+ + \int_{I_n} \rho_v \varphi dt, \\ \\ \int_{I_n} a(t) \theta_v \psi dt + \int_{I_n} b(t) D_{0,t}^{\alpha-1} \theta_v \psi dt = \int_{I_n} (f(U) - f(u)) \psi dt \\ \qquad - \int_{I_n} a(t) \rho_v \psi dt + \int_{I_n} b(t) D_{0,t}^{\alpha-1} \rho_v \psi dt. \end{cases} \tag{6.25}$$

In the first equation of (6.25), set $\varphi = \theta_u$, then we have

$$\begin{aligned} &\int_{I_n} \theta_u' \theta_u dt + \theta_{u,n-1}^+ \theta_{u,n-1}^+ \\ &= \frac{1}{2}(\|\theta_{u,n}^-\|^2 - \|\theta_{u,n-1}^+\|^2) + \|\theta_{u,n-1}^+\| = \frac{1}{2}(\|\theta_{u,n}^-\|^2 + \|\theta_{u,n-1}^+\|^2), \end{aligned}$$

and

$$\begin{aligned} &\theta_{u,n-1}^- \theta_{u,n-1}^+ + \int_{I_n} (\theta_v + \rho_v - \rho_u') dt + [\rho_u]_{n-1} \theta_{u,n-1}^+ \\ &\le \|\theta_{u,n-1}^-\|^2 + \frac{1}{2}\|\theta_{u,n-1}^+\|^2 + \|\theta_u\|_{I_n} (\|\theta_v\|_{I_n} + \|\rho_v\|_{I_n} + \|\rho_u'\|_{I_n}) + [\rho_u]_{n-1} \theta_{u,n-1}^+. \end{aligned}$$

So

$$\|\theta_{u,n}^-\|^2 \le \|\theta_{u,n-1}^-\|^2 + c\|\theta_u\|_{I_n}^2 + c(\|\theta_v\|_{I_n}^2 + \|\rho_v\|_{I_n}^2 + \|\rho'\|_{I_n}^2). \tag{6.26}$$

Integration the first equation of (6.25) by parts, we have

$$-\int_{I_n} \theta_u \frac{d\varphi}{dt} dt + \theta_{u,n}^- \varphi_n^- = \theta_{u,n-1}^- \varphi_{n-1}^+ + \int_{I_n} (\theta_v + \rho_v - \rho_u') \varphi dt. \tag{6.27}$$

Set $\varphi = \theta_u(t_n - t)$, and putting it into (6.27), one has

$$\begin{aligned} &- \int_{I_n} (\theta_u \theta_u' (t_n - t) + \theta_u^2 (-1)) \\ &= \int_{I_n} (t_n - t) \frac{d\theta_u^2}{dt} dt + \int_{I_n} \theta_u^2 dt \\ &= \frac{\theta_u}{2} (t_n - t)|_{I_n} - \int_{I_n} \frac{\theta_u^2}{2} dt + \int_{I_n} \theta_u^2 dt \\ &= \frac{k_n}{2} \|\theta_{u,n-1}^+\|^2 + \frac{1}{2} \|\theta_u\|_{I_n}^2, \end{aligned} \tag{6.28}$$

and

$$\begin{aligned}
&\theta^-_{u,n-1}\theta^+_{u,n-1}(t_n - t_{n-1}) + \textstyle\int_{I_n}(\theta_v + \rho_v - \rho'_u)\theta_u(t_n - t)dt \\
&\quad + [\rho_u]_{n-1}\theta^+_{u,n-1}(t_n - t_{n-1}) \\
&= k_n\theta^-_{u,n-1}\theta^+_{u,n-1} + k_n\textstyle\int_{I_n}(\theta_v + \rho_v - \rho'_u)\theta_u dt + k_n[\rho_u]_{n-1}\theta^+_{u,n-1} \\
&\le \frac{k_n}{2}\|\theta^-_{u,n-1}\|^2 + \frac{k_n}{2}\|\theta^+_{u,n-1}\|^2 + ck_n\|\theta_u\|^2_{I_n} + ck_n(\|\theta_v\|^2_{I_n} + \|\rho_v\|^2_{I_n} + \|\rho'_u\|^2_{I_n}) \\
&\quad + |k_n[\rho_u]_{n-1}\theta^+_{u,n-1}|.
\end{aligned}$$

With the definition of $P^{\pm}_h$, one gets

$$(1 - ck_n)\|\theta_u\|^2 \le \frac{k_n}{2}\|\theta^-_{u,n-1}\|^2 + ck_n(\|\theta_v\|^2_{I_n} + \|\rho_v\|^2_{I_n} + \|\rho'_u\|^2_{I_n}).$$

When k_n is sufficient small, one can conclude that

$$\|\theta_u\|^2_{I_n} \le ck_n(\|\theta^-_{u,n-1}\|^2 + \|\theta_v\|^2_{I_n} + \|\rho_v\|^2_{I_n} + \|\rho'_u\|^2_{I_n}). \tag{6.29}$$

Putting (6.29) into (6.26) leads to

$$\|\theta^-_{u,n}\|^2 \le c(1 + k_n)(\|\theta^-_{u,n-1}\|^2 + \|\theta_v\|^2_{I_n} + \|\rho_v\|^2_{I_n} + \|\rho'_u\|^2_{I_n}). \tag{6.30}$$

Iteration from 1 to n gives

$$\begin{aligned}
\|\theta^-_{u,n}\|^2 \le c\Sigma^n_{m=0}(\Pi^n_{j=m+1}c(1 + k_j))(\|\theta_v\|^2_{I_j} + \|\rho_v\|^2_{I_j} \\
+\|\rho'_u\|^2_{I_j} + \Pi^n_{j=0}c(1 + k_j)\|\theta^-_0\|^2.
\end{aligned} \tag{6.31}$$

When $\|\theta_0\| = 0$, it gives

$$\|\theta^-_{u,n}\|^2 \le c\Sigma^n_{j=0}(\|\theta_v\|^2_{I_j} + \|\rho_v\|^2_{I_j} + \|\rho'_u\|^2_{I_j}). \tag{6.32}$$

Putting (6.32) into (6.29), one has that

$$\|\theta_u\|^2_{(0,t_n)} \le ck(\|\theta_v\|^2_{(0,t_n)} + \|\rho_v\|^2_{(0,t_n)} + \|\frac{d\rho}{dt}\|^2_{(0,t_n)}). \tag{6.33}$$

Setting $\psi = \theta_v$, putting it into the second equation of (6.25) and summing from 1 to n, one has

$$\begin{aligned}
&\sum_{j=1}^{n}\textstyle\int_{I_j} a(t)\theta^2_v dt + \sum^n_{j=1}\int_{I_j} b(t)D^{\alpha-1}_{0,t}\theta_v\theta_v dt \\
&\ge m\textstyle\int_0^{t_n}\theta^2_v dt + m\int_)^{t_n} D^{\alpha-1}_{0,t}\theta_v\theta_v dt \\
&= m\|\theta_v\|^2_{(0,t_n)} + m\textstyle\int_0^{t_n} D^{\frac{\alpha-1}{2}}_{0,t}\theta_v D^{\frac{\alpha-1}{2}}_{t,t_n}\theta_v dt \\
&= m\|\theta_v\|^2_{(0,t_n)} + mcos\frac{1-\alpha}{2}\textstyle\int_0^{t_n}(D^{\frac{\alpha-1}{2}}_{0,t}\theta_v)^2 dt \\
&\ge m\|\theta_v\|^2_{(0,t_n)},
\end{aligned} \tag{6.34}$$

and

$$\begin{aligned}
&\sum_{j=1}^{n}\{\textstyle\int_{I_j}(f(U) - f(u))\theta_v dt - \int_{I_j} a\rho_v\theta_v dt - \int_{I_j} bD^{\alpha-1}_{0,t_n}\rho_v\theta_v dt\} \\
&\le L\textstyle\int_0^{t_n}\|\theta_u + \rho_u\|\theta_v dt + M\int_0^{t_n}\rho_v\theta_v dt +_0^{t_n} bD^{\alpha-1}_{0,t_n}\rho_v\theta_v dt\} \\
&\le L\|\theta_v\|_{(0,t_n)}(\|\theta_u\|_{(0,t_n)} + \|\rho_u\|_{(0,t_n)}) + M\|\theta_v\|_{(0,t_n)}(\|\rho_v\|_{(0,t_n)} + \|D^{\alpha-1}_{0,t}\rho_v\|_{(0,t_n)}).
\end{aligned} \tag{6.35}$$

Combining (6.34) with (6.35) yields

$$\|\theta_v\|^2_{(0,t_n)} \leq \frac{L}{M_1}\|(\|\theta_u\|_{(0,t_n)}+\|\rho_u\|_{(0,t_n)})+\frac{M_2}{M_1}(\|\rho_v\|_{(0,t_n)}+\|D^{\alpha-1}_{0,t}\rho_v\|_{(0,t_n)}). \quad (6.36)$$

Taking the estimation of $\|\theta_u\|_{(0,t_n)}$ into (6.36) gives

$$\begin{aligned}(1-c\frac{Lk^{\frac{1}{2}}}{M_1})\|\theta_v\|_{(0,t_n)} \leq (\frac{Lk^{\frac{1}{2}}}{M_1}+\frac{M_2}{M_1})\|\rho_v\|_{(0,t_n)}+\frac{M_2}{M_1}\|D^{\alpha-1}_{0,t}\rho_v\|_{(0,t_n)}\\ +\frac{L}{M_1}\|\rho_u\|_{(0,t_n)}+\frac{Lk^{\frac{1}{2}}}{M_1}\|\rho'_u\|_{(0,t_n)}.\end{aligned}$$

When k is sufficiently small, $\frac{Lk^{\frac{1}{2}}}{M_1} < 1$, the above inequality can be rewritten as follows.

$$\|\theta_v\|_{(0,t_n)} \leq c(\|\rho_v\|_{(0,t_n)}+\|D^{\alpha-1}_{0,t}\rho_v\|_{(0,t_n)}+\|\rho_u\|_{(0,t_n)}+\|\rho'_u\|). \quad (6.37)$$

Combing the estimation of $\|\theta_v\|_{(0,t_n)}$ with the right hand side of (6.33), one gets

$$\begin{aligned}\|\theta_u\|^2_{(0,t_n)} &\leq ck(\|\rho_v\|^2_{(0,t_n)}+\|D^{\alpha-1}_{0,t_n}\rho_v\|^2_{(0,t_n)}+\|\rho_u\|^2_{(0,t_n)}+\|\rho'_u\|^2_{(0,t_n)})\\ &\leq ck^{2r+1}(\|v\|^2_{(0,t_n)}+\|D^{\alpha-1}_{0,t_n}v\|^2_{(0,t_n)}+\|u\|^2_{(0,t_n)}+\|u'\|^2_{(0,t_n)}).\end{aligned}$$

As to the max norm, we have the inverse estimate

$$\max_{I_n}\|y(t)\|^2 \leq ck_n^{-1}\int_{I_n}\|y(t)\|^2dt.$$

So the proof of the estimation of the max $\|U-u\|_{(0,t_n)}$ is thus finished. □

6.5 Numerical examples

In this section, we present numerical results. Let S_h denote a uniform partition of spacial domain $[0,a]$, and X_h the space of continuous piecewise linear functions on S_h, i.e., $r=1$. We associate with the shape function of space X_h with the standard basis of the constant functions on the uniform grid of size $h=\frac{1}{n}$ along time axis. In our scheme, the finite element trial and test spaces for Problem 6.1 are chosen to be the same.

Example 6.1. Our first example deals with the linear equation

$$\begin{cases} D^{0.5}u = t^{1.5}/\Gamma(2.5), & 0\leq t\leq 1,\\ u(0)=0.\end{cases}$$

The exact solution of the equation is $u(x,t)=t^2$.

Table 6.1 includes the numerical calculations over a regular partition of [0,1]. We can see that when the size of grid is smaller, and the finite element approximation is better. And we can also observe that the experimental rates of convergence agree well with the theoretical convergence rates for the numerical solution.

Table 6.1 The error estimation of numerical Example 6.1.

h	$\|u - u_h\|_\infty$	cvge. rate
1/10	0.021186	—
1/20	0.009546	1.1627
1/40	0.004714	1.0050
1/80	0.002353	1.0029
1/160	0.001175	1.0065
1/320	0.000587	1.0009

Example 6.2. We consider another space-fractional non-linear Fokker-Planck differential equation whose exact solution is $u(t) = t^2 - t$,

$$\begin{cases} D^\alpha u = \dfrac{2.0t^{2-\alpha}}{\Gamma(3.0-\alpha)} - \dfrac{t^{1-\alpha}}{\Gamma(2.0-\alpha)} + t^4 - 2t^3 + t^2 - u^2(t), \quad 0 \le t \le 2, \\ u(t, 0) = 0, \end{cases}$$

Table 6.2 shows that the error results at different size of space grid. We can still observe that the experimental convergence rates agree with the theoretical rates.

Table 6.2 Numerical error result for Example 6.2.

h	$\|u - u_h\|_\infty$	cvge. rate
1/10	0.021186	—
1/20	0.009546	1.1627
1/40	0.004714	1.0050
1/80	0.002353	1.0029
1/160	0.001175	1.0065
1/320	0.000587	1.0009

6.6 Conclusion

In this chapter, we present a new discontinuous Galerkin numerical method for a type of nonlinear fractional Cauchy problem. The usual approximate method is to change the fractional differential equation into an integral equation, then to solve the equivalent integral equation. Our numerical method is different. By introducing an auxiliary function, we change the fractional differential equation into a system of equations. With the help of modified discontinuous Galerkin method, a new numerical scheme is derived. The existence and uniqueness of the solution to the discrete system are proved. The error estimates are also derived. Numerical results are in line with the theoretical results.

Bibliography

Adolfsson, Enelund, K. M. and Larsson, S. (2004). Adaptive discretization of fractional order viscoelasticity using sparse time history, *Comput. Methods Appl. Mech. Engrg.* **193**, pp. 4567-4590.

Agrawal, O. P. (2008). A general finite element formulation for fractional variational problem, *J. Math. Anal. Appl.* **337**, 1, pp. 1–12.

Alberty, J. and Carstensen, C. (2002). Discontinuous Galerkin time discretization in elastoplasticity: motivation, numerical algorithms, and applications, *Computer Methods in Applied Mechanics and Engineering* **191**, 43, pp. 4949–4968.

Baeumer, B., Kovács, M. and Meerschaert, M. M.(2008). Numerical solutions for fractional reaction-diffusion equations, *Comput. Math. Appl.* **55**, pp. 2212–2226.

Brockmann, D. and sokolov, I. M. (2002). Levy flights in external force fields: from models to equations, *Chemical Physics* **284**, 1, pp. 409–421.

Brunner, H. and Schötzau, D. (2006). *hp*-discontinuous Galerkin time-stepping for Volterra Integrodifferential equations, *SIAM J. Numer. Anal.* **44**, 1, pp. 224–245.

Delfour, M., Hager, W. and Trochu, F. (1981). Discontinuous Galerkin methods for ordinary differential equations, *Math. Comput.* **36**, 154, pp. 455–473.

Denisov, S. I., Horsthemke, W. and Hänggi, P. (2009). Generalized Fokker-Planck equation: derivation and exact solutions, *The European Physical Journal B* **68**, pp. 567–575.

Diethelm, K. Ford, N. J. and Freed, A. D. (2004). Detailed error analysis for a fractional Adams method, *Numer. Algor.* **36**, pp. 31–52.

Ervin, V. J. and Roop, J. P. (2006). Variational formulation for the stationary fractional advection dispersion equation, *Numer. Methods P. D. E.* **22**, pp. 558–576.

Fix, G. J. and Roop, J. P. (2004). Least Ssquares finite element solution of a fractional order two-Point boundary value problem, *Comput. Math. Appl.* **48**, pp. 1017–1033.

Hänggi, P. and Thomas, H. (1982). Stochastic-prcesses time evolution symmetries ans linear response, *Physics Report* **88**, 4, pp. 207–319.

Jumarie, G. (2004). Fractional Brownian motions via random walk in the complex plane and via fractional derivative, comparison and further results on their Fokker–Planck equations, *Chaos, Solitons and Fractals* **22**, pp. 907–925.

Kumar, P. and Agrawal, O. P. (2006). An approximate method for numerical solution of fractional differential equations, *Signal Processing* **86**, pp. 2602–2610.

Larsson, S., Thomée, V. and Wahlbin L. B. (1998). Numerical solution of parabolic integrao-differential equations by the discontinous Galerkin method, *Math. Comput.* **67**, 221, pp. 45–71.

Lasis, A. and Süli, E. (2007). Hp-version discontinuous Galerkin finite element method for semilinear parabolic problems, *SIAM J. Numer. Anal.* **45**, 4, pp. 1544–1569.

Lesaint, P. and Raviart, P. A. (1974). *Mathematical aspects of finite elements in partial differential equations* (Academic Press, NewYork).

Li, H. and Liu, R. X. (2001). The space-time finite element method for nonlinear parabolic problems, *Appl. math. Mech.* **22**, 6, pp. 613–623.

Lin, R. and Liu, F.(2007). Fractional high order methods for the nonlinear fractional ordinary differential equation, *Nonlinear Analysis* **66**, pp. 856–869.

Liu, F., Anh, V. and Turner, I. (2004). Numerical solution of the space fractional Fokker-Planck equation, *J. Comput. Appl. Math.* **166**, pp. 209–219.

Liu, F., Zhuang, P., Anh, V., Turner, I. and Burrage, K. (2007). Stability and convergence of the difference methods for the space-time fractional advection-diffusion equation, *Appl. Math. Comput.* **191**, pp. 12–20.

Metzler, R. and Nonnenmacher, T. F. (2002). Space- and time-fractional diffusion and wave equations, fractional Fokker–Planck equations, and physical motivation, *Chemical Physics* **284**, pp. 67–90.

Odibat, Z. and Momani, S. (2007). Numerical solution of Fokker-Planck equation with space- and time-fractional derivatives, *Physics Letters A* **369**, pp. 349-358.

Roop, J. P. (2006). Computational aspects of FEM approximation of fractional advection dispersion equations on bounded domains in R^2, *J. Comput. Appl. Math.* **193**, pp. 243–268.

Schötzau, D. and Schwab, C. (2000). An hp a priori error analysis of the DG time-stepping method for initial value problems, *Calcolo* **37**, pp. 207–232.

Shaw, S. and Whiteman, J. R. (1996). Discontinuous Galerkin method with a posteriori $L_p(0, t_i)$ error estimate for second-dind Volterra problems, *Mumer. Math.* **74**, pp. 361–383.

Sun, T. and Ma, K. (2009). A space-time discontinuous Galerkin mmethod for linear convection-dominated Sobolev equations, *Appl. Math. Comput.* **210**, pp. 490–503.

Thomée, V. (2006). *Galerkin finite element methods for parabolic problems* (Springer Verlag).

Tsallis, C. and Lenzia, E. K. (2002). Anomalous diffusion: Nonlinear fractional Fokker-Planck equation, *Chemical Physics* **284**, pp. 341–347.

Wang, K., Wang, H., Sun, S. and Wheeler, M. F. (2009). An optimal-order L_2 error estimate for nonsymmetric discontinuous Galerkin methods for a parabolic equation in multiple space dimensions, *Computer Methods in Applied Mechanics and Engineering* **198**, pp. 2190–2197.

Zheng, Y. Y., Li, C. P. and Zhao, Z. G. (2010a). A note on the finite element method for the space-fractional advection diffusion equation, *Comput. Math. Appl.* **59**, pp. 1718–1726.

Zheng, Y. Y., Li, C. P. and Zhao, Z. G. (2010b). A fully discrete discontinuous Galerkin method for nonlinear fractional Fokker–Planck equation, *Mathematical Problems in Engineering* **2010**, Article ID 279038, 26 pages.

Zheng, Y. Y. and Zhao, Z. G. (2011). A fully discrete Galerkin method for a nonlinear space-fractional diffusion equation, *Mathematical Problems in Engineering* **2011**, Article ID 171620, 20 pages.

Chapter 7

Asymptotic analysis of a singularly perturbed parabolic problem in a general smooth domain[1]

Yu-Jiang Wu*, Na Zhang and Lun-Ji Song

School of Mathematics and Statistics, Lanzhou University, Lanzhou 730000, Gansu, PR China

**myjaw@lzu.edu.cn*

The asymptotic expansion of the solution of a singularly perturbed parabolic problem in a general, bounded, and smooth domain is considered when the diffusivity parameter ε is small by using techniques of curvilinear coordinates and Laplace transformation. The boundedness of asymptotic expansion is estimated. Convergence results with respect to ε are carefully proved. Actual errors of all orders are performed, especially the optimal error order is obtained.

Keywords: Singularly perturbed problem, asymptotic analysis, curvilinear coordinates.

7.1 Introduction

In the work of this chapter, we consider the following singularly perturbed parabolic problem

$$\begin{cases} \partial_t u^\epsilon(\mathbf{x},t) - \epsilon \Delta u^\epsilon(\mathbf{x},t) + u^\epsilon(\mathbf{x},t) = f(\mathbf{x},t), \quad (\mathbf{x},t) \in \Omega \times (0,T], \\ u^\epsilon(\mathbf{x},t) = 0, \quad \text{on } \partial\Omega, \\ u^\epsilon|_{t=0} = u_0(\mathbf{x}), \end{cases} \tag{7.1}$$

where Ω is a bounded domain in $\mathbb{R}^d\,(d \geq 2)$ with smooth boundary, f and u_0 are given smooth functions, and $0 < \epsilon \ll 1$ is a constant. Since we are not concerned with the exact regularities of $\partial\Omega$ and functions f, u_0, we assume that they are sufficiently smooth for the construction of the asymptotic expansion of the solution

[1]This project is supported by National Basic Research Program of China, 973 Program (grant no. 2011CB706903) and Tianyuan Youth Foundation of Mathematics (NSFC no. 11026065).

and the actual error analysis. Besides, we impose the consistency condition on data, i.e.

$$u_0 = 0, \quad \text{on } \partial\Omega.$$

In [Gie (2009); Gie, Hamouda and Temam (2010a,b)], the authors pointed out that, due to the curvature of the boundary of the domain under consideration, the usual expansion in powers of ϵ can not give a suitable approximation (see, for example, [Hamouda and Temam (2008); Jung and Temam (2006); Roos, Stynes and Tobiska (2008)]). To handle the difficulty, Gie *et al.* first proposed to add terms of order $\epsilon^{j+\frac{1}{2}}$ in the asymptotic expansion. They initially considered a singularly perturbed reaction-diffusion equation in a general smooth bounded domain and, using the techniques of differential geometry, proposed the new asymptotic expansion and recovered the optimal convergence rate of the remainder at all orders. By a standard technique in ordinary differential equations, correctors at all orders are obtained. And the error analysis there guarantees the optimality of the convergence rate. Secondly, the authors considered the boundary layer problem of the Stokes equations with small viscosity (also known as the linearized Navier-Stokes equations (LNSE)), in a smoothly bounded domain in $\mathbb{R}^3$ with a characteristic boundary. And the authors showed that the solution of LNSE behaves like the corresponding Euler solution except in a thin region near the boundary. Subsequently, the authors considered a heat equation in a smoothly bounded domain in $\mathbb{R}^d$, $d \geq 2$. And the authors, take advantage of the curvilinear coordinates and the Green function for the differential operator, obtained the asymptotic expansion with respect to ϵ, $\epsilon^{\frac{1}{2}}$ at all orders and also proved the convergence rate which seems optimal.

In our work, we consider a more general problem with a reaction term compared with that in [Gie, Hamouda and Temam (2010b)], which gives rise to difficulty in finding the explicit expression of the Green's function of the differential operator in the boundary layer equations. To surmount the difficulty, we resort to the Laplace transformation to solve the boundary layer equations, instead. Besides, it turns out that the exploitation of the Laplace transformation makes both the resolvent of the boundary corrector and the error analysis much simpler and intuitive. Moreover, we have finally proved the optimal convergence rate. Numerical computation for the singularly perturbed problem is another important topic, which will appear elsewhere. Newly similar studies can refer to [Sun *et al.* (2011)] and references therein.

The rest of this chapter is organized as follows. In Section 7.2, we introduce the curvilinear coordinates and give some basic notions that would be used in the following. In Section 7.3, the asymptotic expansion of the solution of problem (7.1) at all orders is obtained by means of the Laplace transformation. And we estimate the bound of the asymptotic expansion. In Section 7.4, we perform the error analysis at all orders, and by means of a simple technique, prove the optimal error order. In Section 7.5, an example is shown to confirm our theoretical analysis.

7.2 The curvilinear coordinates

We assume that $(x_1, \cdots, x_d)$ is the Cartesian coordinates of a point $\mathbf{x} \in \mathbb{R}^d$ and Δ is the Laplace operator with respect to the space variable $\mathbf{x}$. Now, as in [Gie (2009)], we introduce the curvilinear coordinate system $\boldsymbol{\xi} = (\xi', \xi_d)$ satisfying

$$\partial\Omega = \{\boldsymbol{\xi} \text{ in } \mathbb{R}^d | \xi_d = 0 \text{ and } \Omega \text{ is located on the side } \xi_d > 0\}. \tag{7.2}$$

Moreover, since Ω is a smooth bounded domain in $\mathbb{R}^d_{\xi}$, we can write

$$\Omega = \{\boldsymbol{\xi} = (\xi', \xi_d) \in \mathbb{R}^d_{\boldsymbol{\xi}} | \xi' \in \omega',\ 0 \le \xi_d \le l(\xi')\},$$

where ω' is an open bounded domain in $\mathbb{R}^{d-1}_{\xi'}$. Without loss of generality, let us assume that the curvilinear coordinate system is orthogonal and satisfies

$$\frac{\partial \mathbf{x}}{\partial \xi_d} \cdot \frac{\partial \mathbf{x}}{\partial \xi_d} = \sum_{j=1}^{d} (\frac{\partial x_j}{\partial \xi_d})^2 = 1. \tag{7.3}$$

From the classical differential geometry, it gives that

$$\begin{cases} \mathbf{g}_\alpha = \frac{\partial \mathbf{x}}{\partial \xi_\alpha} = (\frac{\partial x_1}{\partial \xi_\alpha}, \cdots, \frac{\partial x_d}{\partial \xi_\alpha}), \quad \alpha = 1, \cdots, d, \\ (g_{\alpha\beta})_{1 \le \alpha, \beta \le d} = (\mathbf{g}_\alpha, \mathbf{g}_\beta) = \operatorname{diag}(g_{11}, \cdots, g_{d-1,d-1}, 1), \\ g = \det(g_{\alpha\beta})_{1 \le \alpha, \beta \le d}, \\ (g^{\alpha\beta})_{1 \le \alpha, \beta \le d} = (g_{\alpha\beta})^{-1} = \operatorname{diag}(\frac{1}{g_{11}}, \cdots, \frac{1}{g_{d-1,d-1}}, 1). \end{cases} \tag{7.4}$$

The Laplace operator under the curvilinear coordinate system has the form of

$$\begin{aligned} \Delta u &= \sum_{i,j=1}^{d} \frac{1}{\sqrt{g}} \frac{\partial}{\partial \xi_j} \left(\sqrt{g} g^{ij} \frac{\partial u}{\partial \xi_i} \right) \\ &= \sum_{j=1}^{d-1} \frac{1}{\sqrt{g}} \frac{\partial}{\partial \xi_j} \left(\sqrt{g} g^{jj} \frac{\partial u}{\partial \xi_j} \right) + \frac{1}{\sqrt{g}} \frac{\partial}{\partial \xi_d} \left(\sqrt{g} \frac{\partial u}{\partial \xi_d} \right) \\ &= \sum_{j=1}^{d-1} \frac{1}{h} \frac{\partial}{\partial \xi_j} \left(h g^{jj} \frac{\partial u}{\partial \xi_j} \right) + \frac{h'}{h} \frac{\partial u}{\partial \xi_d} + \frac{\partial^2 u}{\partial \xi_d^2}. \end{aligned} \tag{7.5}$$

In the last equality, we set $h = \sqrt{g}$ and $h' = \frac{\partial h}{\partial \xi_d}$ for simplicity. One can refer to [Ciarlet (2005)] to learn more about the curvilinear coordinates.

7.3 Asymptotic expansion

Usually one can refer to Refs. [Eckhaus (1979); Holmes (1995); Lions (1973); O'Malley (1991)] to learn more details about the asymptotic expansion methods for singularly perturbed equations, and we will follow the ideas in these references.

pansion

ınsion can approximate u^ϵ very well away from the boundary layer. e is to try to find

$$u_g = \sum_{j=0}^{\infty} \epsilon^j u^j, \tag{7.6}$$

ere u^j, $j \geq 0$, are yet to be determined.

Substituting u_g into problem (7.1) and collecting the coefficients of ϵ^j, $j = 0, 1, \ldots$, we obtain

$$\begin{cases} \partial_t u^0(\mathbf{x},t) + u^0(\mathbf{x},t) = f(\mathbf{x},t), \quad u^0(\mathbf{x},0) = u_0(\mathbf{x}), \\ \partial_t u^j(\mathbf{x},t) + u^j(\mathbf{x},t) = \Delta u^{j-1}(\mathbf{x},t), \quad u^j(\mathbf{x},0) = 0, \quad j \geq 1. \end{cases} \tag{7.7}$$

Equation $(7.7)_1$ is usually regarded as the reduced problem corresponding to (7.1).

Solving the above system (7.7), we obtain

$$\begin{cases} u^0(\mathbf{x},t) = \int_0^t e^{-(t-s)} f(\mathbf{x},s)ds + u_0(\mathbf{x})e^{-t}, \\ u^j(\mathbf{x},t) = \frac{t^j}{j!} e^{-t} \Delta^j u_0(\mathbf{x}) + \int_0^t \frac{(t-s)^j}{j!} e^{-(t-s)} \Delta^j f(\mathbf{x},s)ds. \end{cases} \tag{7.8}$$

7.3.2 *Boundary corrector*

Note that, u_g cannot be an approximation to u^ϵ validly for all $x \in \Omega$. We are going to look for a boundary corrector. Let us introduce the inner expansion

$$v_{\text{loc}} = \sum_{j=0}^{\infty} (\epsilon^j \theta^j + \epsilon^{j+\frac{1}{2}} \theta^{j+\frac{1}{2}}), \tag{7.9}$$

where θ^j and $\theta^{j+\frac{1}{2}}$, $j \geq 0$, are yet to be determined.

To make v_{loc} a good approximation of $w = u - u_g$, by $(7.1)_1$ and (7.7), the function v_{loc} should satisfy

$$\begin{cases} \partial_t v_{\text{loc}}(\mathbf{x},t) - \epsilon \Delta v_{\text{loc}}(\mathbf{x},t) + v_{\text{loc}}(\mathbf{x},t) = 0, \quad (\mathbf{x},t) \in \Omega \times (0,T), \\ v_{\text{loc}}(\mathbf{x},t) = -u_g(\mathbf{x},t), \quad \text{on } \partial\Omega, \\ v_{\text{loc}}|_{t=0} = 0. \end{cases} \tag{7.10}$$

Inserting (7.9) into $(7.10)_1$ makes it easy to obtain

$$\sum_{j=0}^{\infty} (\epsilon^j \partial_t \theta^j + \epsilon^{j+\frac{1}{2}} \partial_t \theta^{j+\frac{1}{2}}) - \epsilon \sum_{j=0}^{\infty} (\epsilon^j \Delta\theta^j + \epsilon^{j+\frac{1}{2}} \Delta\theta^{j+\frac{1}{2}}) + \sum_{j=0}^{\infty} (\epsilon^j \theta^j + \epsilon^{j+\frac{1}{2}} \theta^{j+\frac{1}{2}}) = 0. \tag{7.11}$$

One can refer to [Holmes (1995); Roos, Stynes and Tobiska (2008)] to learn more details.

Let us again introduce a stretched variable $\bar{\xi}_d$ of ξ_d, i.e.

$$\bar{\xi}_d = \frac{\xi_d}{\sqrt{\epsilon}}. \tag{7.12}$$

By means of the new coordinate introduced in (7.12), the Laplace operator expressed in (7.5) can be rewritten in the following compact form

$$\begin{cases} \Delta = S + \epsilon^{-\frac{1}{2}} L + \epsilon^{-1} \frac{\partial^2}{\partial \overline{\xi}_d^2}, \\ S = \sum\limits_{j=1}^{d-1} \frac{1}{h} \frac{\partial}{\partial \xi_j} \left(h g^{jj} \frac{\partial}{\partial \xi_j} \right), \quad L = \frac{h'}{h} \frac{\partial}{\partial \overline{\xi}_d}. \end{cases} \tag{7.13}$$

Note that, the operators S, L in (7.13) are in fact dependent on ϵ. To address the dependence on ϵ, we use the Taylor expansion in terms of the new variable $\overline{\xi}_d$

$$\begin{cases} \frac{1}{h} = \sum\limits_{j=0}^{\infty} \epsilon^{\frac{j}{2}} \overline{\xi}_d^j (\frac{1}{h})_j, \\ h g^{jj} = \sum\limits_{j=0}^{\infty} \epsilon^{\frac{j}{2}} \overline{\xi}_d^j (h g^{jj})_j, \\ \frac{h'}{h} = \sum\limits_{j=0}^{\infty} \epsilon^{\frac{j}{2}} \overline{\xi}_d^j \left(\frac{h'}{h} \right)_j. \end{cases}$$

Here we use the notation ϕ_j to denote $\frac{1}{j!}\phi^{(j)}(0)$. Thus, one can obtain

$$\begin{cases} S = \sum\limits_{k=1}^{d-1} \sum\limits_{j=0}^{\infty} \epsilon^{\frac{j}{2}} \overline{\xi}_d^j \sum\limits_{j_1+j_2=j} (\frac{1}{h})_{j_1} \frac{\partial}{\partial \xi_k} \left\{ (h g^{kk})_{j_2} \frac{\partial}{\partial \xi_k} \right\} = \sum\limits_{j=0}^{\infty} \epsilon^{\frac{j}{2}} \overline{\xi}_d^j S_{\frac{j}{2}}, \\ L = \sum\limits_{j=0}^{\infty} \epsilon^{\frac{j}{2}} \overline{\xi}_d^j \left(\frac{h'}{h} \right)_j \frac{\partial}{\partial \overline{\xi}_d} = \sum\limits_{j=0}^{\infty} \epsilon^{\frac{j}{2}} \overline{\xi}_d^j L_{\frac{j}{2}}, \end{cases}$$

where

$$\begin{cases} S_{\frac{j}{2}} = \sum\limits_{k=1}^{d-1} \sum\limits_{j_1+j_2=j} (\frac{1}{h})_{j_1} \frac{\partial}{\partial \xi_k} \left\{ (h g^{kk})_{j_2} \frac{\partial}{\partial \xi_k} \right\}, \\ L_{\frac{j}{2}} = \left(\frac{h'}{h} \right)_j \frac{\partial}{\partial \overline{\xi}_d}. \end{cases}$$

Inserting the above expression of the operators S, L into $(7.13)_1$ leads us to obtain

$$\Delta = \sum_{j=0}^{\infty} \epsilon^{\frac{j}{2}} \overline{\xi}_d^j S_{\frac{j}{2}} + \epsilon^{-\frac{1}{2}} \sum_{j=0}^{\infty} \epsilon^{\frac{j}{2}} \overline{\xi}_d^j L_{\frac{j}{2}} + \epsilon^{-1} \frac{\partial^2}{\partial \overline{\xi}_d^2}. \tag{7.14}$$

Remark 7.1. Note that $S_{\frac{j}{2}}$, $T_{\frac{j}{2}}$, $L_{\frac{j}{2}}$ are no longer dependent on ϵ.

By inserting the expression (7.14) into (7.11), we derive

$$\begin{aligned} 0 = & \sum_{j=0}^{\infty} (\epsilon^j \partial_t \theta^j + \epsilon^{j+\frac{1}{2}} \partial_t \theta^{j+\frac{1}{2}}) \\ & - \epsilon \sum_{j=0}^{\infty} \epsilon^j \left[\sum_{i=0}^{\infty} \epsilon^{i/2} \overline{\xi}_2^i S_{\frac{i}{2}} + \epsilon^{-\frac{1}{2}} \sum_{i=0}^{\infty} \epsilon^{i/2} \overline{\xi}_2^i L_{\frac{i}{2}} + \epsilon^{-1} \frac{\partial^2}{\partial \overline{\xi}_d^2} \right] \theta^j \\ & - \epsilon \sum_{j=0}^{\infty} \epsilon^{j+\frac{1}{2}} \left[\sum_{i=0}^{\infty} \epsilon^{i/2} \overline{\xi}_2^i S_{\frac{i}{2}} + \epsilon^{-\frac{1}{2}} \sum_{i=0}^{\infty} \epsilon^{i/2} \overline{\xi}_2^i L_{\frac{i}{2}} + \epsilon^{-1} \frac{\partial^2}{\partial \overline{\xi}_d^2} \right] \theta^{j+\frac{1}{2}} \end{aligned}$$

$$+\sum_{j=0}^{\infty}(\epsilon^j\theta^j+\epsilon^{j+\frac{1}{2}}\theta^{j+\frac{1}{2}}). \tag{7.15}$$

Thus our task now is to find the local correctors θ^N, $\theta^{N+\frac{1}{2}}$ for $N \geq 0$.

Collecting the coefficients of ϵ^N in (7.15), we obtain the equation for θ^N

$$\begin{aligned}\partial_t\theta^N-\frac{\partial^2\theta^N}{\partial\bar{\xi}_d^2}+\theta^N=&\sum_{j=0}^{N-1}\Big(\bar{\xi}_d^{2N-2j-2}S_{N-j-1}\theta^j+\bar{\xi}_d^{2N-2j-1}L_{N-j-\frac{1}{2}}\theta^j\\&+\bar{\xi}_d^{2N-2j-2}L_{N-j-1}\theta^{j+\frac{1}{2}}\Big)+\sum_{j=0}^{N-2}\bar{\xi}_d^{2N-2j-3}S_{N-j-\frac{3}{2}}\theta^{j+\frac{1}{2}}.\end{aligned} \tag{7.16}$$

In accordance with (7.10), we supplement (7.16) with boundary and initial conditions

$$\begin{cases}\theta^N|_{\bar{\xi}_d=0}=-u^N|_{\bar{\xi}_d=0},\\ \theta^N|_{\bar{\xi}_d\to\infty}=0,\\ \theta^N|_{t=0}=0.\end{cases} \tag{7.17}$$

Similarly, we can obtain the equation for $\theta^{N+\frac{1}{2}}$ by collecting the coefficients of $\epsilon^{N+\frac{1}{2}}$

$$\begin{aligned}\partial_t\theta^{N+\frac{1}{2}}-\frac{\partial^2\theta^{N+\frac{1}{2}}}{\partial\bar{\xi}_d^2}+\theta^{N+\frac{1}{2}}=&\sum_{j=0}^{N-1}\Big[\bar{\xi}_d^{2N-2j-1}S_{N-j-\frac{1}{2}}\theta^j+\bar{\xi}_d^{2N-2j-2}S_{N-j-1}\theta^{j+\frac{1}{2}}\\&+\bar{\xi}_d^{2N-2j-1}L_{N-j-\frac{1}{2}}\theta^{j+\frac{1}{2}}\Big]+\sum_{j=0}^{N}\bar{\xi}_d^{2N-2j}L_{N-j}\theta^j,\end{aligned} \tag{7.18}$$

and, supplement (7.18) with boundary and initial conditions

$$\begin{cases}\theta^{N+\frac{1}{2}}|_{\bar{\xi}_d=0}=0,\\ \theta^{N+\frac{1}{2}}|_{\bar{\xi}_d\to\infty}=0,\\ \theta^{N+\frac{1}{2}}|_{t=0}=0.\end{cases} \tag{7.19}$$

As a matter of fact, Eq.s (7.16)-(7.19) give the explicit expressions for θ^N and $\theta^{N+\frac{1}{2}}$, $N=0,1,\ldots$. This means that we have obtained the inner expansion of v_{loc}.

7.3.3 *Estimates of the solutions of boundary layer equations*

Let $u^\epsilon \simeq u_g + v_{\text{loc}}$. Then Subsections 7.3.1 and 7.3.2 provide the possibility to obtain the asymptotic expansion of u^ϵ. In order to give the error analysis in the subsequent section, we need to consider the boundary correctors.

At first, we consider the lower order terms θ^0, $\theta^{\frac{1}{2}}$. From (7.15), the equation for θ^0 is

$$\partial_t \theta^0 - \frac{\partial^2 \theta^0}{\partial \bar{\xi}_d^2} + \theta^0 = 0, \tag{7.20}$$

associated with boundary and initial conditions

$$\begin{cases} \theta^0|_{\bar{\xi}_d=0} = -u^0|_{\bar{\xi}_d=0}, \\ \theta^0|_{\bar{\xi}_d\to\infty} = 0, \\ \theta^0|_{t=0} = 0. \end{cases} \tag{7.21}$$

Similarly, the equation for $\theta^{\frac{1}{2}}$ is

$$\partial_t \theta^{\frac{1}{2}} - \frac{\partial^2 \theta^{\frac{1}{2}}}{\partial \bar{\xi}_d^2} + \theta^{\frac{1}{2}} = L_0 \theta^0, \tag{7.22}$$

associated with boundary and initial conditions

$$\begin{cases} \theta^{\frac{1}{2}}|_{\bar{\xi}_d=0} = 0, \\ \theta^{\frac{1}{2}}|_{\bar{\xi}_d\to\infty} = 0, \\ \theta^{\frac{1}{2}}|_{t=0} = 0. \end{cases} \tag{7.23}$$

Then we try to use the Laplace transformation method to solve the above two problems. Hereafter, we use the notation $\hat{f}$ to denote the Laplace transformation of a function f with respect to the variable t. (One can refer to [Schiff (1999)] to learn more about the Laplace transformation.) Due to Laplace transformation, the above two problems are changed into

$$\begin{cases} \frac{\partial^2}{\partial \bar{\xi}_d^2} \hat{\theta}^0 - (p+1)\hat{\theta}^0 = 0, \\ \hat{\theta}^0|_{\bar{\xi}_d=0} = -\hat{u}^0|_{\bar{\xi}_d=0}, \\ \hat{\theta}^0|_{\bar{\xi}_d\to\infty} = 0, \end{cases} \tag{7.24}$$

and

$$\begin{cases} \frac{\partial^2}{\partial \bar{\xi}_d^2} \hat{\theta}^{\frac{1}{2}} - (p+1)\hat{\theta}^{\frac{1}{2}} = -L_0 \hat{\theta}^0, \\ \hat{\theta}^{\frac{1}{2}}|_{\bar{\xi}_d=0} = 0, \\ \hat{\theta}^{\frac{1}{2}}|_{\bar{\xi}_d\to\infty} = 0, \end{cases} \tag{7.25}$$

respectively. Note that

$$\hat{u}^0(\xi', 0, p) = \textstyle\int_0^{+\infty} e^{-pt} \int_0^t e^{-(t-s)} f(\xi', 0, s)\, ds dt = \frac{1}{p+1} \hat{f}(\xi', 0, s).$$

After solving problems (7.24) and (7.25), we obtain

$$\hat{\theta}^0(\xi, p) = -\frac{1}{p+1} \hat{f}(\xi', 0, p) e^{-\sqrt{p+1}\bar{\xi}_d}, \tag{7.26}$$

$$\hat{\theta}^{\frac{1}{2}} = \frac{1}{2}\frac{h'(0)}{h(0)}\frac{1}{p+1}\hat{f}(\xi',0,p)e^{-\sqrt{p+1}\bar{\xi}_d}\bar{\xi}_d. \tag{7.27}$$

Afterwards, taking Laplace transformation for both sides of the equation (7.16) and the boundary conditions in (7.17), we get

$$\begin{cases} -\frac{\partial^2\hat{\theta}^N}{\partial\bar{\xi}_d^2} + (p+1)\hat{\theta}^N = \sum_{j=0}^{N-1}\Big[\bar{\xi}_d^{2N-2j-2}S_{N-j-1}\hat{\theta}^j + \bar{\xi}_d^{2N-2j-1}L_{N-j-1/2}\hat{\theta}^j \\ \qquad +\bar{\xi}_d^{2N-2j-2}L_{N-j-1}\hat{\theta}^{j+1/2}\Big] + \sum_{j=0}^{N-2}\bar{\xi}_d^{2N-2j-3}S_{N-j-\frac{3}{2}}\hat{\theta}^{j+1/2}, \\ \hat{\theta}^N|_{\bar{\xi}_d=0} = -\hat{u}^N|_{\bar{\xi}_d=0}, \\ \hat{\theta}^N|_{\bar{\xi}_d\to\infty} = 0, \end{cases} \tag{7.28}$$

where

$$\begin{aligned} \hat{u}^N(\xi',0,p) &= \int_0^\infty e^{-(p+1)t}\int_0^t \frac{(t-s)^N}{N!}\cdot e^s\cdot\Delta^N f(\xi',0,s)dsdt \\ &= \frac{1}{p+1}\int_0^\infty e^{-(p+1)t}\int_0^t\frac{(t-s)^{N-1}}{(N-1)!}\cdot e^s\cdot\Delta^N f(\xi',0,s)dsdt \\ &\;\vdots \\ &= \frac{1}{(p+1)^N}\int_0^\infty e^{-(p+1)t}\int_0^t e^s\cdot\Delta^N f(\xi',0,s)dsdt \\ &= \frac{1}{(p+1)^{N+1}}\int_0^\infty e^{-pt}\cdot\Delta^N f(\xi',0,t)dt \\ &= \frac{1}{(p+1)^{N+1}}\Delta^N\hat{f}(\xi',0,p). \end{aligned}$$

Similarly, from (7.18), (7.19), we get

$$\begin{cases} -\frac{\partial^2\hat{\theta}^{N+1/2}}{\partial\bar{\xi}_d^2} + (p+1)\hat{\theta}^{N+1/2} = \sum_{j=0}^{N-1}\Big[\bar{\xi}_d^{2N-2j-1}S_{N-j-1/2}\hat{\theta}^j \\ \qquad +\bar{\xi}_d^{2N-2j-2}S_{N-j-1}\hat{\theta}^{j+1/2} + \bar{\xi}_d^{2N-2j-1}L_{N-j-1/2}\hat{\theta}^{j+1/2}\Big] \\ \qquad + \sum_{j=0}^{N}\bar{\xi}_d^{2N-2j}L_{N-j}\hat{\theta}^j, \\ \hat{\theta}^{N+\frac{1}{2}}|_{\bar{\xi}_d=0} = 0, \\ \hat{\theta}^{N+\frac{1}{2}}|_{\bar{\xi}_d\to\infty} = 0. \end{cases} \tag{7.29}$$

Let us introduce some basic lemmas.

Lemma 7.1. *A particular solution of the following differential equation*

$$-f_{xx}(x) + \alpha f(x) = \beta_j x^j e^{-\sqrt{\alpha}x},$$

is given by

$$f(x) = \sum_{k=1}^{j+1} C_k x^k e^{-\sqrt{\alpha}x},$$

where

$$C_k = \frac{1}{2\sqrt{\alpha}}\beta_j\left(\frac{1}{2\sqrt{\alpha}}\right)^{j-k+1}\frac{j!}{k!}.$$

The proof of the Lemma 7.1 is omitted, and one can refer to, for example, [Ding and Li (2004)] for details.

Lemma 7.2. *Assume that the boundary of Ω and the data u_0 and f are sufficiently smooth. Then, the local correctors have the following forms*

$$\hat{\theta}^N = \frac{1}{p+1}\sum_{k=0}^{2N}\hat{C}_k^N(\xi',p)\left(\frac{1}{p+1}\right)^{N-\frac{k}{2}}\bar{\xi}_d^k e^{-\sqrt{p+1}\bar{\xi}_d}, \tag{7.30}$$

$$\hat{\theta}^{N+\frac{1}{2}} = \frac{1}{p+1}\sum_{k=1}^{2N+1}\hat{D}_k^N(\xi',p)\left(\frac{1}{p+1}\right)^{N+\frac{1}{2}-\frac{k}{2}}\bar{\xi}_d^k e^{-\sqrt{p+1}\bar{\xi}_d}, \tag{7.31}$$

where $\hat{C}_k^N(\xi',p)$ and $\hat{D}_k^N(\xi',p)$ are coefficients which only depend on derivatives about ξ' of the coefficients of Taylor expansion of h', $\frac{1}{h}$ and functions $\Delta^i\hat{f}(\xi',0,p)$, $i \le N$.

Proof. We proceed by induction on N. By (7.26) and (7.27), it is easy to get the conclusion for $N=0$.

Now, suppose the conclusion is valid for $N = 0, 1, 2 \ldots, M-1$, we find that, for $N = M$,

$$\begin{aligned}
&\text{(R.H.S. of (7.28))}\\
&= \sum_{j=0}^{M-1}\Bigg\{\bar{\xi}_d^{2M-2j-2}S_{N-j-1}\left[\frac{1}{p+1}\sum_{k=0}^{2j}\hat{C}_k^j(\xi',p)\left(\frac{1}{p+1}\right)^{j-\frac{k}{2}}\bar{\xi}_d^k e^{-\sqrt{p+1}\bar{\xi}_d}\right]\\
&\quad+\bar{\xi}_d^{2M-2j-1}L_{M-j-1/2}\left[\frac{1}{p+1}\sum_{k=0}^{2j}\hat{C}_k^j(\xi',p)\left(\frac{1}{p+1}\right)^{j-\frac{k}{2}}\bar{\xi}_d^k e^{-\sqrt{p+1}\bar{\xi}_d}\right]\\
&\quad+\bar{\xi}_d^{2M-2j-2}L_{N-j-1}\left[\frac{1}{p+1}\sum_{k=1}^{2j+1}\hat{D}_k^j(\xi',p)\left(\frac{1}{p+1}\right)^{j+\frac{1}{2}-\frac{k}{2}}\bar{\xi}_d^k e^{-\sqrt{p+1}\bar{\xi}_d}\right]\Bigg\}\\
&\quad+\sum_{j=0}^{M-2}\bar{\xi}_d^{2M-2j-3}S_{M-j-\frac{3}{2}}\left[\frac{1}{p+1}\sum_{k=1}^{2j+1}\hat{D}_k^j(\xi',p)\left(\frac{1}{p+1}\right)^{j+\frac{1}{2}-\frac{k}{2}}\bar{\xi}_d^k e^{-\sqrt{p+1}\bar{\xi}_d}\right],\\
&= \frac{1}{p+1}e^{-\sqrt{p+1}\bar{\xi}_d}\Bigg\{\sum_{j=0}^{M-1}\Bigg[\sum_{k=0}^{2j}\hat{C}_{k,1}^j(\xi',p)\left(\frac{1}{p+1}\right)^{j-\frac{k}{2}}\bar{\xi}_d^{2M-2j+k-2}\\
&\quad+\sum_{k=0}^{2j}\hat{C}_{k,2}^j(\xi',p)\left(\frac{1}{p+1}\right)^{j-\frac{k}{2}}\bar{\xi}_d^{2N-2j+k-2}\\
&\quad-\sum_{k=0}^{2j}\hat{C}_{k,2}^j(\xi',p)\left(\frac{1}{p+1}\right)^{j-\frac{1}{2}-\frac{k}{2}}\bar{\xi}_d^{2M-2j+k-1}
\end{aligned}$$

$$
\begin{aligned}
&+\sum_{k=1}^{2j+1}\hat{D}^j_{k,1}(\xi',p)\left(\frac{1}{p+1}\right)^{j+\frac{1}{2}-\frac{k}{2}}\bar{\xi}_d^{2M-2j+k-3}\\
&-\sum_{k=1}^{2j+1}\hat{D}^j_{k,1}(\xi',p)\left(\frac{1}{p+1}\right)^{j-\frac{k}{2}}\bar{\xi}_d^{2N-2j+k-2}\Bigg]\\
&+\sum_{j=0}^{M-2}\sum_{k=1}^{2j+1}\hat{D}^j_{k,2}(\xi',p)\left(\frac{1}{p+1}\right)^{j+\frac{1}{2}-\frac{k}{2}}\bar{\xi}_d^{2M-2j+k-3}\Bigg\},\\
&=\frac{1}{p+1}e^{-\sqrt{p+1}\bar{\xi}_d}\Bigg\{\sum_{j=0}^{M-1}\Bigg[\sum_{k=0}^{2j}\left(\hat{C}^j_{k,1}(\xi',p)+\hat{C}^j_{k,2}(\xi',p)\right)\left(\frac{1}{p+1}\right)^{\frac{k}{2}}\bar{\xi}_d^{2M-k-2}\\
&-\sum_{k=0}^{2j}\hat{C}^j_{k,2}(\xi',p)\left(\frac{1}{p+1}\right)^{\frac{k}{2}-\frac{1}{2}}\bar{\xi}_d^{2M-k-1}\\
&+\sum_{k=0}^{2j}\hat{D}^j_{k,1}(\xi',p)\left(\frac{1}{p+1}\right)^{\frac{k}{2}}\bar{\xi}_d^{2M-k-2}\\
&-\sum_{k=0}^{2j}\hat{D}^j_{k,1}(\xi',p)\left(\frac{1}{p+1}\right)^{\frac{k}{2}-\frac{1}{2}}\bar{\xi}_d^{2M-k-1}\Bigg]\\
&+\sum_{j=0}^{M-2}\sum_{k=0}^{2j}\hat{D}^j_{k,2}(\xi',p)\left(\frac{1}{p+1}\right)^{\frac{k}{2}}\bar{\xi}_d^{2M-k-2}\Bigg\}\\
&=\frac{1}{p+1}e^{-\sqrt{p+1}\bar{\xi}_d}\sum_{j=0}^{2M-1}\hat{Q}^M_j(\xi',p)\left(\frac{1}{p+1}\right)^{M-1-\frac{j}{2}}\bar{\xi}_d^j,
\end{aligned}
$$

where $\hat{Q}^M_j(\xi',p),\forall j$ are only dependent on derivatives about ξ' of the coefficients of Taylor expansion of h', $\frac{1}{h}$ and functions $\Delta^i\hat{f}(\xi',0,p)$, $i\le M-1$. By Lemma 7.2, problem (7.28) has a particular solution of the form

$$
\begin{aligned}
\hat{\theta}^N &= \frac{1}{p+1}e^{-\sqrt{p+1}\bar{\xi}_d}\\
&\quad\times\sum_{j=0}^{2M-1}\hat{Q}^M_j(\xi',p)\left(\frac{1}{p+1}\right)^{M-1-\frac{j}{2}}\sum_{k=1}^{j+1}\frac{1}{2\sqrt{p+1}}\left(\frac{1}{2\sqrt{p+1}}\right)^{j-k+1}\frac{j!}{k!}\bar{\xi}_d^k\\
&=\frac{1}{p+1}e^{-\sqrt{p+1}\bar{\xi}_d}\sum_{k=1}^{2M}\hat{C}^M_k(\xi',p)\left(\frac{1}{p+1}\right)^{M-\frac{k}{2}}\bar{\xi}_d^k. \qquad (7.32)
\end{aligned}
$$

Combine with the boundary conditions in (7.28), we obtain the conclusion for (7.30) when $N=M$. And by the same argument, we can also get the conclusion for (7.31) when $N=M$. Therefore, the proof is completed. □

Taking the Laplace inverse transformation, one can get

$$
\theta^N=\sum_{k=0}^{2N-1}\left(C^N_k(\xi',t)*\frac{t^{N-\frac{k}{2}-1}e^{-t}}{\Gamma(N-\frac{k}{2})}*\operatorname{erfc}\left(\frac{\bar{\xi}_d}{2\sqrt{t}}\right)e^{-t}\right)\bar{\xi}_d^k
$$

$$
\begin{aligned}
&+\left(C_{2N}^N(\xi',t)*\mathrm{erfc}(\frac{\overline{\xi}_d}{2\sqrt{t}})e^{-t}\right)\overline{\xi}_d^{2N},\\
\theta^{N+\frac{1}{2}} &= \sum_{k=1}^{2N}\left(C_k^N(\xi',t)*\frac{t^{N-\frac{k}{2}-\frac{1}{2}}e^{-t}}{\Gamma(N-\frac{k}{2}+\frac{1}{2})}*\mathrm{erfc}(\frac{\overline{\xi}_d}{2\sqrt{t}})e^{-t}\right)\overline{\xi}_d^{k}\\
&+\left(C_{2N+1}^N(\xi',t)*\mathrm{erfc}(\frac{\overline{\xi}_d}{2\sqrt{t}})e^{-t}\right)\overline{\xi}_d^{2N+1},
\end{aligned}
\tag{7.33}
$$

where $*$ denotes the convolution of two functions, $C_k^N(\xi',t)$ and $D_k^N(\xi',t)$ are coefficients which only depend on derivatives about ξ' of the Taylor expansions of h', $\frac{1}{h}$ and functions $\Delta^i f(\xi',0,t)$, $i \le N$. Moreover, $\mathrm{erfc}(z) = \frac{2}{\sqrt{\pi}}\int_z^\infty e^{-s^2}ds$.

Note that the following inequality holds true (see also [Gie, Hamouda and Temam (2010b)])

$$
\begin{aligned}
(\mathrm{erfc}(z))^2 &= \frac{4}{\pi}\int_z^\infty\int_z^\infty e^{-(x^2+y^2)}dxdy\\
&\le \int_z^\infty e^{-r^2}rdr\\
&= \frac{1}{2}e^{-z^2}.
\end{aligned}
$$

Thus,

$$
\left|\mathrm{erfc}(\frac{\overline{\xi}_d}{2\sqrt{t}})\right| \le \exp(-\frac{\overline{\xi}_d^2}{8t}). \tag{7.34}
$$

Lemma 7.3. *Assume that the boundary of Ω and the data u_0 and f are sufficiently smooth. Then for all $t \in [0,T]$*

$$
|\theta^N| \le P_{2N}(\overline{\xi}_d)\exp(-\frac{\overline{\xi}_d^2}{8t}), \tag{7.35}
$$

and

$$
|\theta^{N+\frac{1}{2}}| \le P_{2N+1}(\overline{\xi}_d)\exp(-\frac{\overline{\xi}_d^2}{8t}), \tag{7.36}
$$

where the coefficients in $P_{2N}(\overline{\xi}_d)$ and $P_{2N+1}(\overline{\xi}_d)$ are the corresponding upper bounds of the convolutions in (7.33).

Proof. By $(7.33)_1$, we have

$$
\begin{aligned}
\theta^N = &\sum_{k=0}^{2N-1}\int_0^t\int_0^{t-s} C_k^N(\xi',t-s-\tau)\frac{\tau^{N-1-\frac{k}{2}}e^{-\tau}}{\Gamma(N-\frac{k}{2})}d\tau\,\mathrm{erfc}\left(\frac{\overline{\xi}_d}{2\sqrt{s}}\right)e^{-s}ds\,\overline{\xi}_d^k\\
&+\int_0^t C_{2N}^N(\xi',t-s)\mathrm{erfc}\left(\frac{\overline{\xi}_d}{2\sqrt{s}}\right)e^{-s}ds\,\overline{\xi}_d^k.
\end{aligned}
$$

By the definition of C_k^N, $k=0,\ldots,2N$, it is easy to check that

$$
\left|\int_0^{t-s} C_k^N(\xi',t-s-\tau)\frac{\tau^{N-1-\frac{k}{2}}e^{-\tau}}{\Gamma(N-\frac{k}{2})}d\tau\right| \le a_k, \quad k=0,\ldots,2N-1,
$$

and

$$\left|C_{2N}^N(\xi', t-s)\right| \le a_{2N}.$$

Therefore, we obtain

$$\begin{aligned}|\theta^N| &\le \sum_{k=0}^{2N-1} a_k \int_0^t \operatorname{erfc}\left(\frac{\bar{\xi}_d}{2\sqrt{s}}\right) e^{-s} ds\, \bar{\xi}_d^k + a_{2N} \int_0^t \operatorname{erfc}\left(\frac{\bar{\xi}_d}{2\sqrt{s}}\right) e^{-s} ds\, \bar{\xi}_d^{2N} \\ &\le \sum_{k=0}^{2N} a_k \bar{\xi}_d^k \exp\left(-\frac{\bar{\xi}_d^2}{8t}\right) \quad \text{(by (7.34))}.\end{aligned}$$

Following the same argument, we obtain the conclusion for $\theta^{N+\frac{1}{2}}$. □

Lemma 7.4. *For any polynomial in* $\bar{\xi}_d$, $P_N(\bar{\xi}_d) = \sum_{k=0}^{N} a_k(\xi', t)\bar{\xi}_d^k$, $N \ge 0$, *with coefficients* $a_k(\xi', t)$ *bounded for all* $\xi' \in \omega'$ *and* $t \in (0, T]$, *we have for all* $t \in [0, T]$

$$\left|P_N e^{-\frac{\bar{\xi}_d^2}{8t}}\right|_{L^2(\Omega)} = O(\epsilon^{\frac{1}{4}}). \tag{7.37}$$

Proof. Note that, there exist a constant κ, such that

$$P_N(\bar{\xi}_d) \le \kappa_1 e^{\frac{\bar{\xi}_d^2}{16t}},$$

then

$$P_N(\bar{\xi}_d) e^{-\frac{\bar{\xi}_d^2}{8t}} \le \kappa_1 e^{-\frac{\bar{\xi}_d^2}{16t}}.$$

Therefore, by the boundary condition given in (7.1), we have

$$\begin{aligned}\left|P_N e^{-\frac{\bar{\xi}_d^2}{8t}}\right|_{L^2(\Omega)}^2 &= \int_\Omega \left|P_N e^{-\frac{\bar{\xi}_d^2}{8t}}\right|^2 dx = \int_\Omega \left|P_N e^{-\frac{\bar{\xi}_d^2}{8t}}\right|^2 \sqrt{g} d\xi \\ &\le \kappa_2 \int_{\omega'} \int_0^{l(\xi')} \kappa_1^2 e^{-\left(\frac{\bar{\xi}_d}{\sqrt{8t}}\right)^2} d\xi_d d\xi' \\ &= \kappa_2 \sqrt{8\epsilon t} \int_{\omega'} \int_0^\infty \kappa_1^2 e^{-z^2} dz d\xi' \quad (\text{let } z = \frac{\bar{\xi}_d}{\sqrt{8t}}) \\ &\le \kappa_3 \epsilon^{\frac{1}{2}},\end{aligned}$$

where κ_i, $i = 1, 2, 3$, are positive constants independent of ϵ. □

Now we are about to get the main conclusion of this section.

Theorem 7.1. *Assume that the boundary of* Ω *and the data* u_0 *and* f *are sufficiently smooth. Then we have the following estimates for the local boundary correctors, for all* $t \in [0, T]$

$$\left|\theta^{N+\frac{k}{2}}\right|_{L^2(\Omega)} \le c_1 \epsilon^{\frac{1}{4}}, \tag{7.38}$$

$$\left| \frac{\partial^m}{\partial \xi'^m} \theta^{N+\frac{k}{2}} \right|_{L^2(\Omega)} \leq c_2 \epsilon^{\frac{1}{4}}, \quad m = 1, 2, \tag{7.39}$$

$$\left| \frac{\partial}{\partial \bar{\xi}_d} \theta^{N+\frac{k}{2}} \right|_{L^2(\Omega)} \leq c_3 \epsilon^{\frac{1}{4}}, \tag{7.40}$$

where $k = 0,\ 1,\ c_i,\ i = 1,\ 2,\ 3,$ *are constants independent of* ϵ.

Proof. By Lemmas 7.3 and 7.4, it is clear that the inequality (7.38) holds.

For (7.39), one deduces by means of the expression of θ^N in $(7.33)_1$ that

$$\begin{aligned} \frac{\partial^m}{\partial \xi'^m} \theta^N &= \sum_{k=0}^{2N-1} \left(\frac{\partial^m}{\partial \xi'^m} C_k^N(\xi', t) * \frac{t^{N-\frac{k}{2}-1} e^{-t}}{\Gamma(N - \frac{k}{2})} * \operatorname{erfc}(\frac{\bar{\xi}_d}{2\sqrt{t}}) e^{-t} \right) \bar{\xi}_d^k \\ &+ \left(\frac{\partial^m}{\partial \xi'^m} C_{2N}^N(\xi', t) * \operatorname{erfc}(\frac{\bar{\xi}_d}{2\sqrt{t}}) e^{-t} \right) \bar{\xi}_d^{2N}. \end{aligned}$$

Denoting again $\frac{\partial^m}{\partial \xi'^m} C_k^N(\xi', t)$, $k = 0, \ldots, 2N$, by $C_k^N(\xi', t)$ would not cause any ambiguity. Since Ω and f are sufficiently smooth, by Lemmas 7.3 and 7.4, thus one has (7.39).

For (7.40), the direct calculation to $(7.33)_1$ with respect to $\bar{\xi}_d$ deduces that

$$\begin{aligned} \frac{\partial}{\partial \bar{\xi}_d} \theta^N &= \sum_{k=0}^{2N-2} \left(k \cdot C_k^N(\xi', t) * \frac{t^{N-\frac{k}{2}-1} e^{-t}}{\Gamma(N - \frac{k}{2})} * \operatorname{erfc}(\frac{\bar{\xi}_d}{2\sqrt{t}}) e^{-t} \right) \bar{\xi}_d^k \\ &+ \left(2N \cdot C_{2N}^N(\xi', t) * \operatorname{erfc}(\frac{\bar{\xi}_d}{2\sqrt{t}}) e^{-t} \right) \bar{\xi}_d^{2N-1} \\ &+ \sum_{k=0}^{2N-1} \left(C_k^N(\xi', t) * \frac{t^{N-\frac{k}{2}-1} e^{-t}}{\Gamma(N - \frac{k}{2})} * \frac{1}{\sqrt{\pi t}} \exp(-\frac{\bar{\xi}_d^2}{4t}) e^{-t} \right) \bar{\xi}_d^k \\ &+ \left(C_{2N}^N(\xi', t) * \frac{1}{\sqrt{\pi t}} \exp(-\frac{\bar{\xi}_d^2}{4t}) e^{-t} \right) \bar{\xi}_d^{2N}. \end{aligned}$$

By Lemmas 7.3 and 7.4, the inequality (7.40) is obtained. The proof is completed. □

7.4 Error estimate

To perform the error analysis, it needs to introduce the remainder at order ϵ^N and $\epsilon^{N+\frac{1}{2}}$

$$\begin{cases} r^N = u^\epsilon - \sum_{j=0}^{N} (\epsilon^j u^j + \epsilon^j \theta^j) - \sum_{j=0}^{N-1} \epsilon^{j+\frac{1}{2}} \theta^{j+\frac{1}{2}}, \\ r^{N+\frac{1}{2}} = u^\epsilon - \sum_{j=0}^{N} (\epsilon^j u^j + \epsilon^j \theta^j) - \sum_{j=0}^{N} \epsilon^{j+\frac{1}{2}} \theta^{j+\frac{1}{2}}. \end{cases} \tag{7.41}$$

Since

$$
\begin{aligned}
\partial_t r^N - \epsilon\Delta r^N + r^N &= \epsilon^{N+1}\Delta u^N - \sum_{j=0}^{N}\epsilon^j(\partial_t\theta^j - \epsilon\Delta\theta^j + \theta^j)\\
&\quad - \sum_{j=0}^{N-1}\epsilon^{j+\frac{1}{2}}(\partial_t\theta^{j+\frac{1}{2}} - \epsilon\Delta\theta^{j+\frac{1}{2}} + \theta^{j+\frac{1}{2}})\\
&= \epsilon^{N+1}\Delta u^N + \sum_{j=0}^{N}\epsilon^{j+1}S\theta^j + \sum_{j=0}^{N}\epsilon^{j+\frac{1}{2}}L\theta^j\\
&\quad - \sum_{j=0}^{N}\epsilon^j\left[\partial_t\theta^j - \frac{\partial^2}{\partial\bar{\xi}_d^2}\theta^j + \theta^j\right]\\
&\quad + \sum_{j=0}^{N-1}\epsilon^{j+\frac{3}{2}}S\theta^{j+\frac{1}{2}} + \sum_{j=0}^{N-1}\epsilon^{j+1}L\theta^{j+\frac{1}{2}}\\
&\quad - \sum_{j=0}^{N-1}\epsilon^{j+\frac{1}{2}}\left[\partial_t\theta^{j+\frac{1}{2}} - \frac{\partial^2}{\partial\bar{\xi}_d^2}\theta^{j+\frac{1}{2}} + \theta^{j+\frac{1}{2}}\right]\\
&= \epsilon^{N+1}\Delta u^N\\
&\quad - \sum_{j=0}^{N}\epsilon^j\left[\sum_{k=0}^{2j-2}\bar{\xi}_d^{2j-2-k}S_{j-1-\frac{k}{2}}\theta^{\frac{k}{2}} + \sum_{k=0}^{2j-1}\bar{\xi}_d^{2j-1-k}L_{j-\frac{1}{2}-\frac{k}{2}}\theta^{\frac{k}{2}}\right]\\
&\quad - \sum_{j=0}^{N-1}\epsilon^{j+\frac{1}{2}}\left[\sum_{k=0}^{2j-1}\bar{\xi}_d^{2j-1-k}S_{j-\frac{1}{2}-\frac{k}{2}}\theta^{\frac{k}{2}} + \sum_{k=0}^{2j}\bar{\xi}_d^{2j-k}L_{j-\frac{k}{2}}\theta^{\frac{k}{2}}\right]\\
&\quad + \sum_{j=0}^{2N}\epsilon^{\frac{j}{2}+1}S\theta^{\frac{j}{2}} + \sum_{j=0}^{2N}\epsilon^{\frac{j}{2}+\frac{1}{2}}L\theta^{\frac{j}{2}}\\
&= \epsilon^{N+1}\Delta u^N + \sum_{k=0}^{2N-2}\epsilon^{\frac{k}{2}+1}\left[S - \sum_{j=0}^{2N-k-2}\epsilon^{\frac{j}{2}}\bar{\xi}_d^{j}S_{\frac{j}{2}}\right]\theta^{\frac{k}{2}}\\
&\quad + \epsilon^{N+\frac{1}{2}}S\theta^{N-\frac{1}{2}} + \epsilon^{N+1}S\theta^N\\
&\quad + \sum_{k=0}^{2N-1}\epsilon^{\frac{k+1}{2}}\left[L - \sum_{j=0}^{2N-k-1}\epsilon^{\frac{j}{2}}\bar{\xi}_d^{j}L_{\frac{j}{2}}\right]\theta^{\frac{k}{2}} + \epsilon^{N+\frac{1}{2}}L\theta^N\\
&= \epsilon^{N+1}\Delta u^N + \sum_{k=0}^{2N-2}\epsilon^{\frac{k}{2}+1}\left[S - \sum_{j=0}^{2N-k-2}\epsilon^{\frac{j}{2}}\bar{\xi}_d^{j}S_{\frac{j}{2}}\right]\theta^{\frac{k}{2}}\\
&\quad + \epsilon^{N+\frac{1}{2}}S\theta^{N-\frac{1}{2}} + \epsilon^{N+1}S\theta^N\\
&\quad + \sum_{k=0}^{2N-1}\epsilon^{\frac{k+1}{2}}\left[L - \sum_{j=0}^{2N-k-1}\epsilon^{\frac{j}{2}}\bar{\xi}_d^{j}L_{\frac{j}{2}}\right]\theta^{\frac{k}{2}} + \epsilon^{N+\frac{1}{2}}L\theta^N.
\end{aligned}
$$

Therefore, r^N satisfies the equation

$$\begin{aligned}\partial_t r^N - \epsilon\Delta r^N + r^N = \epsilon^{N+1}\Delta u^N + \sum_{k=0}^{2N-2} \epsilon^{\frac{k}{2}+1}\left(S - \sum_{j=0}^{2N-k-2} \epsilon^{\frac{j}{2}}\bar{\xi}_d^j S_{\frac{j}{2}}\right)\theta^{\frac{k}{2}} \\ +\epsilon^{N+\frac{1}{2}}S\theta^{N-\frac{1}{2}} + \epsilon^{N+1}S\theta^N \\ + \sum_{k=0}^{2N-1} \epsilon^{\frac{k+1}{2}}\left(L - \sum_{j=0}^{2N-k-1} \epsilon^{\frac{j}{2}}\bar{\xi}_d^j L_{\frac{j}{2}}\right)\theta^{\frac{k}{2}} + \epsilon^{N+\frac{1}{2}}L\theta^N.\end{aligned}$$

Similarly, the equation satisfied by $r^{N+\frac{1}{2}}$ is the following

$$\begin{aligned}\partial_t r^{N+\frac{1}{2}} - \epsilon\Delta r^{N+\frac{1}{2}} + r^{N+\frac{1}{2}} = \epsilon^{N+1}\Delta u^N + \sum_{k=0}^{2N-1} \epsilon^{\frac{k}{2}+1}\left(S - \sum_{j=0}^{2N-k-1} \epsilon^{\frac{j}{2}}\bar{\xi}_d^j S_{\frac{j}{2}}\right)\theta^{\frac{k}{2}} \\ +\epsilon^{N+1}S\theta^N + \epsilon^{N+\frac{3}{2}}S\theta^{N+\frac{1}{2}} \\ +\sum_{k=0}^{2N} \epsilon^{\frac{k+1}{2}}\left(L - \sum_{j=0}^{2N-k} \epsilon^{\frac{j}{2}}\bar{\xi}_d^j L_{\frac{j}{2}}\right)\theta^{\frac{k}{2}} + \epsilon^{N+1}L\theta^{N+\frac{1}{2}}.\end{aligned}$$

Here, for simplicity, we use notations

$$\begin{aligned}R^N &= \sum_{k=0}^{2N-2} \epsilon^{\frac{k}{2}+1}\left(S - \sum_{j=0}^{2N-k-2} \epsilon^{\frac{j}{2}}\bar{\xi}_d^j S_{\frac{j}{2}}\right)\theta^{\frac{k}{2}} + \epsilon^{N+\frac{1}{2}}S\theta^{N-\frac{1}{2}} + \epsilon^{N+1}S\theta^N \\ &\quad + \sum_{k=0}^{2N-1} \epsilon^{\frac{k+1}{2}}\left(L - \sum_{j=0}^{2N-k-1} \epsilon^{\frac{j}{2}}\bar{\xi}_d^j L_{\frac{j}{2}}\right)\theta^{\frac{k}{2}} + \epsilon^{N+\frac{1}{2}}L\theta^N,\end{aligned} \tag{7.42}$$

$$\begin{aligned}R^{N+\frac{1}{2}} &= \sum_{k=0}^{2N-1} \epsilon^{\frac{k}{2}+1}\left(S - \sum_{j=0}^{2N-k-1} \epsilon^{\frac{j}{2}}\bar{\xi}_d^j S_{\frac{j}{2}}\right)\theta^{\frac{k}{2}} + \epsilon^{N+1}S\theta^N + \epsilon^{N+\frac{3}{2}}S\theta^{N+\frac{1}{2}} \\ &\quad +\sum_{k=0}^{2N} \epsilon^{\frac{k+1}{2}}\left(L - \sum_{j=0}^{2N-k} \epsilon^{\frac{j}{2}}\bar{\xi}_d^j L_{\frac{j}{2}}\right)\theta^{\frac{k}{2}} + \epsilon^{N+1}L\theta^{N+\frac{1}{2}}.\end{aligned} \tag{7.43}$$

Thus, the equations for r^N and $r^{N+\frac{1}{2}}$ can be written in the following compact form,

$$\begin{cases} \partial_t r^{N+\frac{m}{2}} - \epsilon\Delta r^{N+\frac{m}{2}} + r^{N+\frac{m}{2}} = \epsilon^{N+1}\Delta u^N + R^{N+\frac{m}{2}}, \\ r^{N+\frac{m}{2}}(\xi', \xi_d = 0, t) = 0, \\ r^{N+\frac{m}{2}}|_{t=0} = 0, \end{cases} \tag{7.44}$$

where $m = 0$, 1.

Multiplying both sides of $(7.44)_1$ with $r^{N+\frac{m}{2}}$, and integrating by part over Ω, we obtain

$$\frac{1}{2}\frac{d}{dt}\left|r^{N+\frac{m}{2}}\right|^2_{L^2(\Omega)} + \epsilon\left|r^{N+\frac{m}{2}}\right|^2_{H^1(\Omega)} + \left|r^{N+\frac{m}{2}}\right|^2_{L^2(\Omega)}$$

$$
\begin{aligned}
&= \epsilon^{N+1} \int_\Omega \Delta u^N r^{N+\frac{m}{2}} dx + \int_\Omega R^{N+\frac{m}{2}} r^{N+\frac{m}{2}} dx \\
&\leq \frac{\epsilon^{2N+2}}{2} |\Delta u^N|^2_{L^2(\Omega)} + \frac{1}{2} \left|R^{N+\frac{m}{2}}\right|^2_{L^2(\Omega)} + \left|r^{N+\frac{m}{2}}\right|^2_{L^2(\Omega)}.
\end{aligned}
$$

Thus

$$
\frac{1}{2}\frac{d}{dt} \left|r^{N+\frac{m}{2}}\right|^2_{L^2(\Omega)} + \epsilon \left|r^{N+\frac{m}{2}}\right|^2_{H^1(\Omega)} \leq \frac{\epsilon^{2N+2}}{2} |\Delta u^N|^2_{L^2(\Omega)} + \frac{1}{2} \left|R^{N+\frac{m}{2}}\right|^2_{L^2(\Omega)}.
$$

Therefore, the estimates for $r^{N+\frac{m}{2}}$ are

$$
\left|r^{N+\frac{m}{2}}\right|^2_{L^2(\Omega)} \leq \int_0^T (\epsilon^{2N+2} |\Delta u^N|^2_{L^2(\Omega)} + |R^{N+\frac{m}{2}}|^2_{L^2(\Omega)}) dt, \tag{7.45}
$$

$$
\int_0^T \left|r^{N+\frac{m}{2}}\right|^2_{H^1(\Omega)} dt \leq \frac{1}{2}\epsilon^{-1} \int_0^T (\epsilon^{2N+2} |\Delta u^N|^2_{L^2(\Omega)} + |R^{N+\frac{m}{2}}|^2_{L^2(\Omega)}) dt. \tag{7.46}
$$

According to estimates (7.45) and (7.46), it is necessary to give furthermore estimates for $|R^{N+\frac{m}{2}}|_{L^2(\Omega)}, m = 0, 1$.

Theorem 7.2. *Suppose that the boundary of Ω and the data u_0 and f are sufficiently smooth. Then we have the following estimates,*

$$
\left|R^N\right|_{L^\infty(0,T;L^2(\Omega))} \leq C\epsilon^{N+\frac{3}{4}}, \tag{7.47}
$$

$$
\left|R^{N+\frac{1}{2}}\right|_{L^\infty(0,T;L^2(\Omega))} \leq C\epsilon^{N+\frac{5}{4}}, \tag{7.48}
$$

where $N \geq 0$.

Proof. By (7.42), it follows that

$$
\begin{aligned}
\left|R^N\right|_{L^2(\Omega)} &\leq \epsilon^{N+\frac{1}{2}} \left|\bar{\xi}_d^{2N-k-1} S_{N-\frac{k}{2}-\frac{1}{2}} \theta^{\frac{k}{2}}\right|_{L^2(\Omega)} \\
&+ \epsilon^{N+\frac{1}{2}} \left|S_0 \theta^{N-\frac{1}{2}}\right|_{L^2(\Omega)} + \epsilon^{N+1} \left|S_0 \theta^N\right|_{L^2(\Omega)} \\
&+ \epsilon^{N+\frac{1}{2}} \left|\bar{\xi}_d^{2N-k} L_{N-\frac{k}{2}} \theta^{\frac{k}{2}}\right|_{L^2(\Omega)} + \epsilon^{N+\frac{1}{2}} \left|L_0 \theta^N\right|_{L^2(\Omega)}.
\end{aligned} \tag{7.49}
$$

Similarly, for $R^{N+\frac{1}{2}}$, we have

$$
\begin{aligned}
\left|R^{N+\frac{1}{2}}\right|_{L^2(\Omega)} &\leq \epsilon^{N+1} \left|\bar{\xi}_d^{2N-k} S_{N-\frac{k}{2}} \theta^{\frac{k}{2}}\right|_{L^2(\Omega)} + \epsilon^{N+1} \left|S_0 \theta^N\right|_{L^2(\Omega)} \\
&+ \epsilon^{N+\frac{3}{2}} \left|S \theta^{N+\frac{1}{2}}\right|_{L^2(\Omega)} + \epsilon^{N+1} \left|\bar{\xi}_d^{2N-k+1} L_{N-\frac{k}{2}+\frac{1}{2}} \theta^{\frac{k}{2}}\right|_{L^2(\Omega)} \\
&+ \epsilon^{N+1} \left|L_0 \theta^{N+\frac{1}{2}}\right|_{L^2(\Omega)}.
\end{aligned} \tag{7.50}
$$

Then by Theorem 7.1, we go to the conclusion. □

The following theorem gives the error estimates.

Theorem 7.3. *Assume that the boundary of Ω and the data u_0 and f are sufficiently smooth. Then, for the remainders defined in* (7.41)*, we have the following estimates*

$$\left|r^N\right|_{L^\infty(0,T;L^2(\Omega))} \leq C\epsilon^{N+\frac{3}{4}}, \quad \left|r^N\right|_{L^\infty(0,T;H^1(\Omega))} \leq C\epsilon^{N+\frac{1}{4}}, \tag{7.51}$$

$$\left|r^{N+\frac{1}{2}}\right|_{L^\infty(0,T;L^2(\Omega))} \leq C\epsilon^{N+1}, \quad \left|r^{N+\frac{1}{2}}\right|_{L^\infty(0,T;H^1(\Omega))} \leq C\epsilon^{N+\frac{1}{2}}, \tag{7.52}$$

where $N \geq 0$.

7.5 An example

This section gives the application of the asymptotic expansion described in Section 7.3. In [Gie (2009)], the steady state of problem (7.1) is discussed with a domain $\Omega = \{x | x_1^2 + x_2^2 \leq R\}$, where $R > 0$. While without loss of generality, we consider the problem (7.1) with the domain being a unit circle in $\mathbb{R}^2$, i.e.

$$\Omega = \{\mathbf{x} | x_1^2 + x_2^2 \leq 1\}. \tag{7.53}$$

Introduce the curvilinear coordinates $\boldsymbol{\xi} = (\xi_1, \xi_2)$,

$$\begin{cases} x_1 = (1 - \xi_2)\cos(\xi_1), \\ x_2 = (1 - \xi_2)\sin(\xi_1), \end{cases} \tag{7.54}$$

note that the conditions in (7.2) and (7.3) are satisfied.

We seek for the low order expansion of u^ϵ,

$$u^\epsilon \simeq u^0 + \theta^0 + \theta^{\frac{1}{2}}. \tag{7.55}$$

By $(7.7)_1$, we have the expression of u^0. We will devote ourselves to find θ^0 and $\theta^{\frac{1}{2}}$ in successive steps.

First of all, as in [Gie (2009)], it is easy to obtain

$$h = \sqrt{g} = 1 - \xi_2. \tag{7.56}$$

Then, by (7.25) and (7.26), we have

$$\begin{cases} \theta^0(\boldsymbol{\xi}, p) = -\frac{1}{p+1}\hat{f}(\xi', 0, p)e^{-\sqrt{p+1}\overline{\xi}_2}, \\ \theta^{\frac{1}{2}} = \frac{1}{2}\frac{1}{p+1}\hat{f}(\xi', 0, p)e^{-\sqrt{p+1}\overline{\xi}_2}\overline{\xi}_2. \end{cases} \tag{7.57}$$

Finally, by taking the Laplace inverse transformation, we achieve

$$\begin{cases} \theta^0(\boldsymbol{\xi}, p) = -\int_0^t e^{-s}\text{erfc}(\frac{\xi_2}{2\sqrt{\epsilon s}})f(\xi_1, 0, t-s)ds, \\ \theta^{\frac{1}{2}} = \frac{\xi_2}{2\sqrt{\epsilon}}\int_0^t e^{-s}\text{erfc}(\frac{\xi_2}{2\sqrt{\epsilon s}})f(\xi_1, 0, t-s)ds. \end{cases} \tag{7.58}$$

For $f = 1$, $u_0 = 0$, the two low order expansions are as follows.

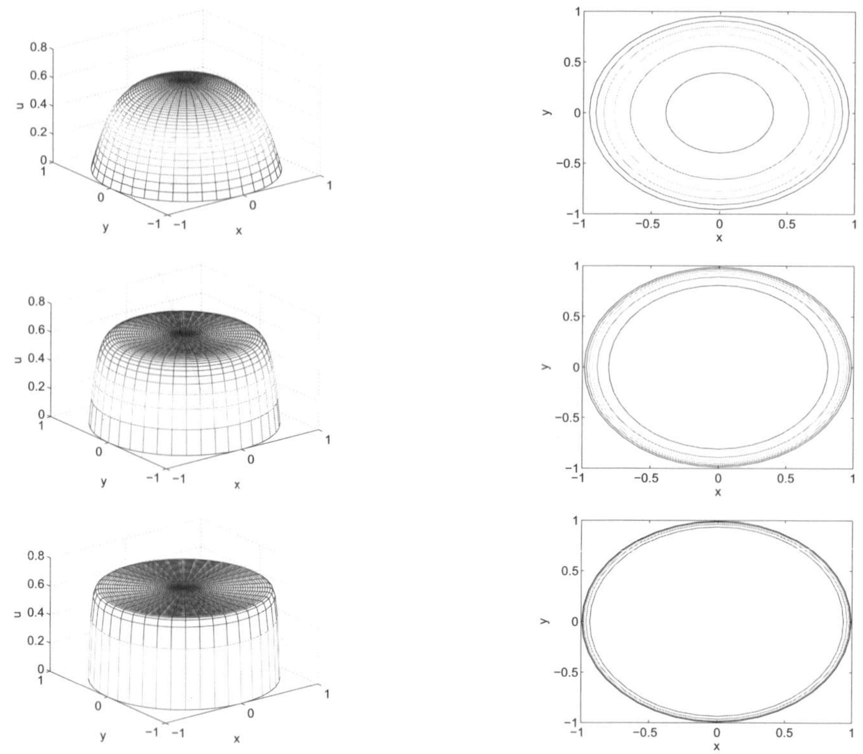

Fig. 7.1 $u^\epsilon \simeq u^0 + \theta^0$: the asymptotic solution (the left) and the contour (the right) with different ϵ.

$$\begin{cases} u^\epsilon \simeq u^0 + \theta^0 = 1 - e^{-t} - \int_0^t e^{-(t-s)} \mathrm{erfc}(\frac{\xi_2}{2\sqrt{\epsilon(t-s)}}) ds, \\ u^\epsilon \simeq u^0 + \theta^0 + \theta^{\frac{1}{2}} = 1 - e^{-t} + \left(\frac{\xi_2}{2\sqrt{\epsilon}} - 1\right) \int_0^t e^{-(t-s)} \mathrm{erfc}(\frac{\xi_2}{2\sqrt{\epsilon(t-s)}}) ds. \end{cases} \tag{7.59}$$

In Fig. 7.1, the low order expansion in $(7.59)_1$ of the solution of problem (7.1) and its contour, with $\epsilon = 0.1$, $\epsilon = 0.01$, $\epsilon = 0.001$, are plotted respectively. One can see that the layer occurs along all of the boundary. While in Fig. 7.2, the low order expansion in $(7.59)_2$, with an extra term $\theta^{\frac{1}{2}}$, of the solution of the problem (7.1) and its contour, with $\epsilon = 0.1$, $\epsilon = 0.01$, $\epsilon = 0.001$, are plotted respectively. And comparing Fig. 7.2 with Fig. 7.1, we find that, with introduction of the terms $\epsilon^{j+\frac{1}{2}}\theta^{j+\frac{1}{2}}$, the asymptotic expansion can better reproduce the boundary layer. Therefore, we can recover the optimal convergence rate.

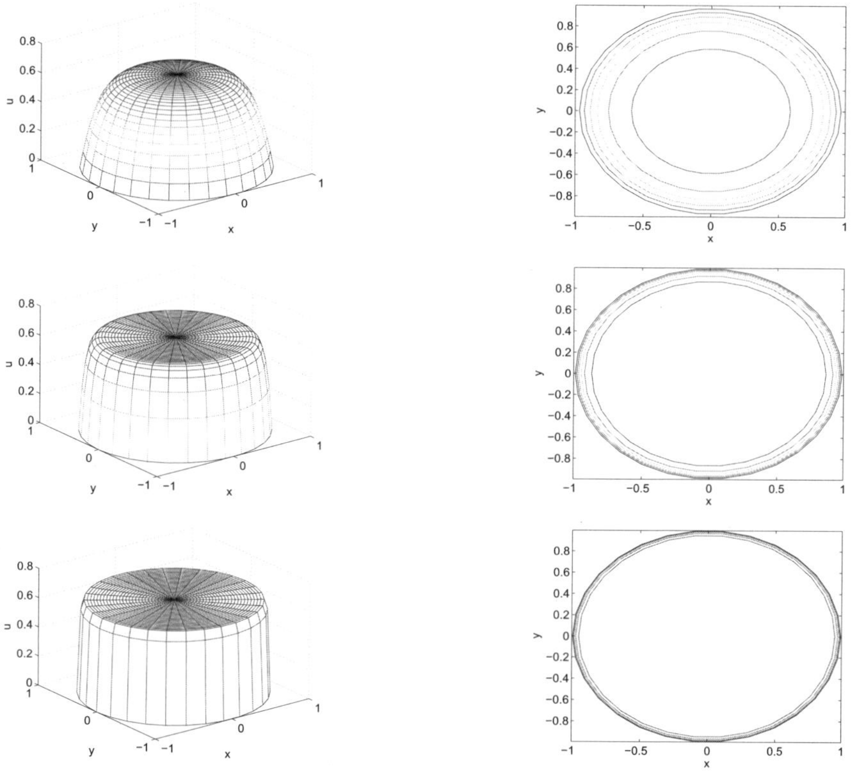

Fig. 7.2 $u^\epsilon \simeq u^0 + \theta^0 + \theta^{\frac{1}{2}}$: the asymptotic solution (the left) and the contour (the right) with different ϵ.

Bibliography

Ciarlet, P. G. (2005). *An Introduction to Differential Geometry with Applications to Elasticity* (Springer Press, Dordrecht).

Ding, T. R. and Li, C. Z. (2004). *Ordinary Differential Equations* (Higher Education Press, Beijing).

Eckhaus, W. (1979). *Asymptotic Analysis of Singular Perturbations* (North Holland Publishing Company, Amsterdam).

Gie, G. M. (2009). Singular perturbation problems in a general smooth domain, *Asymptot. Anal.* **62**, pp. 227–249.

Gie, G. M., Hamouda, M. and Temam, R. (2010a). Asymptotic analysis of the Stokes problems on general bounded domains: the case of a characteristic boundary, *Appl. Anal.* **89**, pp. 49–66.

Gie, G. M., Hamouda, M. and Temam, R. (2010b). Boundary layers in smooth curvilinear domains: parabolic problems, *Discrete Contin. Dyn. Syst.-Series A* **26**, pp. 1213–1240.

Hamouda, M. and Temam, R. (2008). Boundary layers for the Navier-Stokes equations, The case of a characteristic boundary, *Georgian Math. J.* **15**, pp. 517–530.

Holmes, M. H. (1995). *Introduction to Perturbation Methods* (Springer Press, New York).

Jung, C. Y. and Temam, R. (2006). On parabolic boundary layers for convection-diffusion equations in a channel: analysis and numerical applications, *J. Sci. Comput.* **28**, pp. 361–410.

Lions, J. L. (1973). *Perturbations Singulières dans Les problèmes aux Limites et en Contrôle Optimal, Lecture Notes in Math, Vol. 323* (Springer Press, New York).

O'Malley, R. E. (1991). *Singular Perturbation Methods for Ordinary Differential Equations* (Springer Press, New York).

Roos, H. G., Stynes, M. and Tobiska, L. (2008). *Robust Numerical Methods for Singularly Perturbed Differential Equations* (Springer Press, Berlin).

Schiff, J. L. (1999). *The Laplace Transform: Theory and Applications* (Springer Press, New York).

Sun, L. N., Wu, Y. J. and Yang, A. L. (2011). Uniform convergence analysis of finite difference approximations for advection-reaction-diffusion problem on adaptive grids, *Int. J. Comput. Math.* **88**, pp. 3293–3307.

Chapter 8

Incremental unknowns methods for the alternating directional implicit and semi-implicit schemes[1]

Ai-Li Yang[a,*], Yu-Jiang Wu[a] and Zhong-Hua Yang[b]
[a]*School of Mathematics and Statistics, Lanzhou University, Lanzhou 730000, Gansu, PR China*
[*]*yangaili@lzu.edu.cn*
[b]*School of Mathematical Science, Shanghai Normal University, Shanghai 200234, PR China*

For two-dimensional heat equation with time-dependent coefficients, we construct an alternating directional scheme using finite difference method. That is an alternating directional incremental unknowns (IUs) semi-implicit scheme. The stability of this scheme is carefully studied with the Fourier method. Numerical results show that the new scheme is more efficient than the classical alternating directional implicit (ADI) scheme for many problems.
Keywords: Alternating direction implicit scheme, incremental unknowns, two-dimensional heat equation

8.1 Introduction

Incremental unknowns (IUs) method was first introduced by Temam [Temam (1990)] to study long time integration of dissipative evolutionary equations when finite difference approximations are used, see also [Chen and Temam (1991, 1993a,b); Pouit (1996)]. It is well known, IUs method usually yields a very good conditioned matrix in linear algebraic equations [Chen and Temam (1993c); Garcia (1998)]. Hence, many articles have been developed to analyze the properties and efficiencies of the IUs method when it is used for solving many kinds of differential equations.

[1]This research is partially supported by the National Basic Research Program of China, 973 Program (grant no. 2011CB706903), the Mathematical Tianyuan Foundation of China (grant no. 11026064), the Fundamental Research Fund for Physics and Mathematics of Lanzhou University (grant no. LZULL200904) and the Fundamental Research Fund for the Central Universities (grant no. lzujbky-2010-169.

For one-dimensional heat equation with constant or variable coefficient ν,

$$\frac{\partial u}{\partial t} - \nu \frac{\partial^2 u}{\partial x^2} = f(x,t), \tag{8.1}$$

several kinds of IUs-type semi-implicit schemes have been proposed to approximate the solution of the equation (8.1) [Huang and Wu (2005); Pouit (1996); Wu and Yang (2007)]. The stability conditions of these schemes have been improved efficiently. Moreover, the condition numbers of the coefficient matrices of these schemes have also been reduced.

For the two dimensional heat equation with time-dependent coefficients, if we construct the semi-implicit IUs-scheme and analyze it with the similar technique used in the one dimensional case [Huang and Wu (2005); Pouit (1996); Wu and Yang (2007)], we have to find the exact eigenvalues and eigenvectors of a tri-diagonal block matrix. This is very difficult, even impossible.

In this work, we construct an alternating directional IUs-type semi-implicit (ADIUSI) scheme based on the classical alternating directional (AD) scheme [Quarteroni *et al.* (2000); Thomas (1995)]. The stability analysis of this new scheme only needs us to find the exact eigenvalues and eigenvectors of a simple tri-diagonal matrix instead of a tri-diagonal block matrix. The stability condition for each of the sub-schemes of the ADIUSI scheme is improved by the IUs method comparing with that of the AD scheme. Moreover, the numerical results presented in Section 8.4 verifies the efficiency of the ADIUSI scheme.

8.2 Two dimensional heat equation and the AD scheme

We consider the two-dimensional heat equation

$$\begin{cases} \dfrac{\partial u}{\partial t} - \nu(t)\left(\dfrac{\partial^2 u}{\partial x^2} + \dfrac{\partial^2 u}{\partial y^2}\right) = f(x,y,t), & (x,y) \in \Omega, \quad t \in (0,T], \\ u(x,y,0) = u_0(x,y), & (x,y) \in \Omega, \\ u(x,y,t) = 0, & (x,y) \in \partial\Omega, \quad t \in (0,T], \end{cases} \tag{8.2}$$

where $\Omega = (0,1) \times (0,1)$, $\partial\Omega$ is the boundary of Ω, the time-dependent coefficient function $\nu(t)$ is non-increasing and bounded on $[0,T]$ with $\nu(T) > 0$.

Let $h = 1/2N$ and $\tau = T/M$ be the spacial and temporal mesh steps, respectively, with positive integers N and M. Denote $x_i = ih$, $y_j = jh$, $t_k = k\tau$, the discretized step function of u can be defined as

$$u_{ij}^k = u(x_i, y_j, t_k), \quad 0 \le i,j \le 2N, \quad 0 \le k \le M.$$

Let

$$\begin{aligned} & v_{ij}^{k+1/2} = (v_{ij}^k + v_{ij}^{k+1})/2, \\ & \delta_x v_{i+1/2,j}^k = (v_{i+1,j}^k - v_{ij}^k)/h, \quad \delta_x^2 v_{ij}^k = (\delta_x v_{i+1/2,j}^k - \delta_x v_{i-1/2,j}^k)/h, \\ & \delta_y v_{i,j+1/2}^k = (v_{i,j+1}^k - v_{ij}^k)/h, \quad \delta_y^2 v_{ij}^k = (\delta_y v_{i,j+1/2}^k - \delta_y v_{i,j-1/2}^k)/h. \end{aligned}$$

Using finite difference method and introducing an intermediate variable $\overline{u}_{ij}$, the AD scheme [Quarteroni *et al.* (2000); Thomas (1995)] of equation (8.2) can be derived as follows:

$$(I - \frac{\tau}{2}\nu^{k+1/2}\delta_x^2)\overline{u}_{ij} = (I + \frac{\tau}{2}\nu^k\delta_y^2)u_{ij}^k + \frac{\tau}{2}f_{ij}^{k+\frac{1}{2}}, \tag{8.3}$$

$$(I - \frac{\tau}{2}\nu^{k+1}\delta_y^2)u_{ij}^{k+1} = (I + \frac{\tau}{2}\nu^{k+1/2}\delta_x^2)\overline{u}_{ij} + \frac{\tau}{2}f_{ij}^{k+\frac{1}{2}}. \tag{8.4}$$

The initial and boundary conditions are

$$u(x_i, y_j, 0) = (u_0)_{ij}, \quad i, j = 1, 2, \ldots, 2N-1,$$
$$u(x_i, y_j, t_k) = 0, \quad i, j = 0 \text{ or } 2N, \quad 0 < k \leq M.$$

8.3 ADIUSI scheme and stability

8.3.1 *ADIUSI scheme*

First, we write the scheme (8.3) in matrix form with $j = 1, 2, \ldots, 2N-1$,

$$\overline{U}_j^* + \frac{\tau}{2h^2}\nu^{k+1/2}A^*\overline{U}_j^* = \frac{\tau}{2h^2}\nu^k U_{j-1}^{k*} + (1 - \frac{\tau}{h^2}\nu^k)U_j^{k*} + \frac{\tau}{2h^2}\nu^k U_{j+1}^{k*} + \frac{\tau}{2}F_j^{k+\frac{1}{2},*}, \tag{8.5}$$

where

$$A_{2N-1}^* = \begin{pmatrix} 2 & -1 & & \\ -1 & \ddots & \ddots & \\ & \ddots & \ddots & -1 \\ & & -1 & 2 \end{pmatrix}, \qquad U_j^* = \begin{pmatrix} u_{1j} \\ u_{2j} \\ \vdots \\ u_{2N-1,j} \end{pmatrix}.$$

The eigenvalues and the corresponding eigenvectors of matrix A_{2N-1}^* can be found in [Pouit (1996)], which are

$$\lambda_p = 4\sin^2\left(\frac{p\pi}{4N}\right), \quad p = 1, 2, \ldots, 2N-1,$$

and

$$(W_p^*)_i = \sqrt{2}\sin\left(\frac{p\pi i}{2N}\right), \quad i, p = 1, 2, \ldots, 2N-1.$$

Denote $r = \tau/h^2$ and reorder the vector U_j^*, we obtain the following equivalent form of scheme (8.5), that is

$$\overline{U}_j + \frac{r}{2}\nu^{k+1/2}A\overline{U}_j = \frac{r}{2}\nu^k U_{j-1}^k + (1 - r\nu^k)U_j^k + \frac{r}{2}\nu^k U_{j+1}^k + \frac{\tau}{2}F_j^{k+\frac{1}{2}}, \tag{8.6}$$

where

$$A = \begin{pmatrix} 2I_{N-1} & -2G^T \\ -2G & 2I_N \end{pmatrix}, \quad U_j = \begin{pmatrix} (u_{2i,j})_{i=1,\ldots,N-1} \\ (u_{2i+1,j})_{i=0,\ldots,N-1} \end{pmatrix},$$

and $G = (g_{ij})$ is given by $g_{ij} = 0$ except that $g_{ii} = g_{i+1,i} = 1/2$. It is easy to find that the eigenvalues of matrix A are same with those of the matrix A^*, the eigenvectors of which are changed to be

$$W_p = \begin{pmatrix} (W_p^*)_{2i}\ , & i = 1, 2, \ldots, N-1 \\ (W_p^*)_{2i+1}\ , & i = 0, 1, \ldots, N-1 \end{pmatrix},$$

with $p = 1, 2, \ldots, 2N-1$. Now we expand the vectors $\overline{U}_j$, U_s^k, and $F_j^{k+1/2}$ as:

$$\overline{U}_j = \sum_{p=1}^{2N-1} \overline{\mathcal{U}}_{jp} W_p, \quad U_s^k = \sum_{p=1}^{2N-1} \mathcal{U}_{sp}^k W_p, \quad F_j^{k+\frac{1}{2}} = \sum_{p=1}^{2N-1} \mathcal{F}_{jp}^{k+\frac{1}{2}} W_p. \tag{8.7}$$

After introducing a new basis $(\varphi_p)_{p=1,2,\ldots,2N-1}$ with $\varphi_p = S^{-1}W_p$, we obtain an alternating directional IUs-type (ADIU) scheme of (8.3), which is

$$\begin{aligned} &\left(S^T S + \frac{r}{2}\nu^{k+1/2} S^T A S\right) \widetilde{U}_j \\ &= \frac{r}{2}\nu^k S^T S \widetilde{U}_{j-1}^k + \left(1 - r\nu^k\right) S^T S \widetilde{U}_j^k + \frac{r}{2}\nu^k S^T S \widetilde{U}_{j+1}^k + \frac{\tau}{2} S^T S \widetilde{F}_j^{k+\frac{1}{2}}, \end{aligned} \tag{8.8}$$

where

$$\widetilde{U}_j = S^{-1}\overline{U}_j = \begin{pmatrix} \overline{Y}_j \\ \overline{Z}_j \end{pmatrix}, \quad \widetilde{U}_j^k = S^{-1}U_j^k = \begin{pmatrix} Y_j^k \\ Z_j^k \end{pmatrix}, \quad S = \begin{pmatrix} I_{N-1} & 0 \\ G & I_N \end{pmatrix}.$$

Remark 8.1. The ADIU-scheme of (8.4) is

$$\begin{aligned} &\left(S^T S + \frac{r}{2}\nu^{k+1} S^T A S\right) \widetilde{V}_i \\ &= \frac{r}{2}\nu^{k+1/2} S^T S \widetilde{V}_{i-1}^k + \left(1 - r\nu^{k+1/2}\right) S^T S \widetilde{V}_i^k \\ &\quad + \frac{r}{2}\nu^{k+1/2} S^T S \widetilde{V}_{i+1}^k + \frac{\tau}{2} S^T S \widetilde{G}_i^{k+\frac{1}{2}}, \end{aligned} \tag{8.9}$$

where

$$V_i = S^{-1}\begin{pmatrix} (u_{i,2j})_{j=1,\ldots,N-1} \\ (u_{i,2j+1})_{j=0,\ldots,N-1} \end{pmatrix}.$$

Remark 8.2. The incremental unknowns method used here is the one-dimensional case [Chen and Temam (1993a); Pouit (1996)].

Now, we treat Y and Z differently in the second term of the left hand side of scheme (8.8), the ADIUSI scheme of (8.3) is obtained, that is

$$\begin{aligned} &S^T S \begin{pmatrix} \overline{Y}_j \\ \overline{Z}_j \end{pmatrix} + \frac{r}{2}\nu^{k+1/2} S^T A S \begin{pmatrix} Y_j^k \\ \overline{Z}_j \end{pmatrix} \\ &= \left(1 - r\nu^k\right) S^T S \begin{pmatrix} Y_j^k \\ Z_j^k \end{pmatrix} + \frac{r}{2}\nu^k S^T S \left[\begin{pmatrix} Y_{j-1}^k \\ Z_{j-1}^k \end{pmatrix} + \begin{pmatrix} Y_{j+1}^k \\ Z_{j+1}^k \end{pmatrix}\right] + \frac{\tau}{2} S^T S \widetilde{F}_j^{k+\frac{1}{2}}. \end{aligned} \tag{8.10}$$

Remark 8.3. With the same method, we get the ADIUSI scheme of (8.4), that is

$$\begin{aligned} &S^T S \begin{pmatrix} \overline{Y}_i \\ \overline{Z}_i \end{pmatrix} + \frac{r}{2}\nu^{k+1} S^T A S \begin{pmatrix} Y_i^k \\ \overline{Z}_i \end{pmatrix} \\ &= \left(1 - r\nu^{k+\frac{1}{2}}\right) S^T S \begin{pmatrix} Y_i^k \\ Z_i^k \end{pmatrix} + \frac{r}{2}\nu^{k+\frac{1}{2}} S^T S \left[\begin{pmatrix} Y_{i-1}^k \\ Z_{i-1}^k \end{pmatrix} + \begin{pmatrix} Y_{i+1}^k \\ Z_{i+1}^k \end{pmatrix}\right] + \frac{\tau}{2} S^T S \widetilde{F}_i^{k+\frac{1}{2}}. \end{aligned} \tag{8.11}$$

8.3.2 Stability study of the ADIUSI scheme

Denote $\hat{U}_j = S\begin{pmatrix} Y_j^k \\ Z_j \end{pmatrix}$, then

$$\hat{U}_j = S\begin{pmatrix} I_{N-1} & 0 \\ 0 & 0 \end{pmatrix} S^{-1}U_j^k + S\begin{pmatrix} 0 & 0 \\ 0 & I_N \end{pmatrix} S^{-1}\overline{U}_j$$
$$= \sum_{p=1}^{2N-1}\left[\mathfrak{U}_{jp}^k S\begin{pmatrix} I_{N-1} & 0 \\ 0 & 0 \end{pmatrix} S^{-1}W_p + \overline{\mathfrak{U}}_{jp} S\begin{pmatrix} 0 & 0 \\ 0 & I_N \end{pmatrix} S^{-1}W_p\right].$$

The scheme (8.10) becomes

$$\sum_{p=1}^{2N-1}\overline{\mathfrak{U}}_{jp}\left(S^T + \frac{r}{2}\nu^{k+1/2}S^T AS\begin{pmatrix} 0 & 0 \\ 0 & I_N \end{pmatrix} S^{-1}\right)W_p$$
$$= \sum_{p=1}^{2N-1}\mathfrak{U}_{jp}^k\left(\left(1 - r\nu^k\right)S^T - \frac{r}{2}\nu^{k+1/2}S^T AS\begin{pmatrix} I_{N-1} & 0 \\ 0 & 0 \end{pmatrix} S^{-1}\right)W_p \tag{8.12}$$
$$+ \sum_{p=1}^{2N-1}(\mathfrak{U}_{j-1,p}^k + \mathfrak{U}_{j+1,p}^k)\frac{r}{2}\nu^k S^T W_p + \sum_{p=1}^{2N-1}\frac{\tau}{2}S^T \mathfrak{F}_{jp}^{k+\frac{1}{2}}W_p.$$

Now, introducing another basis $(W_p^Y,\ W_p^Z,\ W_N^Z)$ defined by

$$W_p^Y = \begin{pmatrix} (W_p)_{2i}, \quad i = 1, 2, \ldots, N-1 \\ 0 \end{pmatrix}_{p=1,2,\ldots,N-1} \tag{8.13}$$

and

$$W_p^Z = \begin{pmatrix} 0 \\ (W_p)_{2j+1}, \quad j = 0, 1, \ldots, N-1 \end{pmatrix}_{p=1,2,\ldots,N} \tag{8.14}$$

gives

$$S^T W_p = \begin{pmatrix} 2\cos^2\left(\frac{p\pi}{4N}\right)(W_p)_{2i} \\ (W_p)_{2j+1} \end{pmatrix} = 2\cos^2\left(\frac{p\pi}{4N}\right)W_p^Y + W_p^Z,$$
$$S^T AS\begin{pmatrix} 0 & 0 \\ 0 & I_N \end{pmatrix} S^{-1}W_p = \begin{pmatrix} 0 \\ 4\sin^2\left(\frac{p\pi}{4N}\right)\cdot(W_p)_{2i+1} \end{pmatrix} = 4\sin^2\left(\frac{p\pi}{4N}\right)W_p^Z,$$
$$S^T AS\begin{pmatrix} I_{N-1} & 0 \\ 0 & 0 \end{pmatrix} S^{-1}W_p = \begin{pmatrix} 2\sin^2\left(\frac{p\pi}{2N}\right)\cdot(W_p)_{2i} \\ 0 \end{pmatrix} = 2\sin^2\left(\frac{p\pi}{2N}\right)W_p^Y.$$

Then, the scheme (8.12) is simplified as

$$\sum_{p=1}^{2N-1}\overline{\mathfrak{U}}_{jp}\left(a_p W_p^Y + b_p W_p^Z)\right)$$
$$= \frac{r}{2}\nu^k\sum_{p=1}^{2N-1}\mathfrak{U}_{j-1,p}^k\left(a_p W_p^Y + W_p^Z\right) + \frac{r}{2}\nu^k\sum_{p=1}^{2N-1}\mathfrak{U}_{j+1,p}^k\left(a_p W_p^Y + W_p^Z\right)$$

$$+\sum_{p=1}^{2N-1}\mathcal{U}_{jp}^{k}\left(c_pW_p^Y+d_pW_p^Z\right)+\frac{\tau}{2}\mathcal{F}_{jp}^{k+\frac{1}{2}}\left(a_pW_p^Y+W_p^Z\right)\right), \tag{8.15}$$

where

$$\begin{aligned} a_p &= \frac{1}{2}\lambda_{2N-p}, \quad b_p = 1+\frac{r}{2}\nu^{k+1/2}\lambda_p, \quad d_p = 1-r\nu^k \\ c_p &= \frac{1}{2}\left(1-r\nu^k\right)\lambda_{2N-p}-\frac{r}{4}\nu^{k+1/2}\lambda_p\lambda_{2N-p}. \end{aligned} \tag{8.16}$$

Since $W_{2N-p}^Y=-W_p^Y$ and $W_{2N-p}^Z=W_p^Z$ for $p=1,2,\ldots,N$, we have

$$\begin{aligned} &\sum_{p=1}^{N-1}\left(\Gamma_{p,j}^{-}(a,\overline{\mathcal{U}})W_p^Y+\Gamma_{p,j}^{+}(b,\overline{\mathcal{U}})W_p^Z\right)+b_N\overline{\mathcal{U}}_{jN}W_N^Z \\ &=\frac{r}{2}\nu^k\sum_{p=1}^{N-1}\left(\Gamma_{p,j-1}^{-}(a,\mathcal{U}^k)W_p^Y+\Gamma_{p,j-1}^{+}(1,\mathcal{U}^k)W_p^Z\right)+\frac{r}{2}\nu^k\mathcal{U}_{j-1,N}^kW_N^Z \\ &+\frac{r}{2}\nu^k\sum_{p=1}^{N-1}\left(\Gamma_{p,j+1}^{-}(a,\mathcal{U}^k)W_p^Y+\Gamma_{p,j+1}^{+}(1,\mathcal{U}^k)W_p^Z\right)+\frac{r}{2}\nu^k\mathcal{U}_{j+1,N}^kW_N^Z \\ &+\sum_{p=1}^{N-1}\left(\Gamma_{p,j}^{-}(c,\mathcal{U}^k)W_p^Y+\Gamma_{p,j}^{+}(d,\mathcal{U}^k)W_p^Z\right)+d_N\mathcal{U}_{jN}^kW_N^Z \\ &+\frac{\tau}{2}\sum_{p=1}^{N-1}\left(\Gamma_{p,j}^{-}(a,\mathcal{F}^{k+\frac{1}{2}})W_p^Y+\Gamma_{p,j}^{+}(1,\mathcal{F}^{k+\frac{1}{2}})W_p^Z\right)+\frac{\tau}{2}\mathcal{F}_{jN}^{k+\frac{1}{2}}W_N^Z, \end{aligned} \tag{8.17}$$

where $\Gamma_{p,j}^{\pm}(a,\varphi)=a_p\varphi_{jp}\pm a_{2N-p}\varphi_{j,2N-p}$. The matrix form of scheme (8.17) can be written as

$$\mathcal{A}\overline{\mathcal{U}}_j=\frac{r}{2}\nu^k\mathcal{B}\mathcal{U}_{j-1}^k+\mathcal{C}\mathcal{U}_j^k+\frac{r}{2}\nu^k\mathcal{B}\mathcal{U}_{j+1}^k+\frac{\tau}{2}\mathcal{B}\mathcal{F}_j^{k+\frac{1}{2}}, \tag{8.18}$$

where matrix $\mathcal{A}=(\alpha_{ij})$ is given by $\alpha_{ij}=0$ except that

$$\alpha_{ii}=\begin{cases} a_i, \ i=1,2,\ldots,N-1, \\ b_i, \ i=N,\ldots,2N-1, \end{cases} \quad \text{and} \quad \alpha_{i,2N-i}=\begin{cases} -a_{2N-i}, \ i=1,2,\ldots,N-1, \\ b_{2N-i}, \quad i=N,\ldots,2N-1, \end{cases}$$

and matrix $\mathcal{B}$ is obtained by replacing b_i $(i=1,\ldots,2N-1)$ in matrix $\mathcal{A}$ with 1. If we replace a_i and b_i in matrix $\mathcal{A}$ by c_i and d_i respectively for $i=1,\ldots,2N-1$, the matrix $\mathcal{A}$ becomes matrix $\mathcal{C}$.

Solving the row N of the linear system (8.18) gives

$$\overline{\mathcal{U}}_{jN}=\frac{\frac{r}{2}\nu^k\mathcal{U}_{j-1,N}^k+\left(1-r\nu^k\right)\mathcal{U}_{jN}^k+\frac{r}{2}\nu^k\mathcal{U}_{j+1,N}^k+\frac{\tau}{2}\mathcal{F}_{jN}^{k+\frac{1}{2}}}{1+\frac{r}{2}\nu^{k+1/2}\lambda_N}. \tag{8.19}$$

Hence, the amplification factor is

$$G=\left(1+\frac{r}{2}\nu^{k+1/2}\lambda_N\right)^{-1}\left(1-2r\nu^k\sin^2\left(\frac{nh}{2}\right)\right),$$

then we have $|G| < 1$, if r satisfies the following condition

$$r < \frac{2}{2\nu(0) - \nu(T)}. \tag{8.20}$$

For the rest of the system (8.18), we have $N-1$ systems of two equations with two unknowns:

$$\begin{cases} \Gamma^-_{p,j}(a,\overline{\mathcal{U}}) = \frac{r}{2}\nu^k \Gamma^-_{p,j-1}(a,\mathcal{U}^k) + \Gamma^-_{p,j}(c,\mathcal{U}^k) \\ \qquad + \frac{r}{2}\nu^k \Gamma^-_{p,j+1}(a,\mathcal{U}^k) + \frac{\tau}{2}\Gamma^-_{p,j}(a,\mathcal{F}^{k+\frac{1}{2}}), \\ \Gamma^+_{p,j}(b,\overline{\mathcal{U}}) = \frac{r}{2}\nu^k \Gamma^+_{p,j-1}(1,\mathcal{U}^k) + \Gamma^+_{p,j}(d,\mathcal{U}^k) \\ \qquad + \frac{r}{2}\nu^k \Gamma^+_{p,j+1}(1,\mathcal{U}^k) + \frac{\tau}{2}\Gamma^+_{p,j}(1,\mathcal{F}^{k+\frac{1}{2}}). \end{cases} \tag{8.21}$$

Denote

$$\begin{aligned} &\rho_1 = a_{2N-p} - a_{2N-p}b_{2N-p}, && \rho_2 = a_p + a_{2N-p}b_p, \\ &\rho_3 = a_p b_{2N-p} + a_{2N-p}, && \rho_4 = c_p b_{2N-p} + a_{2N-p}d_p, \\ &\rho_5 = a_p - a_p b_p, && \rho_6 = a_{2N-p}d_{2N-p} - c_{2N-p}b_{2N-p}, \\ &\rho_7 = a_p d_{2N-p} + c_{2N-p}b_p, && \rho_8 = a_p d_p - c_p b_p, \\ &\rho_9 = a_p b_{2N-p} + a_{2N-p}b_p. \end{aligned}$$

The matrix form of (8.21) becomes

$$\begin{aligned} \overline{\xi}_j = {} & \frac{r\nu^k}{2\rho_9}\begin{pmatrix} \rho_3 & \rho_1 \\ \rho_5 & \rho_2 \end{pmatrix}\xi^k_{j-1} + \frac{1}{\rho_9}\begin{pmatrix} \rho_4 & \rho_6 \\ \rho_8 & \rho_7 \end{pmatrix}\xi^k_j \\ & + \frac{r\nu^k}{2\rho_9}\begin{pmatrix} \rho_3 & \rho_1 \\ \rho_5 & \rho_2 \end{pmatrix}\xi^k_{j+1} + \frac{\tau}{2\rho_9}\begin{pmatrix} \rho_3 & \rho_1 \\ \rho_5 & \rho_2 \end{pmatrix}\eta_j^{k+\frac{1}{2}}, \end{aligned} \tag{8.22}$$

where

$$\overline{\xi}_j = \begin{pmatrix} \overline{\mathcal{U}}_{jp} \\ \overline{\mathcal{U}}_{j,2N-p} \end{pmatrix}, \quad \xi^k_j = \begin{pmatrix} \mathcal{U}^k_{jp} \\ \mathcal{U}^k_{j,2N-p} \end{pmatrix}, \quad \eta_j^{k+\frac{1}{2}} = \begin{pmatrix} \mathcal{F}^{k+\frac{1}{2}}_{jp} \\ \mathcal{F}^{k+\frac{1}{2}}_{j,2N-p} \end{pmatrix}.$$

The amplification matrix of (8.22) is

$$G = \frac{r\nu^k}{\rho_9}\begin{pmatrix} \rho_3 & \rho_1 \\ \rho_5 & \rho_2 \end{pmatrix}\cos(nh) + \frac{1}{\rho_9}\begin{pmatrix} \rho_4 & \rho_6 \\ \rho_8 & \rho_7 \end{pmatrix}.$$

So the characteristic polynomial of G is

$$P(\alpha_p) = \alpha_p^2 + \Gamma_0\left(\Gamma_1 r^2 + \Gamma_2 r - 8\left(\lambda_p + \lambda_{2N-p}\right)\right)\alpha_p + \Gamma_0\Gamma_3, \tag{8.23}$$

where

$$\begin{aligned} \Gamma_0 = {} & \left[4\lambda_{2N-p} + 2r\nu^{k+1/2}\lambda^2_{2N-p} + 2\lambda_p(2 + r\nu^{k+1/2}\lambda_p)\right]^{-1}, \\ \Gamma_1 = {} & 2\nu^k\nu^{k+1/2}(1-\cos(hn))\lambda^2_{2N-p} + (\nu^{k+1/2})^2\lambda_p\lambda^2_{2N-p} + 2\nu^k\nu^{k+1/2}\lambda_p^2 \\ & - 2\nu^k\nu^{k+1/2}\cos(hn)\lambda_p^2 + (\nu^{k+1/2})^2\lambda_p^2\lambda_{2N-p}, \\ \Gamma_2 = {} & 8\nu^k(1-\cos(hn))\lambda_{2N-p} - 2\nu^{k+1/2}\lambda^2_{2N-p} + 8\nu^k(1-\cos(hn))\lambda_p \\ & + 4\nu^{k+1/2}\lambda_p\lambda_{2N-p} - 2\nu^{k+1/2}\lambda_p^2, \\ \Gamma_3 = {} & 4\left(1 - r\nu^k + r\nu^k\cos(hn)\right)\Big(\left(r\nu^k\cos(hn) - r\nu^k + 1\right)\lambda_p \\ & + \lambda_{2N-p}\left(1 - r\nu^k + r\nu^k\cos(hn) - r\nu^{k+1/2}\lambda_p\right)\Big). \end{aligned} \tag{8.24}$$

The discriminant of (8.23) is

$$\Delta = r^2\Gamma_0^2 \cdot \Psi(r), \tag{8.25}$$

with

$$\begin{aligned}\Psi(r) = &\left[2\nu^k\left(\cos(hn)-1\right)\lambda_p^2 + \lambda_{2N-p}^2\left(2\nu^k(1-\cos(nh)) + \nu^{k+1/2}\lambda_p\right)\right]^2 r^2\\ &+ \nu^{k+1/2}\lambda_p^2\lambda_{2N-p} - 16(\nu^{k+1/2})^2(\lambda_p - \lambda_{2N-p})^2\left[-2\nu^k\left(-1+\cos(nh)\right)\lambda_p\right.\\ &\left.+\lambda_{2N-p}\left(2\nu^k(1-\cos(nh)) + \nu^{k+1/2}\lambda_p\right)\right] r + 4(\nu^{k+1/2})^2(\lambda_p^2 - \lambda_{2N-p}^2)^2.\end{aligned}$$

Inasmuch as the coefficient of r^2 in $\Psi(r)$ is positive and the discriminant of $\Psi(r)$ satisfies

$$\begin{aligned}\delta = &-64(\nu^{k+1/2})^4\lambda_p^2\lambda_{2N-p}^2\left((2-2\cos(nh))\nu^k + \nu^{k+1/2}\lambda_{2N-p}\right)\\ &\times(\lambda_p^2 - \lambda_{2N-p}^2)^2\left((2-2\cos(nh))\nu^k + \nu^{k+1/2}\lambda_p\right) < 0,\end{aligned}$$

the discriminant Δ defined in (8.25) is positive. Hence the equation $P(\alpha_p) = 0$ has two real roots denoted by ϱ_1 and ϱ_2. Using (8.20), we obtain the following stability condition of ADIUSI scheme (8.10):

Proposition 8.1. *the ADIUSI scheme* (8.10) *is stable if*

- $r < 2/\left(2\nu(0) - \nu(T)\right)$,
- ϱ_1 *and* ϱ_2 *are belong to* $(-1,1)$ *for any* $p = 1, 2, \ldots, N-1$.

Note that the second condition can be interpreted as: the minimum of $P(\alpha_p)$ is obtained in $(-1,1)$, $P(-1) > 0$ and $P(1) > 0$. In the immediately following, we will consider this condition in three steps:

8.3.2.1 *The minimum of $P(\alpha_p)$ is obtained in $(-1,1)$*

The minimum of $P(\alpha_p)$ is obtained at

$$\begin{aligned}\alpha_{min} = &-\frac{1}{2}\Gamma_0\left(-32 + 8r\nu^k(1-\cos(nh))\lambda_{2N-p} - 2r\nu^{k+1/2}\lambda_{2N-p}^2\right.\\ &+2r^2\nu^k\nu^{k+1/2}(1-\cos(nh))\lambda_{2N-p}^2 + 8r\nu^k(1-\cos(nh))\lambda_p\\ &+4r\nu^{k+1/2}\lambda_{2N-p}\lambda_p + r^2(\nu^{k+1/2})^2\lambda_{2N-p}^2\lambda_p - 2r\nu^{k+1/2}\lambda_p^2\\ &\left.+2r^2\nu^k\nu^{k+1/2}(1-\cos(nh))\lambda_p^2 + r^2(\nu^{k+1/2})^2\lambda_{2N-p}\lambda_p^2\right).\end{aligned} \tag{8.26}$$

Obviously, $\alpha_{min} < 1$ is equivalent to

$$\begin{aligned}&r\left[\lambda_{2N-p}^2\nu^{k+1/2}(2 + 2r\nu^k(1-\cos(nh)) + r\nu^{k+1/2}\lambda_p)\right.\\ &+2\lambda_p(4\nu^k(1-\cos(nh)) + (\nu^{k+1/2} + r\nu^k\nu^{k+1/2}(1-\cos(nh)))\lambda_p)\\ &\left.+\lambda_{2N-p}(8\nu^k(1-\cos(nh)) + 4\nu^{k+1/2}\lambda_p + r(\nu^{k+1/2})^2\lambda_p^2)\right] > 0.\end{aligned}$$

The above inequality is unconditionally true.

Denote by $\sigma = \cos(nh)$ ($\sigma \in [-1, 1]$), the condition $\alpha_{min} > -1$ becomes

$$\begin{aligned}\phi(\sigma) = & -64 + r(32\nu^k(1-\sigma) - 6\nu^{k+1/2}\lambda_{2N-p}^2 + 4\nu^{k+1/2}\lambda_{2N-p}\lambda_p - 6\nu^{k+1/2}\lambda_p^2) \\ & + r^2\left(2\nu^k\nu^{k+1/2}\lambda_{2N-p}^2 - 2\nu^k\nu^{k+1/2}\sigma\lambda_{2N-p}^2 + 4(\nu^{k+1/2})^2\lambda_{2N-p}\lambda_p\right. \\ & \left. +2\nu^k\nu^{k+1/2}\lambda_p^2 - 2\nu^k\nu^{k+1/2}\sigma\lambda_p^2\right) < 0.\end{aligned}$$

Since $\phi'(\sigma) = -32r\nu^k - 2r^2\nu^k\nu^{k+1/2}\lambda_{2N-p}^2 - 2r^2\nu^k\nu^{k+1/2}\lambda_p^2 < 0$, function $\phi(\sigma)$ is decreasing about the variable σ. Hence

$$\begin{aligned}\phi(\sigma) \le \phi(-1) \le & -64 + r\left(64\nu^k - 6\nu^{k+1/2}\left(\lambda_{2N-p}^2 + \lambda_p^2\right) + 4\nu^{k+1/2}\lambda_{2N-p}\lambda_p\right) \\ & + r^2\left(4\nu^k\nu^{k+1/2}\lambda_{2N-p}^2 + 4\nu^k\nu^{k+1/2}\lambda_p^2 + 4(\nu^{k+1/2})^2\lambda_{2N-p}\lambda_p\right).\end{aligned}$$

From the right hand side of above inequality, we know that the condition $\phi(\sigma) < 0$ holds if

$$r < \frac{3\nu^{k+1/2} - 2\nu^k - 2\nu^{k+1/2}\sin^2\frac{p\pi}{2N} + \sqrt{\Delta}}{4\nu^k\nu^{k+1/2} - \nu^{k+1/2}(2\nu^k - \nu^{k+1/2})\sin^2\left(\frac{p\pi}{2N}\right)},$$

with

$$\Delta = (\nu^{k+1/2})^2(4\sin^4\left(\frac{p\pi}{2N}\right) - 8\sin^2\left(\frac{p\pi}{2N}\right) + 8) + (\nu^{k+1/2} + 2\nu^k)^2.$$

After simplification, it is sufficient to set

$$r < \frac{\nu^{k+1/2} - 2\nu^k + \sqrt{4(\nu^{k+1/2})^2 + (2\nu^k + \nu^{k+1/2})^2}}{4\nu^k\nu^{k+1/2}}.$$

Owing to the non-increasing property of function $\nu(t)$ on interval $[0, T]$, the condition of $\alpha_{\min} > -1$ can be further simplified as

$$r < \frac{\nu(T) - 2\nu(0) + \sqrt{4(\nu(T))^2 + (2\nu(0) + \nu(T))^2}}{4\nu(0)\nu(T)}. \tag{8.27}$$

8.3.2.2 *The condition $P(1) > 0$*

From the characteristic polynomial (8.23), we get

$$P(1) = 4r^2\Gamma_0(2(1-\cos(nh))\nu^k + \lambda_{2N-p}\nu^{k+1/2})(2(1-\cos(nh))\nu^k + \lambda_p\nu^{k+1/2}).$$

It is evident that $P(1) > 0$.

8.3.2.3 *The condition $P(-1) > 0$*

Letting $\alpha_p = -1$ in (8.23) gives

$$P(-1) = \Gamma_0 Q(\cos(nh)),$$

where

$$\begin{aligned}Q(\sigma) = & 64 + r(64\nu^k(\sigma - 1) + 4\nu^{k+1/2}(\lambda_{2N-p} - \lambda_p)^2) \\ & + r^2\left(16(1-\sigma)^2(\nu^k)^2 - 4\lambda_{2N-p}\lambda_p(\nu^{k+1/2})^2\right. \\ & \left. + \left(2\sigma(\lambda_{2N-p} - \lambda_p)^2 - 2(\lambda_p^2 + \lambda_{2N-p}^2)\right)\nu^k\nu^{k+1/2}\right).\end{aligned}$$

Inasmuch as

$$Q'(\sigma) = 32r\nu^k(2 - r\nu^k - r\sigma\nu^k) + 8r^2\nu^k\nu^{k+1/2}(2 - \lambda_p)^2,$$

we find that $Q'(\sigma) \geq 0$ holds if

$$r \leq (\nu(0))^{-1}. \tag{8.28}$$

Hence

$$Q(\cos(nh)) \geq Q(-1) = 16\Phi(r),$$

with

$$\begin{aligned}\Phi(r) = &\left(4(\nu^k)^2 - 4\nu^k\nu^{k+1/2}\cos^2\left(\frac{p\pi}{2N}\right) - (\nu^{k+1/2})^2\sin^2\left(\frac{p\pi}{2N}\right)\right)r^2 \\ &- 4\left(2\nu^k - \nu^{k+1/2}\cos^2\left(\frac{p\pi}{2N}\right)\right)r + 4.\end{aligned}$$

Now, it is easy to find that $P(-1) > 0$ is equivalent to $\Phi(r) > 0$. Since the coefficient of r^2 in $\Phi(r)$ is positive and the discriminant of $\Phi(r)$ satisfies

$$\delta = 16(\nu^{k+1/2})^2\left(1 - \cos^2\left(\frac{p\pi}{2N}\right) + \cos^4\left(\frac{p\pi}{2N}\right)\right) > 0,$$

the equation $\Phi(r) = 0$ has two positive roots. Therefore $\Phi(r) > 0$ holds if

$$r < \frac{4\nu^k - 2\nu^{k+1/2}\kappa - 2\nu^{k+1/2}\sqrt{1-\kappa+\kappa^2}}{4(\nu^k)^2 - 4\nu^k\nu^{k+1/2}\kappa - (\nu^{k+1/2})^2(1-\kappa)} =: \Upsilon(\kappa). \tag{8.29}$$

Here $\kappa = \cos^2\left(\frac{p\pi}{2N}\right) \in [0, 1]$. Note that

$$\Upsilon'(\kappa) = \frac{\nu^{k+1/2}\left(\zeta(\kappa) + 2\nu^{k+1/2}(\nu^k - \nu^{k+1/2})^2\sqrt{1-\kappa+\kappa^2}\right)}{\nu^{k+1/2}\sqrt{1-\kappa+\kappa^2}\left(4(\nu^k)^2 - 4\nu^k\nu^{k+1/2}\kappa - (\nu^{k+1/2})^2(1-\kappa)\right)^2}, \tag{8.30}$$

with

$$\begin{aligned}\zeta(\kappa) = &\, 4(1-2\kappa)(\nu^k)^2\nu^{k+1/2} - 4(2-\kappa)\nu^k(\nu^{k+1/2})^2 \\ &+ (1+\kappa)(\nu^{k+1/2})^3 + 6(\nu^k)^2\nu^{k+1/2}\sqrt{1-\kappa+\kappa^2},\end{aligned}$$

we have with $\nu^k \geq \nu^{k+1/2}$ that

$$\begin{aligned}\zeta'(\kappa) = &- 8(\nu^k)^2\nu^{k+1/2} + 4\nu^k(\nu^{k+1/2})^2 + (\nu^{k+1/2})^3 + \frac{3(2\kappa-1)(\nu^k)^2\nu^{k+1/2}}{\sqrt{1-\kappa+\kappa^2}} \\ \leq &- 5(\nu^k)^2\nu^{k+1/2} + 4\nu^k(\nu^{k+1/2})^2 + (\nu^{k+1/2})^3 \leq 0.\end{aligned}$$

Thus

$$\zeta(\kappa) \geq \zeta(1) = 2\nu^{k+1/2}(\nu^k - \nu^{k+1/2})^2 \geq 0.$$

From (8.30), we have $\Upsilon'(\kappa) \geq 0$. The inequality (8.29) holds if

$$r < \Upsilon(0) = \frac{4\nu^k - 2\nu^{k+1/2}}{4(\nu^k)^2 - (\nu^{k+1/2})^2}. \tag{8.31}$$

Combining (8.28) and (8.31), the sufficient condition of $P(-1) > 0$ is obtained, that is

$$r < \min\left\{\frac{1}{\nu(0)}, \frac{2\nu(0)}{4\nu(0)^2 - \nu(T)^2}\right\} = \frac{2\nu(0)}{4\nu(0)^2 - \nu(T)^2}. \tag{8.32}$$

From Proposition 8.1, conditions (8.27), (8.32) and

$$\frac{2}{2\nu(0) - \nu(T)} > \frac{2\nu(0)}{4\nu(0)^2 - \nu(T)^2},$$

we finally obtain the following result:

Lemma 8.1. *The stability condition of the ADIUSI scheme* (8.10) *is*

$$r < \min\left\{\frac{\nu(T) - 2\nu(0) + \sqrt{4\nu(T)^2 + (2\nu(0) + \nu(T))^2}}{4\nu(0)\nu(T)}, \frac{2\nu(0)}{4\nu(0)^2 - \nu(T)^2}\right\}. \tag{8.33}$$

We can attain the exactly same result for the ADIUSI scheme (8.11) with the same method. Therefore, we get the main result of this work, that is

Theorem 8.1. *The ADIUSI scheme* (8.10) *and* (8.11) *is stable if* r *satisfies the condition* (8.33).

Remark 8.4. If the coefficient function $\nu(t)$ is a constant ν, the stability condition of the ADIUSI scheme (8.10) and (8.11) becomes

$$r < \frac{-1 + \sqrt{13}}{4\nu}.$$

8.4 Numerical results

In this section, we will perform numerical tests to examine the efficiency of the ADIUSI scheme and the correctness of the stability condition derived in Section 8.3. Now, consider the following problem

$$\begin{cases} \dfrac{\partial u}{\partial t} - e^{1-t}\left(\dfrac{\partial^2 u}{\partial x^2} + \dfrac{\partial^2 u}{\partial y^2}\right) = f(x,y,t), & (x,y) \in \Omega, \quad t \in (0,1], \\ u(x,y,0) = e^2(1-x)(1-y)\sin(x)\sin(y), & (x,y) \in \Omega, \\ u(x,y,t) = 0, & (x,y) \in \partial\Omega, \quad t \in (0,1]. \end{cases} \tag{8.34}$$

where

$$\begin{aligned} f(x,y,t) = {} & 2e^{3-2t}(\cos(x)\sin(y)(1-y) + \sin(x)\cos(y)(1-x)) \\ & + \sin(x)\sin(y)(1-x)(1-y)e^{2-t}(2e^{1-t} - 1). \end{aligned}$$

The exact solution is $u = e^{2-t}(1-x)(1-y)\sin(x)\sin(y)$.

From (8.33) and $\nu(t) = e^{1-t}$, the sufficient condition of the stability of the ADIUSI scheme is

$$r < \min\left\{\frac{1 - 2e + \sqrt{4 + (2e+1)^2}}{4e}, \frac{2e}{4e^2 - 1}\right\} = \frac{2e}{4e^2 - 1} \approx 0.1904. \tag{8.35}$$

Table 8.1 Numerical solutions and errors for AD and ADIUSI schemes in the case $h = 1/10, \tau = 1/1000, r = 0.1$

(x, y, t)	exact	AD	|error|	ADIUSI	|error|
(0.2, 0.2, 0.5)	0.1132	0.1134	1.858e-4	0.1134	1.545e-4
(0.5, 0.2, 0.5)	0.1707	0.1710	2.797e-4	0.1710	2.360e-4
(0.8, 0.2, 0.5)	0.1022	0.1024	1.675e-4	0.1023	1.409e-4
(0.5, 0.5, 0.5)	0.2575	0.2579	4.210e-4	0.2579	3.605e-4
(0.8, 0.5, 0.5)	0.1541	0.1544	2.521e-4	0.1543	2.152e-4
(0.8, 0.8, 0.5)	0.0923	0.0924	1.510e-4	0.0924	1.286e-4
(0.2, 0.2, 1)	0.0687	0.0688	1.145e-4	0.0688	0.957e-4
(0.5, 0.2, 1)	0.1036	0.1037	1.726e-4	0.1037	1.455e-4
(0.8, 0.2, 1)	0.0620	0.0621	1.033e-4	0.0621	0.867e-4
(0.5, 0.5, 1)	0.1562	0.1565	2.603e-4	0.1564	2.213e-4
(0.8, 0.5, 1)	0.0935	0.0936	1.558e-4	0.0936	1.319e-4
(0.8, 0.8, 1)	0.0560	0.0560	0.933e-4	0.0560	0.786e-4

Denote by

$$E_\infty(h,\tau) = \max_{\substack{1 \le i,j \le 2N-1 \\ 0 < k \le M}} \left| u(x_i, y_j, t_k) - u_{ij}^k \right|,$$

and

$$e(t_k) = \|u(x,y,t_k) - u_{h\tau}(x,y,t_k)\|_2,$$

the maximal error of the numerical solution and the L^2-norm of the error at time t_k, respectively.

For the problem (8.34), we construct AD and ADIUSI schemes based on the finite difference method. The numerical results of the two schemes are presented in Tables 8.1, 8.2 and Figs. 8.1, 8.2. The results in Table 8.1 show that, under the same termination condition for each linear system obtained on each time-level t_k with $k = 1, 2, \ldots, M$, the error of ADIUSI scheme is obviously smaller than that of the AD scheme when r satisfies the stability condition (8.35). More intuitive comparison of the errors derived by the two schemes at $t = 1$ can be seen in Fig. 8.1.

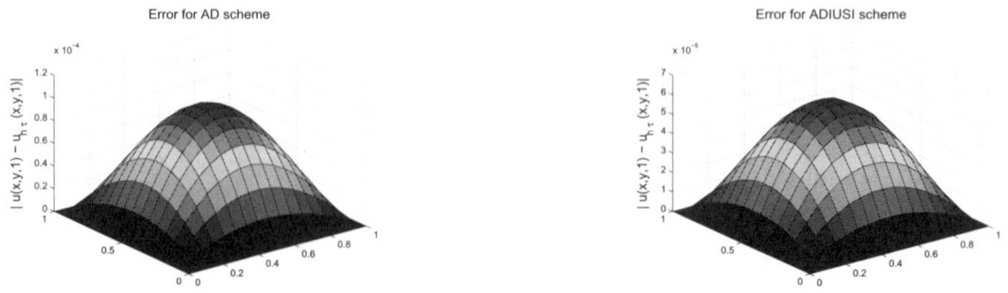

(a) (b)

Fig. 8.1 Error surfaces for AD (left) and ADIUSI (right) schemes.

Table 8.2 $E_\infty(h,\tau)$ for AD and ADIUSI schemes

h	τ	r	AD	ADIUSI
1/10	1/1000	0.1	2.6028e-04	2.2126e-04
1/10	1/500	0.2	2.6136e-04	1.8470e-04
1/20	1/4000	0.1	6.4948e-05	5.6366e-05
1/20	1/2000	0.2	6.5017e-05	4.3848e-05

From Table 8.2, we find that the ADIUSI scheme is stable if r satisfies the condition (8.33) (or (8.35)). Hence, the stability condition (8.33) derived in Section 8.3 is sufficient for the problem (8.34).

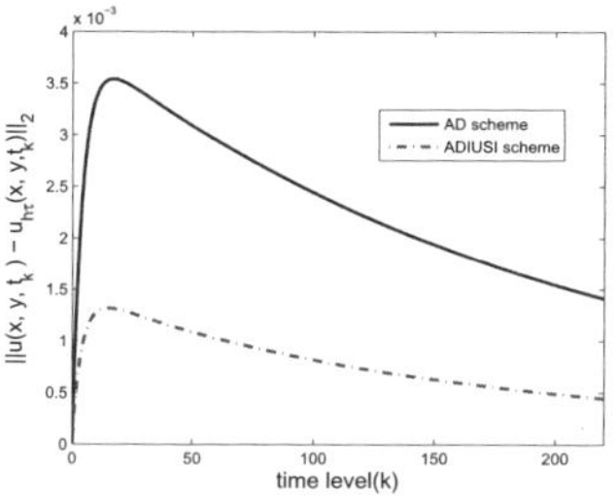

Fig. 8.2 Graphs of $e(t_k)$ for AD and ADIUSI schemes.

The L^2-norms of the errors $e(t_k)$ at time t_k with $k = 1, 2, \ldots, M$ for the AD and ADIUSI schemes are given in Fig. 8.2. It is obviously that the error of the ADIUSI scheme at each time level t_k is smaller than that of the AD scheme. Therefore, the ADIUSI scheme is efficient than the AD scheme for many problems.

Bibliography

Chen, M. and Temam, R. (1991). Incremental unknowns for solving partial differential equations, *Numer. Math.* **59**, pp. 255–271.

Chen, M. and Temam, R. (1993a). Incremental unknowns for convection-diffusion equations, *Appl. Numer. Math.* **11**, pp. 365–383.

Chen, M. and Temam, R. (1993b). Nonlinear Galerkin method in the finite difference case and wavelet-like incremental unknowns, *Numer. Math.* **64**, pp. 271–294.

Chen, M. and Temam, R. (1993c). Incremental unknowns in finite differences: condition number of the matrix, *SIAM J. Matrix Anal. Appl.* **14**, pp. 432–455.

Garcia, S. (1998). lgebraic conditioning analysis of the incremental unknowns preconditioner, *Appl. Math. Model.* **22**, pp. 351–366.

Huang, J. Q. and Wu, Y. J. (2005). On stability and error estimates for a class of weighted semi-implicit schemes of the incremental unknowns methods, *Math. Numer. Sinica (Chinese)* **27**, pp. 183–198.

Pouit, F. (1996). Stability study, error estimation, and condition number for semi-implicit schemes using incremental unknowns, *Numer. Meth. Part. Diff. Equ.* **12**, pp. 743–766.

Quarteroni, A., Sacco, R. and Saleri, F. (2000). *Numerical Partial Differential Equations: Finite Difference Methods* (Springer-Verlag, New York).

Temam, R. (1990). Inertial manifolds and multigrid methods, *SIAM J. Math. Anal.* **21**, pp. 154–178.

Thomas, J. W. (1995). *Numerical Partial Differential Equations: Finite Difference Methods* (Springer-Verlag, New York).

Wu, Y. J. and Yang, A. L. (2007). Incremental unknowns for the heat equation with time-dependent coefficients: semi-implicit θ-schemes and their stability, *J. Comput. Math.* **25**, pp. 573–582.

Chapter 9

Stability of a colocated finite volume scheme for the 3D incompressible Navier-Stokes equations[1]

Xu Li[a,b,*] and Shu-qin Wang[a]

[a] *School of Mathematics and Statistics, Lanzhou University, Lanzhou 730000, Gansu, PR China*

[*] *mathlixu@163.com*

[b] *Department of Mathematics, Gansu Lianhe University, Lanzhou 730000, Gansu, PR China*

For the numerical solution of 2D incompressible Navier-Stokes equations, authors as represented by S. Faure and R. Temam described a colocated finite volume scheme and studied its stability. In this article, the extension to the three-dimensional case is presented, and stability of the scheme is discussed. First, we give the full discretization of the 3D incompressible Navier-Stokes equations, the time discretization used in the scheme is a projection method. Then, we state the stability result and prove it following the methods of Marion and Temam.

Keywords: Navier-Stokes equations, colocated finite volume scheme, stability

9.1 Introduction

We consider the three-dimensional Navier-Stokes equations in their velocity-pressure formulation and the continuity equation written for an incompressible viscous fluid. Let Ω denote an open bounded set of $\mathbb{R}^3$ with boundary $\partial\Omega$. Let $T > 0$ be fixed and let $\mathbf{f} \in L^2(\Omega \times [0,T])^3$ be a given vector function. We are looking for a vector function $\mathbf{u}=(u,v,w)$ and a scalar function p, representing the velocity and the pressure of

[1]This research is supported by National Basic Research Program of China, 973 Program (grant no. 2011CB706903).

the fluid, which are defined on $\Omega \times [0,T]$ and satisfy the following equations:

$$\begin{cases} \dfrac{\partial \mathbf{u}}{\partial t} - \nu\Delta\mathbf{u} + (\mathbf{u}\cdot\nabla)\mathbf{u} + \nabla p = \mathbf{f}, & \text{in } \Omega\times(0,T], \\ \text{div } \mathbf{u} = 0, & \text{in } \Omega\times[0,T], \\ \mathbf{u} = \mathbf{0}, & \text{on } \partial\Omega\times[0,T], \\ \mathbf{u}(x,y,z,0) = \mathbf{u}_0(x,y,z), & \text{in } \Omega, \end{cases} \tag{9.1}$$

where $\nu > 0$ is the viscosity and $\mathbf{u}_0 : \Omega \to \mathbb{R}^3$ is given.

We define the function spaces $\mathbf{W}=H_0^1(\Omega)^3$, $\mathbf{V}=\{\boldsymbol{\varphi} \in H_0^1(\Omega)^3; \text{div } \boldsymbol{\varphi}=0\}$ and $\mathbf{H}=\{\boldsymbol{\varphi} \in L^2(\Omega)^3; \text{div } \boldsymbol{\varphi}=0, \boldsymbol{\varphi}\cdot\mathbf{n}=0 \text{ on } \partial\Omega\}$. We denote by $(\cdot,\cdot)$, $|\cdot|$ the inner product and norm of the space $L^2(\Omega)^3$, and $((\cdot,\cdot))$, $\|\cdot\|$ the inner product and norm of the space $H_0^1(\Omega)^3$. The weak formulation of the problems (9.1) is as follows (see [Temam (2001)]): for given $\mathbf{f} \in L^2(0,T;\mathbf{V}')$ and $\mathbf{u}_0 \in \mathbf{H}$, we look for $\mathbf{u} \in L^2(0,T;\mathbf{V})$ satisfying

$$\frac{d}{dt}(\mathbf{u},\boldsymbol{\varphi}) + \nu((\mathbf{u},\boldsymbol{\varphi})) + \mathbf{b}(\mathbf{u},\mathbf{u},\boldsymbol{\varphi}) = (\mathbf{f},\boldsymbol{\varphi}), \qquad \forall\, \boldsymbol{\varphi} \in \mathbf{V}, \tag{9.2}$$

where, the definition of the trilinear form $\mathbf{b}$ is

$$\mathbf{b}(\mathbf{u},\boldsymbol{\varphi},\boldsymbol{\xi}) = \sum_{i,j=1}^{3} \int_\Omega u_i(x)\frac{\partial \varphi_j(x)}{\partial x_i}\xi_j(x)dx. \tag{9.3}$$

The aim of this work is to introduce a finite volume scheme adapted to this problem and prove its stability. For the two-dimensional incompressible Navier-Stokes equations, S. Faure in [Faure (2005); Faure *et al.* (2008)] proposed a colocated finite volume approximation and studied its stability. In [Li *et al.* (2009)], new estimates on the discrete nonlinear term and two kinds of new *a priori* estimates on approximate solutions were proposed, which led to an improved stability theorem.

In this chapter, an extension to the three-dimensional case is presented. As the two-dimensional case in [Faure (2005)], our motivation for colocated schemes comes from the fact that it is easier to develop a multilevel method with them than with staggered grid schemes. Indeed, as the unknowns share the same location, we can make a simple hierarchical space discretization, see, e.g., [Faure *et al.* (2005)].

The chapter is organized as follows. In Section 9.2, we describe the projection method used for the time discretization and the finite volume scheme used for the space discretization. Then, in Section 9.3, we introduce the notations, give the discrete weak formulation and state the main stability result of our scheme. Section 9.4 contains the technical lemmas used in Section 9.3, and Section 9.5 contains the *a priori* estimates. In Section 9.6, we give the proof of stability.

9.2 Full discretization: finite volume scheme in space and projection method in time

In this section, we give the time and space discretizations in order to solve the Navier-Stokes equations.

First we use a projection method for the time discretization. For the sake of simplicity in the presentation, we will describe here the original projection method of [Chorin (1968)]. Here we use three steps to compute the new velocity vector $\mathbf{u}^{n+1}$ from velocity vector $\mathbf{u}^{n-1}$ and $\mathbf{u}^n$:

• The first step: compute the intermediate velocity vector $\mathbf{u}^{n+\frac{1}{2}}$ with

$$\frac{\mathbf{u}^{n+\frac{1}{2}} - \mathbf{u}^n}{\Delta t} = \nu\Delta(\frac{\mathbf{u}^{n+\frac{1}{2}} + \mathbf{u}^n}{2}) - \frac{3}{2}\mathbf{C}^n + \frac{1}{2}\mathbf{C}^{n-1} + \frac{1}{2}(\mathbf{f}^{n+1} + \mathbf{f}^n), \tag{9.4}$$

$$\mathbf{u}^{n+\frac{1}{2}}|_{\partial\Omega} = 0, \tag{9.5}$$

where $\Delta t{=}T/N_t$ is the time step, N_t is an integer; $\mathbf{u}^n(x, y, z) \simeq \mathbf{u}(x, y, z, n\Delta t)$ if $t \in [n\Delta t, (n+1)\Delta t]$; and $\mathbf{C}^n{=}(\mathbf{u}^n \cdot \nabla)\mathbf{u}^n$ is the nonlinear term.

• The second step: compute the pressure p^{n+1}

$$\frac{\mathbf{u}^{n+1} - \mathbf{u}^{n+\frac{1}{2}}}{\Delta t} = -\nabla p^{n+1}, \tag{9.6}$$

$$\text{div } \mathbf{u}^{n+1} = 0, \tag{9.7}$$

$$\mathbf{u}^{n+1} \cdot \mathbf{n}|_{\partial\Omega} = 0, \tag{9.8}$$

where $p^n(x, y, z) \simeq p(x, y, z, n\Delta t)$ if $t \in [n\Delta t, (n+1)\Delta t]$, and $\mathbf{n}$ is the unit outward normal on $\partial\Omega$.

Equations (9.6) and (9.7) imply the projection equation:

$$\Delta p^{n+1} = \frac{1}{\Delta t} \text{div } \mathbf{u}^{n+\frac{1}{2}}. \tag{9.9}$$

Due to (9.5), (9.6), and (9.8), we have a Neumann boundary condition for p^{n+1}:

$$\frac{\partial p^{n+1}}{\partial \mathbf{n}} = 0. \tag{9.10}$$

Consequently, we are able to compute p^{n+1} from (9.9) and (9.10), up to an additive constant.

• The third step: compute the new velocity vector $\mathbf{u}^{n+1}$ from (9.6).

Now, we describe the space discretization. Here, the results are presented for a cuboid domain $\Omega = (0, L_1) \times (0, L_2) \times (0, L_3)$. In view of extending these results to more general situations, we use convenient notations and concepts of [Eymard *et al.* (2000)]. In order to discretize the domain Ω, we use cuboid finite volumes. We assume that all the volumes have the same dimensions $\Delta x \Delta y \Delta z$ with $M\Delta x{=}L_1$, $N\Delta y{=}L_2$ and $Q\Delta z{=}L_3$ where M, N and Q are given integers. Hence, we have MNQ volumes defined by

$$(K_{ijk}{=}[x_{i-\frac{1}{2}}, x_{i+\frac{1}{2}}] \times [y_{j-\frac{1}{2}}, y_{j+\frac{1}{2}}] \times [z_{k-\frac{1}{2}}, z_{k+\frac{1}{2}}])_{i=1,\ldots,M, j=1,\ldots,N, k=1,\ldots,Q},$$

where

$$x_{i+\frac{1}{2}}{=}i\Delta x \qquad \text{for } i{=}0, \ldots, M,$$

$$y_{j+\frac{1}{2}}{=}j\Delta y \qquad \text{for } j{=}0, \ldots, N,$$

$$z_{k+\frac{1}{2}}{=}k\Delta z \qquad \text{for } k{=}0, \ldots, Q.$$

As the scheme studied uses a colocated grid, all the unknowns are located at the centers of the volumes. So, for i=1, ..., M, j=1, ..., N, and k=1, ..., Q, the unknowns are meant to be the approximations of the cell averages:

- for the velocity:

$$\mathbf{u}_{ijk} \simeq \frac{1}{\Delta x \Delta y \Delta z} \int_{x_{i-\frac{1}{2}}}^{x_{i+\frac{1}{2}}} \int_{y_{j-\frac{1}{2}}}^{y_{j+\frac{1}{2}}} \int_{z_{k-\frac{1}{2}}}^{z_{k+\frac{1}{2}}} \mathbf{u}(x,y,z,t) \mathrm{d}x\mathrm{d}y\mathrm{d}z,$$

- for the pressure:

$$p_{ijk} \simeq \frac{1}{\Delta x \Delta y \Delta z} \int_{x_{i-\frac{1}{2}}}^{x_{i+\frac{1}{2}}} \int_{y_{j-\frac{1}{2}}}^{y_{j+\frac{1}{2}}} \int_{z_{k-\frac{1}{2}}}^{z_{k+\frac{1}{2}}} p(x,y,z,t) \mathrm{d}x\mathrm{d}y\mathrm{d}z.$$

To further simplify the notations, we introduce the velocities at the edges obtained by linear interpolation:

$$\mathbf{u}_{i+\frac{1}{2}jk} = \frac{\mathbf{u}_{i+1jk} + \mathbf{u}_{ijk}}{2}, \quad \mathbf{u}_{ij+\frac{1}{2}k} = \frac{\mathbf{u}_{ij+1k} + \mathbf{u}_{ijk}}{2}, \quad \mathbf{u}_{ijk+\frac{1}{2}} = \frac{\mathbf{u}_{ijk+1} + \mathbf{u}_{ijk}}{2}. \tag{9.11}$$

We integrate Equation (9.4) over a control volume K_{ijk}, for i=1, ..., M, j=1, ..., N, and k=1, ..., Q. This suggests to consider the following scheme (first half step) to compute the intermediate velocity vector $\mathbf{u}^{n+\frac{1}{2}}$:

$$\begin{aligned}
\Delta x \Delta y \Delta z \frac{\mathbf{u}_{ijk}^{n+\frac{1}{2}} - \mathbf{u}_{ijk}^{n}}{\Delta t} =& \frac{\nu}{2}\Bigg[\Delta y \Delta z \frac{\mathbf{u}_{i+1jk}^{n+\frac{1}{2}} - \mathbf{u}_{ijk}^{n+\frac{1}{2}}}{\Delta x} + \Delta y \Delta z \frac{\mathbf{u}_{i-1jk}^{n+\frac{1}{2}} - \mathbf{u}_{ijk}^{n+\frac{1}{2}}}{\Delta x} \\
&+ \Delta z \Delta x \frac{\mathbf{u}_{ij+1k}^{n+\frac{1}{2}} - \mathbf{u}_{ijk}^{n+\frac{1}{2}}}{\Delta y} + \Delta z \Delta x \frac{\mathbf{u}_{ij-1k}^{n+\frac{1}{2}} - \mathbf{u}_{ijk}^{n+\frac{1}{2}}}{\Delta y} \\
&+ \Delta x \Delta y \frac{\mathbf{u}_{ijk+1}^{n+\frac{1}{2}} - \mathbf{u}_{ijk}^{n+\frac{1}{2}}}{\Delta z} + \Delta x \Delta y \frac{\mathbf{u}_{ijk-1}^{n+\frac{1}{2}} - \mathbf{u}_{ijk}^{n+\frac{1}{2}}}{\Delta z} \\
&+ \Delta y \Delta z \frac{\mathbf{u}_{i+1jk}^{n} - \mathbf{u}_{ijk}^{n}}{\Delta x} + \Delta y \Delta z \frac{\mathbf{u}_{i-1jk}^{n} - \mathbf{u}_{ijk}^{n}}{\Delta x} \\
&+ \Delta z \Delta x \frac{\mathbf{u}_{ij+1k}^{n} - \mathbf{u}_{ijk}^{n}}{\Delta y} + \Delta z \Delta x \frac{\mathbf{u}_{ij-1k}^{n} - \mathbf{u}_{ijk}^{n}}{\Delta y} \\
&+ \Delta x \Delta y \frac{\mathbf{u}_{ijk+1}^{n} - \mathbf{u}_{ijk}^{n}}{\Delta z} + \Delta x \Delta y \frac{\mathbf{u}_{ijk-1}^{n} - \mathbf{u}_{ijk}^{n}}{\Delta z}\Bigg] \\
&- \frac{3}{2}\Big[\Delta y \Delta z (u_{i+\frac{1}{2}jk}^{n} \mathbf{u}_{i+\frac{1}{2}jk}^{n} - u_{i-\frac{1}{2}jk}^{n} \mathbf{u}_{i-\frac{1}{2}jk}^{n}) \\
&+ \Delta z \Delta x (v_{ij+\frac{1}{2}k}^{n} \mathbf{u}_{ij+\frac{1}{2}k}^{n} - v_{ij-\frac{1}{2}k}^{n} \mathbf{u}_{ij-\frac{1}{2}k}^{n}) \\
&+ \Delta x \Delta y (w_{ijk+\frac{1}{2}}^{n} \mathbf{u}_{ijk+\frac{1}{2}}^{n} - w_{ijk-\frac{1}{2}}^{n} \mathbf{u}_{ijk-\frac{1}{2}}^{n})\Big] \\
&+ \frac{1}{2}\Big[\Delta y \Delta z (u_{i+\frac{1}{2}jk}^{n-1} \mathbf{u}_{i+\frac{1}{2}jk}^{n-1} - u_{i-\frac{1}{2}jk}^{n-1} \mathbf{u}_{i-\frac{1}{2}jk}^{n-1}) \\
&+ \Delta z \Delta x (v_{ij+\frac{1}{2}k}^{n-1} \mathbf{u}_{ij+\frac{1}{2}k}^{n-1} - v_{ij-\frac{1}{2}k}^{n-1} \mathbf{u}_{ij-\frac{1}{2}k}^{n-1}) \\
&+ \Delta x \Delta y (w_{ijk+\frac{1}{2}}^{n-1} \mathbf{u}_{ijk+\frac{1}{2}}^{n-1} - w_{ijk-\frac{1}{2}}^{n-1} \mathbf{u}_{ijk-\frac{1}{2}}^{n-1})\Big] \\
&+ \frac{1}{2} \Delta x \Delta y \Delta z (\mathbf{f}_{ijk}^{n+1} + \mathbf{f}_{ijk}^{n}).
\end{aligned} \tag{9.12}$$

Because of (9.5), we propose the following Dirichlet boundary condition at step $n+\frac{1}{2}$:

$$\mathbf{u}_{\frac{1}{2}jk}^{n+\frac{1}{2}} = \mathbf{u}_{M+\frac{1}{2}jk}^{n+\frac{1}{2}} = \mathbf{u}_{i\frac{1}{2}k}^{n+\frac{1}{2}} = \mathbf{u}_{iN+\frac{1}{2}k}^{n+\frac{1}{2}} = \mathbf{u}_{ij\frac{1}{2}}^{n+\frac{1}{2}} = \mathbf{u}_{ijQ+\frac{1}{2}}^{n+\frac{1}{2}} = 0 \tag{9.13}$$

which also means, in accordance with the definition (9.11), that

$$\begin{aligned} &\mathbf{u}_{0jk}^{n+\frac{1}{2}} = -\mathbf{u}_{1jk}^{n+\frac{1}{2}}, \quad \mathbf{u}_{M+1jk}^{n+\frac{1}{2}} = -\mathbf{u}_{Mjk}^{n+\frac{1}{2}}, \quad \mathbf{u}_{i0k}^{n+\frac{1}{2}} = -\mathbf{u}_{i1k}^{n+\frac{1}{2}}, \\ &\mathbf{u}_{iN+1k}^{n+\frac{1}{2}} = -\mathbf{u}_{iNk}^{n+\frac{1}{2}}, \quad \mathbf{u}_{ij0}^{n+\frac{1}{2}} = -\mathbf{u}_{ij1}^{n+\frac{1}{2}}, \quad \mathbf{u}_{ijQ+1}^{n+\frac{1}{2}} = -\mathbf{u}_{ijQ}^{n+\frac{1}{2}}. \end{aligned} \tag{9.14}$$

Notice that when i=0 or i=M+1 or j=0 or j=N+1 or k=0 or k=Q+1, the control volume K_{ijk} is a fictitious control volume outside of the domain Ω.

Similarly, we integrate Equation (9.7) over a control volume K_{ijk}, for $i=1,\ldots,M$, $j=1,\ldots,N$, and $k=1,\ldots,Q$. This suggests the following incompressibility condition:

$$\begin{aligned} \frac{\Delta y \Delta z}{2}(u_{i+1jk}^{n+1} - u_{i-1jk}^{n+1}) + \frac{\Delta z \Delta x}{2}(v_{ij+1k}^{n+1} - v_{ij-1k}^{n+1}) \\ + \frac{\Delta x \Delta y}{2}(w_{ijk+1}^{n+1} - w_{ijk-1}^{n+1}) = 0 \end{aligned} \tag{9.15}$$

with

$$\begin{aligned} &u_{0jk}^{n+1} = -u_{1jk}^{n+1}, \quad u_{M+1jk}^{n+1} = -u_{Mjk}^{n+1}, \quad v_{i0k}^{n+1} = -v_{i1k}^{n+1}, \\ &v_{iN+1k}^{n+1} = -v_{iNk}^{n+1}, \quad w_{ij0}^{n+1} = -w_{ij1}^{n+1}, \quad w_{ijQ+1}^{n+1} = -w_{ijQ}^{n+1}. \end{aligned} \tag{9.16}$$

Furthermore, we integrate Equation (9.6) over a control volume K_{ijk}, for $i=1,\ldots,M$, $j=1,\ldots,N$, and $k=1,\ldots,Q$. This leads us to propose the following discrete equations for the second half step:

$$\Delta x \Delta y \Delta z \frac{\mathbf{u}_{ijk}^{n+1} - \mathbf{u}_{ijk}^{n+\frac{1}{2}}}{\Delta t} = -\begin{pmatrix} \Delta y \Delta z [p_{i+\frac{1}{2}jk}^{n+1} - p_{i-\frac{1}{2}jk}^{n+1}] \\ \Delta z \Delta x [p_{ij+\frac{1}{2}k}^{n+1} - p_{ij-\frac{1}{2}k}^{n+1}] \\ \Delta x \Delta y [p_{ijk+\frac{1}{2}}^{n+1} - p_{ijk-\frac{1}{2}}^{n+1}] \end{pmatrix}. \tag{9.17}$$

To proceed with (9.17), we must interpolate the pressures $p_{i+\frac{1}{2}jk}^{n+1}$, $p_{ij+\frac{1}{2}k}^{n+1}$, $p_{ijk+\frac{1}{2}}^{n+1}$ at the edges. By analogy with the fluxes, we define these pressures by linear interpolation:

$$p_{i+\frac{1}{2}jk}^{n+1} = \frac{p_{i+1jk}^{n+1} + p_{ijk}^{n+1}}{2} \qquad \text{for } i=1,\ldots,M, \tag{9.18}$$

$$p_{ij+\frac{1}{2}k}^{n+1} = \frac{p_{ij+1k}^{n+1} + p_{ijk}^{n+1}}{2} \qquad \text{for } j=1,\ldots,N, \tag{9.19}$$

$$p_{ijk+\frac{1}{2}}^{n+1} = \frac{p_{ijk+1}^{n+1} + p_{ijk}^{n+1}}{2} \qquad \text{for } k=1,\ldots,Q, \tag{9.20}$$

where the pressure on the fictitious control volumes is defined by $p_{0jk}^{n+1}=p_{1jk}^{n+1}$, $p_{M+1jk}^{n+1}=p_{Mjk}^{n+1}$, $p_{i0k}^{n+1}=p_{i1k}^{n+1}$, $p_{iN+1k}^{n+1}=p_{iNk}^{n+1}$, $p_{ij0}^{n+1}=p_{ij1}^{n+1}$, $p_{ijQ+1}^{n+1}=p_{ijQ}^{n+1}$. These definitions are consistent with the pressure Neumann boundary condition (9.10) and allow us to rewrite the Equation (9.17) as follows: for $i=1,\ldots,M$, $j=1,\ldots,N$, and $k=1,\ldots,Q$,

$$\Delta x\Delta y\Delta z\frac{\mathbf{u}_{ijk}^{n+1}-\mathbf{u}_{ijk}^{n+\frac{1}{2}}}{\Delta t}=-\begin{pmatrix}\frac{\Delta y\Delta z}{2}\left[p_{i+1jk}^{n+1}-p_{i-1jk}^{n+1}\right]\\ \frac{\Delta z\Delta x}{2}\left[p_{ij+1k}^{n+1}-p_{ij-1k}^{n+1}\right]\\ \frac{\Delta x\Delta y}{2}\left[p_{ijk+1}^{n+1}-p_{ijk-1}^{n+1}\right]\end{pmatrix}. \tag{9.21}$$

Now, to obtain the pressure equation, we take the discrete divergence [defined as for (9.15)] of Equation (9.21). First, inside the domain, i.e., for $i=2,\ldots,M-1$, $j=2,\ldots,N-1$, and $k=2,\ldots,Q-1$, we have

$$\begin{aligned}&\frac{1}{2\Delta x}\left[\frac{\Delta x\Delta y\Delta z}{\Delta t}(u_{i+1jk}^{n+1}-u_{i+1jk}^{n+\frac{1}{2}})+\frac{\Delta y\Delta z}{2}(p_{i+2jk}^{n+1}-p_{ijk}^{n+1})\right.\\&\qquad\left.-\frac{\Delta x\Delta y\Delta z}{\Delta t}(u_{i-1jk}^{n+1}-u_{i-1jk}^{n+\frac{1}{2}})-\frac{\Delta y\Delta z}{2}(p_{ijk}^{n+1}-p_{i-2jk}^{n+1})\right]\\&+\frac{1}{2\Delta y}\left[\frac{\Delta x\Delta y\Delta z}{\Delta t}(v_{ij+1k}^{n+1}-v_{ij+1k}^{n+\frac{1}{2}})+\frac{\Delta z\Delta x}{2}(p_{ij+2k}^{n+1}-p_{ijk}^{n+1})\right.\\&\qquad\left.-\frac{\Delta x\Delta y\Delta z}{\Delta t}(v_{ij-1k}^{n+1}-v_{ij-1k}^{n+\frac{1}{2}})-\frac{\Delta z\Delta x}{2}(p_{ijk}^{n+1}-p_{ij-2k}^{n+1})\right]\\&+\frac{1}{2\Delta z}\left[\frac{\Delta x\Delta y\Delta z}{\Delta t}(w_{ijk+1}^{n+1}-w_{ijk+1}^{n+\frac{1}{2}})+\frac{\Delta x\Delta y}{2}(p_{ijk+2}^{n+1}-p_{ijk}^{n+1})\right.\\&\qquad\left.-\frac{\Delta x\Delta y\Delta z}{\Delta t}(w_{ijk-1}^{n+1}-w_{ijk-1}^{n+\frac{1}{2}})-\frac{\Delta x\Delta y}{2}(p_{ijk}^{n+1}-p_{ijk-2}^{n+1})\right]=0.\end{aligned} \tag{9.22}$$

Hence, using the incompressibility condition (9.15):

$$\begin{aligned}&\Delta y\Delta z\frac{p_{i+2jk}^{n+1}-p_{ijk}^{n+1}}{2\Delta x}+\Delta y\Delta z\frac{p_{i-2jk}^{n+1}-p_{ijk}^{n+1}}{2\Delta x}+\Delta z\Delta x\frac{p_{ij+2k}^{n+1}-p_{ijk}^{n+1}}{2\Delta y}\\&+\Delta z\Delta x\frac{p_{ij-2k}^{n+1}-p_{ijk}^{n+1}}{2\Delta y}+\Delta x\Delta y\frac{p_{ijk+2}^{n+1}-p_{ijk}^{n+1}}{2\Delta z}+\Delta x\Delta y\frac{p_{ijk-2}^{n+1}-p_{ijk}^{n+1}}{2\Delta z}\\&=\frac{1}{\Delta t}\left[\Delta y\Delta z(u_{i+1jk}^{n+\frac{1}{2}}-u_{i-1jk}^{n+\frac{1}{2}})+\Delta z\Delta x(v_{ij+1k}^{n+\frac{1}{2}}-v_{ij-1k}^{n+\frac{1}{2}})\right.\\&\quad\left.+\Delta x\Delta y(w_{ijk+1}^{n+\frac{1}{2}}-w_{ijk-1}^{n+\frac{1}{2}})\right].\end{aligned} \tag{9.23}$$

Then, near the boundary, for example, when $i=1$, (9.15) becomes in accordance with the boundary conditions (9.16):

$$\begin{aligned}&\frac{\Delta y\Delta z}{2}(u_{2jk}^{n+1}+u_{1jk}^{n+1})+\frac{\Delta z\Delta x}{2}(v_{1j+1k}^{n+1}-v_{1j-1k}^{n+1})\\&\qquad+\frac{\Delta x\Delta y}{2}(w_{1jk+1}^{n+1}-w_{1jk-1}^{n+1})=0.\end{aligned} \tag{9.24}$$

So, when we apply this discrete divergence to Equation (9.21), we find, for $i=1$,

j=2, ..., $N-1$, and k=2, ..., $Q-1$:

$$\begin{aligned}
&\Delta y\Delta z\frac{p^{n+1}_{3jk}-p^{n+1}_{1jk}}{2\Delta x}+\Delta y\Delta z\frac{p^{n+1}_{2jk}-p^{n+1}_{1jk}}{2\Delta x}+\Delta z\Delta x\frac{p^{n+1}_{1j+2k}-p^{n+1}_{1jk}}{2\Delta y}\\
&+\Delta z\Delta x\frac{p^{n+1}_{1j-2k}-p^{n+1}_{1jk}}{2\Delta y}+\Delta x\Delta y\frac{p^{n+1}_{1jk+2}-p^{n+1}_{1jk}}{2\Delta z}+\Delta x\Delta y\frac{p^{n+1}_{1jk-2}-p^{n+1}_{1jk}}{2\Delta z}\\
&=\frac{1}{\Delta t}\Big[\Delta y\Delta z(u^{n+\frac12}_{2jk}+u^{n+\frac12}_{1jk})+\Delta z\Delta x(v^{n+\frac12}_{1j+1k}-v^{n+\frac12}_{1j-1k})\\
&\quad+\Delta x\Delta y(w^{n+\frac12}_{1jk+1}-w^{n+\frac12}_{1jk-1})\Big].
\end{aligned}\tag{9.25}$$

The pressure equation can be finally written like, for i=1, ..., M, j=1, ..., N, and k=1, ..., Q:

$$\begin{aligned}
&\Delta y\Delta z\frac{p^{n+1}_{i+2jk}-p^{n+1}_{ijk}}{2\Delta x}+\Delta y\Delta z\frac{p^{n+1}_{i-2jk}-p^{n+1}_{ijk}}{2\Delta x}+\Delta z\Delta x\frac{p^{n+1}_{ij+2k}-p^{n+1}_{ijk}}{2\Delta y}\\
&+\Delta z\Delta x\frac{p^{n+1}_{ij-2k}-p^{n+1}_{ijk}}{2\Delta y}+\Delta x\Delta y\frac{p^{n+1}_{ijk+2}-p^{n+1}_{ijk}}{2\Delta z}+\Delta x\Delta y\frac{p^{n+1}_{ijk-2}-p^{n+1}_{ijk}}{2\Delta z}\\
&=\frac{1}{\Delta t}\Big[\Delta y\Delta z(u^{n+\frac12}_{i+1jk}-u^{n+\frac12}_{i-1jk})+\Delta z\Delta x(v^{n+\frac12}_{ij+1k}-v^{n+\frac12}_{ij-1k})\\
&\quad+\Delta x\Delta y(w^{n+\frac12}_{ijk+1}-w^{n+\frac12}_{ijk-1})\Big],
\end{aligned}\tag{9.26}$$

with the following definitions for the pressure on the fictitious control volumes:

$$\begin{aligned}
&p^{n+1}_{M+1jk}=p^{n+1}_{Mjk},\quad p^{n+1}_{0jk}=p^{n+1}_{1jk},\quad p^{n+1}_{iN+1k}=p^{n+1}_{iNk},\quad p^{n+1}_{i0k}=p^{n+1}_{i1k},\\
&p^{n+1}_{ijQ+1}=p^{n+1}_{ijQ},\quad p^{n+1}_{ij0}=p^{n+1}_{ij1},\quad p^{n+1}_{M+2jk}=p^{n+1}_{M-1jk},\quad p^{n+1}_{-1jk}=p^{n+1}_{2jk},\\
&p^{n+1}_{iN+2k}=p^{n+1}_{iN-1k},\quad p^{n+1}_{i-1k}=p^{n+1}_{i2k},\quad p^{n+1}_{ijQ+2}=p^{n+1}_{ijQ-1},\quad p^{n+1}_{ij-1}=p^{n+1}_{ij2}.
\end{aligned}\tag{9.27}$$

9.3 The main result: stability of the scheme

9.3.1 *Notations*

The mesh considered here is an admissible finite mesh in the sense of [Eymard *et al.* (2000)]. For a function $\boldsymbol{\varphi}\in L^2(\Omega)^3$, we can define its approximate function $\boldsymbol{\varphi}_h$, which is constant over each control volume K_{ijk} of the mesh:

$$\boldsymbol{\varphi}_h(x,y,z)=\boldsymbol{\varphi}_{ijk}\qquad\text{if}\ \ (x,y,z)\in K_{ijk},$$

where $(i,j,k)\in\{(i,j,k),(\frac12,j,k),(M+\frac12,j,k),(i,\frac12,k),(i,N+\frac12,k),(i,j,\frac12),(i,j,Q+\frac12);i{=}1,\dots,M,j{=}1,\dots,N,\text{and }k{=}1,\dots,Q\}$. In the following, we write h for $(\Delta x,\Delta y,\Delta z)$.

We define an approximation $\mathbf{W}_h$ of $\mathbf{W}=H^1_0(\Omega)^3$:

$$\begin{aligned}
\mathbf{W}_h&=\{\boldsymbol{\varphi}_h;\boldsymbol{\varphi}_{\frac12jk}=\boldsymbol{\varphi}_{M+\frac12jk}=\boldsymbol{\varphi}_{i\frac12k}=\boldsymbol{\varphi}_{iN+\frac12k}=\boldsymbol{\varphi}_{ij\frac12}=\boldsymbol{\varphi}_{ijQ+\frac12}=0,\\
&\qquad i{=}1,\dots,M,j{=}1,\dots,N,k{=}1,\dots,Q\}\\
&=\{\boldsymbol{\varphi}_h;\boldsymbol{\varphi}_{0jk}{=}-\boldsymbol{\varphi}_{1jk},\boldsymbol{\varphi}_{M+1jk}{=}-\boldsymbol{\varphi}_{Mjk},\boldsymbol{\varphi}_{i0k}{=}-\boldsymbol{\varphi}_{i1k},\boldsymbol{\varphi}_{iN+1k}{=}-\boldsymbol{\varphi}_{iNk},\\
&\qquad\boldsymbol{\varphi}_{ij0}{=}-\boldsymbol{\varphi}_{ij1},\boldsymbol{\varphi}_{ijQ+1}{=}-\boldsymbol{\varphi}_{ijQ},i{=}1,\dots,M,j{=}1,\dots,N,k{=}1,\dots,Q\}
\end{aligned}\tag{9.28}$$

and an approximation $\mathbf{V}_h \subset \mathbf{W}_h$ of $\mathbf{V}=\{\boldsymbol{\varphi} \in H_0^1(\Omega)^3; \text{div}\, \boldsymbol{\varphi} = 0\}$:

$$\begin{aligned}\mathbf{V}_h =\Big\{\boldsymbol{\varphi}_h \in \mathbf{W}_h; \frac{\Delta y\Delta z}{2}(\boldsymbol{\varphi}_{i+1jk} - \boldsymbol{\varphi}_{i-1jk}) + \frac{\Delta z\Delta x}{2}(\boldsymbol{\varphi}_{ij+1k} - \boldsymbol{\varphi}_{ij-1k})\\ + \frac{\Delta x\Delta y}{2}(\boldsymbol{\varphi}_{ijk+1} - \boldsymbol{\varphi}_{ijk-1})=0\Big\}.\end{aligned} \tag{9.29}$$

Then, for $\boldsymbol{\varphi}_h$ and $\boldsymbol{\psi}_h$ in $\mathbf{W}_h$, we introduce the discrete inner products and norms of the function spaces used:

• Discrete inner product and norm corresponding to the inner product $(\cdot,\cdot)$ and norm $|\cdot|$ of the space $L^2(\Omega)$ are written:

$$(\boldsymbol{\varphi}_h, \boldsymbol{\psi}_h)_h = \sum_{i=1}^{M}\sum_{j=1}^{N}\sum_{k=1}^{Q} \Delta x\Delta y\Delta z \boldsymbol{\varphi}_{ijk} \cdot \boldsymbol{\psi}_{ijk}, \tag{9.30}$$

$$|\boldsymbol{\varphi}_h|_h^2 = (\boldsymbol{\varphi}_h, \boldsymbol{\varphi}_h)_h. \tag{9.31}$$

• Discrete inner product and norm corresponding to the inner product $((\cdot,\cdot))$ and norm $\|\cdot\|$ of the space $\mathbf{H}_0^1(\Omega)$ are written:

$$\begin{aligned}((\boldsymbol{\varphi}_h, \boldsymbol{\psi}_h))_h =&\sum_{j=1}^{N}\sum_{k=1}^{Q}\sum_{i=1}^{M-1} \Delta x\Delta y\Delta z \frac{\boldsymbol{\varphi}_{i+1jk} - \boldsymbol{\varphi}_{ijk}}{\Delta x} \cdot \frac{\boldsymbol{\psi}_{i+1jk} - \boldsymbol{\psi}_{ijk}}{\Delta x}\\
&+\sum_{k=1}^{Q}\sum_{i=1}^{M}\sum_{j=1}^{N-1} \Delta x\Delta y\Delta z \frac{\boldsymbol{\varphi}_{ij+1k} - \boldsymbol{\varphi}_{ijk}}{\Delta y} \cdot \frac{\boldsymbol{\psi}_{ij+1k} - \boldsymbol{\psi}_{ijk}}{\Delta y}\\
&+\sum_{i=1}^{M}\sum_{j=1}^{N}\sum_{k=1}^{Q-1} \Delta x\Delta y\Delta z \frac{\boldsymbol{\varphi}_{ijk+1} - \boldsymbol{\varphi}_{ijk}}{\Delta z} \cdot \frac{\boldsymbol{\psi}_{ijk+1} - \boldsymbol{\psi}_{ijk}}{\Delta z}\\
&+2\sum_{j=1}^{N}\sum_{k=1}^{Q} \Delta x\Delta y\Delta z \Big[\frac{\boldsymbol{\varphi}_{Mjk}}{\Delta x} \cdot \frac{\boldsymbol{\psi}_{Mjk}}{\Delta x} + \frac{\boldsymbol{\varphi}_{1jk}}{\Delta x} \cdot \frac{\boldsymbol{\psi}_{1jk}}{\Delta x}\Big]\\
&+2\sum_{k=1}^{Q}\sum_{i=1}^{M} \Delta x\Delta y\Delta z \Big[\frac{\boldsymbol{\varphi}_{iNk}}{\Delta y} \cdot \frac{\boldsymbol{\psi}_{iNk}}{\Delta y} + \frac{\boldsymbol{\varphi}_{i1k}}{\Delta y} \cdot \frac{\boldsymbol{\psi}_{i1k}}{\Delta y}\Big]\\
&+2\sum_{i=1}^{M}\sum_{j=1}^{N} \Delta x\Delta y\Delta z \Big[\frac{\boldsymbol{\varphi}_{ijQ}}{\Delta z} \cdot \frac{\boldsymbol{\psi}_{ijQ}}{\Delta z} + \frac{\boldsymbol{\varphi}_{ij1}}{\Delta z} \cdot \frac{\boldsymbol{\psi}_{ij1}}{\Delta z}\Big],\end{aligned} \tag{9.32}$$

$$\|\boldsymbol{\varphi}_h\|_h^2 = ((\boldsymbol{\varphi}_h, \boldsymbol{\varphi}_h))_h. \tag{9.33}$$

Finally, the discrete operator $\mathbf{G}_h$ which will be used for the pressure gradient is defined for $i=1,\ldots,M$, $j=1,\ldots,N$, and $k=1,\ldots,Q$ by

$$\mathbf{G}_h p_h(x,y,z) = \begin{pmatrix} \frac{p_{i+1jk}-p_{i-1jk}}{2\Delta x} \\ \frac{p_{ij+1k}-p_{ij-1k}}{2\Delta y} \\ \frac{p_{ijk+1}-p_{ijk-1}}{2\Delta z} \end{pmatrix}, \qquad \text{if } (x,y,z) \in K_{ijk} \tag{9.34}$$

with

$$p_{0jk}=p_{1jk}, \quad p_{M+1jk}=p_{Mjk}, \quad p_{i0k}=p_{i1k},$$

$$p_{iN+1k}=p_{iNk}, \quad p_{ij0}=p_{ij1}, \quad p_{ijQ+1}=p_{ijQ}.$$

9.3.2 *Discrete weak formulation*

First, we introduce the discrete form corresponding to the nonlinear term $\mathbf{b}(\mathbf{u}, \boldsymbol{\varphi}, \boldsymbol{\xi})$:

$$\begin{aligned}\mathbf{b}_h(\mathbf{u}_h, \boldsymbol{\varphi}_h, \boldsymbol{\xi}_h) = \sum_{i=1}^{M}\sum_{j=1}^{N}\sum_{k=1}^{Q} \Big[&\Delta y \Delta z (u_{i+\frac{1}{2}jk}\boldsymbol{\varphi}_{i+\frac{1}{2}jk} \cdot \boldsymbol{\xi}_{ijk} - u_{i-\frac{1}{2}jk}\boldsymbol{\varphi}_{i-\frac{1}{2}jk} \cdot \boldsymbol{\xi}_{ijk}) \\ &+ \Delta z \Delta x (v_{ij+\frac{1}{2}k}\boldsymbol{\varphi}_{ij+\frac{1}{2}k} \cdot \boldsymbol{\xi}_{ijk} - v_{ij-\frac{1}{2}k}\boldsymbol{\varphi}_{ij-\frac{1}{2}k} \cdot \boldsymbol{\xi}_{ijk}) \\ &+ \Delta x \Delta y (w_{ijk+\frac{1}{2}}\boldsymbol{\varphi}_{ijk+\frac{1}{2}} \cdot \boldsymbol{\xi}_{ijk} - w_{ijk-\frac{1}{2}}\boldsymbol{\varphi}_{ijk-\frac{1}{2}} \cdot \boldsymbol{\xi}_{ijk}) \Big].\end{aligned} \tag{9.35}$$

Then, we write the weak formulation.

Lemma 9.1 (Discrete weak formulation). *With the previous notations, Equations (9.12) and (9.21) can be reinterpreted in the following way:*

$$\begin{aligned}\boldsymbol{u}_h^{n+\frac{1}{2}} \in \boldsymbol{W}_h, \quad &\frac{1}{\Delta t}(\boldsymbol{u}_h^{n+\frac{1}{2}} - \boldsymbol{u}_h^n, \boldsymbol{\varphi}_h)_h = -\nu((\frac{\boldsymbol{u}_h^{n+\frac{1}{2}} + \boldsymbol{u}_h^n}{2}, \boldsymbol{\varphi}_h))_h - \frac{3}{2}\boldsymbol{b}_h(\boldsymbol{u}_h^n, \boldsymbol{u}_h^n, \boldsymbol{\varphi}_h) \\ &+ \frac{1}{2}\boldsymbol{b}_h(\boldsymbol{u}_h^{n-1}, \boldsymbol{u}_h^{n-1}, \boldsymbol{\varphi}_h) + (\frac{\boldsymbol{f}^{n+1} + \boldsymbol{f}^n}{2}, \boldsymbol{\varphi}_h)_h, \quad \forall \boldsymbol{\varphi}_h \in \boldsymbol{W}_h,\end{aligned} \tag{9.36}$$

and

$$\boldsymbol{u}_h^{n+1} \in \boldsymbol{V}_h, \quad \frac{1}{\Delta t}(\boldsymbol{u}_h^{n+1} - \boldsymbol{u}_h^{n+\frac{1}{2}}, \boldsymbol{\varphi}_h)_h = -(\boldsymbol{G}_h p_h^{n+1}, \boldsymbol{\varphi}_h)_h, \quad \forall \boldsymbol{\varphi}_h \in \boldsymbol{V}_h. \tag{9.37}$$

Proof. In order to find the weak formulation (9.36), we multiply our finite volume scheme (9.12) by $\boldsymbol{\varphi}_{ijk}$ and sum for i=1, ..., M, j=1, ..., N, and k=1, ..., Q. Regarding each term, we have

- For the time derivative:

$$\Delta x \Delta y \Delta z \sum_{i=1}^{M}\sum_{j=1}^{N}\sum_{k=1}^{Q} \frac{\mathbf{u}_{ijk}^{n+\frac{1}{2}} - \mathbf{u}_{ijk}^n}{\Delta t} \cdot \boldsymbol{\varphi}_{ijk} = \frac{1}{\Delta t}(\mathbf{u}_h^{n+\frac{1}{2}} - \mathbf{u}_h^n, \boldsymbol{\varphi}_h)_h.$$

- For the diffusive term:

$$\begin{aligned}\sum_{i=1}^{M}\sum_{j=1}^{N}\sum_{k=1}^{Q} \Big[&\Delta y \Delta z \frac{\mathbf{u}_{i+1jk} - \mathbf{u}_{ijk}}{\Delta x} + \Delta y \Delta z \frac{\mathbf{u}_{i-1jk} - \mathbf{u}_{ijk}}{\Delta x} \\ &+ \Delta z \Delta x \frac{\mathbf{u}_{ij+1k} - \mathbf{u}_{ijk}}{\Delta y} + \Delta z \Delta x \frac{\mathbf{u}_{ij-1k} - \mathbf{u}_{ijk}}{\Delta y} \\ &+ \Delta x \Delta y \frac{\mathbf{u}_{ijk+1} - \mathbf{u}_{ijk}}{\Delta z} + \Delta x \Delta y \frac{\mathbf{u}_{ijk-1} - \mathbf{u}_{ijk}}{\Delta z} \Big] \cdot \boldsymbol{\varphi}_{ijk} \\ = -\sum_{j=1}^{N}\sum_{k=1}^{Q}\sum_{i=1}^{M-1} &\Delta x \Delta y \Delta z \frac{\mathbf{u}_{i+1jk} - \mathbf{u}_{ijk}}{\Delta x} \cdot \frac{\boldsymbol{\varphi}_{i+1jk} - \boldsymbol{\varphi}_{ijk}}{\Delta x} \\ -\sum_{k=1}^{Q}\sum_{i=1}^{M}\sum_{j=1}^{N-1} &\Delta x \Delta y \Delta z \frac{\mathbf{u}_{ij+1k} - \mathbf{u}_{ijk}}{\Delta y} \cdot \frac{\boldsymbol{\varphi}_{ij+1k} - \boldsymbol{\varphi}_{ijk}}{\Delta y} \\ -\sum_{i=1}^{M}\sum_{j=1}^{N}\sum_{k=1}^{Q-1} &\Delta x \Delta y \Delta z \frac{\mathbf{u}_{ijk+1} - \mathbf{u}_{ijk}}{\Delta z} \cdot \frac{\boldsymbol{\varphi}_{ijk+1} - \boldsymbol{\varphi}_{ijk}}{\Delta z}\end{aligned}$$

$$+\sum_{j=1}^{N}\sum_{k=1}^{Q}\Delta y\Delta z\Big[\frac{\mathbf{u}_{M+1jk}-\mathbf{u}_{Mjk}}{\Delta x}\cdot\boldsymbol{\varphi}_{Mjk}+\frac{\mathbf{u}_{0jk}-\mathbf{u}_{1jk}}{\Delta x}\cdot\boldsymbol{\varphi}_{1jk}\Big]$$

$$+\sum_{k=1}^{Q}\sum_{i=1}^{M}\Delta z\Delta x\Big[\frac{\mathbf{u}_{iN+1k}-\mathbf{u}_{iNk}}{\Delta y}\cdot\boldsymbol{\varphi}_{iNk}+\frac{\mathbf{u}_{i0k}-\mathbf{u}_{i1k}}{\Delta y}\cdot\boldsymbol{\varphi}_{i1k}\Big]$$

$$+\sum_{i=1}^{M}\sum_{j=1}^{N}\Delta x\Delta y\Big[\frac{\mathbf{u}_{ijQ+1}-\mathbf{u}_{ijQ}}{\Delta z}\cdot\boldsymbol{\varphi}_{ijQ}+\frac{\mathbf{u}_{ij0}-\mathbf{u}_{ij1}}{\Delta z}\cdot\boldsymbol{\varphi}_{ij1}\Big]$$

(due to the boundary conditions (9.13))

$$=-\sum_{j=1}^{N}\sum_{k=1}^{Q}\sum_{i=1}^{M-1}\Delta x\Delta y\Delta z\frac{\mathbf{u}_{i+1jk}-\mathbf{u}_{ijk}}{\Delta x}\cdot\frac{\boldsymbol{\varphi}_{i+1jk}-\boldsymbol{\varphi}_{ijk}}{\Delta x}$$

$$-\sum_{k=1}^{Q}\sum_{i=1}^{M}\sum_{j=1}^{N-1}\Delta x\Delta y\Delta z\frac{\mathbf{u}_{ij+1k}-\mathbf{u}_{ijk}}{\Delta y}\cdot\frac{\boldsymbol{\varphi}_{ij+1k}-\boldsymbol{\varphi}_{ijk}}{\Delta y}$$

$$-\sum_{i=1}^{M}\sum_{j=1}^{N}\sum_{k=1}^{Q-1}\Delta x\Delta y\Delta z\frac{\mathbf{u}_{ijk+1}-\mathbf{u}_{ijk}}{\Delta z}\cdot\frac{\boldsymbol{\varphi}_{ijk+1}-\boldsymbol{\varphi}_{ijk}}{\Delta z}$$

$$-\sum_{j=1}^{N}\sum_{k=1}^{Q}\Delta x\Delta y\Delta z\Big[\frac{2\mathbf{u}_{Mjk}}{\Delta x}\cdot\frac{\boldsymbol{\varphi}_{Mjk}}{\Delta x}+\frac{2\mathbf{u}_{1jk}}{\Delta x}\cdot\frac{\boldsymbol{\varphi}_{1jk}}{\Delta x}\Big]$$

$$-\sum_{k=1}^{Q}\sum_{i=1}^{M}\Delta x\Delta y\Delta z\Big[\frac{2\mathbf{u}_{iNk}}{\Delta y}\cdot\frac{\boldsymbol{\varphi}_{iNk}}{\Delta y}+\frac{2\mathbf{u}_{i1k}}{\Delta y}\cdot\frac{\boldsymbol{\varphi}_{i1k}}{\Delta y}\Big]$$

$$-\sum_{i=1}^{M}\sum_{j=1}^{N}\Delta x\Delta y\Delta z\Big[\frac{2\mathbf{u}_{ijQ}}{\Delta z}\cdot\frac{\boldsymbol{\varphi}_{ijQ}}{\Delta z}+\frac{2\mathbf{u}_{ij1}}{\Delta z}\cdot\frac{\boldsymbol{\varphi}_{ij1}}{\Delta z}\Big]$$

$$=-\,((\mathbf{u}_h,\boldsymbol{\varphi}_h))_h.$$

- For the nonlinear term:

$$\sum_{i=1}^{M}\sum_{j=1}^{N}\sum_{k=1}^{Q}\Big[\Delta y\Delta z(u_{i+\frac{1}{2}jk}\mathbf{u}_{i+\frac{1}{2}jk}\cdot\boldsymbol{\varphi}_{ijk}-u_{i-\frac{1}{2}jk}\mathbf{u}_{i-\frac{1}{2}jk}\cdot\boldsymbol{\varphi}_{ijk})$$
$$+\Delta z\Delta x(v_{ij+\frac{1}{2}k}\mathbf{u}_{ij+\frac{1}{2}k}\cdot\boldsymbol{\varphi}_{ijk}-v_{ij-\frac{1}{2}k}\mathbf{u}_{ij-\frac{1}{2}k}\cdot\boldsymbol{\varphi}_{ijk})$$
$$+\Delta x\Delta y(w_{ijk+\frac{1}{2}}\mathbf{u}_{ijk+\frac{1}{2}}\cdot\boldsymbol{\varphi}_{ijk}-w_{ijk-\frac{1}{2}}\mathbf{u}_{ijk-\frac{1}{2}}\cdot\boldsymbol{\varphi}_{ijk})\Big]$$
$$=\mathbf{b}_h(\mathbf{u}_h,\mathbf{u}_h,\boldsymbol{\varphi}_h).$$

- For the right-hand side:

$$\Delta x\Delta y\Delta z\sum_{i=1}^{M}\sum_{j=1}^{N}\sum_{k=1}^{Q}\frac{\mathbf{f}_{ijk}^{n+1}+\mathbf{f}_{ijk}^{n}}{2}\cdot\boldsymbol{\varphi}_{ijk}=\frac{1}{2}(\mathbf{f}^{n+1}+\mathbf{f}^{n},\boldsymbol{\varphi}_h)_h.$$

Finally, we obtain the following weak formulation (9.36).

To find the weak formulation (9.37), we multiply our finite volume scheme (9.21) by $\boldsymbol{\varphi}_{ijk} = ((\varphi_x)_{ijk}, (\varphi_y)_{ijk}, (\varphi_z)_{ijk})$ and sum for i=1, ..., M, j=1, ..., N, and k=1, ..., Q. Regarding each term, we have

- For the left-hand side:

$$\sum_{i=1}^{M}\sum_{j=1}^{N}\sum_{k=1}^{Q} \Delta x \Delta y \Delta z \frac{\mathbf{u}_{ijk}^{n+1} - \mathbf{u}_{ijk}^{n+\frac{1}{2}}}{\Delta t} \cdot \boldsymbol{\varphi}_{ijk} = \frac{1}{\Delta t}(\mathbf{u}_h^{n+1} - \mathbf{u}_h^{n+\frac{1}{2}}, \boldsymbol{\varphi}_h)_h.$$

- For the right-hand side:

$$-\sum_{i=1}^{M}\sum_{j=1}^{N}\sum_{k=1}^{Q} \begin{pmatrix} \frac{\Delta y \Delta z}{2}[p_{i+1jk}^{n+1} - p_{i-1jk}^{n+1}] \\ \frac{\Delta z \Delta x}{2}[p_{ij+1k}^{n+1} - p_{ij-1k}^{n+1}] \\ \frac{\Delta x \Delta y}{2}[p_{ijk+1}^{n+1} - p_{ijk-1}^{n+1}] \end{pmatrix} \cdot \boldsymbol{\varphi}_{ijk}$$

$$= -\sum_{i=1}^{M}\sum_{j=1}^{N}\sum_{k=1}^{Q} \Big[\frac{\Delta y \Delta z}{2}(p_{i+1jk}^{n+1} - p_{i-1jk}^{n+1})(\varphi_x)_{ijk} + \frac{\Delta z \Delta x}{2}(p_{ij+1k}^{n+1} - p_{ij-1k}^{n+1})(\varphi_y)_{ijk}$$

$$+ \frac{\Delta x \Delta y}{2}(p_{ijk+1}^{n+1} - p_{ijk-1}^{n+1})(\varphi_z)_{ijk}\Big]$$

$$= -(\mathbf{G}_h p_h^{n+1}, \boldsymbol{\varphi}_h)_h.$$

So, we have obtained the second weak formulation (9.37). □

9.3.3 *Stability result*

Our aim in this section is to perform the stability result of this scheme. We know from [Eymard *et al.* (2000)] that there exist four constants K_1, K_2, K_3, and $\tilde{K}_3$, which are independent of h and Δt, such that

$$|\mathbf{u}_h^0|_h \leq K_1, \tag{9.38}$$

$$|\mathbf{f}_h^n|_h^2 \leq K_2, \quad \text{for } n{=}1, \ldots, N_t, \tag{9.39}$$

$$|\mathbf{u}_h^1|_h^2 \leq K_3, \tag{9.40}$$

$$|\mathbf{u}_h^{1+\frac{1}{2}}|_h^2 \leq \tilde{K}_3. \tag{9.41}$$

Note that these conditions are easily verified in practice.

Theorem 9.1 (Stability). *We assume that* (9.38)–(9.41) *hold. If*

$$\Delta t \leq \min\left(\frac{1}{16\nu S_h^2}, \frac{\nu}{384 K_4 S_h^3}\right), \tag{9.42}$$

where S_h is the constant of the inverse inequality, then there exist K_4 and K_5 independent of h and Δt, such that,

$$|\boldsymbol{u}_h^n|_h^2 \leq K_4, \quad \text{for } n{=}1, \ldots, N_t, \tag{9.43}$$

$$\sum_{n=0}^{N_t-1} |\boldsymbol{u}_h^{n+1} - \boldsymbol{u}_h^n|_h^2 \leq K_5, \tag{9.44}$$

$$\nu \Delta t \sum_{n=0}^{N_t-1} \|\boldsymbol{u}_h^{n+1}\|_h^2 \leq K_5. \tag{9.45}$$

Proof. As the results of Sections 9.4 and 9.5 are needed to show this stability result, we have postponed the proof of theorem 9.1 to Section 9.6. The proof follows the methods of [Temam and Marion (1998)]. □

9.4 Technical lemmas

This section contains the proofs of lemmas used to show the stability theorem. We start by the Poincaré inequality and an inverse inequality. Then, we prove a usual lemma for the Navier-Stokes equations: the discrete nonlinear term properties. At last, we provide an estimate for the discrete nonlinear term and prove a specific lemma required for the proof of the discrete nonlinear term estimation.

9.4.1 *The Poincaré inequality and an inverse inequality*

Lemma 9.2 (Poincaré inequality). *There exists a constant* $c_1 = diam(\Omega)$, *such that, for every* $\boldsymbol{u}_h \in \boldsymbol{W}_h$:

$$|\boldsymbol{u}_h|_h \leq c_1 \|\boldsymbol{u}_h\|_h.$$

Proof. As these discrete norms are usual in the finite volume context, the reader can find the proof of this Poincaré inequality in, e.g., [Eymard *et al.* (2000)]. Here, due to our discretization, the constant c_1 is equal to $\max(N\Delta x, M\Delta y, Q\Delta z) = \max(L_1, L_2, L_3)$. □

Lemma 9.3 (Inverse inequality). *There exists a constant* S_h *depending on* h, *such that, for every* $\boldsymbol{u}_h \in \boldsymbol{W}_h$:

$$\|\boldsymbol{u}_h\|_h \leq S_h |\boldsymbol{u}_h|_h.$$

Proof. We recall the definition of the discrete norm corresponding to the continuous norm $H_0^1(\Omega)^3$:

$$\begin{aligned}
\|\mathbf{u}_h\|_h^2 = & \sum_{j=1}^{N}\sum_{k=1}^{Q}\sum_{i=1}^{M-1} \Delta x \Delta y \Delta z \left(\frac{\mathbf{u}_{i+1jk} - \mathbf{u}_{ijk}}{\Delta x}\right)^2 \\
& + \sum_{k=1}^{Q}\sum_{i=1}^{M}\sum_{j=1}^{N-1} \Delta x \Delta y \Delta z \left(\frac{\mathbf{u}_{ij+1k} - \mathbf{u}_{ijk}}{\Delta y}\right)^2 \\
& + \sum_{i=1}^{M}\sum_{j=1}^{N}\sum_{k=1}^{Q-1} \Delta x \Delta y \Delta z \left(\frac{\mathbf{u}_{ijk+1} - \mathbf{u}_{ijk}}{\Delta z}\right)^2 \\
& + 2\sum_{j=1}^{N}\sum_{k=1}^{Q} \Delta x \Delta y \Delta z \left[\left(\frac{\mathbf{u}_{Mjk}}{\Delta x}\right)^2 + \left(\frac{\mathbf{u}_{1jk}}{\Delta x}\right)^2\right]
\end{aligned}$$

$$+2\sum_{k=1}^{Q}\sum_{i=1}^{M}\Delta x\Delta y\Delta z\Big[(\frac{\mathbf{u}_{iNk}}{\Delta y})^2+(\frac{\mathbf{u}_{i1k}}{\Delta y})^2\Big]$$
$$+2\sum_{i=1}^{M}\sum_{j=1}^{N}\Delta x\Delta y\Delta z\Big[(\frac{\mathbf{u}_{ijQ}}{\Delta z})^2+(\frac{\mathbf{u}_{ij1}}{\Delta z})^2\Big].$$

Then we use the inequality $(a-b)^2 \le 2a^2+2b^2$ and the boundary conditions (9.13), we can find

$$\begin{aligned}\|\mathbf{u}_h\|_h^2 \le& \frac{2}{\Delta x^2}\sum_{j=1}^{N}\sum_{k=1}^{Q}\sum_{i=1}^{M-1}\Delta x\Delta y\Delta z(\mathbf{u}_{i+1jk}^2+\mathbf{u}_{ijk}^2)\\
&+\frac{2}{\Delta y^2}\sum_{k=1}^{Q}\sum_{i=1}^{M}\sum_{j=1}^{N-1}\Delta x\Delta y\Delta z(\mathbf{u}_{ij+1k}^2+\mathbf{u}_{ijk}^2)\\
&+\frac{2}{\Delta z^2}\sum_{i=1}^{M}\sum_{j=1}^{N}\sum_{k=1}^{Q-1}\Delta x\Delta y\Delta z(\mathbf{u}_{ijk+1}^2+\mathbf{u}_{ijk}^2)\\
&+\frac{2}{\Delta x^2}\sum_{j=1}^{N}\sum_{k=1}^{Q}\Delta x\Delta y\Delta z(\mathbf{u}_{Mjk}^2+\mathbf{u}_{1jk}^2)\\
&+\frac{2}{\Delta y^2}\sum_{k=1}^{Q}\sum_{i=1}^{M}\Delta x\Delta y\Delta z(\mathbf{u}_{iNk}^2+\mathbf{u}_{i1k}^2)\\
&+\frac{2}{\Delta z^2}\sum_{i=1}^{M}\sum_{j=1}^{N}\Delta x\Delta y\Delta z(\mathbf{u}_{ijQ}^2+\mathbf{u}_{ij1}^2)\\
=&4(\frac{1}{\Delta x^2}+\frac{1}{\Delta y^2}+\frac{1}{\Delta z^2})|\mathbf{u}_h|_h^2.\end{aligned}$$

So, the inequality is proved with

$$S_h = 2\sqrt{\frac{1}{\Delta x^2}+\frac{1}{\Delta y^2}+\frac{1}{\Delta z^2}}.$$

□

9.4.2 *Standard lemma*

Lemma 9.4 (Properties of the discrete nonlinear term). *For all approximate functions $\boldsymbol{u}_h \in \boldsymbol{V}_h$ and $\boldsymbol{\varphi}_h$, $\boldsymbol{\xi}_h \in \boldsymbol{W}_h$, the discrete nonlinear term satisfies the properties:*

$$\boldsymbol{b}_h(\boldsymbol{u}_h,\boldsymbol{\varphi}_h,\boldsymbol{\varphi}_h)=0, \tag{9.46}$$

$$\boldsymbol{b}_h(\boldsymbol{u}_h,\boldsymbol{\varphi}_h,\boldsymbol{\xi}_h)=-\boldsymbol{b}_h(\boldsymbol{u}_h,\boldsymbol{\xi}_h,\boldsymbol{\varphi}_h). \tag{9.47}$$

Proof. We recall the definition of the discrete nonlinear term:

$$\begin{aligned}\mathbf{b}_h(\mathbf{u}_h,\boldsymbol{\varphi}_h,\boldsymbol{\xi}_h) =& \sum_{i=1}^{M}\sum_{j=1}^{N}\sum_{k=1}^{Q}\Big[\Delta y\Delta z(u_{i+\frac{1}{2}jk}\boldsymbol{\varphi}_{i+\frac{1}{2}jk}\cdot\boldsymbol{\xi}_{ijk} - u_{i-\frac{1}{2}jk}\boldsymbol{\varphi}_{i-\frac{1}{2}jk}\cdot\boldsymbol{\xi}_{ijk})\\
&+\Delta z\Delta x(v_{ij+\frac{1}{2}k}\boldsymbol{\varphi}_{ij+\frac{1}{2}k}\cdot\boldsymbol{\xi}_{ijk} - v_{ij-\frac{1}{2}k}\boldsymbol{\varphi}_{ij-\frac{1}{2}k}\cdot\boldsymbol{\xi}_{ijk})\\
&+\Delta x\Delta y(w_{ijk+\frac{1}{2}}\boldsymbol{\varphi}_{ijk+\frac{1}{2}}\cdot\boldsymbol{\xi}_{ijk} - w_{ijk-\frac{1}{2}}\boldsymbol{\varphi}_{ijk-\frac{1}{2}}\cdot\boldsymbol{\xi}_{ijk})\Big].\end{aligned}$$

If we set $\boldsymbol{\xi}_h = \boldsymbol{\varphi}_h$, then

$$\begin{aligned}\mathbf{b}_h(\mathbf{u}_h,\boldsymbol{\varphi}_h,\boldsymbol{\varphi}_h) =& \sum_{i=1}^{M}\sum_{j=1}^{N}\sum_{k=1}^{Q}\Big[\Delta y\Delta z(u_{i+\frac{1}{2}jk}\boldsymbol{\varphi}_{i+\frac{1}{2}jk}\cdot\boldsymbol{\varphi}_{ijk} - u_{i-\frac{1}{2}jk}\boldsymbol{\varphi}_{i-\frac{1}{2}jk}\cdot\boldsymbol{\varphi}_{ijk})\\
&+\Delta z\Delta x(v_{ij+\frac{1}{2}k}\boldsymbol{\varphi}_{ij+\frac{1}{2}k}\cdot\boldsymbol{\varphi}_{ijk} - v_{ij-\frac{1}{2}k}\boldsymbol{\varphi}_{ij-\frac{1}{2}k}\cdot\boldsymbol{\varphi}_{ijk})\\
&+\Delta x\Delta y(w_{ijk+\frac{1}{2}}\boldsymbol{\varphi}_{ijk+\frac{1}{2}}\cdot\boldsymbol{\varphi}_{ijk} - w_{ijk-\frac{1}{2}}\boldsymbol{\varphi}_{ijk-\frac{1}{2}}\cdot\boldsymbol{\varphi}_{ijk})\Big].\end{aligned}$$

The fluxes being linearly interpolated, it follows:

$$\begin{aligned}&\mathbf{b}_h(\mathbf{u}_h,\boldsymbol{\varphi}_h,\boldsymbol{\varphi}_h)\\
&=\sum_{i=1}^{M}\sum_{j=1}^{N}\sum_{k=1}^{Q}\Big[\Delta y\Delta z(u_{i+\frac{1}{2}jk}\frac{\boldsymbol{\varphi}_{i+1jk}+\boldsymbol{\varphi}_{ijk}}{2}\cdot\boldsymbol{\varphi}_{ijk} - u_{i-\frac{1}{2}jk}\frac{\boldsymbol{\varphi}_{i-1jk}+\boldsymbol{\varphi}_{ijk}}{2}\cdot\boldsymbol{\varphi}_{ijk})\\
&+\Delta z\Delta x(v_{ij+\frac{1}{2}k}\frac{\boldsymbol{\varphi}_{ij+1k}+\boldsymbol{\varphi}_{ijk}}{2}\cdot\boldsymbol{\varphi}_{ijk} - v_{ij-\frac{1}{2}k}\frac{\boldsymbol{\varphi}_{ij-1k}+\boldsymbol{\varphi}_{ijk}}{2}\cdot\boldsymbol{\varphi}_{ijk})\\
&+\Delta x\Delta y(w_{ijk+\frac{1}{2}}\frac{\boldsymbol{\varphi}_{ijk+1}+\boldsymbol{\varphi}_{ijk}}{2}\cdot\boldsymbol{\varphi}_{ijk} - w_{ijk-\frac{1}{2}}\frac{\boldsymbol{\varphi}_{ijk-1}+\boldsymbol{\varphi}_{ijk}}{2}\cdot\boldsymbol{\varphi}_{ijk})\Big].\end{aligned}$$

which can be rewritten as

$$\begin{aligned}&\mathbf{b}_h(\mathbf{u}_h,\boldsymbol{\varphi}_h,\boldsymbol{\varphi}_h)\\
&=\sum_{j=1}^{N}\sum_{k=1}^{Q}\Big[\sum_{i=1}^{M-1}\Delta y\Delta z u_{i+\frac{1}{2}jk}\frac{\boldsymbol{\varphi}_{i+1jk}+\boldsymbol{\varphi}_{ijk}}{2}\cdot\boldsymbol{\varphi}_{ijk}\\
&-\sum_{i=2}^{M}\Delta y\Delta z u_{i-\frac{1}{2}jk}\frac{\boldsymbol{\varphi}_{i-1jk}+\boldsymbol{\varphi}_{ijk}}{2}\cdot\boldsymbol{\varphi}_{ijk}\Big]\\
&+\sum_{k=1}^{Q}\sum_{i=1}^{M}\Big[\sum_{j=1}^{N-1}\Delta z\Delta x v_{ij+\frac{1}{2}k}\frac{\boldsymbol{\varphi}_{ij+1k}+\boldsymbol{\varphi}_{ijk}}{2}\cdot\boldsymbol{\varphi}_{ijk}\\
&-\sum_{j=2}^{N}\Delta z\Delta x v_{ij-\frac{1}{2}k}\frac{\boldsymbol{\varphi}_{ij-1k}+\boldsymbol{\varphi}_{ijk}}{2}\cdot\boldsymbol{\varphi}_{ijk}\Big]\\
&+\sum_{i=1}^{M}\sum_{j=1}^{N}\Big[\sum_{k=1}^{Q-1}\Delta x\Delta y w_{ijk+\frac{1}{2}}\frac{\boldsymbol{\varphi}_{ijk+1}+\boldsymbol{\varphi}_{ijk}}{2}\cdot\boldsymbol{\varphi}_{ijk}\\
&-\sum_{k=2}^{Q}\Delta x\Delta y w_{ijk-\frac{1}{2}}\frac{\boldsymbol{\varphi}_{ijk-1}+\boldsymbol{\varphi}_{ijk}}{2}\cdot\boldsymbol{\varphi}_{ijk}\Big].\end{aligned}$$

which can be also rewritten as

$$
\begin{aligned}
&\mathbf{b}_h(\mathbf{u}_h, \boldsymbol{\varphi}_h, \boldsymbol{\varphi}_h)\\
=&\sum_{j=1}^{N}\sum_{k=1}^{Q}\Big[\sum_{i=1}^{M}\frac{\Delta y\Delta z}{2}(u_{i+\frac12 jk}-u_{i-\frac12 jk})\boldsymbol{\varphi}_{ijk}^2+\sum_{i=1}^{M-1}\frac{\Delta y\Delta z}{2}u_{i+\frac12 jk}\boldsymbol{\varphi}_{i+1jk}\cdot\boldsymbol{\varphi}_{ijk}\\
&-\sum_{i=2}^{M}\frac{\Delta y\Delta z}{2}u_{i-\frac12 jk}\boldsymbol{\varphi}_{i-1jk}\cdot\boldsymbol{\varphi}_{ijk}\Big]\\
&+\sum_{k=1}^{Q}\sum_{i=1}^{M}\Big[\sum_{j=1}^{N}\frac{\Delta z\Delta x}{2}(v_{ij+\frac12 k}-v_{ij-\frac12 k})\boldsymbol{\varphi}_{ijk}^2+\sum_{j=1}^{N-1}\frac{\Delta z\Delta x}{2}v_{ij+\frac12 k}\boldsymbol{\varphi}_{ij+1k}\cdot\boldsymbol{\varphi}_{ijk}\\
&-\sum_{j=2}^{N}\frac{\Delta z\Delta x}{2}v_{ij-\frac12 k}\boldsymbol{\varphi}_{ij-1k}\cdot\boldsymbol{\varphi}_{ijk}\Big]\\
&+\sum_{i=1}^{M}\sum_{j=1}^{N}\Big[\sum_{k=1}^{Q}\frac{\Delta x\Delta y}{2}(w_{ijk+\frac12}-w_{ijk-\frac12})\boldsymbol{\varphi}_{ijk}^2+\sum_{k=1}^{Q-1}\frac{\Delta x\Delta y}{2}w_{ijk+\frac12}\boldsymbol{\varphi}_{ijk+1}\cdot\boldsymbol{\varphi}_{ijk}\\
&-\sum_{k=2}^{Q}\frac{\Delta x\Delta y}{2}w_{ijk-\frac12}\boldsymbol{\varphi}_{ijk-1}\cdot\boldsymbol{\varphi}_{ijk}\Big].
\end{aligned}
$$

In each square brackets, the two last terms cancel out and we obtain

$$
\begin{aligned}
&\mathbf{b}_h(\mathbf{u}_h, \boldsymbol{\varphi}_h, \boldsymbol{\varphi}_h)\\
=&\frac12\sum_{i=1}^{M}\sum_{j=1}^{N}\sum_{k=1}^{Q}\Big[\Delta y\Delta z(u_{i+\frac12 jk}-u_{i-\frac12 jk})+\Delta z\Delta x(v_{ij+\frac12 k}-v_{ij-\frac12 k})\\
&+\Delta x\Delta y(w_{ijk+\frac12}-w_{ijk-\frac12})\Big]\boldsymbol{\varphi}_{ijk}^2\\
=&\frac14\sum_{i=1}^{M}\sum_{j=1}^{N}\sum_{k=1}^{Q}\Big[\Delta y\Delta z(u_{i+1jk}-u_{i-1jk})+\Delta z\Delta x(v_{ij+1k}-v_{ij-1k})\\
&+\Delta x\Delta y(w_{ijk+1}-w_{ijk-1})\Big]\boldsymbol{\varphi}_{ijk}^2.
\end{aligned}
$$

Finally, according to the incompressibility condition (9.15), the right-hand side is equal to zero, so

$$\mathbf{b}_h(\mathbf{u}_h, \boldsymbol{\varphi}_h, \boldsymbol{\varphi}_h) = 0.$$

Moreover, considering that $\mathbf{b}_h(\mathbf{u}_h, \boldsymbol{\varphi}_h + \boldsymbol{\xi}_h, \boldsymbol{\varphi}_h + \boldsymbol{\xi}_h)$, we see that (9.46) classically implies (9.47). □

9.4.3 *Specific lemmas for the Navier-Stokes equations*

Lemma 9.5 (Estimates on the discrete nonlinear term). *For all approximate functions* $\boldsymbol{u}_h=(u_h, v_h, w_h) \in \boldsymbol{V}_h$, $\boldsymbol{\varphi}_h=((\varphi_x)_h, (\varphi_y)_h, (\varphi_z)_h) \in \boldsymbol{W}_h$ *and* $\boldsymbol{\xi}_h=$

$((\xi_x)_h, (\xi_y)_h, (\xi_z)_h) \in \boldsymbol{W}_h,$

$$|\boldsymbol{b}_h(\boldsymbol{u}_h, \boldsymbol{\varphi}_h, \boldsymbol{\xi}_h)| \leq 2|\boldsymbol{u}_h|_h^{\frac{1}{4}} \|\boldsymbol{u}_h\|_h^{\frac{3}{4}} |\boldsymbol{\varphi}_h|_h^{\frac{1}{4}} \|\boldsymbol{\varphi}_h\|_h^{\frac{3}{4}} \|\boldsymbol{\xi}_h\|_h,$$
$$|\boldsymbol{b}_h(\boldsymbol{u}_h, \boldsymbol{\varphi}_h, \boldsymbol{\xi}_h)| \leq 2|\boldsymbol{u}_h|_h^{\frac{1}{4}} \|\boldsymbol{u}_h\|_h^{\frac{3}{4}} |\boldsymbol{\xi}_h|_h^{\frac{1}{4}} \|\boldsymbol{\xi}_h\|_h^{\frac{3}{4}} \|\boldsymbol{\varphi}_h\|_h.$$

Remark 9.1. (Notation). In the proof of this lemma, we will use the same notation for the discrete norms of $L^2(\Omega)$ and $L^2(\Omega)^3$. So, we will write for $\mathbf{u}_h=(u_h, v_h, w_h)$, $|u_h|_h$ for the discrete $L^2(\Omega)$-norm of u_h and $|\mathbf{u}_h|_h$ for the discrete $L^2(\Omega)^3$-norm of $\mathbf{u}_h$.

Proof. We want to bound properly the discrete nonlinear term:

$$\begin{aligned}|\mathbf{b}_h(\mathbf{u}_h, \boldsymbol{\varphi}_h, \boldsymbol{\xi}_h)| = |\sum_{i=1}^{M}\sum_{j=1}^{N}\sum_{k=1}^{Q} \Big[&\Delta y \Delta z (u_{i+\frac{1}{2}jk} \boldsymbol{\varphi}_{i+\frac{1}{2}jk} \cdot \boldsymbol{\xi}_{ijk} - u_{i-\frac{1}{2}jk} \boldsymbol{\varphi}_{i-\frac{1}{2}jk} \cdot \boldsymbol{\xi}_{ijk}) \\ &+ \Delta z \Delta x (v_{ij+\frac{1}{2}k} \boldsymbol{\varphi}_{ij+\frac{1}{2}k} \cdot \boldsymbol{\xi}_{ijk} - v_{ij-\frac{1}{2}k} \boldsymbol{\varphi}_{ij-\frac{1}{2}k} \cdot \boldsymbol{\xi}_{ijk}) \\ &+ \Delta x \Delta y (w_{ijk+\frac{1}{2}} \boldsymbol{\varphi}_{ijk+\frac{1}{2}} \cdot \boldsymbol{\xi}_{ijk} - w_{ijk-\frac{1}{2}} \boldsymbol{\varphi}_{ijk-\frac{1}{2}} \cdot \boldsymbol{\xi}_{ijk}) \Big] | .\end{aligned}$$

We simplify it using the Dirichlet boundary conditions $\mathbf{u}_{\frac{1}{2}jk}^{n+\frac{1}{2}} = \mathbf{u}_{M+\frac{1}{2}jk}^{n+\frac{1}{2}} = \mathbf{u}_{i\frac{1}{2}k}^{n+\frac{1}{2}} = \mathbf{u}_{iN+\frac{1}{2}k}^{n+\frac{1}{2}} = \mathbf{u}_{ij\frac{1}{2}}^{n+\frac{1}{2}} = \mathbf{u}_{ijQ+\frac{1}{2}}^{n+\frac{1}{2}} = 0$:

$$\begin{aligned}&|\ \mathbf{b}_h(\mathbf{u}_h, \boldsymbol{\varphi}_h, \boldsymbol{\xi}_h)\ | \\ &= |\sum_{j=1}^{N}\sum_{k=1}^{Q}(\sum_{i=1}^{M-1} \Delta y \Delta z u_{i+\frac{1}{2}jk} \boldsymbol{\varphi}_{i+\frac{1}{2}jk} \cdot \boldsymbol{\xi}_{ijk} - \sum_{i=2}^{M} \Delta y \Delta z u_{i-\frac{1}{2}jk} \boldsymbol{\varphi}_{i-\frac{1}{2}jk} \cdot \boldsymbol{\xi}_{ijk}) \\ &\quad + \sum_{k=1}^{Q}\sum_{i=1}^{M}(\sum_{j=1}^{N-1} \Delta z \Delta x v_{ij+\frac{1}{2}k} \boldsymbol{\varphi}_{ij+\frac{1}{2}k} \cdot \boldsymbol{\xi}_{ijk} - \sum_{j=2}^{N} \Delta z \Delta x v_{ij-\frac{1}{2}k} \boldsymbol{\varphi}_{ij-\frac{1}{2}k} \cdot \boldsymbol{\xi}_{ijk}) \\ &\quad + \sum_{i=1}^{M}\sum_{j=1}^{N}(\sum_{k=1}^{Q-1} \Delta x \Delta y w_{ijk+\frac{1}{2}} \boldsymbol{\varphi}_{ijk+\frac{1}{2}} \cdot \boldsymbol{\xi}_{ijk} - \sum_{k=2}^{Q} \Delta x \Delta y w_{ijk-\frac{1}{2}} \boldsymbol{\varphi}_{ijk-\frac{1}{2}} \cdot \boldsymbol{\xi}_{ijk})\ |,\end{aligned}$$

and we arrange this expression and write:

$$\begin{aligned}&|\mathbf{b}_h(\mathbf{u}_h, \boldsymbol{\varphi}_h, \boldsymbol{\xi}_h)| \\ &= | \underbrace{\sum_{j=1}^{N}\sum_{k=1}^{Q}\sum_{i=1}^{M-1} \Delta x \Delta y \Delta z u_{i+\frac{1}{2}jk} \boldsymbol{\varphi}_{i+\frac{1}{2}jk} \cdot \frac{\boldsymbol{\xi}_{ijk} - \boldsymbol{\xi}_{i+1jk}}{\Delta x}}_{I}\end{aligned}$$

$$+\underbrace{\sum_{k=1}^{Q}\sum_{i=1}^{M}\sum_{j=1}^{N-1}\Delta x\Delta y\Delta z v_{ij+\frac{1}{2}k}\boldsymbol{\varphi}_{ij+\frac{1}{2}k}\cdot\frac{\boldsymbol{\xi}_{ijk}-\boldsymbol{\xi}_{ij+1k}}{\Delta y}}_{II}$$

$$+\underbrace{\sum_{i=1}^{M}\sum_{j=1}^{N}\sum_{k=1}^{Q-1}\Delta x\Delta y\Delta z w_{ijk+\frac{1}{2}}\boldsymbol{\varphi}_{ijk+\frac{1}{2}}\cdot\frac{\boldsymbol{\xi}_{ijk}-\boldsymbol{\xi}_{ijk+1}}{\Delta z}}_{III}\,|\,.$$

Then, the method used to bound these three terms I, II and III is very similar, we explain how to bound the first term I, then we will briefly do the same for the second term II and the third term III. For I, we have

$$|I| = |\sum_{j=1}^{N}\sum_{k=1}^{Q}\sum_{i=1}^{M-1}\Delta x\Delta y\Delta z u_{i+\frac{1}{2}jk}\Big[(\varphi_x)_{i+\frac{1}{2}jk}\cdot\frac{(\xi_x)_{ijk}-(\xi_x)_{i+1jk}}{\Delta x}$$
$$+(\varphi_y)_{i+\frac{1}{2}jk}\cdot\frac{(\xi_y)_{ijk}-(\xi_y)_{i+1jk}}{\Delta x}+(\varphi_z)_{i+\frac{1}{2}jk}\cdot\frac{(\xi_z)_{ijk}-(\xi_z)_{i+1jk}}{\Delta x}\Big]\,|,$$

which can be bounded as

$$|I| \le |\sum_{j=1}^{N}\sum_{k=1}^{Q}\sum_{i=1}^{M-1}\Delta x\Delta y\Delta z u_{i+\frac{1}{2}jk}(\varphi_x)_{i+\frac{1}{2}jk}\cdot\frac{(\xi_x)_{ijk}-(\xi_x)_{i+1jk}}{\Delta x}\,|$$
$$+|\sum_{j=1}^{N}\sum_{k=1}^{Q}\sum_{i=1}^{M-1}\Delta x\Delta y\Delta z u_{i+\frac{1}{2}jk}(\varphi_y)_{i+\frac{1}{2}jk}\cdot\frac{(\xi_y)_{ijk}-(\xi_y)_{i+1jk}}{\Delta x}\,|$$
$$+|\sum_{j=1}^{N}\sum_{k=1}^{Q}\sum_{i=1}^{M-1}\Delta x\Delta y\Delta z u_{i+\frac{1}{2}jk}(\varphi_z)_{i+\frac{1}{2}jk}\cdot\frac{(\xi_z)_{ijk}-(\xi_z)_{i+1jk}}{\Delta x}\,|\,.$$

Using the Cauchy-Schwarz inequality, we obtain

$$|I| \le \Big[\sum_{j=1}^{N}\sum_{k=1}^{Q}\sum_{i=1}^{M-1}\Delta x\Delta y\Delta z u^2_{i+\frac{1}{2}jk}(\varphi_x)^2_{i+\frac{1}{2}jk}\Big]^{\frac{1}{2}}$$
$$\times\Big[\sum_{j=1}^{N}\sum_{k=1}^{Q}\sum_{i=1}^{M-1}\Delta x\Delta y\Delta z(\frac{(\xi_x)_{i+1jk}-(\xi_x)_{ijk}}{\Delta x})^2\Big]^{\frac{1}{2}}$$
$$+\Big[\sum_{j=1}^{N}\sum_{k=1}^{Q}\sum_{i=1}^{M-1}\Delta x\Delta y\Delta z u^2_{i+\frac{1}{2}jk}(\varphi_y)^2_{i+\frac{1}{2}jk}\Big]^{\frac{1}{2}}$$
$$\times\Big[\sum_{j=1}^{N}\sum_{k=1}^{Q}\sum_{i=1}^{M-1}\Delta x\Delta y\Delta z(\frac{(\xi_y)_{i+1jk}-(\xi_y)_{ijk}}{\Delta x})^2\Big]^{\frac{1}{2}}$$
$$+\Big[\sum_{j=1}^{N}\sum_{k=1}^{Q}\sum_{i=1}^{M-1}\Delta x\Delta y\Delta z u^2_{i+\frac{1}{2}jk}(\varphi_z)^2_{i+\frac{1}{2}jk}\Big]^{\frac{1}{2}}$$

$$\times\Big[\sum_{j=1}^{N}\sum_{k=1}^{Q}\sum_{i=1}^{M-1}\Delta x\Delta y\Delta z(\frac{(\xi_z)_{i+1jk}-(\xi_z)_{ijk}}{\Delta x})^2\Big]^{\frac{1}{2}}.$$

Using again the Cauchy-Schwarz inequality, we find

$$\begin{aligned}|I|\le\ &\Big[\sum_{j=1}^{N}\sum_{k=1}^{Q}\sum_{i=1}^{M-1}\Delta x\Delta y\Delta z u^4_{i+\frac{1}{2}jk}\Big]^{\frac{1}{4}}\\&\times\Big\{\Big[\sum_{j=1}^{N}\sum_{k=1}^{Q}\sum_{i=1}^{M-1}\Delta x\Delta y\Delta z(\varphi_x)^4_{i+\frac{1}{2}jk}\Big]^{\frac{1}{4}}\\&\quad\times\Big[\sum_{j=1}^{N}\sum_{k=1}^{Q}\sum_{i=1}^{M-1}\Delta x\Delta y\Delta z(\frac{(\xi_x)_{i+1jk}-(\xi_x)_{ijk}}{\Delta x})^2\Big]^{\frac{1}{2}}\\&+\Big[\sum_{j=1}^{N}\sum_{k=1}^{Q}\sum_{i=1}^{M-1}\Delta x\Delta y\Delta z(\varphi_y)^4_{i+\frac{1}{2}jk}\Big]^{\frac{1}{4}}\\&\quad\times\Big[\sum_{j=1}^{N}\sum_{k=1}^{Q}\sum_{i=1}^{M-1}\Delta x\Delta y\Delta z(\frac{(\xi_y)_{i+1jk}-(\xi_y)_{ijk}}{\Delta x})^2\Big]^{\frac{1}{2}}\\&+\Big[\sum_{j=1}^{N}\sum_{k=1}^{Q}\sum_{i=1}^{M-1}\Delta x\Delta y\Delta z(\varphi_z)^4_{i+\frac{1}{2}jk}\Big]^{\frac{1}{4}}\\&\quad\times\Big[\sum_{j=1}^{N}\sum_{k=1}^{Q}\sum_{i=1}^{M-1}\Delta x\Delta y\Delta z(\frac{(\xi_z)_{i+1jk}-(\xi_z)_{ijk}}{\Delta x})^2\Big]^{\frac{1}{2}}\Big\}.\end{aligned}$$

Moreover, due to the inequality $(\frac{a+b}{2})^4\le\frac{a^4+b^4}{2}$ and the definition of the fluxes by linear interpolation, we obtain the following upper bound:

$$\begin{aligned}&\Big[\sum_{j=1}^{N}\sum_{k=1}^{Q}\sum_{i=1}^{M-1}\Delta x\Delta y\Delta z u^4_{i+\frac{1}{2}jk}\Big]^{\frac{1}{4}}\\&=\Big[\sum_{j=1}^{N}\sum_{k=1}^{Q}\sum_{i=1}^{M-1}\Delta x\Delta y\Delta z(\frac{u_{i+1jk}+u_{ijk}}{2})^4\Big]^{\frac{1}{4}}\\&\le\Big[\sum_{j=1}^{N}\sum_{k=1}^{Q}\sum_{i=1}^{M-1}\Delta x\Delta y\Delta z(\frac{u^4_{i+1jk}+u^4_{ijk}}{2})\Big]^{\frac{1}{4}}\\&=(\sum_{j=1}^{N}\sum_{k=1}^{Q}\sum_{i=2}^{M}\Delta x\Delta y\Delta z\frac{u^4_{ijk}}{2}+\sum_{j=1}^{N}\sum_{k=1}^{Q}\sum_{i=1}^{M-1}\Delta x\Delta y\Delta z\frac{u^4_{ijk}}{2})^{\frac{1}{4}}\\&\le(\sum_{j=1}^{N}\sum_{k=1}^{Q}\sum_{i=1}^{M}\Delta x\Delta y\Delta z u^4_{ijk})^{\frac{1}{4}}\\&=|u_h^2|_h^{\frac{1}{2}}.\end{aligned}$$

So, from this inequality, we can majorize the first term as follows:

$$
\begin{aligned}
|I| \leq & |u_h^2|_h^{\frac{1}{2}} \Bigg\{ |(\varphi_x)_h^2|_h^{\frac{1}{2}} \Big[\sum_{j=1}^{N} \sum_{k=1}^{Q} \sum_{i=1}^{M-1} \Delta x \Delta y \Delta z (\frac{(\xi_x)_{i+1jk} - (\xi_x)_{ijk}}{\Delta x})^2 \Big]^{\frac{1}{2}} \\
& + |(\varphi_y)_h^2|_h^{\frac{1}{2}} \Big[\sum_{j=1}^{N} \sum_{k=1}^{Q} \sum_{i=1}^{M-1} \Delta x \Delta y \Delta z (\frac{(\xi_y)_{i+1jk} - (\xi_y)_{ijk}}{\Delta x})^2 \Big]^{\frac{1}{2}} \\
& + |(\varphi_z)_h^2|_h^{\frac{1}{2}} \Big[\sum_{j=1}^{N} \sum_{k=1}^{Q} \sum_{i=1}^{M-1} \Delta x \Delta y \Delta z (\frac{(\xi_z)_{i+1jk} - (\xi_z)_{ijk}}{\Delta x})^2 \Big]^{\frac{1}{2}} \Bigg\}.
\end{aligned}
$$

Proceeding similarly for the second term II and the third term III, we find

$$
\begin{aligned}
|II| \leq & |v_h^2|_h^{\frac{1}{2}} \Bigg\{ |(\varphi_x)_h^2|_h^{\frac{1}{2}} \Big[\sum_{k=1}^{Q} \sum_{i=1}^{M} \sum_{j=1}^{N-1} \Delta x \Delta y \Delta z (\frac{(\xi_x)_{ij+1k} - (\xi_x)_{ijk}}{\Delta y})^2 \Big]^{\frac{1}{2}} \\
& + |(\varphi_y)_h^2|_h^{\frac{1}{2}} \Big[\sum_{k=1}^{Q} \sum_{i=1}^{M} \sum_{j=1}^{N-1} \Delta x \Delta y \Delta z (\frac{(\xi_y)_{ij+1k} - (\xi_y)_{ijk}}{\Delta y})^2 \Big]^{\frac{1}{2}} \\
& + |(\varphi_z)_h^2|_h^{\frac{1}{2}} \Big[\sum_{k=1}^{Q} \sum_{i=1}^{M} \sum_{j=1}^{N-1} \Delta x \Delta y \Delta z (\frac{(\xi_z)_{ij+1k} - (\xi_z)_{ijk}}{\Delta y})^2 \Big]^{\frac{1}{2}} \Bigg\}.
\end{aligned}
$$

$$
\begin{aligned}
|III| \leq & |w_h^2|_h^{\frac{1}{2}} \Bigg\{ |(\varphi_x)_h^2|_h^{\frac{1}{2}} \Big[\sum_{i=1}^{M} \sum_{j=1}^{N} \sum_{k=1}^{Q-1} \Delta x \Delta y \Delta z (\frac{(\xi_x)_{ijk+1} - (\xi_x)_{ijk}}{\Delta z})^2 \Big]^{\frac{1}{2}} \\
& + |(\varphi_y)_h^2|_h^{\frac{1}{2}} \Big[\sum_{i=1}^{M} \sum_{j=1}^{N} \sum_{k=1}^{Q-1} \Delta x \Delta y \Delta z (\frac{(\xi_y)_{ijk+1} - (\xi_y)_{ijk}}{\Delta z})^2 \Big]^{\frac{1}{2}} \\
& + |(\varphi_z)_h^2|_h^{\frac{1}{2}} \Big[\sum_{i=1}^{M} \sum_{j=1}^{N} \sum_{k=1}^{Q-1} \Delta x \Delta y \Delta z (\frac{(\xi_z)_{ijk+1} - (\xi_z)_{ijk}}{\Delta z})^2 \Big]^{\frac{1}{2}} \Bigg\}.
\end{aligned}
$$

Due to the Cauchy-Schwarz inequality:

$$
\begin{aligned}
& | \mathbf{b}_h(\mathbf{u}_h, \boldsymbol{\varphi}_h, \boldsymbol{\xi}_h) | \\
\leq & \ \big\{ |u_h^2|_h + |v_h^2|_h + |w_h^2|_h \big\}^{\frac{1}{2}} \big\{ |(\varphi_x)_h^2|_h + |(\varphi_y)_h^2|_h + |(\varphi_z)_h^2|_h \big\}^{\frac{1}{2}} \\
& \Bigg\{ \sum_{j=1}^{N} \sum_{k=1}^{Q} \sum_{i=1}^{M-1} \Delta x \Delta y \Delta z \Big[(\frac{(\xi_x)_{i+1jk} - (\xi_x)_{ijk}}{\Delta x})^2 + (\frac{(\xi_y)_{i+1jk} - (\xi_y)_{ijk}}{\Delta x})^2 \\
& + (\frac{(\xi_z)_{i+1jk} - (\xi_z)_{ijk}}{\Delta x})^2 \Big] + \sum_{k=1}^{Q} \sum_{i=1}^{M} \sum_{j=1}^{N-1} \Delta x \Delta y \Delta z \Big[(\frac{(\xi_x)_{ij+1k} - (\xi_x)_{ijk}}{\Delta y})^2 \\
& + (\frac{(\xi_y)_{ij+1k} - (\xi_y)_{ijk}}{\Delta y})^2 + (\frac{(\xi_z)_{ij+1k} - (\xi_z)_{ijk}}{\Delta y})^2 \Big] \\
& + \sum_{i=1}^{M} \sum_{j=1}^{N} \sum_{k=1}^{Q-1} \Delta x \Delta y \Delta z \Big[(\frac{(\xi_x)_{ijk+1} - (\xi_x)_{ijk}}{\Delta z})^2 + (\frac{(\xi_y)_{ijk+1} - (\xi_y)_{ijk}}{\Delta z})^2
\end{aligned}
$$

$$+(\frac{(\xi_z)_{ijk+1}-(\xi_z)_{ijk}}{\Delta z})^2\Big]\Big\}^{\frac{1}{2}}$$

$$\leq \{|u_h^2|_h+|v_h^2|_h+|w_h^2|_h\}^{\frac{1}{2}}\{|(\varphi_x)_h^2|_h+|(\varphi_y)_h^2|_h+|(\varphi_z)_h^2|_h\}^{\frac{1}{2}}\|\boldsymbol{\xi}_h\|_h.$$

Thanks to Lemma 9.6:

$$| \mathbf{b}_h(\mathbf{u}_h,\boldsymbol{\varphi}_h,\boldsymbol{\xi}_h) |$$

$$\leq 2\Big\{|u_h|_h^{\frac{1}{2}}\|u_h\|_h^{\frac{3}{2}}+|v_h|_h^{\frac{1}{2}}\|v_h\|_h^{\frac{3}{2}}+|w_h|_h^{\frac{1}{2}}\|w_h\|_h^{\frac{3}{2}}\Big\}^{\frac{1}{2}}$$

$$\times\Big\{|(\varphi_x)_h|_h^{\frac{1}{2}}\|(\varphi_x)_h\|_h^{\frac{3}{2}}+|(\varphi_y)_h|_h^{\frac{1}{2}}\|(\varphi_y)_h\|_h^{\frac{3}{2}}+|(\varphi_z)_h|_h^{\frac{1}{2}}\|(\varphi_z)_h\|_h^{\frac{3}{2}}\Big\}^{\frac{1}{2}}\|\boldsymbol{\xi}_h\|_h.$$

Using Hölder inequality, we find

$$|\mathbf{b}_h(\mathbf{u}_h,\boldsymbol{\varphi}_h,\boldsymbol{\xi}_h)|$$

$$\leq 2\Big\{|u_h|_h^2+|v_h|_h^2+|w_h|_h^2\Big\}^{\frac{1}{8}}\Big\{\|u_h\|_h^2+\|v_h\|_h^2+\|w_h\|_h^2\Big\}^{\frac{3}{8}}$$

$$\times\Big\{|(\varphi_x)_h|_h^2+|(\varphi_y)_h|_h^2+|(\varphi_z)_h|_h^2\Big\}^{\frac{1}{8}}\Big\{\|(\varphi_x)_h\|_h^2+\|(\varphi_y)_h\|_h^2+\|(\varphi_z)_h\|_h^2\Big\}^{\frac{3}{8}}\|\boldsymbol{\xi}_h\|_h$$

$$=2|\mathbf{u}_h|_h^{\frac{1}{4}}\|\mathbf{u}_h\|_h^{\frac{3}{4}}|\boldsymbol{\varphi}_h|_h^{\frac{1}{4}}\|\boldsymbol{\varphi}_h\|_h^{\frac{3}{4}}\|\boldsymbol{\xi}_h\|_h.$$

Also, Lemma 9.4 implies

$$|\mathbf{b}_h(\mathbf{u}_h,\boldsymbol{\varphi}_h,\boldsymbol{\xi}_h)|=|\mathbf{b}_h(\mathbf{u}_h,\boldsymbol{\xi}_h,\boldsymbol{\varphi}_h)|\leq 2|\mathbf{u}_h|_h^{\frac{1}{4}}\|\mathbf{u}_h\|_h^{\frac{3}{4}}|\boldsymbol{\xi}_h|_h^{\frac{1}{4}}\|\boldsymbol{\xi}_h\|_h^{\frac{3}{4}}\|\boldsymbol{\varphi}_h\|_h.$$ □

Lemma 9.6.

$$|u_h^2|_h\leq 2|u_h|_h^{\frac{1}{2}}\|u_h\|_h^{\frac{3}{2}}.$$

Proof. First, we note that

$$u_{ijk}^2=u_{1jk}^2+\sum_{\alpha=1}^{i-1}(u_{\alpha+1jk}^2-u_{\alpha jk}^2)$$

$$\leq u_{1jk}^2+\sum_{\alpha=1}^{i-1}\Big[|u_{\alpha+1jk}||u_{\alpha+1jk}-u_{\alpha jk}|+|u_{\alpha jk}||u_{\alpha+1jk}-u_{\alpha jk}|\Big],$$

which yields for i=1, …, M

$$u_{ijk}^2\leq u_{1jk}^2+\sum_{\alpha=1}^{M-1}|u_{\alpha+1jk}||u_{\alpha+1jk}-u_{\alpha jk}|+\sum_{\alpha=1}^{M-1}|u_{\alpha jk}||u_{\alpha+1jk}-u_{\alpha jk}|.$$

Proceeding similarly in other direction, we find

$$u_{ijk}^2\leq u_{i1k}^2+\sum_{\beta=1}^{N-1}|u_{i\beta+1k}||u_{i\beta+1k}-u_{i\beta k}|+\sum_{\beta=1}^{N-1}|u_{i\beta k}||u_{i\beta+1k}-u_{i\beta k}|,$$

$$u_{ijk}^2\leq u_{ij1}^2+\sum_{\gamma=1}^{Q-1}|u_{ij\gamma+1}||u_{ij\gamma+1}-u_{ij\gamma}|+\sum_{\gamma=1}^{Q-1}|u_{ij\gamma}||u_{ij\gamma+1}-u_{ij\gamma}|.$$

So

$$
\begin{aligned}
&\sum_{j=1}^{N}\sum_{k=1}^{Q}\Delta y\Delta z u_{ijk}^4\\
&\le \underbrace{\sum_{k=1}^{Q}\Delta z\Big[u_{i1k}^2+\sum_{j=1}^{N-1}|u_{ij+1k}||u_{ij+1k}-u_{ijk}|+\sum_{j=1}^{N-1}|u_{ijk}||u_{ij+1k}-u_{ijk}|\Big]}_{I}\\
&\quad\times\underbrace{\sum_{j=1}^{N}\Delta y\Big[u_{ij1}^2+\sum_{k=1}^{Q-1}|u_{ijk+1}||u_{ijk+1}-u_{ijk}|+\sum_{k=1}^{Q-1}|u_{ijk}||u_{ijk+1}-u_{ijk}|\Big]}_{II}.
\end{aligned}
$$

We explain how to bound the first term I, then we will briefly do the same for the second term II. For I, using the Cauchy-Schwarz inequality, we have

$$
\begin{aligned}
I \le&\sum_{k=1}^{Q}\Delta z\Big[(\sum_{j=1}^{N}u_{ijk}^2)^{\frac12}(\sum_{j=1}^{N-1}|u_{ij+1k}-u_{ijk}|^2+u_{i1k}^2)^{\frac12}\\
&+(\sum_{j=1}^{N-1}u_{ijk}^2)^{\frac12}(\sum_{j=1}^{N-1}|u_{ij+1k}-u_{ijk}|^2)^{\frac12}\Big]\\
\le&2\sum_{k=1}^{Q}\Delta z\Big[(\sum_{j=1}^{N}u_{ijk}^2)^{\frac12}(\sum_{j=1}^{N-1}|u_{ij+1k}-u_{ijk}|^2+u_{i1k}^2)^{\frac12}\Big],
\end{aligned}
$$

then we find

$$
I\le 2\Big[\sum_{j=1}^{N}\sum_{k=1}^{Q}\Delta y\Delta z u_{ijk}^2\Big]^{\frac12}\cdot\Big[\sum_{k=1}^{Q}\sum_{j=1}^{N-1}\Delta y\Delta z(\frac{u_{ij+1k}-u_{ijk}}{\Delta y})^2+\sum_{k=1}^{Q}\Delta y\Delta z(\frac{u_{i1k}}{\Delta y})^2\Big]^{\frac12}.
$$

Proceeding similarly for the second term II, we find

$$
II\le 2\Big[\sum_{j=1}^{N}\sum_{k=1}^{Q}\Delta y\Delta z u_{ijk}^2\Big]^{\frac12}\cdot\Big[\sum_{j=1}^{N}\sum_{k=1}^{Q-1}\Delta y\Delta z(\frac{u_{ijk+1}-u_{ijk}}{\Delta z})^2+\sum_{j=1}^{N}\Delta y\Delta z(\frac{u_{ij1}}{\Delta z})^2\Big]^{\frac12}.
$$

Hence the inequality:

$$
\begin{aligned}
&\sum_{j=1}^{N}\sum_{k=1}^{Q}\Delta y\Delta z u_{ijk}^4\\
\le&4\Big[\sum_{j=1}^{N}\sum_{k=1}^{Q}\Delta y\Delta z u_{ijk}^2\Big]\cdot\Big[\sum_{k=1}^{Q}\sum_{j=1}^{N-1}\Delta y\Delta z(\frac{u_{ij+1k}-u_{ijk}}{\Delta y})^2+\sum_{k=1}^{Q}\Delta y\Delta z(\frac{u_{i1k}}{\Delta y})^2\Big]^{\frac12}\\
&\times\Big[\sum_{j=1}^{N}\sum_{k=1}^{Q-1}\Delta y\Delta z(\frac{u_{ijk+1}-u_{ijk}}{\Delta z})^2+\sum_{j=1}^{N}\Delta y\Delta z(\frac{u_{ij1}}{\Delta z})^2\Big]^{\frac12}.
\end{aligned}
$$

Due to the well-known inequality $2ab \leq a^2 + b^2$, we have

$$\begin{aligned}&\sum_{j=1}^{N}\sum_{k=1}^{Q}\Delta y\Delta z u_{ijk}^4\\ \leq & 2\Big[\sum_{j=1}^{N}\sum_{k=1}^{Q}\Delta y\Delta z u_{ijk}^2\Big]\cdot\Big[\sum_{k=1}^{Q}\sum_{j=1}^{N-1}\Delta y\Delta z(\frac{u_{ij+1k}-u_{ijk}}{\Delta y})^2+\sum_{k=1}^{Q}\Delta y\Delta z(\frac{u_{i1k}}{\Delta y})^2\\ &+\sum_{j=1}^{N}\sum_{k=1}^{Q-1}\Delta y\Delta z(\frac{u_{ijk+1}-u_{ijk}}{\Delta z})^2+\sum_{j=1}^{N}\Delta y\Delta z(\frac{u_{ij1}}{\Delta z})^2\Big].\end{aligned}$$

Let $|u_{Rjk}| = \max_{1\leq i\leq M}\{|u_{ijk}|\}$, hence we have the inequality:

$$\begin{aligned}|u_h^2|_h^2 &= \sum_{i=1}^{M}\sum_{j=1}^{N}\sum_{k=1}^{Q}\Delta x\Delta y\Delta z u_{ijk}^4 = \sum_{i=1}^{M}\Delta x(\sum_{j=1}^{N}\sum_{k=1}^{Q}\Delta y\Delta z u_{ijk}^4)\\ &\leq 2\sum_{i=1}^{M}\Bigg\{\Big[\sum_{j=1}^{N}\sum_{k=1}^{Q}\Delta y\Delta z u_{ijk}^2\Big]\\ &\quad\times\Big[\sum_{k=1}^{Q}\sum_{j=1}^{N-1}\Delta x\Delta y\Delta z(\frac{u_{ij+1k}-u_{ijk}}{\Delta y})^2+\sum_{k=1}^{Q}\Delta x\Delta y\Delta z(\frac{u_{i1k}}{\Delta y})^2\\ &\quad+\sum_{j=1}^{N}\sum_{k=1}^{Q-1}\Delta x\Delta y\Delta z(\frac{u_{ijk+1}-u_{ijk}}{\Delta z})^2+\sum_{j=1}^{N}\Delta x\Delta y\Delta z(\frac{u_{ij1}}{\Delta z})^2\Big]\Bigg\}\\ &\leq 2\Big[\sum_{j=1}^{N}\sum_{k=1}^{Q}\Delta y\Delta z u_{Rjk}^2\Big]\\ &\quad\times\Big[\sum_{i=1}^{M}\sum_{k=1}^{Q}\sum_{j=1}^{N-1}\Delta x\Delta y\Delta z(\frac{u_{ij+1k}-u_{ijk}}{\Delta y})^2+\sum_{i=1}^{M}\sum_{k=1}^{Q}\Delta x\Delta y\Delta z(\frac{u_{i1k}}{\Delta y})^2\\ &\quad+\sum_{i=1}^{M}\sum_{j=1}^{N}\sum_{k=1}^{Q-1}\Delta x\Delta y\Delta z(\frac{u_{ijk+1}-u_{ijk}}{\Delta z})^2+\sum_{i=1}^{M}\sum_{j=1}^{N}\Delta x\Delta y\Delta z(\frac{u_{ij1}}{\Delta z})^2\Big]\\ &\leq 2\Big[\sum_{j=1}^{N}\sum_{k=1}^{Q}\Delta y\Delta z u_{Rjk}^2\Big]\cdot\|u_h\|_h^2.\end{aligned}$$

Similar to the method of bounding the terms I and II, we also have

$$\begin{aligned}&\sum_{j=1}^{N}\sum_{k=1}^{Q}\Delta y\Delta z u_{Rjk}^2\\ \leq\ & 2\Big[\sum_{j=1}^{N}\sum_{k=1}^{Q}\sum_{i=1}^{N}\Delta x\Delta y\Delta z u_{ijk}^2\Big]^{\frac{1}{2}}\\ &\times\Big[\sum_{j=1}^{N}\sum_{k=1}^{Q}\sum_{i=1}^{M-1}\Delta x\Delta y\Delta z(\frac{u_{i+1jk}-u_{ijk}}{\Delta x})^2+\sum_{j=1}^{N}\sum_{k=1}^{Q}\Delta x\Delta y\Delta z(\frac{u_{1jk}}{\Delta x})^2\Big]^{\frac{1}{2}}\end{aligned}$$

$\leq 2|u_h|_h\|u_h\|_h.$

Finally, we obtain

$$|u_h^2|_h^2 \leq 4|u_h|_h\|u_h\|_h^3.$$

The above inequality implies that

$$|u_h^2|_h \leq 2|u_h|_h^{\frac{1}{2}}\|u_h\|_h^{\frac{3}{2}}.$$

□

9.5 *Apriori* Estimate

In this section, we derive *a priori* estimates required for the proof of the stability theorem. We start by a lemma that contains the properties deduced from (9.37). Then, we will use (9.36) and this lemma in order to obtain two *a priori* estimates: one for the intermediate velocities $\mathbf{u}_h^{n+\frac{1}{2}} \in \mathbf{W}_h$ and one for the new velocities $\mathbf{u}_h^{n+1} \in \mathbf{V}_h$.

Lemma 9.7.

$$|\boldsymbol{u}_h^{n+1} - \boldsymbol{u}_h^{n+\frac{1}{2}}|_h^2 = |\boldsymbol{u}_h^{n+\frac{1}{2}}|_h^2 - |\boldsymbol{u}_h^{n+1}|_h^2, \tag{9.48}$$

$$|\boldsymbol{u}_h^{n+1}|_h \leq |\boldsymbol{u}_h^{n+\frac{1}{2}}|_h, \tag{9.49}$$

$$|\boldsymbol{u}_h^{n+1}|_h^2 - (\boldsymbol{u}_h^{n+\frac{1}{2}}, \boldsymbol{u}_h^{n+1})_h = 0. \tag{9.50}$$

Proof. We note that (9.48) is just a rewriting of (9.50) and that (9.49) is an easy consequence of (9.48). Hence, we just need to show (9.50). To do so, we replace $\boldsymbol{\varphi}_h$ by $\mathbf{u}_h^{n+1}$ in (9.37):

$$\frac{1}{\Delta t}(\mathbf{u}_h^{n+1} - \mathbf{u}_h^{n+\frac{1}{2}}, \mathbf{u}_h^{n+1})_h = -(\mathbf{G}_h p_h^{n+1}, \mathbf{u}_h^{n+1})_h,$$

and we reorganize the right-hand side as follows:

$$\begin{aligned}
&(\mathbf{G}_h p_h^{n+1}, \mathbf{u}_h^{n+1})_h \\
&= \sum_{i=1}^{M}\sum_{j=1}^{N}\sum_{k=1}^{Q} \frac{\Delta y \Delta z}{2}[p_{i+1jk}^{n+1} - p_{i-1jk}^{n+1}]u_{ijk}^{n+1} \\
&\quad + \sum_{i=1}^{M}\sum_{j=1}^{N}\sum_{k=1}^{Q} \frac{\Delta z \Delta x}{2}[p_{ij+1k}^{n+1} - p_{ij-1k}^{n+1}]v_{ijk}^{n+1} \\
&\quad + \sum_{i=1}^{M}\sum_{j=1}^{N}\sum_{k=1}^{Q} \frac{\Delta x \Delta y}{2}[p_{ijk+1}^{n+1} - p_{ijk-1}^{n+1}]w_{ijk}^{n+1} \\
&= \sum_{j=1}^{N}\sum_{k=1}^{Q} \frac{\Delta y \Delta z}{2}[\sum_{i=2}^{M+1} p_{ijk}^{n+1}u_{i-1jk}^{n+1} - \sum_{i=0}^{M-1} p_{ijk}^{n+1}u_{i+1jk}^{n+1}]
\end{aligned}$$

$$+\sum_{k=1}^{Q}\sum_{i=1}^{M}\frac{\Delta z\Delta x}{2}[\sum_{j=2}^{N+1}p_{ijk}^{n+1}v_{ij-1k}^{n+1}-\sum_{j=0}^{N-1}p_{ijk}^{n+1}v_{ij+1k}^{n+1}]$$
$$+\sum_{i=1}^{M}\sum_{j=1}^{N}\frac{\Delta x\Delta y}{2}[\sum_{k=2}^{Q+1}p_{ijk}^{n+1}w_{ijk-1}^{n+1}-\sum_{k=0}^{Q-1}p_{ijk}^{n+1}w_{ijk+1}^{n+1}]$$
$$=\sum_{i=1}^{M}\sum_{j=1}^{N}\sum_{k=1}^{Q}p_{ijk}^{n+1}\Big(\frac{\Delta y\Delta z}{2}[u_{i-1jk}^{n+1}-u_{i+1jk}^{n+1}]+\frac{\Delta z\Delta x}{2}[v_{ij-1k}^{n+1}-v_{ij+1k}^{n+1}]$$
$$+\frac{\Delta x\Delta y}{2}[w_{ijk-1}^{n+1}-w_{ijk+1}^{n+1}]\Big)$$
$$+\sum_{j=1}^{N}\sum_{k=1}^{Q}\frac{\Delta y\Delta z}{2}[-p_{1jk}^{n+1}u_{0jk}^{n+1}+p_{M+1jk}^{n+1}u_{Mjk}^{n+1}-p_{0jk}^{n+1}u_{1jk}^{n+1}+p_{Mjk}^{n+1}u_{M+1jk}^{n+1}]$$
$$+\sum_{k=1}^{Q}\sum_{i=1}^{M}\frac{\Delta z\Delta x}{2}[-p_{i1k}^{n+1}v_{i0k}^{n+1}+p_{iN+1k}^{n+1}v_{iNk}^{n+1}-p_{i0k}^{n+1}v_{i1k}^{n+1}+p_{iNk}^{n+1}v_{iN+1k}^{n+1}]$$
$$+\sum_{i=1}^{M}\sum_{j=1}^{N}\frac{\Delta x\Delta y}{2}[-p_{ij1}^{n+1}w_{ij0}^{n+1}+p_{ijQ+1}^{n+1}w_{ijQ}^{n+1}-p_{ij0}^{n+1}w_{ij1}^{n+1}+p_{ijQ}^{n+1}w_{ijQ+1}^{n+1}].$$

Now, using the incompressibility condition (9.15), we find

$$(\mathbf{G}_hp_h^{n+1},\mathbf{u}_h^{n+1})_h$$
$$=\sum_{j=1}^{N}\sum_{k=1}^{Q}\frac{\Delta y\Delta z}{2}[-p_{1jk}^{n+1}u_{0jk}^{n+1}+p_{M+1jk}^{n+1}u_{Mjk}^{n+1}-p_{0jk}^{n+1}u_{1jk}^{n+1}+p_{Mjk}^{n+1}u_{M+1jk}^{n+1}]$$
$$+\sum_{k=1}^{Q}\sum_{i=1}^{M}\frac{\Delta z\Delta x}{2}[-p_{i1k}^{n+1}v_{i0k}^{n+1}+p_{iN+1k}^{n+1}v_{iNk}^{n+1}-p_{i0k}^{n+1}v_{i1k}^{n+1}+p_{iNk}^{n+1}v_{iN+1k}^{n+1}]$$
$$+\sum_{i=1}^{M}\sum_{j=1}^{N}\frac{\Delta x\Delta y}{2}[-p_{ij1}^{n+1}w_{ij0}^{n+1}+p_{ijQ+1}^{n+1}w_{ijQ}^{n+1}-p_{ij0}^{n+1}w_{ij1}^{n+1}+p_{ijQ}^{n+1}w_{ijQ+1}^{n+1}],$$

and using the boundary conditions (9.16) and (9.27), we have $(\mathbf{G}_hp_h^{n+1},\mathbf{u}_h^{n+1})_h=0$ and (9.50) is proved. □

Now, we show two *a priori* estimates used in the stability section.

Lemma 9.8 (*A priori* estimate). *We have two a priori estimates, one for the intermediate velocities:*

$$\frac{3}{4}|\boldsymbol{u}_h^{n+\frac{1}{2}}-\boldsymbol{u}_h^n|_h^2+|\boldsymbol{u}_h^{n+\frac{1}{2}}|_h^2+\nu\Delta t\|\boldsymbol{u}_h^{n+\frac{1}{2}}\|_h^2\Big\{1-[2\nu\Delta tS_h^2+\frac{\Delta tS_h^3}{\nu}(42|\boldsymbol{u}_h^n|_h^2$$
$$+10|\boldsymbol{u}_h^{n-1}|_h^2)]\Big\}\leq\frac{c_1^2\Delta t}{2\nu}(|\boldsymbol{f}^{n+1}|_h^2+|\boldsymbol{f}^n|_h^2)+|\boldsymbol{u}_h^{n-\frac{1}{2}}|_h^2+\frac{1}{4}|\boldsymbol{u}_h^{n-\frac{1}{2}}-\boldsymbol{u}_h^{n-1}|_h^2, \tag{9.51}$$

and one for the new velocities:

$$\frac{3}{4}|\boldsymbol{u}_h^{n+1}-\boldsymbol{u}_h^n|_h^2+|\boldsymbol{u}_h^{n+1}|_h^2+\frac{\nu\Delta t}{2}\|\boldsymbol{u}_h^{n+1}\|_h^2\Big\{1-[4\nu\Delta t S_h^2+\frac{\Delta t S_h^3}{\nu}(80|\boldsymbol{u}_h^n|_h^2+ \\ 16|\boldsymbol{u}_h^{n-1}|_h^2)]\Big\}\leq\frac{2c_1^2\Delta t}{\nu}(|\boldsymbol{f}^{n+1}|_h^2+|\boldsymbol{f}^n|_h^2)+|\boldsymbol{u}_h^n|_h^2+\frac{1}{4}|\boldsymbol{u}_h^n-\boldsymbol{u}_h^{n-1}|_h^2+\nu\Delta t\|\boldsymbol{u}_h^{n+\frac{1}{2}}\|_h^2. \tag{9.52}$$

Proof. To find the first *a priori* estimate, for the intermediate velocity, we replace φ_h by $2\Delta t\mathbf{u}_h^{n+\frac{1}{2}}$ in (9.36):

$$\begin{aligned}&(\mathbf{u}_h^{n+\frac{1}{2}}-\mathbf{u}_h^n,2\mathbf{u}_h^{n+\frac{1}{2}})_h\\ &=-\nu((\mathbf{u}_h^{n+\frac{1}{2}}+\mathbf{u}_h^n,\Delta t\mathbf{u}_h^{n+\frac{1}{2}}))_h-3\Delta t\mathbf{b}_h(\mathbf{u}_h^n,\mathbf{u}_h^n,\mathbf{u}_h^{n+\frac{1}{2}})\\ &\quad+\Delta t\mathbf{b}_h(\mathbf{u}_h^{n-1},\mathbf{u}_h^{n-1},\mathbf{u}_h^{n+\frac{1}{2}})+(\mathbf{f}^{n+1}+\mathbf{f}^n,\Delta t\mathbf{u}_h^{n+\frac{1}{2}})_h.\end{aligned}$$

Moreover, we can modify each term as follows:

$$\begin{aligned}&(\mathbf{u}_h^{n+\frac{1}{2}}-\mathbf{u}_h^n,2\mathbf{u}_h^{n+\frac{1}{2}})_h=(\mathbf{u}_h^{n+\frac{1}{2}}-\mathbf{u}_h^n,\mathbf{u}_h^{n+\frac{1}{2}}-\mathbf{u}_h^n)_h+(\mathbf{u}_h^{n+\frac{1}{2}}-\mathbf{u}_h^n,\mathbf{u}_h^{n+\frac{1}{2}}+\mathbf{u}_h^n)_h\\ =&|\mathbf{u}_h^{n+\frac{1}{2}}-\mathbf{u}_h^n|_h^2+|\mathbf{u}_h^{n+\frac{1}{2}}|_h^2-|\mathbf{u}_h^n|_h^2,((\mathbf{u}_h^{n+\frac{1}{2}}+\mathbf{u}_h^n,\Delta t\mathbf{u}_h^{n+\frac{1}{2}}))_h\\ =&2\Delta t\|\mathbf{u}_h^{n+\frac{1}{2}}\|_h^2-\Delta t((\mathbf{u}_h^{n+\frac{1}{2}}-\mathbf{u}_h^n,\mathbf{u}_h^{n+\frac{1}{2}}))_h,\\ &-3\Delta t\mathbf{b}_h(\mathbf{u}_h^n,\mathbf{u}_h^n,\mathbf{u}_h^{n+\frac{1}{2}})+\Delta t\mathbf{b}_h(\mathbf{u}_h^{n-1},\mathbf{u}_h^{n-1},\mathbf{u}_h^{n+\frac{1}{2}})\\ =&2\Delta t\mathbf{b}_h(\mathbf{u}_h^n,\mathbf{u}_h^{n+\frac{1}{2}}-\mathbf{u}_h^n,\mathbf{u}_h^{n+\frac{1}{2}})-\Delta t[\mathbf{b}_h(\mathbf{u}_h^n,\mathbf{u}_h^n,\mathbf{u}_h^{n+\frac{1}{2}})-\mathbf{b}_h(\mathbf{u}_h^{n-1},\mathbf{u}_h^{n-1},\mathbf{u}_h^{n+\frac{1}{2}})],\end{aligned}$$

according to $\mathbf{b}_h(\mathbf{u}_h^n,\mathbf{u}_h^{n+\frac{1}{2}},\mathbf{u}_h^{n+\frac{1}{2}})=0$ (Lemma 9.4).

So, we have the equality:

$$\begin{aligned}&|\mathbf{u}_h^{n+\frac{1}{2}}-\mathbf{u}_h^n|_h^2+|\mathbf{u}_h^{n+\frac{1}{2}}|_h^2-|\mathbf{u}_h^n|_h^2+2\nu\Delta t\|\mathbf{u}_h^{n+\frac{1}{2}}\|_h^2\\ &=\underbrace{\nu\Delta t((\mathbf{u}_h^{n+\frac{1}{2}}-\mathbf{u}_h^n,\mathbf{u}_h^{n+\frac{1}{2}}))_h}_{I}+\underbrace{2\Delta t\mathbf{b}_h(\mathbf{u}_h^n,\mathbf{u}_h^{n+\frac{1}{2}}-\mathbf{u}_h^n,\mathbf{u}_h^{n+\frac{1}{2}})}_{II}\\ &\quad\underbrace{-\Delta t[\mathbf{b}_h(\mathbf{u}_h^n,\mathbf{u}_h^n,\mathbf{u}_h^{n+\frac{1}{2}})-\mathbf{b}_h(\mathbf{u}_h^{n-1},\mathbf{u}_h^{n-1},\mathbf{u}_h^{n+\frac{1}{2}})]}_{III}+\underbrace{\Delta t(\mathbf{f}^{n+1}+\mathbf{f}^n,\mathbf{u}_h^{n+\frac{1}{2}})_h}_{IV}.\end{aligned} \tag{9.53}$$

To find the *a priori* estimate, we need to bound the terms I–IV which are in the right-hand side:

- For the first term:

$$\begin{aligned}I&=\nu\Delta t((\mathbf{u}_h^{n+\frac{1}{2}}-\mathbf{u}_h^n,\mathbf{u}_h^{n+\frac{1}{2}}))_h\\ &\leq\nu\Delta t\|\mathbf{u}_h^{n+\frac{1}{2}}-\mathbf{u}_h^n\|_h\|\mathbf{u}_h^{n+\frac{1}{2}}\|_h \qquad \text{(with Cauchy}-\text{Schwarz inequality)}\\ &\leq\nu\Delta t S_h|\mathbf{u}_h^{n+\frac{1}{2}}-\mathbf{u}_h^n|_h\|\mathbf{u}_h^{n+\frac{1}{2}}\|_h \qquad \text{(with the inverse inequality)}\\ &\leq\frac{1}{8}|\mathbf{u}_h^{n+\frac{1}{2}}-\mathbf{u}_h^n|_h^2+2\nu^2\Delta t^2S_h^2\|\mathbf{u}_h^{n+\frac{1}{2}}\|_h^2. \qquad \text{(with Young inequality)}\end{aligned}$$

• For the second term:

$$
\begin{aligned}
II &= 2\Delta t \mathbf{b}_h(\mathbf{u}_h^n, \mathbf{u}_h^{n+\frac{1}{2}} - \mathbf{u}_h^n, \mathbf{u}_h^{n+\frac{1}{2}}) \\
&\leq 4\Delta t |\mathbf{u}_h^n|_h^{\frac{1}{4}} \|\mathbf{u}_h^n\|_h^{\frac{3}{4}} |\mathbf{u}_h^{n+\frac{1}{2}} - \mathbf{u}_h^n|_h^{\frac{1}{4}} \|\mathbf{u}_h^{n+\frac{1}{2}} - \mathbf{u}_h^n\|_h^{\frac{3}{4}} \|\mathbf{u}_h^{n+\frac{1}{2}}\|_h \quad \text{(with Lemma 9.5)} \\
&\leq 4\Delta t S_h^{\frac{3}{2}} |\mathbf{u}_h^n|_h |\mathbf{u}_h^{n+\frac{1}{2}} - \mathbf{u}_h^n|_h \|\mathbf{u}_h^{n+\frac{1}{2}}\|_h \quad \text{(with the inverse inequality)} \\
&\leq \frac{1}{8} |\mathbf{u}_h^{n+\frac{1}{2}} - \mathbf{u}_h^n|_h^2 + 32\Delta t^2 S_h^3 \|\mathbf{u}_h^{n+\frac{1}{2}}\|_h^2 |\mathbf{u}_h^n|_h^2. \quad \text{(with Young inequality)}
\end{aligned}
$$

• For the third term:

$$
\begin{aligned}
III &= -\Delta t[\mathbf{b}_h(\mathbf{u}_h^n, \mathbf{u}_h^n, \mathbf{u}_h^{n+\frac{1}{2}}) - \mathbf{b}_h(\mathbf{u}_h^{n-1}, \mathbf{u}_h^{n-1}, \mathbf{u}_h^{n+\frac{1}{2}})] \\
&= -\Delta t[\mathbf{b}_h(\mathbf{u}_h^n - \mathbf{u}_h^{n-1}, \mathbf{u}_h^n, \mathbf{u}_h^{n+\frac{1}{2}}) + \mathbf{b}_h(\mathbf{u}_h^{n-1}, \mathbf{u}_h^n - \mathbf{u}_h^{n-1}, \mathbf{u}_h^{n+\frac{1}{2}})] \\
&\leq 2\Delta t |\mathbf{u}_h^n - \mathbf{u}_h^{n-1}|_h^{\frac{1}{4}} \|\mathbf{u}_h^n - \mathbf{u}_h^{n-1}\|_h^{\frac{3}{4}} \|\mathbf{u}_h^{n+\frac{1}{2}}\|_h \\
&\quad \times \left[|\mathbf{u}_h^n|_h^{\frac{1}{4}} \|\mathbf{u}_h^n\|_h^{\frac{3}{4}} + |\mathbf{u}_h^{n-1}|_h^{\frac{1}{4}} \|\mathbf{u}_h^{n-1}\|_h^{\frac{3}{4}} \right] \quad \text{(with Lemma 9.5)} \\
&\leq 2\Delta t S_h^{\frac{3}{2}} |\mathbf{u}_h^n - \mathbf{u}_h^{n-1}|_h \|\mathbf{u}_h^{n+\frac{1}{2}}\|_h [\, |\mathbf{u}_h^n|_h + |\mathbf{u}_h^{n-1}|_h] \quad \text{(with the inverse inequality)} \\
&\leq 2\Delta t S_h^{\frac{3}{2}} [\, |\mathbf{u}_h^n - \mathbf{u}_h^{n-\frac{1}{2}}|_h + |\mathbf{u}_h^{n-\frac{1}{2}} - \mathbf{u}_h^{n-1}|_h \,] \|\mathbf{u}_h^{n+\frac{1}{2}}\|_h [\, |\mathbf{u}_h^n|_h + |\mathbf{u}_h^{n-1}|_h] \\
&\leq |\mathbf{u}_h^n - \mathbf{u}_h^{n-\frac{1}{2}}|_h^2 + \frac{1}{4} |\mathbf{u}_h^{n-\frac{1}{2}} - \mathbf{u}_h^{n-1}|_h^2 \\
&\quad + 10\Delta t^2 S_h^3 \|\mathbf{u}_h^{n+\frac{1}{2}}\|_h^2 [\, |\mathbf{u}_h^n|_h^2 + |\mathbf{u}_h^{n-1}|_h^2 \,] \quad \text{(with Young inequality)} \\
&\leq |\mathbf{u}_h^{n-\frac{1}{2}}|_h^2 - |\mathbf{u}_h^n|_h^2 + \frac{1}{4} |\mathbf{u}_h^{n-\frac{1}{2}} - \mathbf{u}_h^{n-1}|_h^2 \\
&\quad + 10\Delta t^2 S_h^3 \|\mathbf{u}_h^{n+\frac{1}{2}}\|_h^2 [\, |\mathbf{u}_h^n|_h^2 + |\mathbf{u}_h^{n-1}|_h^2 \,]. \quad \text{(with (9.48))}
\end{aligned}
$$

• For the fourth term:

$$
\begin{aligned}
IV &= \Delta t(\mathbf{f}^{n+1} + \mathbf{f}^n, \mathbf{u}_h^{n+\frac{1}{2}})_h \\
&\leq \Delta t(|\mathbf{f}^{n+1}|_h + |\mathbf{f}^n|_h) |\mathbf{u}_h^{n+\frac{1}{2}}|_h \quad \text{(with Cauchy – Schwarz inequality)} \\
&\leq c_1 \Delta t(|\mathbf{f}^{n+1}|_h + |\mathbf{f}^n|_h) \|\mathbf{u}_h^{n+\frac{1}{2}}\|_h \quad \text{(with Poincaré inequality)} \\
&\leq \nu \Delta t \|\mathbf{u}_h^{n+\frac{1}{2}}\|_h^2 + \frac{c_1^2 \Delta t}{2\nu} (|\mathbf{f}^{n+1}|_h^2 + |\mathbf{f}^n|_h^2). \quad \text{(with Young inequality)}
\end{aligned}
$$

So, due to these previous inequalities we obtain the first *a priori* estimate.

To get the second *a priori* estimate about the new velocities $\mathbf{u}_h^{n+1}$, we apply the same method except that we replace $\boldsymbol{\varphi}_h$ by $2\Delta t \mathbf{u}_h^{n+1}$ and not by $2\Delta t \mathbf{u}_h^{n+\frac{1}{2}}$ in (9.36):

$$
\begin{aligned}
&(\mathbf{u}_h^{n+\frac{1}{2}} - \mathbf{u}_h^n, 2\mathbf{u}_h^{n+1})_h \\
&= -\nu((\mathbf{u}_h^{n+\frac{1}{2}} + \mathbf{u}_h^n, \Delta t \mathbf{u}_h^{n+1}))_h - 3\Delta t \mathbf{b}_h(\mathbf{u}_h^n, \mathbf{u}_h^n, \mathbf{u}_h^{n+1}) \\
&\quad + \Delta t \mathbf{b}_h(\mathbf{u}_h^{n-1}, \mathbf{u}_h^{n-1}, \mathbf{u}_h^{n+1}) + (\mathbf{f}^{n+1} + \mathbf{f}^n, \Delta t \mathbf{u}_h^{n+1})_h.
\end{aligned}
$$

Moreover, if we follow the same way that for the previous estimate, we rewrite each term:

$$\begin{aligned}(\mathbf{u}_h^{n+\frac{1}{2}} - \mathbf{u}_h^n, 2\mathbf{u}_h^{n+1})_h =& 2(\mathbf{u}_h^{n+\frac{1}{2}}, \mathbf{u}_h^{n+1})_h - (\mathbf{u}_h^n, 2\mathbf{u}_h^{n+1})_h \\ =& 2(\mathbf{u}_h^{n+1}, \mathbf{u}_h^{n+1})_h - (\mathbf{u}_h^n, 2\mathbf{u}_h^{n+1})_h \quad \text{with (9.50)}, \\ =& (\mathbf{u}_h^{n+1} - \mathbf{u}_h^n, 2\mathbf{u}_h^{n+1}) \\ =& |\mathbf{u}_h^{n+1} - \mathbf{u}_h^n|_h^2 + |\mathbf{u}_h^{n+1}|_h^2 - |\mathbf{u}_h^n|_h^2,\end{aligned}$$

$$\begin{aligned}((\mathbf{u}_h^{n+\frac{1}{2}} + \mathbf{u}_h^n, \mathbf{u}_h^{n+1}))_h =& ((\mathbf{u}_h^{n+\frac{1}{2}} + \mathbf{u}_h^{n+1} - \mathbf{u}_h^{n+1} + \mathbf{u}_h^n, \mathbf{u}_h^{n+1}))_h \\ =& ((\mathbf{u}_h^{n+\frac{1}{2}}, \mathbf{u}_h^{n+1}))_h + \|\mathbf{u}_h^{n+1}\|_h^2 - ((\mathbf{u}_h^{n+1} - \mathbf{u}_h^n, \mathbf{u}_h^{n+1}))_h,\end{aligned}$$

$$\begin{aligned}&- 3\Delta t \mathbf{b}_h(\mathbf{u}_h^n, \mathbf{u}_h^n, \mathbf{u}_h^{n+1}) + \Delta t \mathbf{b}_h(\mathbf{u}_h^{n-1}, \mathbf{u}_h^{n-1}, \mathbf{u}_h^{n+1}) \\ &= 2\Delta t \mathbf{b}_h(\mathbf{u}_h^n, \mathbf{u}_h^{n+1} - \mathbf{u}_h^n, \mathbf{u}_h^{n+1}) - \Delta t[\mathbf{b}_h(\mathbf{u}_h^n, \mathbf{u}_h^n, \mathbf{u}_h^{n+1}) - \mathbf{b}_h(\mathbf{u}_h^{n-1}, \mathbf{u}_h^{n-1}, \mathbf{u}_h^{n+1})].\end{aligned}$$

Hence, the equality obtained from the weak formulation becomes

$$\begin{aligned}&|\mathbf{u}_h^{n+1} - \mathbf{u}_h^n|_h^2 + |\mathbf{u}_h^{n+1}|_h^2 - |\mathbf{u}_h^n|_h^2 + \nu\Delta t\|\mathbf{u}_h^{n+1}\|_h^2 \\ &= \underbrace{\nu\Delta t((\mathbf{u}_h^{n+1} - \mathbf{u}_h^n, \mathbf{u}_h^{n+1}))_h}_{I} \underbrace{-\nu\Delta t((\mathbf{u}_h^{n+1}, \mathbf{u}_h^{n+\frac{1}{2}}))_h}_{II} + \underbrace{2\Delta t \mathbf{b}_h(\mathbf{u}_h^n, \mathbf{u}_h^{n+1} - \mathbf{u}_h^n, \mathbf{u}_h^{n+1})}_{III} \\ &\quad \underbrace{-\Delta t[\mathbf{b}_h(\mathbf{u}_h^n, \mathbf{u}_h^n, \mathbf{u}_h^{n+1}) - \mathbf{b}_h(\mathbf{u}_h^{n-1}, \mathbf{u}_h^{n-1}, \mathbf{u}_h^{n+1})]}_{IV} + \underbrace{\Delta t(\mathbf{f}^{n+1} + \mathbf{f}^n, \mathbf{u}_h^{n+1})_h}_{V}\end{aligned} \tag{9.54}$$

and we have to bound the terms I–V which are in the right-hand side.

- For the first term:

$$\begin{aligned}I = \nu\Delta t((\mathbf{u}_h^{n+1} - \mathbf{u}_h^n, \mathbf{u}_h^{n+1}))_h &\le \nu\Delta t\|\mathbf{u}_h^{n+1} - \mathbf{u}_h^n\|_h\|\mathbf{u}_h^{n+1}\|_h \\ &\le \nu\Delta t S_h|\mathbf{u}_h^{n+1} - \mathbf{u}_h^n|_h\|\mathbf{u}_h^{n+1}\|_h \le \frac{1}{8}|\mathbf{u}_h^{n+1} - \mathbf{u}_h^n|_h^2 + 2\nu^2\Delta t^2 S_h^2\|\mathbf{u}_h^{n+1}\|_h^2.\end{aligned}$$

- For the second term:

$$II = -\nu\Delta t((\mathbf{u}_h^{n+1}, \mathbf{u}_h^{n+\frac{1}{2}}))_h \le \nu\Delta t\|\mathbf{u}_h^{n+1}\|_h\|\mathbf{u}_h^{n+\frac{1}{2}}\|_h \le \frac{1}{4}\nu\Delta t\|\mathbf{u}_h^{n+1}\|_h^2 + \nu\Delta t\|\mathbf{u}_h^{n+\frac{1}{2}}\|_h^2.$$

- For the third term:

$$\begin{aligned}III &= 2\Delta t \mathbf{b}_h(\mathbf{u}_h^n, \mathbf{u}_h^{n+1} - \mathbf{u}_h^n, \mathbf{u}_h^{n+1}) \\ &\le 4\Delta t|\mathbf{u}_h^n|_h^{\frac{1}{4}}\|\mathbf{u}_h^n\|_h^{\frac{3}{4}}|\mathbf{u}_h^{n+1} - \mathbf{u}_h^n|_h^{\frac{1}{4}}\|\mathbf{u}_h^{n+1} - \mathbf{u}_h^n\|_h^{\frac{3}{4}}\|\mathbf{u}_h^{n+1}\|_h \\ &\le 4\Delta t S_h^{\frac{3}{2}}|\mathbf{u}_h^n|_h|\mathbf{u}_h^{n+1} - \mathbf{u}_h^n|_h\|\mathbf{u}_h^{n+1}\|_h \le \frac{1}{8}|\mathbf{u}_h^{n+1} - \mathbf{u}_h^n|_h^2 + 32\Delta t^2 S_h^3\|\mathbf{u}_h^{n+1}\|_h^2|\mathbf{u}_h^n|_h^2.\end{aligned}$$

- For the fourth term:

$$\begin{aligned}IV &= -\Delta t[\mathbf{b}_h(\mathbf{u}_h^n, \mathbf{u}_h^n, \mathbf{u}_h^{n+1}) - \mathbf{b}_h(\mathbf{u}_h^{n-1}, \mathbf{u}_h^{n-1}, \mathbf{u}_h^{n+1})] \\ &= -\Delta t[\mathbf{b}_h(\mathbf{u}_h^n - \mathbf{u}_h^{n-1}, \mathbf{u}_h^n, \mathbf{u}_h^{n+1}) + \mathbf{b}_h(\mathbf{u}_h^{n-1}, \mathbf{u}_h^n - \mathbf{u}_h^{n-1}, \mathbf{u}_h^{n+1})] \\ &\le 2\Delta t|\mathbf{u}_h^n - \mathbf{u}_h^{n-1}|_h^{\frac{1}{4}}\|\mathbf{u}_h^n - \mathbf{u}_h^{n-1}\|_h^{\frac{3}{4}}\|\mathbf{u}_h^{n+1}\|_h \cdot [\, |\mathbf{u}_h^n|_h^{\frac{1}{4}}\|\mathbf{u}_h^n\|_h^{\frac{3}{4}} + |\mathbf{u}_h^{n-1}|_h^{\frac{1}{4}}\|\mathbf{u}_h^{n-1}\|_h^{\frac{3}{4}}] \\ &\le 2\Delta t S_h^{\frac{3}{2}}|\mathbf{u}_h^n - \mathbf{u}_h^{n-1}|_h\|\mathbf{u}_h^{n+1}\|_h[\, |\mathbf{u}_h^n|_h + |\mathbf{u}_h^{n-1}|_h] \\ &\le \frac{1}{4}|\mathbf{u}_h^n - \mathbf{u}_h^{n-1}|_h^2 + 8\Delta t^2 S_h^3\|\mathbf{u}_h^{n+1}\|_h^2[\, |\mathbf{u}_h^n|_h^2 + |\mathbf{u}_h^{n-1}|_h^2\,].\end{aligned}$$

• For the fifth term:

$$V = \Delta t(\mathbf{f}^{n+1} + \mathbf{f}^n, \mathbf{u}_h^{n+1})_h \leq \Delta t(|\mathbf{f}^{n+1}|_h + |\mathbf{f}^n|_h)|\mathbf{u}_h^{n+1}|_h$$
$$\leq c_1\Delta t(|\mathbf{f}^{n+1}|_h + |\mathbf{f}^n|_h)\|\mathbf{u}_h^{n+1}\|_h \leq \frac{\nu\Delta t}{4}\|\mathbf{u}_h^{n+1}\|_h^2 + \frac{2c_1^2\Delta t}{\nu}(|\mathbf{f}^{n+1}|_h^2 + |\mathbf{f}^n|_h^2).$$

So, due to these previous inequalities we obtain the second *apriori* estimate. □

9.6 Proof of stability

In this last section, we present the proof of the Stability Theorem 9.1, which is stated in Section 9.3.

First, we show the inequality (9.43) step by step. We know that the inequality is true for n=0 and n=1. We suppose that $|\mathbf{u}_h^p|_h^2 \leq K_4$ for all p=0, ..., n and we show that $|\mathbf{u}_h^{n+1}|_h^2 \leq K_4$.

The hypothesis (9.42) and the recurrence hypothesis imply that

$$1 - [2\nu\Delta t S_h^2 + \frac{\Delta t S_h^3}{\nu}(42|\mathbf{u}_h^n|_h^2 + 10|\mathbf{u}_h^{n-1}|_h^2)] \geq 1 - [2\nu\Delta t S_h^2 + \frac{52\Delta t S_h^3}{\nu}K_4] \geq \frac{1}{2}.$$

So, the *a priori* estimate (9.51) can be rewritten as

$$\begin{aligned}&\frac{3}{4}|\mathbf{u}_h^{n+\frac{1}{2}} - \mathbf{u}_h^n|_h^2 + |\mathbf{u}_h^{n+\frac{1}{2}}|_h^2 + \frac{\nu\Delta t}{2}\|\mathbf{u}_h^{n+\frac{1}{2}}\|_h^2\\ &\leq \frac{c_1^2\Delta t}{2\nu}(|\mathbf{f}^{n+1}|_h^2 + |\mathbf{f}^n|_h^2) + |\mathbf{u}_h^{n-\frac{1}{2}}|_h^2 + \frac{1}{4}|\mathbf{u}_h^{n-\frac{1}{2}} - \mathbf{u}_h^{n-1}|_h^2.\end{aligned} \tag{9.55}$$

Due to the positive of the third term in the left of (9.55) and $\frac{1}{4} \leq \frac{3}{4}$, we obtain

$$|\mathbf{u}_h^{n+\frac{1}{2}}|_h^2 + \frac{1}{4}|\mathbf{u}_h^{n+\frac{1}{2}} - \mathbf{u}_h^n|_h^2 \leq |\mathbf{u}_h^{1+\frac{1}{2}}|_h^2 + \frac{1}{4}|\mathbf{u}_h^{1+\frac{1}{2}} - \mathbf{u}_h^1|_h^2 + \frac{c_1^2\Delta t}{\nu}\sum_{j=2}^{n+1}|\mathbf{f}^j|_h^2. \tag{9.56}$$

Then, due to the hypotheses (9.38)–(9.41):

$$\begin{aligned}|\mathbf{u}_h^{n+\frac{1}{2}}|_h^2 + \frac{1}{4}|\mathbf{u}_h^{n+\frac{1}{2}} - \mathbf{u}_h^n|_h^2 &\leq |\mathbf{u}_h^{1+\frac{1}{2}}|_h^2 + \frac{1}{2}|\mathbf{u}_h^{1+\frac{1}{2}}|_h^2 + \frac{1}{2}|\mathbf{u}_h^1|_h^2 + \frac{c_1^2}{\nu}TK_2\\ &\leq \underbrace{\frac{3}{2}\tilde{K_3} + \frac{1}{2}K_3 + \frac{c_1^2}{\nu}TK_2}_{K_4}.\end{aligned}$$

Finally, using (9.49), we have shown that $|\mathbf{u}_h^{n+1}|_h^2 \leq K_4$ for all n=0, ..., $N_t - 1$, so we have proved that (9.43) is satisfied.

Now, we want to prove (9.44) and (9.45).

First, we sum the inequality (9.55) for n=2, ..., $N_t - 1$, we obtain

$$\begin{aligned}&\frac{3}{4}\sum_{n=2}^{N_t-1}|\mathbf{u}_h^{n+\frac{1}{2}} - \mathbf{u}_h^n|_h^2 + \sum_{n=2}^{N_t-1}|\mathbf{u}_h^{n+\frac{1}{2}}|_h^2 + \frac{\nu\Delta t}{2}\sum_{n=2}^{N_t-1}\|\mathbf{u}_h^{n+\frac{1}{2}}\|_h^2\\ &\leq \frac{c_1^2\Delta t}{2\nu}\sum_{n=2}^{N_t-1}(|\mathbf{f}^{n+1}|_h^2 + |\mathbf{f}^n|_h^2) + \sum_{n=2}^{N_t-1}|\mathbf{u}_h^{n-\frac{1}{2}}|_h^2 + \frac{1}{4}\sum_{n=2}^{N_t-1}|\mathbf{u}_h^{n-\frac{1}{2}} - \mathbf{u}_h^{n-1}|_h^2.\end{aligned}$$

Simplify the above inequality and use the stability theorem hypotheses (9.38)–(9.41), we write

$$\frac{1}{2}\sum_{n=2}^{N_t-2}|\mathbf{u}_h^{n+\frac{1}{2}}-\mathbf{u}_h^n|_h^2+\frac{3}{4}|\mathbf{u}_h^{N_t-\frac{1}{2}}-\mathbf{u}_h^{N_t-1}|_h^2+|\mathbf{u}_h^{N_t-\frac{1}{2}}|_h^2+\frac{\nu\Delta t}{2}\sum_{n=2}^{N_t-1}\|\mathbf{u}_h^{n+\frac{1}{2}}\|_h^2$$
$$\leq\frac{c_1^2\Delta t}{2\nu}\sum_{n=2}^{N_t-1}(|\mathbf{f}^{n+1}|_h^2+|\mathbf{f}^n|_h^2)+|\mathbf{u}_h^{1+\frac{1}{2}}|_h^2+\frac{1}{4}|\mathbf{u}_h^{1+\frac{1}{2}}-\mathbf{u}_h^1|_h^2$$
$$\leq\frac{c_1^2}{\nu}TK_2+|\mathbf{u}_h^{1+\frac{1}{2}}|_h^2+\frac{1}{4}|\mathbf{u}_h^{1+\frac{1}{2}}-\mathbf{u}_h^1|_h^2\leq K_4.$$

Then, we obtain the following inequality:

$$\nu\Delta t\sum_{n=1}^{N_t-1}\|\mathbf{u}_h^{n+\frac{1}{2}}\|_h^2\leq 2K_4+\tilde{K}_3. \tag{9.57}$$

The hypothesis (9.42) and the stability theorem inequality (9.43) imply that

$$1-[4\nu\Delta tS_h^2+\frac{\Delta tS_h^3}{\nu}(80|\mathbf{u}_h^n|_h^2+16|\mathbf{u}_h^{n-1}|_h^2)]\geq 1-[4\nu\Delta tS_h^2+\frac{96\Delta tS_h^3}{\nu}K_4]\geq\frac{1}{2}.$$

So, the *a priori* estimate (9.52) can be rewritten as

$$\begin{aligned}&\frac{3}{4}|\mathbf{u}_h^{n+1}-\mathbf{u}_h^n|_h^2+|\mathbf{u}_h^{n+1}|_h^2+\frac{\nu\Delta t}{4}\|\mathbf{u}_h^{n+1}\|_h^2\\&\leq\frac{2c_1^2\Delta t}{\nu}(|\mathbf{f}^{n+1}|_h^2+|\mathbf{f}^n|_h^2)+|\mathbf{u}_h^n|_h^2+\frac{1}{4}|\mathbf{u}_h^n-\mathbf{u}_h^{n-1}|_h^2+\nu\Delta t\|\mathbf{u}_h^{n+\frac{1}{2}}\|_h^2.\end{aligned} \tag{9.58}$$

We sum the above inequality for $n=1,\ldots,N_t-1$:

$$\frac{3}{4}\sum_{n=1}^{N_t-1}|\mathbf{u}_h^{n+1}-\mathbf{u}_h^n|_h^2+\sum_{n=1}^{N_t-1}|\mathbf{u}_h^{n+1}|_h^2+\frac{\nu\Delta t}{4}\sum_{n=1}^{N_t-1}\|\mathbf{u}_h^{n+1}\|_h^2$$
$$\leq\frac{2c_1^2\Delta t}{\nu}\sum_{n=1}^{N_t-1}(|\mathbf{f}^{n+1}|_h^2+|\mathbf{f}^n|_h^2)+\sum_{n=1}^{N_t-1}|\mathbf{u}_h^n|_h^2+\frac{1}{4}\sum_{n=1}^{N_t-1}|\mathbf{u}_h^n-\mathbf{u}_h^{n-1}|_h^2$$
$$+\nu\Delta t\sum_{n=1}^{N_t-1}\|\mathbf{u}_h^{n+\frac{1}{2}}\|_h^2.$$

We simplify the above inequality and use the hypotheses (9.38)–(9.41), (9.57), we obtain

$$\frac{1}{2}\sum_{n=1}^{N_t-2}|\mathbf{u}_h^{n+1}-\mathbf{u}_h^n|_h^2+\frac{3}{4}|\mathbf{u}_h^{N_t}-\mathbf{u}_h^{N_t-1}|_h^2+|\mathbf{u}_h^{N_t}|_h^2+\frac{\nu\Delta t}{4}\sum_{n=1}^{N_t-1}\|\mathbf{u}_h^{n+1}\|_h^2$$
$$\leq\frac{2c_1^2\Delta t}{\nu}\sum_{n=1}^{N_t-1}(|\mathbf{f}^{n+1}|_h^2+|\mathbf{f}^n|_h^2)+|\mathbf{u}_h^1|_h^2+\frac{1}{4}|\mathbf{u}_h^1-\mathbf{u}_h^0|_h^2+\nu\Delta t\sum_{n=1}^{N_t-1}\|\mathbf{u}_h^{n+\frac{1}{2}}\|_h^2$$
$$\leq\frac{4c_1^2}{\nu}TK_2+\frac{3}{2}K_3+\frac{1}{2}K_1+2K_4+\tilde{K}_3.$$

Finally, we can conclude that there exists a constant K_5 such that

$$\sum_{n=0}^{N_t-1}|\mathbf{u}_h^{n+1}-\mathbf{u}_h^n|_h^2\leq K_5,\qquad \nu\Delta t\sum_{n=0}^{N_t-1}\|\mathbf{u}_h^{n+1}\|_h^2\leq K_5.$$

So, the stability theorem inequalities (9.44) and (9.45) are shown and the proof is completed.

Acknowledgement

We are very grateful to our advisor, Professor Yu-Jiang Wu, whose advice and support are very helpful.

Bibliography

Chorin, A. J. (1968). Numerical solution of the Navier-Stokes equations, *Math. Comp.* **22**, pp. 745–762.

Eymard, R., Gallouet, T. and Herbin, R. (2000). Finite volume methods, *Handbook of numerical analysis* **VII**, (North-Holland, Amsterdam), pp. 713–1020.

Faure, S. (2005). Stability of a colocated finite volume scheme for the Navier-Stokes equations, *Numer. Methods P. D. E.* **21**, pp. 242–271.

Faure, S., Laminie, J. and Temam, R. (2005). Finite volume discretization and multilevel methods in flow problems, *J. Sci. Comput.* **25**, pp. 231–261.

Faure, S., Laminie, J. and Temam, R. (2008). Colocated finite volume schemes for fluid flows, *Commun. Comput. Phys.* **4**, pp. 1–25.

Temam, R. (2001). *Navier-Stokes Equations: Theory and numerical analysis,* 3th edn. (AMS Chelsea Publishing, Providence, RI).

Temam, R. and Marion, M. (1998). Navier-Stokes equations: theory and approximation, *Handbook of numerical analysis* **VI**, (North-Holland, Amsterdam), pp. 503–688.

Li, X., Wu, Y. J. and Zhao X. D. (2009). Improvement of the stability condition of a colocated finite volume scheme for two-dimensional Navier-Stokes equations, *Numer. Math. J. Chinese Univ.* **31**, pp. 349–368.

Chapter 10

Computing the multiple positive solutions to p-Henon equation on the unit square

Zhaoxiang Li* and Zhong-hua Yang

Department of Mathematics, Shanghai Normal University, Shanghai 200234, PR China

**zxli@shnu.edu.cn*

In this chapter, the algorithms are proposed to solve the D_4 symmetric positive solutions to the boundary value problem of p-Henon equation. Taking r in p-Henon equation as a bifurcation parameter, the symmetry-breaking bifurcation points on the branch of the D_4 symmetric positive solutions are found via the extended systems and other symmetric positive solutions are computed by the branch switching method based on the Liapunov-Schmidt reduction.

Keywords: p-Henon equation, symmetry-breaking bifurcation, multiple solutions, extended system, branch switching

10.1 Introduction

In this chapter, We consider the boundary value problem (BVP) of p-Henon equation

$$f(u,r) = \begin{cases} \Delta_p u + |\boldsymbol{x} - \boldsymbol{x}^0|^r |u|^{q-1} u = 0, & \boldsymbol{x} \in \Omega, \\ u|_{\partial\Omega} = 0, \end{cases} \tag{10.1}$$

where $\Delta_p u = div(|\nabla u|^{p-2}\nabla u)$, $q \geq \frac{1}{2}$, $1 < p < q+1$, $r \geq 0$, $\boldsymbol{x} = (x_1, x_2)$, Ω is the unit square $[0,1] \times [0,1]$ in $\mathbb{R}^2$, $\partial\Omega$ is the boundary of Ω, $\boldsymbol{x}^0 = (1/2, 1/2)$.

$\Delta_p u(\boldsymbol{x})$ is the nonlinear p-Laplacian differential operator, which has a variety of applications in physics, engineering, biology and ecology. The p-Laplacian operator also appears in the study of flow in a porous media ($p = \frac{3}{2}$), nonlinear elasticity ($p > 2$) and glaciology ($p \in (1, \frac{4}{3})$) [Diaz (1985)]. When p=2 in p-Henon equation (10.1), the properties of the solutions to the equation (10.1) have been studied by many authors [Byeon and Wang (2005); Cao and Peng (2003)].

Since 60's of the 20th century, the existence and multiplicity of solutions to the boundary value problems of the nonlinear elliptic PDE's such as problem (10.1) are studied by the monotone iterative method in the ordered Banach space [Amann (1976); Amann and Crandall (1978)] and the mountain pass lemma and the min-max theorem in the critical point theory [Chang (1993); Struwe (1996)]. It becomes an important field in PDE study. But what distribution and structure the solutions to the BVP of the nonlinear elliptic equations have and how to compute them have attracted the attention of many mathematicians, physicists and engineers.

There are mainly five numerical methods for computing such kinds of problems: the Monotone Iterative Scheme [Pao (1995); Deng *et al.* (1996)], the Mountain Pass Algorithm (MPA) [Choi and McKenna (1993)], the High Linking Algorithm (HLA) [Ding *et al.* (1999)], the Min-Max Algorithm (MMA) [Li and Zhou (2002); Yao and Zhou (2005)] and the Search Extension Method (SEM) [Chen and Xie (2004)]. MIS is based on the monotone iterative methods in the ordered Banach space. MPA, MMA and HLA are based on the numerical implement of the mountain pass lemma and the min-max theorem in the critical point theory. MPA was proposed by Choi and McKenna to compute the solutions with the Morse Index (MI) 0 or 1. Ding, Costa and Chen established HLA for sign-changing solution (MI=2) of semilinear elliptic problems. Li and Zhou designed a new min-max algorithm (MMA) to find multiple saddle points with any Morse index which is more constructive than the traditional min-max theorem. Chen and Xie proposed SEM, which searches the initial guess based on the linear combination of the eigenfunctions of the linearlized problem and then gets the better initial guess by the continuation method for the discretized problem by the finite element method. The bifurcation method [Yang *et al.* (2008); Li and Yang (2010); Zhu and Li (2010); Li *et al.* (2011b,a)] is applied successfully to solving the BVP of the Henon equation on the unit square. The advantages of the bifurcation method are that it can compute the solutions to problem (10.1) with any Morse index and the different symmetries as many as possible and it can simplify the computation of problem (10.1). On the other hand, the difficulty in searching the initial guess in the other methods can be solved effectively by the bifurcation method.

The most interesting solutions to problems (10.1) are the positive solutions with D_4-symmetry or other symmetry [Li and Yang (2004); Li *et al.* (2005)]. The symmetry group $D_4 = \{I, R_1, R_2, R_3, S_1, S_2, S_1', S_2'\}$, where

$$\begin{aligned} R_1u(x_1,x_2) &= u(1-x_2,x_1), & R_2u(x_1,x_2) &= u(1-x_1,1-x_2),\\ R_3u(x_1,x_2) &= u(x_2,1-x_1), & Iu(x_1,x_2) &= u(x_1,x_2),\\ S_1u(x_1,x_2) &= u(x_1,1-x_2), & S_1'u(x_1,x_2) &= u(1-x_1,x_2),\\ S_2u(x_1,x_2) &= u(x_2,x_1), & S_2'u(x_1,x_2) &= u(1-x_2,1-x_1). \end{aligned}$$

The problem (10.1) is Γ equivariant, namely

$$f(\gamma u, r) = \gamma f(u, r), \quad \forall \gamma \in \Gamma,$$

where $\Gamma = D_4 \times Z_2, Z_2 = \{I, -I\}$. The isotropy subgroups of D_4 are $D_4 = \{I, R_1, R_2, R_3, S_1, S_2, S_1', S_2'\}$; $\Sigma_R = \{I, R_1, R_2, R_3\}$; $\Sigma_r = \{I, R_2\}$; $\Sigma_1 = \{I, S_1\}$; $\Sigma_1' = \{I, S_1'\}$; $\Sigma_2 = \{I, S_2\}$; $\Sigma_2' = \{I, S_2'\}$; $\Sigma_d = \{I, R_2, S_2, S_2'\}$; $\Sigma_M =$

$\{I, R_2, S_1, S_1'\}$. Let Σ be one of the above isotropy subgroups and X^Σ be the invariant subspace of Σ.

In Section 10.2, we give two algorithms, which are applied to computing the D_4 symmetric positive solutions to the problem (10.1). In Section 10.3, we take r in p-Henon equation as a bifurcation parameter and propose an extended systems, which can detect the symmetry-breaking bifurcation points on the branch of the D_4 symmetric positive solutions. In Section 10.4, other symmetric positive solutions are computed by the branch switching method based on the Liapunov-Schmidt reduction. Finally, in Section 10.5 the numerical results are given and the multiple positive solutions to problem (10.1) are visualized.

10.2 Computation of D_4 symmetric positive solutions

The first algorithm: p continuation.

Step 1: Discretization.

Ω is divided homogeneously into $2N^2$ small triangle , where $N = 160, h = 1/160$. The triangulation keeps D_4 symmetry of Ω (see Fig. 10.1). The linear finite element approximation is used to discrete problem (10.1). The discretization equations

$$f(u_h, r) = 0, \tag{10.2}$$

where $u_h = \{u_i | i = 1, 2, \cdots, (N-1)^2\}$ keep the same Γ equivariance as the original problem (10.1).

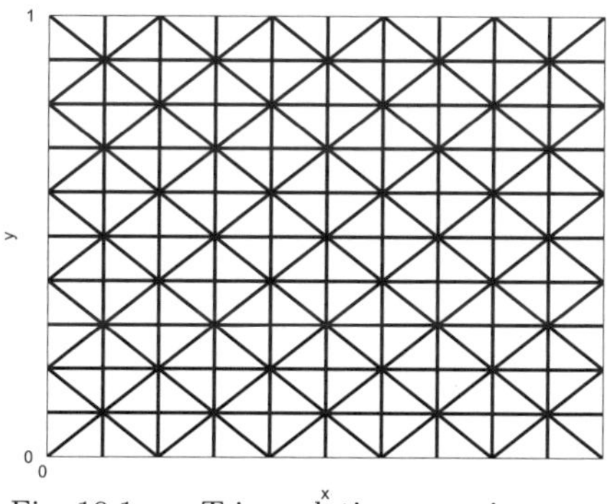

Fig. 10.1 Triangulation on unit square

Step 2: p continuation.

Starting from the D_4 symmetric positive solution to problem (10.1) for $p = 2$, we can get the D_4 symmetric positive solutions to the problem (10.1) for $p = p^*$ (p^*=1.75 or 3) with p continuation and the Newton iteration method (see Fig. 10.2 and Fig. 10.7).

The second algorithm: Taking r as a parameter, the D_4 symmetric positive solutions to problem (10.1) can be computed with r continuation.

The D_4 symmetric positive solution to problem (10.1) for r=0 is used as a starting point on the D_4 symmetric positive solution branch Γ of problem (10.1) with varied r which can be computed effectively with r continuation and the Newton iteration method. While r is continued, the eigenvalues of Jacobian $f_{u_h}(u_h, r)$ are monitored. When $p = 1.75, q = 3$, the eigenvalue with small absolute value of Jacobian $f_{u_h}(u_h, r)$ are found at r near 0.41 and 2.29. The corresponding eigenvector has Σ_1 (Σ_2) and Σ_d symmetry respectively. When $p = 3, q = 7$, the eigenvalue with small absolute value of Jacobian $f_{u_h}(u_h, r)$ are found at r near 0.22, 2.23, 2.56 and 5.19. The corresponding eigenvector has Σ_1 (Σ_2), Σ_d and Σ_M symmetry respectively. These are the potential symmetry-breaking bifurcation point [Golubitsky *et al.* (1988)], [Yang (2007)].

10.3 Computation of the symmetry-breaking bifurcation point

In the following, let Σ be one of Σ_1, Σ_2 , Σ_d and Σ_M, X^Σ be the invariant subspace of Σ. Since

$$f(\gamma u, r) = \gamma f(u, r), \qquad \forall \gamma \in D_4, \tag{10.3}$$

X^Σ can be decomposed into $X^\Sigma = X^{D_4} \oplus W$, where $W = X^\Sigma \cap (X^{D_4})^\perp$, $(X^{D_4})^\perp$ is an orthogonal complement of X^{D_4}.

Lemma 10.1. *If $u \in X^{D_4}$, then the following conclusions hold $\forall\ r \in \mathbb{R}$:*
(i) $f(u, r)$, $f_r(u, r) \in X^{D_4}$,
(ii) X^{D_4} and W are the invariant subspaces of $f_u(u, r)$ and $f_{ur}(u, r)$,
(iii) if $v \in X^{D_4}$, then X^{D_4} and W are the invariant subspaces of $f_{uu}(u, r)v$.

Proof. (i) From equivariance of (10.3), if $u \in X^{D_4}$,

$$f(u, r) = \gamma f(u, r), \quad \forall \gamma \in D_4,$$

$$f_r(u, r) = \gamma f_r(u, r), \quad \forall \gamma \in D_4,$$

which means $f(u, r), f_r(u, r) \in X^{D_4}$.

(ii)

(1) From (10.3), it is known that

$$\gamma f_u(u, r) = f_u(\gamma u, r)\gamma, \quad \forall \gamma \in D_4. \tag{10.4}$$

If $u, v \in X^{D_4}$, we get

$$\gamma f_u(u, r)v = f_u(u, r)v, \quad \forall \gamma \in D_4,$$

which means $f_u(u, r)v \in X^{D_4}$, X^{D_4} is an invariant subspace of $f_u(u, r)$. Similarly, X^{D_4} is an invariant subspace of $f_{ur}(u, r)$, X^Σ also is an invariant subspace of $f_u(u, r)$

and $f_{ur}(u,r)$.

(2) On the other hand, from (10.4) we obtain

$$f_u^{\mathrm{T}}(u,r)\gamma^{\mathrm{T}} = \gamma^{\mathrm{T}} f_u^{\mathrm{T}}(\gamma u, r), \quad \forall \gamma \in D_4. \tag{10.5}$$

If $v \in X^{D_4}$, then

$$f_u^{\mathrm{T}}(u,r)v = \gamma^{\mathrm{T}} f_u^{\mathrm{T}}(u,r)v, \quad \forall \gamma \in D_4,$$

which means $f_u^{\mathrm{T}}(u,r)v \in X^{D_4}$, namely, X^{D_4} is an invariant subspace of $f_u^{\mathrm{T}}(u,r)$. Similarly, X^{D_4} is an invariant subspace of $f_{ur}^{\mathrm{T}}(u,r)$, X^{Σ} also is an invariant subspace of $f_u^{\mathrm{T}}(u,r)$ and $f_{ur}^{\mathrm{T}}(u,r)$.

(3) If $v \in X^{D_4}$, then $f_u^{\mathrm{T}}(u,r)v \in X^{D_4}$,

$$\langle f_u^{\mathrm{T}}(u,r)v,\ w\rangle = 0, \qquad \forall w \in W, \qquad v \in X^{D_4},$$

$$\langle v,\ f_u(u,r)w\rangle = 0, \qquad \forall w \in W, \qquad v \in X^{D_4},$$

therefore $f_u(u,r)w \in W$, namely, W is an invariant subspace of $f_u(u,r)$. Similarly, W is an invariant subspace of $f_{ur}(u,r)$.

(iii) From (10.3), it is true that

$$\gamma f_{uu}(u,r)vw = f_{uu}(\gamma u, r)\gamma v \gamma w, \quad \forall \gamma \in D_4.$$

If $u, v, w \in X^{D_4}$, we have

$$\gamma f_{uu}(u,r)vw = f_u(u,r)vw, \quad \forall \gamma \in D_4,$$

which means $f_{uu}(u,r)vw \in X^{D_4}$, namely X^{D_4} is an invariant subspace of $f_{uu}(u,r)v$. Similarly X^{Σ} also is an invariant subspace of $f_{uu}(u,r)v$.

Also, we have

$$\gamma f_{uu}(u,r)v = f_{uu}(\gamma u, r)\gamma v \gamma.$$

Therefore

$$(f_{uu}(u,r)v)^{\mathrm{T}}\gamma^{\mathrm{T}} = \gamma^{\mathrm{T}}(f_{uu}(\gamma u, r)\gamma v)^{\mathrm{T}},$$

$$(f_{uu}(u,r)v)^{\mathrm{T}}\gamma^{\mathrm{T}} w = \gamma^{\mathrm{T}}(f_{uu}(\gamma u, r)\gamma v)^{\mathrm{T}} w.$$

If $u, v, w \in X^{D_4}$, then $(f_{uu}(u,r)v)^{\mathrm{T}} w \in X^{D_4}$, which means X^{D_4} is an invariant subspace of $(f_{uu}(u,r)v)^{\mathrm{T}}$. If $w \in W$, $v, h \in X^{D_4}$, then

$$\langle h, f_{uu}vw\rangle = \langle (f_{uu}v)^{\mathrm{T}}h, w\rangle = 0,$$

which means $f_{uu}vw \in W$, thus W is an invariant subspace of $f_{uu}v$. □

On the D_4 symmetric positive solution branch, if there is a point (u_0, r_0) at which $f_u^0 = f_u(u_0, r_0)$ is singular and its null space is $N(f_u^0) = span\{\phi_0\}$, its range space is $R(f_u^0) = \{x \in X^{\Sigma} | \langle \psi_0, x\rangle = 0\}$, where $\phi_0 \in W$ and $\psi_0 \in W$ are null eigenvector of f_u^0 and $(f_u^0)^{\mathrm{T}}$ respectively, and if

$$\langle \psi_0, (f_{uu}^0 v_r + f_{ur}^0)\phi_0\rangle \neq 0, \tag{10.6}$$

where $v_r \in X^{D_4}$ is the unique solution to

$$f_u^0 v_r + f_r^0 = 0, \tag{10.7}$$

then the point (u_0, r_0) on the D_4 symmetric positive solution branch of problem (10.1) is called D_4-Σ symmetry-breaking bifurcation point with respect to r.

The following is the extended system for detecting the D_4-Σ symmetry-breaking bifurcation points:

$$F(y) = \begin{pmatrix} f(u,r) \\ f_u(u,r)\phi \\ \langle h_0, \phi\rangle - 1 \end{pmatrix} = 0, \tag{10.8}$$

where $y = (u, \phi, r) \in Y = X^{D_4} \times W \times \mathbb{R}$, $y_0 = (u_0, \phi_0, r_0)$, $h_0 \in W$ is a normalization of ϕ_0.

Theorem 10.1. *The extended system (10.8) is regular at the D_4-Σ symmetry-breaking bifurcation point $y_0 = (u_0, \phi_0, r_0)$.*

Proof. We can know

$$F_y^0 = \begin{pmatrix} f_u^0 & 0 & f_r^0 \\ f_{uu}^0\phi_0 & f_u^0 & f_{ur}^0\phi_0 \\ 0 & \langle h_0, \cdot\rangle & 0 \end{pmatrix}.$$

Next, we prove that $F_y^0 : Y \longrightarrow Y$ is one-to-one. Let

$$F_y^0 Z = 0, \tag{10.9}$$

where $Z = (v, w, \alpha)^{\mathrm{T}}$, $v \in X^{D_4}$, $w \in W$, $\alpha \in \mathbb{R}$. Expanding (10.9) yields

$$f_u^0 v + \alpha f_r^0 = 0, \tag{10.10}$$

$$f_{uu}^0 \phi_0 v + f_u^0 w + \alpha f_{ur}^0 \phi_0 = 0, \tag{10.11}$$

$$\langle h_0, w\rangle = 0. \tag{10.12}$$

From (10.10), we can get $v = \alpha v_r$. Substituting $v = \alpha v_r$ into (10.11) and taking an inner product with ψ_0 lead to $\alpha\langle\psi_0, (f_{uu}^0 v_r \phi_0 + f_{ur}^0 \phi_0)\rangle = 0$. Therefore $\alpha = 0$ due to (10.6). Also $w = 0$ by solving $f_u^0 w = 0$, $\langle h_0, w\rangle = 0$. Similarly we can prove that $F_y^0 : Y \longrightarrow Y$ is onto. Therefore F_y^0 is regular. □

Since F_y^0 is regular, we can solve the extended system (10.8) with Newton iteration method. During continuation of D_4 symmetric positive solutions, we can find some $u^* \in X^{D_4}$, $r^* \in \mathbb{R}$, at which the Jacobian $f_u(u, r)$ has eigenvalues with small absolute value. (u^*, r^*) and the corresponding eigenvector of the eigenvalue with small absolute value can be used as the initial guess for the Newton iteration. The following is the numerical results for $p = 1.75, q = 3$:

Table 10.1 Symmetry-breaking bifurcation points for $p = 1.75, q = 3$.

Bifurcation point	D_4-Σ_1	D_4-Σ_2	D_4-Σ_d
r	0.4140625	0.4140625	2.29960937

We find $r = 0.4140625$ is a double symmetry breaking bifurcation point. The corresponding null space $N(f_u(u,r)) = span\{\phi_0^{(1)}, \phi_0^{(2)}\}$, where $\phi_0^{(1)} \in X^{\Sigma_1}$, $\phi_0^{(2)} \in X^{\Sigma_2}$. The next is the numerical results for $p = 3, q = 7$:

Table 10.2 Symmetry-breaking bifurcation points for $p = 3, q = 7$.

Bifurcation point	D_4-Σ_1	D_4-Σ_2	D_4-Σ_d	D_4-Σ_M	D_4-Σ_M
r	0.221851	0.221851	2.233057	2.561963	5.196680

We again find $r = 0.22185058$ is a double symmetry breaking bifurcation point. The corresponding null space $N(f_u(u,r)) = span\{\phi_0^{(1)}, \phi_0^{(2)}\}$, where $\phi_0^{(1)} \in X^{\Sigma_1}$, $\phi_0^{(2)} \in X^{\Sigma_2}$.

10.4 Branch switching to Σ symmetric solutions

Let the D_4-Σ symmetry-breaking bifurcation point be $r = r_0$, $u = u_0 \in X^{D_4}$, $\psi_0 \in W$, $\phi_0 \in W$. The numerical computation shows

$$\begin{aligned} a &= \langle \psi_0, f_{uu}^0 \phi_0 \phi_0 \rangle = 0, \\ b &= \langle \psi_0, (f_{uu}^0 v_r + f_{ur}^0) \phi_0 \rangle \neq 0, \\ c &= \langle \psi_0, (f_{uu}^0 v_r v_r + 2 f_{ur}^0 v_r + f_{rr}^0) \rangle = 0, \end{aligned}$$

where $v_r \in X^{D_4}$ is the unique solution to equation (10.7). Therefore assumption (10.6) is satisfied. From the algebraic bifurcation equation [Yang (2007)], [Golubitsky *et al.* (1988)], we know that at the symmetry-breaking bifurcation point, the tangent vector along the D_4 symmetric positive solution branch is $(v_r, 1)$, the tangent vector along the Σ symmetric positive solution branch is $(\phi_0, 0)$. We define that

$$G(w,\eta,\varepsilon) = \begin{cases} \frac{1}{\varepsilon} f(u_s(r_0+\eta) + \varepsilon(\phi_0 + w), r_0 + \eta), & \varepsilon \neq 0, \\ f_u(u_s(r_0+\eta), r_0+\eta)(\phi_0 + w), & \varepsilon = 0, \end{cases} \tag{10.13}$$

$$N(w,\eta,\varepsilon) = \langle \phi_0, w \rangle. \tag{10.14}$$

where $u_s(r_0+\eta)$ are the D_4 symmetric positive solutions, $w \in W, \eta, \varepsilon \in \mathbb{R}$. Obviously, $G(0,0,0) = 0$, $N(0,0,0) = 0$. Jacobian of (10.13), (10.14) with respect to w, η at $(w,\eta,\varepsilon) = (0,0,0)$ is that

$$A^0 = \frac{\partial(G,N)}{\partial(w,\eta)}|_{(0,0,0)} = \begin{pmatrix} f_u^0 & B^0 \\ \langle \phi_0, \cdot \rangle & 0 \end{pmatrix},$$

where $B^0 = [f_{uu}^0(u_s(r_0), r_0) u_s'(r_0) + f_{ur}^0(u_s(r_0), r_0)]\phi_0 = [f_{uu}^0 v_r + f_{ur}^0]\phi_0$. Since $\langle \psi_0, B^0 \rangle = b \neq 0, \phi_0 \in N(f_u^0)$, we have $B^0 \notin R(f_u^0)$, $N(f_u^0) \cap N(\langle \phi_0, \cdot \rangle) = \{0\}$. Keller lemma [Yang (2007)], [Golubitsky *et al.* (1988)] ensures that A^0 is nonsingular. The implicit function theorem leads that

$$\begin{cases} G(w,\eta,\varepsilon) = 0, \\ N(w,\eta,\varepsilon) = 0, \end{cases} \tag{10.15}$$

have solutions $(w(\varepsilon), \eta(\varepsilon))$, $\forall\ |\varepsilon| < \varepsilon_0$, which can be solved with the Newton iteration method. Therefore, we obtain the Σ symmetric positive solution branch $(u_s(r_0 + \eta(\varepsilon)) + \varepsilon(\phi_0 + w(\varepsilon)), r_0 + \eta(\varepsilon))$ of problem (10.1) which is switched from the D_4 symmetric positive solution branch.

10.5 Numerical results

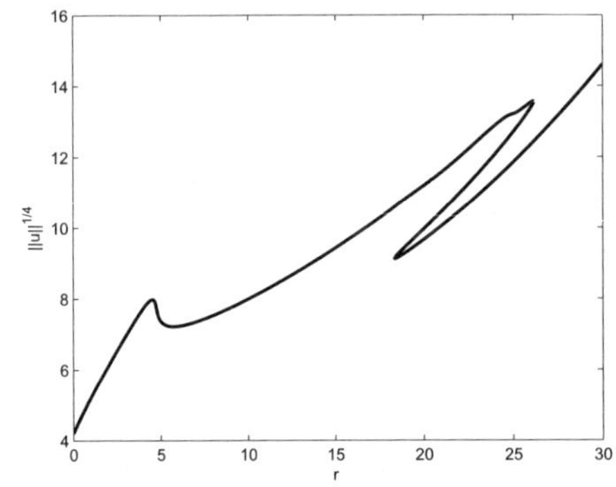

Fig. 10.2 D_4 symmetric positive solutions branch (p=1.75,q=3).

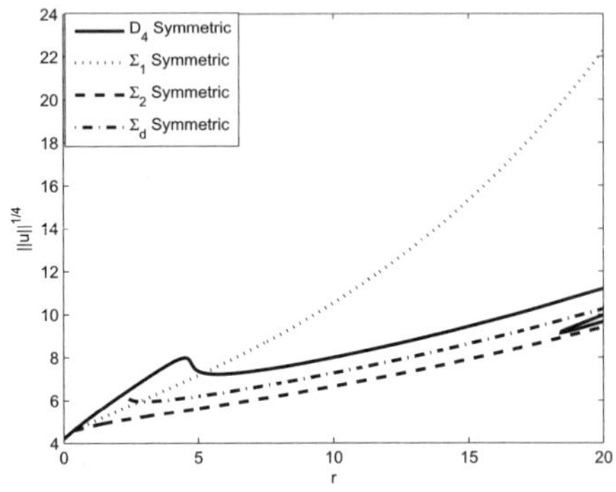

Fig. 10.3 symmetry-breaking bifurcation diagram (p=1.75,q=3).

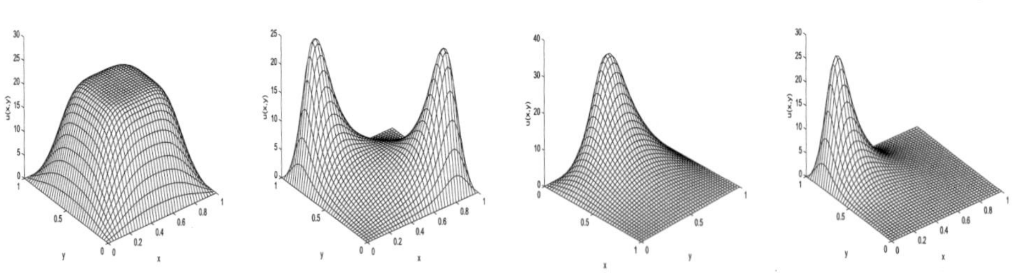

Fig. 10.4 Four positive solutions to (10.1) for $p = 1.75, q = 3, r = 3$ with D_4 , Σ_d, Σ_1 and Σ_2 symmetry.

For the case of $p = 1.75, q = 3$, Fig. 10.2 shows the D_4 symmetric positive solution branch of problem (10.1), on which there are two symmetry-breaking bifurcation points. In Fig. 10.3, the D_4 symmetric positive solution branch, Σ_d symmetric positive solution branch, Σ_1 symmetric positive solution branch and Σ_2 symmetric

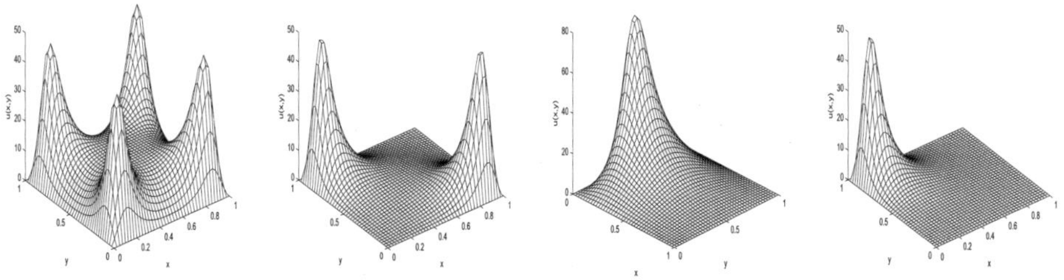

Fig. 10.5 Four positive solutions to (10.1) for $p = 1.75, q = 3, r = 5$ with D_4 , Σ_d, Σ_1 and Σ_2 symmetry.

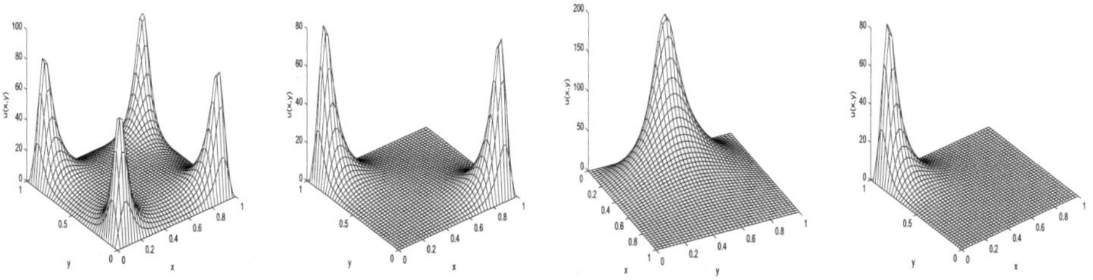

Fig. 10.6 Four positive solutions to (10.1) for $p = 1.75, q = 3, r = 7$ with D_4 , Σ_d, Σ_1 and Σ_2 symmetry.

positive solution branch are respectively drawn by the solid line, the dashed line,the dot-dashed line and the dotted line. When $0 < r < 0.4140625$, problem (10.1) has a D_4 symmetric positive solution. When $0.4140625 < r < 2.29960937$, problem (10.1) has three positive solutions, which are respectively D_4, Σ_1 and Σ_2 symmetric. When $r > 2.29960937$, problem (10.1) has four positive solutions, which are respectively D_4, Σ_1, Σ_2 and Σ_d symmetric. Fig. 10.4 shows four positive solutions to problem (10.1) for $p = 1.75, q = 3, r = 3$ with D_4, Σ_d, Σ_1 and Σ_2 symmetry. Fig. 10.5 shows four positive solutions to (10.1) for $p = 1.75, q = 3, r = 5$ with D_4, Σ_d, Σ_1 and Σ_2 symmetry. Fig. 10.6 shows four positive solutions to (10.1) for $p = 1.75, q = 3, r = 7$ with D_4, Σ_d, Σ_1 and Σ_2 symmetry.

For the case of $p = 3, q = 7$, Fig. 10.7 shows the D_4 symmetric positive solution branch of problem (10.1), on which there are four symmetry-breaking bifurcation points. In Fig. 10.8 the Σ_2 symmetric positive solution branch, Σ_1 symmetric positive solution branch, Σ_d symmetric positive solution branch, Σ_M symmetric positive solution branch, Σ_M symmetric positive solution branch and D_4 symmetric posi-

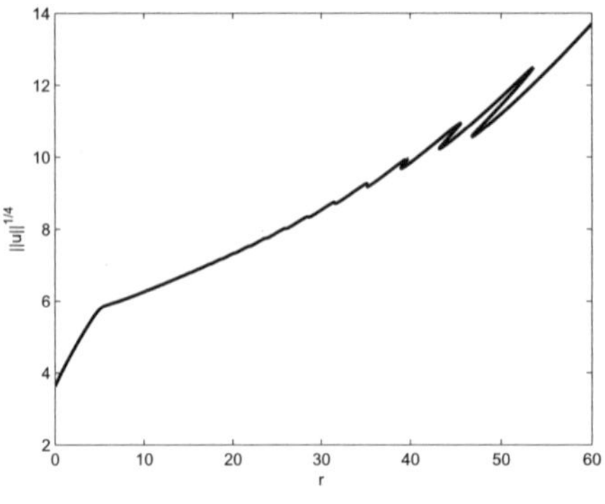

Fig. 10.7 D_4 symmetric positive solutions branch (p=3,q=7).

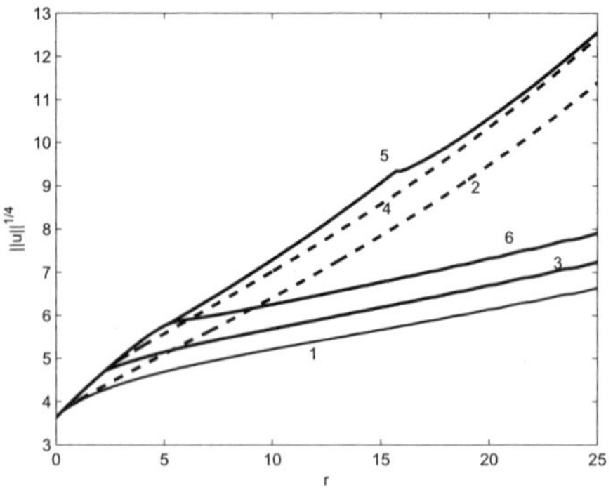

Fig. 10.8 symmetry-breaking bifurcation diagram (p=3,q=7).

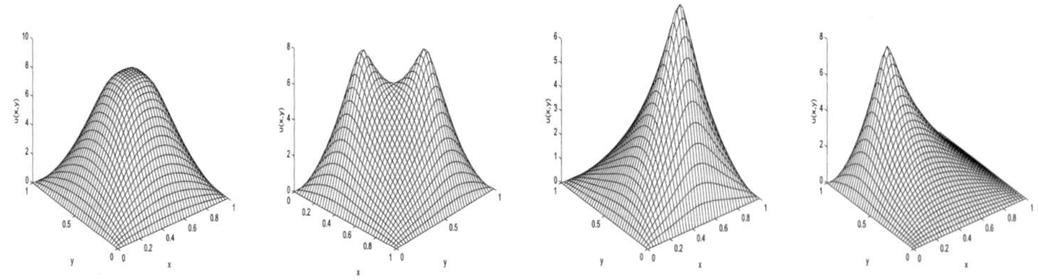

Fig. 10.9 Four positive solutions to (10.1) for $p = 3, q = 7,\ r = 3$ with D_4 , Σ_d, Σ_1 and Σ_2 symmetry.

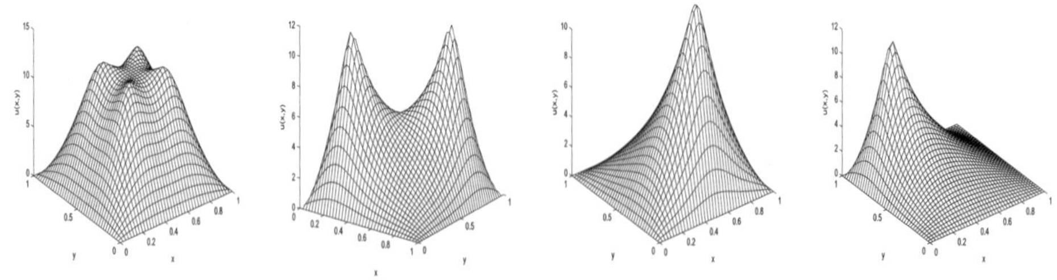

Fig. 10.10 Four positive solutions to (10.1) for $p = 3, q = 7,\ r = 5$ with D_4 , Σ_d, Σ_1 and Σ_2 symmetry.

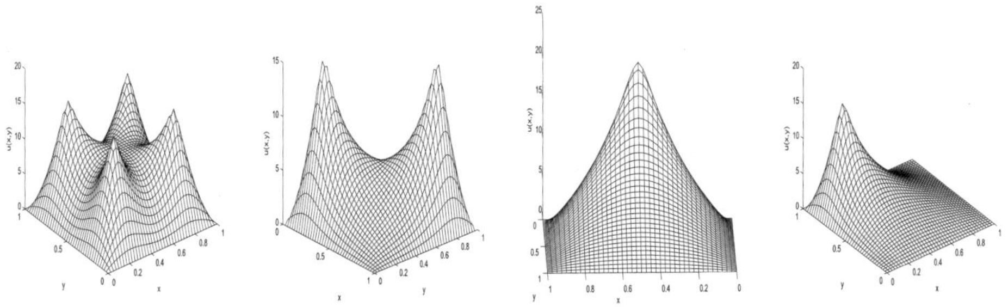

Fig. 10.11 Four positive solutions to (10.1) for $p = 3, q = 7,\ r = 7$ with D_4 , Σ_d, Σ_1 and Σ_2 symmetry.

tive solution branch are respectively drawn with marking 1, 2, 3, 4, 5 and 6. When $0 < r < 0.22185058$, problem (10.1) has a D_4 symmetric positive solution. When $0.22185058 < r < 2.23305664$, problem (10.1) has three positive solutions, which are respectively D_4, Σ_1 and Σ_2 symmetric. When $2.23305664 < r < 2.56196289$, problem (10.1) has four positive solutions, which are respectively D_4, Σ_1, Σ_2 and Σ_d symmetric. When $2.56196289 < r < 5.19667968$, problem (10.1) has five positive solutions, which are respectively D_4, Σ_1, Σ_2, Σ_d and Σ_M symmetric. When $r > 5.19667968$, problem (10.1) has six positive solutions, which are respectively D_4, Σ_1, Σ_2, Σ_d, Σ_M and Σ_M symmetric. Fig. 10.9 shows four positive solutions to problem (10.1) for $p = 3, q = 7, r = 3$ with D_4, Σ_d, Σ_1 and Σ_2 symmetry. Fig. 10.10 shows four positive solutions to problem (10.1) for $p = 3, q = 7, r = 5$ with D_4, Σ_d, Σ_1 and Σ_2 symmetry. Fig. 10.11 shows four positive solutions to (10.1) for $p = 3, q = 7, r = 7$ with D_4, Σ_d, Σ_1 and Σ_2 symmetry.

Bibliography

Amann, H. (1976). Supersolution,monotone iteration and stability, *J. Diff. Eq.* **22**, pp. 367–377.

Amann, H and Crandall, M. G. (1978). On some existence theorems for semilinear elliptic equations, *Indian Univ. Math. J.* **27**, pp. 779–790.

Byeon, J. and Wang, Z. Q. (2005) On the Henon equation: asymptotic profile of ground states II, *J Differential Equations* **216**, pp. 78–108.

Cao, D. M. and Peng, S. J. (2003). The asymptotic behavior of the ground state solutions for Henon equation, *J. Math. Anal. Appl.* **278**, pp. 1–17.

Chang, K. C. (1993). *Variational Methods, A Series of Modern Surveys in Math* (Birkhauser, Boston).

Chen, C. M. and Xie, Z. Q. (2004). Search-extension method for multiple solutions of nonlinear problem, *Comp. Math. Appl.* **47**, pp. 327–343.

Choi, Y. S. and McKenna, P. J. (1993). A mountain pass method for the numerical solutions of semilinear elliptic problems, *Nonlinear Anal.* **20**, pp. 417–437.

Deng, Y., Chen, G. and Ni, W. M. (1996). Boundary element monotone iteration scheme for semilinear elliptic partial differential equations, *Math. Comput.* **65**, pp. 943–982.

Diaz, J. I. (1985). *Nonlinear Partial Differential Equations and Free Boundaries* (Pitman Advanced Pub. Program, Boston).

Ding, Z. H., Costa, D. and Chen, G. (1999). A high-linking algorithm for sign-changing solutions of semilinear elliptic equations, *Nonlinear Anal.* **38**, pp. 151–172.

Golubitsky, M., Stewart, I. and Schaeffer, D. G. (1988). *Singularities and Groups in Bifurcation Theory II* (Springer-Verlag, New York).

Li, C. P. and Yang, Z. H. (2004). Symmetry-breaking bifurcation in O(2)-symmetric nonlinear large problems and its application to the Kuramoto-Sivashinsky equation in two spatial dimensions, *Chaos, Solitons and Fractals* **22**, pp. 451–468.

Li, C. P., Yang, Z. H. and Chen, G. (2005). On bifurcation from steady-state solutions to rotating waves in the Kuramoto-Sivashinsky equation, *J Shanghai Univ.* **9**, pp. 286–291.

Li, Z. X. and Yang, Z. H. (2010). Bifurcation method for solving multiple positive solutions to boundary value problem of p-Henon equation on the unit disk, *Appl. Math. Mech.* **31**, pp. 511–520.

Li, Z. X., Yang, Z. H. and Zhu, H. L. (2011a). A bifurcation method for solving multiple positive solutions to the boundary value problem of the Henon equation on a unit disk, *Comput. Math. Appl.*, in press.

Li, Z. X., Zhu, H. L. and Yang, Z. H. (2011b) Bifurcation method for solving multiple positive solutions to Henon equation on the unit cube, *Commun. Nonlinear Sci.*

Numer. Simulat. **16**, pp. 3673–3683.

Li, Y. and Zhou, J. X. (2002). A minimax method for finding multiple critical points and its applications to semilinear PDEs, *SIAM J. Sci. Comput.* **23**, pp. 840–865.

Pao, C. V. (1995). Block monotone iterative methods for numerical solutions of nonlinear elliptic equations, *Numer. Math.* **72**, pp. 239–262.

Struwe, M. (1996). *Variational Methods, A Series of Modern Surveys in Math* (Springer-Verlag, Berlin).

Yang, Z. H. (2007). *Non-linear Bifurcation: Theory and Computation(in Chinese)* (Science Press, Beijing).

Yang, Z. H., Li, Z. X. and Zhu, H. L. (2008). Bifurcation method for solving multiple positive solutions to Henon equation, *Science in China Series A: Mathematics* **51**, pp. 2330–2342.

Yao, X. D. and Zhou, J. X. (2005). A minimax method for finding multiple critical points in Banach spaces and its application to quasi-linear elliptic PDE, *SIAM J. Sci. Comput.* **26**, pp. 1796–1809.

Zhu, H. L. and Li, Z. X. (2010). Newton method's basin of attraction for sign-changing solutions of concave and convex nonlinearities, *Appl. Math. Comput.* **217**, pp. 2937–2943.

Chapter 11

Multilevel wavelet-like block incremental unknowns methods for a class of reaction-diffusion equations[1]

Yang Wang[a,b], Yu-Jiang Wu[a,*] and Ai-Li Yang[a]

[a] *School of Mathematics and Statistics, Lanzhou University, Lanzhou 730000, Gansu, PR China*

[*] *myjaw@lzu.edu.cn*

[b] *College of Mathematics, Jilin Normal University, Siping 136000, Jilin, PR China*

Based on the wavelet-like block incremental unknowns (WBIUs) method, we propose two WBIUs-type schemes for the anisotropic time-dependent reaction diffusion equations. Some newly established inequalities are given with proof. They play important roles in the process for demonstrating the stability of the schemes. Comparing with the classical wavelet-like incremental unknowns (WIUs) method, the WBIUs method reduces the condition number of the coefficient matrices much more efficiently and thereby spends fewer CPU time in solving the anisotropic problems. Our numerical experiments also confirm these results.

Keywords: Wavelet-like incremental unknowns, multilevel methods, reaction-diffusion equation

11.1 Introduction

Incremental Unknowns (IUs) method has been introduced in [Chen and Temam (1993b); Temam (1990)] in order to implement the Nonlinear Galerkin method [Foias *et al.* (1988)] when finite differences are used for the space discretization. As well known that IUs can reduce the condition number of coefficient matrix [Chen and Temam (1993a); Miranville and Muresan (2002)] and improve the stability of numerical schemes [Pouit (1996); Wu and Yang (2007)]. The Wavelet-like Incremental Unknowns (WIUs) method, as a kind of IUs, has the property of L^2 orthogonality

[1] This work is partially supported by the National Basic Research of China, 973 Program (grant no. 2011CB706903), the Mathematical Tianyuan Foundation of China (grant no. 11026064) and the Fundamental Research Fund for Physics and Mathematics of Lanzhou University (grant no. LZULL200904).

between different levels of unknowns, which makes it particularly appropriate for the approximation of evolution equations [Chen and Temam (1993b,c); Wu *et al.* (2007)]. However, this method is not always effective for all kinds of reaction-diffusion equations.

In the present work, we focus on the following nonlinear reaction-diffusion equations:

$$\begin{cases} \frac{\partial u}{\partial t} - \nu(\epsilon\frac{\partial^2 u}{\partial x^2} + \frac{\partial^2 u}{\partial y^2}) + g(u) = 0, & \text{in } \Omega \times (0,T], \\ u(x,y,0) = u_0(x,y), & \text{in } \Omega, \\ u(x,y,t) = 0, & \text{on } \partial\Omega \times (0,T]. \end{cases} \tag{11.1}$$

Here $\nu > 0$, $0 < \epsilon \ll 1$, Ω is an open bounded set in R^2 with sufficient smooth boundary and

$$g(s) = \sum_{j=0}^{2q-1} b_j s^j, \quad b_{2q-1} > 0. \tag{11.2}$$

For solving the problem (11.1), a natural idea is to use discretization on a uniform grid and then construct and analyze WIUs-type schemes. See, for example, in [Chen and Temam (1993b)]. However, in our case we will see in Section 11.5 that the condition numbers of the coefficient matrices of WIUs-type schemes increase fastly when ϵ becomes small. That is to say, WIUs method is costly for solving equation (11.1) when ϵ is small.

In order to solve (11.1) effectively, we use the WBIUs method introduced in [Wu *et al.* (2009)]. In this way, when ϵ is small, our methods will bring us advantages of reduction of the condition number of coefficient matrix and of usage of less CPU time over the classical WIUs method. Moreover, we set up new inequalities for vector norms and propose two WBIUs-type schemes to solve the equations. The stability for each scheme is proved. Finally, we demonstrate the effectiveness of WBIUs through numerical experiments.

This chapter is organized as follows. In Section 11.2, we define the Wavelet-like Block Incremental Unknowns (WBIUs) and describe their implementation in the space discretization of problem (11.1). In Section 11.3, we propose two different approximate schemes and their equivalent forms. In Section 11.4, we develop the stability analysis of the two schemes. Finally, in Section 11.5, numerical examples show our methods are indeed very efficient.

11.2 Multilevel WBIUs method

For the sake of simplicity, we here let $\Omega = (0,1)^2$. Now, considering spatial discretization by finite difference with mesh size $h_d = 1/(2^d N + 1)$, $N \in \mathbb{N}$, we have

$$\frac{\partial U_d}{\partial t} + \nu A_d U_d + g(U_d) = 0, \tag{11.3}$$

where $U_d \in \mathbb{R}^{(2^d N)^2}$ is the vector that consists of approximate values of u at the grid points. A_d is a matrix of order $(2^d N)^2$, that is

$$A_d = \frac{1}{h_d^2}\begin{pmatrix} B_d & -I & & \\ -I & \ddots & \ddots & \\ & \ddots & \ddots & -I \\ & & -I & B_d \end{pmatrix}, \quad B_d = \begin{pmatrix} 2+2\epsilon & -\epsilon & & \\ -\epsilon & \ddots & \ddots & \\ & \ddots & \ddots & -\epsilon \\ & & -\epsilon & 2+2\epsilon \end{pmatrix},$$

and

$$U_d = \begin{pmatrix} u_1^d \\ u_2^d \\ \vdots \\ u_{2^d N}^d \end{pmatrix}, \quad u_\beta^d = \left(u_{1\beta}^d, u_{2\beta}^d, \dots, u_{2^d N\beta}^d\right)^T.$$

The system (11.3) can be rewritten, in block structure, as

$$\frac{\partial u_\beta^d}{\partial t} + \nu(-u_{\beta-1}^d + B_d u_\beta^d - u_{\beta+1}^d) + g(u_\beta^d) = 0, \quad \beta = 1, 2, ..., 2^d N. \tag{11.4}$$

For $j = 1, 2, ..., 2^{d-1}N$, we define WBIUs as

$$\begin{cases} y_{2j}^d = (u_{2j-1}^d + u_{2j}^d)/2, \\ z_{2j-1}^d = (u_{2j-1}^d - u_{2j}^d)/2. \end{cases} \tag{11.5}$$

The inverse transformation is

$$\begin{cases} u_{2j}^d = y_{2j}^d - z_{2j-1}^d, \\ u_{2j-1}^d = y_{2j}^d + z_{2j-1}^d. \end{cases} \tag{11.6}$$

Denoting

$$\tilde{U}_d = \begin{pmatrix} U_{even}^d \\ U_{odd}^d \end{pmatrix}, \quad \bar{U}_d = \begin{pmatrix} Y^d \\ Z^d \end{pmatrix},$$

with

$$U_{even}^d = \begin{pmatrix} u_2^d \\ u_4^d \\ \vdots \\ u_{2^d N}^d \end{pmatrix}, \quad U_{odd}^d = \begin{pmatrix} u_1^d \\ u_3^d \\ \vdots \\ u_{2^d N-1}^d \end{pmatrix}, \quad Y^d = \begin{pmatrix} y_2^d \\ y_4^d \\ \vdots \\ y_{2^d N}^d \end{pmatrix}, \quad Z^d = \begin{pmatrix} z_1^d \\ z_3^d \\ \vdots \\ z_{2^d N-1}^d \end{pmatrix}.$$

It is obvious to see that

$$U_d = P_d \tilde{U}_d, \quad \tilde{U}_d = S_d \bar{U}_d,$$

where P_d is a permutation matrix of order $(2^d N)^2$

$$P_d = (E_d \ \ V_d) \otimes I_d,$$

and S_d is the transfer matrix of order $(2^d N)^2$,

$$S_d = \begin{pmatrix} I_{d-1} & -I_{d-1} \\ I_{d-1} & I_{d-1} \end{pmatrix} \otimes I_d.$$

I_d is the identity matrix of order $2^d N$, E_d and V_d are matrices of order $2^d N \times 2^{d-1} N$, i.e.,

$$E_d = \begin{pmatrix} 0\,0\,0 \ldots 0\,0 \\ 1\,0\,0 \ldots 0\,0 \\ 0\,0\,0 \ldots 0\,0 \\ 0\,1\,0 \ldots 0\,0 \\ \vdots\,\vdots\,\vdots\,\ddots\,\vdots\,\vdots \\ 0\,0\,0 \ldots 0\,0 \\ 0\,0\,0 \ldots 0\,1 \end{pmatrix}, \quad V_d = \begin{pmatrix} 1\,0\,0 \ldots 0\,0 \\ 0\,0\,0 \ldots 0\,0 \\ 0\,1\,0 \ldots 0\,0 \\ 0\,0\,0 \ldots 0\,0 \\ \vdots\,\vdots\,\vdots\,\ddots\,\vdots\,\vdots \\ 0\,0\,0 \ldots 0\,1 \\ 0\,0\,0 \ldots 0\,0 \end{pmatrix}.$$

Therefore, we get

$$U_d = P_d S_d \bar{U}_d. \tag{11.7}$$

Substituting (11.7) into the finite difference equation (11.3) and multiplying the equation by $(P_d S_d)^T$, we find

$$\frac{\partial (P_d S_d)^T P_d S_d \bar{U}_d}{\partial t} + \nu (P_d S_d)^T A_d P_d S_d \bar{U}_d + (P_d S_d)^T g(P_d S_d \bar{U}_d) = 0. \tag{11.8}$$

Noting that $P_d^T P_d = I_d \otimes I_d$, $S_d^T S_d = 2 I_d \otimes I_d$, P_d^T and g can commute, thus

$$2\frac{\partial \bar{U}_d}{\partial t} + \nu (P_d S_d)^T A_d P_d S_d \bar{U}_d + S_d^T g(S_d \bar{U}_d) = 0, \tag{11.9}$$

which is the 2-level WBIUs-type scheme.

We can again introduce the next level of WBIUs on Y^d by repeating exactly the same procedure. Let

$$\bar{Y}^d = \begin{pmatrix} Y^{d-1} \\ Z^{d-1} \end{pmatrix},$$

with

$$Y^{d-1} = \begin{pmatrix} y_4^{d-1} \\ y_8^{d-1} \\ \vdots \\ y_{2^d N}^{d-1} \end{pmatrix} \quad \text{and} \quad Z^{d-1} = \begin{pmatrix} z_2^{d-1} \\ z_6^{d-1} \\ \vdots \\ z_{2^d N-2}^{d-1} \end{pmatrix}.$$

Similar to (11.5), for $j = 1, 2, ..., 2^{d-2}N$, we define

$$\begin{cases} y_{4j}^d = y_{4j}^{d-1} - z_{4j-2}^{d-1}, \\ y_{4j-2}^d = y_{4j}^{d-1} + z_{4j-2}^{d-1}. \end{cases} \tag{11.10}$$

Therefore

$$Y^d = P_{d-1} S_{d-1} \bar{Y}^d, \tag{11.11}$$

where $P_{d-1} = (E_{d-1}\ V_{d-1}) \otimes I_d$ is a permutation matrix of order $2^{2d-1}N^2$ and

$$S_{d-1} = \begin{pmatrix} I_{d-2} & -I_{d-2} \\ I_{d-2} & I_{d-2} \end{pmatrix} \otimes I_d.$$

Here P_{d-1}, S_{d-1} have the similar structures as P_d, S_d respectively, but they are both matrices of order $(2^{d-1}N)^2$. Let

$$\bar{U}_{d-1} = \begin{pmatrix} Y^{d-1} \\ Z^{d-1} \\ \frac{1}{\sqrt{2}}Z^d \end{pmatrix}$$

and

$$\tilde{P}_{d-1} = \begin{pmatrix} P_{d-1} & 0 \\ 0 & I_{d-1} \otimes I_d \end{pmatrix}, \quad \tilde{S}_{d-1} = \begin{pmatrix} S_{d-1} & 0 \\ 0 & \sqrt{2}I_{d-1} \otimes I_d \end{pmatrix}.$$

We see that

$$\tilde{P}_{d-1}\tilde{P}_{d-1}^T = I_d \otimes I_d, \quad \tilde{S}_{d-1}\tilde{S}_{d-1}^T = 2\begin{pmatrix} I_{d-1} \otimes I_d & 0 \\ 0 & I_{d-1} \otimes I_d \end{pmatrix}$$

and

$$\bar{U}_d = \tilde{P}_{d-1}\tilde{S}_{d-1}\bar{U}_{d-1}. \tag{11.12}$$

Substituting (11.12) into (11.9), and multiply the equation by $(\tilde{P}_{d-1}\tilde{S}_{d-1})^T$, we realize that

$$2^2\frac{\partial \bar{U}_{d-1}}{\partial t} + \nu S^T A_d S \bar{U}_{d-1} + S^T g(S\bar{U}_{d-1}) = 0, \tag{11.13}$$

with $S = P_d S_d \tilde{P}_{d-1}\tilde{S}_{d-1}$. Generally for $l = d-1, d-2, \cdots, 1$, let us define

$$\bar{Y}^{l+1} = \begin{pmatrix} Y^l \\ Z^l \end{pmatrix}. \tag{11.14}$$

We can easily see as previously that

$$Y^{l+1} = P_l S_l \bar{Y}^{l+1}, \tag{11.15}$$

here, P_l, S_l are matrices of the order $2^{d+l-1}N^2$ with

$$P_l = (\, E_l \quad V_l \,) \otimes I_d\,, \qquad S_l = \begin{pmatrix} I_{l-1} & -I_{l-1} \\ I_{l-1} & I_{l-1} \end{pmatrix} \otimes I_d.$$

We can include (11.15) with $l = d$ by letting $Y^{d+1} = U_d, \bar{Y}^{d+1} = \bar{U}_d$. Setting

$$\bar{U}_l = \begin{pmatrix} Y^l \\ Z^l \\ \frac{1}{\sqrt{2}}Z^{l+1} \\ \vdots \\ \frac{1}{\sqrt{2}^{d-l}}Z^d \end{pmatrix} \quad \text{with} \ \ Y^l \in 2^{d+l-1}N^2, \quad Z^l \in 2^{d+l-1}N^2,$$

it can be written as

$$\bar{U}_{l+1} = \tilde{P}_l\tilde{S}_l\bar{U}_l, \tag{11.16}$$

where

$$\tilde{P}_l = \begin{pmatrix} P_l & 0 \\ 0 & I_{k\times k} \otimes I_d \end{pmatrix}, \quad \tilde{S}_l = \begin{pmatrix} S_l & 0 \\ 0 & \sqrt{2}I_{k\times k} \otimes I_d \end{pmatrix},$$

with $k = (2^d - 2^l)N$. Hence

$$\tilde{P}_l\tilde{P}_l^T = \begin{pmatrix} I_l \otimes I_d & 0 \\ 0 & I_{k\times k} \otimes I_d \end{pmatrix}, \quad \tilde{S}_l\tilde{S}_l^T = 2\begin{pmatrix} I_l \otimes I_d & 0 \\ 0 & I_{k\times k} \otimes I_d \end{pmatrix}.$$

Substituting (11.16) into equation (11.13) successively with $l = d-2, d-3, ..., 1$, one can obtain the $(d+1)$-level WBIUs in terms of Y and Z

$$2^d \frac{\partial \bar{U}_1}{\partial t} + \nu S^T A_d S \bar{U}_1 + S^T g(S\bar{U}_1) = 0, \tag{11.17}$$

where

$$S = \tilde{P}_d\tilde{S}_d\tilde{P}_{d-1}\tilde{S}_{d-1}, ..., \tilde{P}_1\tilde{S}_1,$$

$$Y = Y^1 \in 2^d N^2, \quad Z = (Z^1, \frac{1}{\sqrt{2}}Z^2, \cdots, \frac{1}{\sqrt{2}^{d-l}}Z^d)^T.$$

11.3 Approximate schemes and their equivalent forms

11.3.1 *Approximate schemes*

The form of (11.17) with $S = \tilde{P}_d\tilde{S}_d$ is

$$S_d^T S_d \frac{\partial \bar{U}_d}{\partial t} + \nu S_d^T P_d^T A_d P_d S_d \bar{U}_d + S_d^T g(S_d\bar{U}_d) = 0. \tag{11.18}$$

After simple computation, we find

$$S_d^T g(S_d\bar{U}_d) = \begin{pmatrix} 2g(Y^d) + O(|Z^d|^2) \\ O(|Z^d|) \end{pmatrix}.$$

By neglecting the terms $O(|Z_d|^2)$ and $O(|Z_d|)$, Eq. (11.18) becomes

$$2\frac{\partial \bar{U}_d}{\partial t} + \nu S_d^T P_d^T A_d P_d S_d \bar{U}_d + 2\begin{pmatrix} g(Y^d) \\ 0 \end{pmatrix} = 0. \tag{11.19}$$

The form of (11.17) with $S = \tilde{P}_d\tilde{S}_d\tilde{P}_{d-1}\tilde{S}_{d-1}$ is

$$S^T S \frac{\partial \bar{U}_{d-1}}{\partial t} + \nu S^T A_d S \bar{U}_{d-1} + S^T g(S\bar{U}_{d-1}) = 0. \tag{11.20}$$

Using the same approximation, it is clearly

$$2^2 \frac{\partial \bar{U}_{d-1}}{\partial t} + \nu S^T A_d S \bar{U}_{d-1} + 2^2 \begin{pmatrix} g(Y^{d-1}) \\ 0 \end{pmatrix} = 0.$$

In fact, one has

$$\begin{aligned} S^T g(S\bar{U}_{d-1}) &= \tilde{S}_{d-1}^T \tilde{P}_{d-1}^T \tilde{S}_d^T \tilde{P}_d^T g(\tilde{P}_d\tilde{S}_d\tilde{P}_{d-1}\tilde{S}_{d-1}\bar{U}_{d-1}) \\ &= \tilde{S}_{d-1}^T \tilde{P}_{d-1}^T S_d^T P_d^T g(P_d S_d \bar{U}_d) \\ &= \tilde{S}_{d-1}^T \tilde{P}_{d-1}^T S_d^T g(S_d\bar{U}_d). \end{aligned} \tag{11.21}$$

Using the above approximation and the definitions of $\tilde{S}_{d-1}$ and $\tilde{P}_{d-1}$, we deduce that

$$\begin{aligned} S^T g(S\bar{U}_{d-1}) &= 2\begin{pmatrix} S_{d-1}^T & 0 \\ 0 & I \end{pmatrix}\begin{pmatrix} P_{d-1}^T & 0 \\ 0 & I \end{pmatrix}\begin{pmatrix} g(Y^d) \\ 0 \end{pmatrix} \\ &= 2\begin{pmatrix} S_{d-1}^T & 0 \\ 0 & I \end{pmatrix}\begin{pmatrix} P_{d-1}^T & 0 \\ 0 & I \end{pmatrix}\begin{pmatrix} g(P_{d-1}S_{d-1}\bar{Y}^d) \\ 0 \end{pmatrix} \\ &= 2\begin{pmatrix} S_{d-1}^T g(S_{d-1}\bar{Y}^d) \\ 0 \end{pmatrix} \\ &= 2^2\begin{pmatrix} g(Y^{d-1}) \\ 0 \end{pmatrix}. \end{aligned}$$

Therefore, the approximate scheme of $(d+1)$-level WBIUs is

$$2^d\frac{\partial}{\partial t}\begin{pmatrix} Y \\ Z \end{pmatrix} + \nu S^T A_d S\bar{U}_1 + 2^d\begin{pmatrix} g(Y) \\ 0 \end{pmatrix} = 0, \tag{11.22}$$

where $Y \in R^{N\times 2^d N}$, $Z \in R^{(2^d-1)N\times 2^d N}$.

As for time discretization, we propose two kinds of schemes:

Scheme I: Explicit scheme

$$\frac{2^d}{\tau}\begin{pmatrix} Y^{n+1} - Y^n \\ Z^{n+1} - Z^n \end{pmatrix} + \nu S^T A_d S\bar{U}_1^n + 2^d\begin{pmatrix} g(Y^n) \\ 0 \end{pmatrix} = 0.$$

Scheme II: Semi-implicit scheme

$$\frac{2^d}{\tau}\begin{pmatrix} Y^{n+1} - Y^n \\ Z^{n+1} - Z^n \end{pmatrix} + \nu S^T A_d S\bar{U}_1^{n+1} + 2^d\begin{pmatrix} g(Y^n) \\ 0 \end{pmatrix} = 0.$$

11.3.2 *The equivalent forms of approximate schemes*

Let V_{h_l} be the function space spanned by the basis functions $w_{h_l,M}(\mathbf{x})$, $M = (ih_d, jh_l)$, $w_{h_l,M}(\mathbf{x})$ is equal to 1 on the square $K_{ij}^l = [ih_d, (i+1)h_d)\times[jh_l, (j+1)h_l)$ and vanishes outside this square. Let $u_{h_l}(\mathbf{x})$ be a step function in V_{h_l} and $u_{h_l}(\mathbf{x}) = u_{h_l}(M)$ for $\mathbf{x} \in K_{ij}^l$, $i = 1, 2, \cdots, 2^d N$, $j = 1, 2, \cdots, 2^l N$. We can write

$$u_{h_l}(\mathbf{x}) = \sum_{i=1}^{2^d N}\sum_{j=1}^{2^{l-1}N} u_{h_l}(M) w_{h_l,M}(\mathbf{x}), \quad \mathbf{x} \in \Omega.$$

We introduce the finite difference operators ∇_{1h_d} and ∇_{2h_l}:

$$\begin{aligned} \nabla_{1h_d}\Phi(\mathbf{x}) &= \frac{1}{h_d}(\Phi(\mathbf{x} + h_d e_1) - \Phi(\mathbf{x})), \\ \nabla_{2h_l}\Phi(\mathbf{x}) &= \frac{1}{h_l}(\Phi(\mathbf{x} + h_l e_2) - \Phi(\mathbf{x})), \end{aligned}$$

where $e_1 = (1, 0)$, $e_2 = (0, 1)$ and define the discrete scalar product

$$((u_{h_l}, v_{h_l}))_{h_l} = \epsilon(\nabla_{1h_d}u_{h_l}, \nabla_{1h_d}v_{h_l}) + (\nabla_{2h_l}u_{h_l}, \nabla_{2h_l}v_{h_l}),$$

where $(\cdot,\cdot)$ is the scalar product in $L^2(\Omega)$. Let $\|\cdot\|_{h_l} = ((\cdot,\cdot))_{h_l}^{1/2}$, it's easy to find that $\|\cdot\|_{h_l}$ and $|\cdot|$ are Hilbert norms in V_{h_l}.

With the help of step functions, we can write the finite difference discretization scheme (11.3) in variational form

$$(\frac{\partial u_{h_d}}{\partial t}, \tilde{u}) + \nu((u_{h_d}, \tilde{u}))_{h_d} + (g(u_{h_d}), \tilde{u}) = 0, \quad \forall \tilde{u} \in V_{h_d}. \tag{11.23}$$

Scheme (11.3) can be recovered by choosing $\widetilde{u} = w_{h_d,M_{ij}}(\mathbf{x})$. Here, we introduce the notions $M_{\alpha\beta}$(or $M_{\alpha,\beta}) = (\alpha h_d, \beta h_d)$. We define $\mathcal{Y}^d$ as the space spanned by the basis functions $\Psi_{2h_d,M_{i,2j}}(\mathbf{x})$, $\Psi_{2h_d,M_{i,2j}}(\mathbf{x})$ is equal to 1 on the square $[ih_d,(i+1)h_d) \times [(2j-1)h_d,(2j+1)h_d)$ and vanishes outside this square. Thus for every $y^d \in \mathcal{Y}^d$,

$$y^d(\mathbf{x}) = \sum_{i=1}^{2^dN} \sum_{j=1}^{2^{d-1}N} y^d(M_{i,2j})\Psi_{2h_d,M_{i,2j}}(\mathbf{x}), \ \mathbf{x} \in \Omega.$$

We then define $\mathcal{Z}^d$ as the space spanned by

$$\chi_{h_d,M_{i,2j-1}}(\mathbf{x}) = w_{h_d,M_{i,2j-1}}(\mathbf{x}) - w_{h_d,M_{i,2j}}(\mathbf{x}).$$

So, for every $z^d \in \mathcal{Z}^d$, we have

$$z^d(\mathbf{x}) = \sum_{i=1}^{2^dN} \sum_{j=1}^{2^{d-1}N} z^d(M_{i,2j-1})\chi_{h_d,M_{i,2j-1}}(\mathbf{x}), \ \mathbf{x} \in \Omega.$$

From the definition of $\Psi_{2h_d,M_{i,2j}}(\mathbf{x})$ and $\chi_{h_d,M_{i,2j-1}}(\mathbf{x})$, for $\forall\ z^d \in \mathcal{Z}^d, \forall\ y^d \in \mathcal{Y}^d$, we can obtain the following conclusions:

(1) $\displaystyle\int_\Omega z^d(\mathbf{x})\,\mathrm{d}\mathbf{x} = 0,$

(2) $\displaystyle\int_\Omega z^d(\mathbf{x})y^d(\mathbf{x})\,\mathrm{d}\mathbf{x} = 0,$

(3) $\mathcal{Y}^d$ and $\mathcal{Z}^d$ is actually the WIUs space, just as (11.5) defined.

Therefore the space V_{h_d} can be decomposed into

$$V_{h_d} = \mathcal{Y}^d \oplus \mathcal{Z}^d,$$

thus for $\forall\ u_{h_d} \in V_{h_d}$ we have

$$u_{h_d} = y^d + z^d, \quad y^d \in \mathcal{Y}^d,\ z^d \in \mathcal{Z}^d.$$

With above decomposition, we have the following theorem:

Theorem 11.1. *Let* $M = (ih_d, 2jh_d)$, $A_1 = (ih_d,(2j-1)h_d)$. *For* $i = 1,2,\cdots,2^dN$, $j = 1,2,\cdots,2^{d-1}N$, *when* $\tilde{y} = \Psi_{2h_d,M}(\boldsymbol{x})$, $\tilde{z} = \chi_{h_d,A_1}(\boldsymbol{x})$, (11.19) *is identical to the following variational forms*

$$\begin{cases} (\dfrac{\partial y^d}{\partial t}, \tilde{y}) + \nu((y^d + z^d, \tilde{y}))_{h_d} + (g(y^d), \tilde{y}) = 0, & \forall\ \tilde{y} \in \mathcal{Y}^d, \\ (\dfrac{\partial z^d}{\partial t}, \tilde{z}) + \nu((y^d + z^d, \tilde{z}))_{h_d} = 0, & \forall\ \tilde{z} \in \mathcal{Z}^d. \end{cases}$$

Multilevel incremental unknowns can be recovered in a similar fashion, we decompose $\mathcal{Y}^l$, with $l = d, ..., 1$, into

$$\mathcal{Y}^{l+1} = \mathcal{Y}^l \oplus \mathcal{Z}^l.$$

Therefore for any function $u_{h_d} \in V_{h_d}$, we can write it as

$$u_{h_d} = y + z,$$

where $y = y^1 \in \mathcal{Y} = \mathcal{Y}^1$ and $z \in \mathcal{Z} = \mathcal{Z}^1 \oplus \mathcal{Z}^2 \oplus \cdots \oplus \mathcal{Z}^d$. Using the above decomposition, we can prove that (11.22) is identical to the following variational form

$$\begin{cases} (\frac{\partial y}{\partial t}, \tilde{y}) + \nu((y+z, \tilde{y}))_{h_d} + (g(y+z), \tilde{y}) = 0, \quad \forall \tilde{y} \in \mathcal{Y}^1, \\ (\frac{\partial z}{\partial t}, \tilde{z}) + \nu((y+z, \tilde{z}))_{h_d} = 0, \quad \forall \tilde{z} \in \mathcal{Z}. \end{cases} \tag{11.24}$$

Remarks. Due to the space decomposition, these methods resemble the multilevel augmentation methods developed in [Chen, Z. Y. *et al.* (2005, 2006a,b)]. The methods therein were suitable for solving some linear ordinary differential equations, linear integral-differential equations, or even ill-posed such problems. However, our methods are suitable for some nonlinear reaction diffusion equations. Besides, their multilevel augmentation methods are based on essentially the wavelet functions which are considerably more complicated than the simple wavelet-like decompositions we use.

11.4 Stability analysis

11.4.1 *Lemmas for new vector norms*

Before presenting the stability theory, let us first introduce some easy lemmas.

Lemma 11.1 ([Chen and Temam (1993c)]). *For the function g, there exists two constants c_1 and c_2, such that*

$$\begin{aligned} g(s)s \geq & \frac{1}{2} b_{2q-1} s^{2q} - c_1, \\ g(s)^2 \geq & 2b_{2q-1}^2 s^{4q-2} + c_2, \quad \forall s. \end{aligned}$$

Lemma 11.2. *Let l be a integer, for $0 \leq l \leq d$, $\forall u_{h_l} \in V_{h_l}$, we have*

$$\sqrt{2(1+\epsilon)} \cdot |u_{h_l}| \leq \|u_{h_l}\|_{h_l} \leq S_1(l)^{-1} \cdot |u_{h_l}|, \tag{11.25}$$

where $S_1(l) = h_d(4\epsilon + 4^{l-d+1})^{-1/2}$.

Proof. Firstly, we prove the right-hand side of (11.25). Thanks to $u_{\alpha,0} = u_{0,\beta} = u_{\alpha,2^l N+1} = u_{2^d N+1,\beta} = 0$ and $(a-b)^2 \le 2(a^2+b^2)$, it suffices to prove that

$$\begin{aligned}
\|u_{h_l}\|_{h_l}^2 &= \epsilon \sum_{\alpha=0}^{2^d N}\sum_{\beta=1}^{2^l N}\int_\Omega \frac{(u_{\alpha+1,\beta}-u_{\alpha,\beta})^2}{h_d^2}\mathbf{dx} + \sum_{\alpha=1}^{2^d N}\sum_{\beta=0}^{2^l N}\int_\Omega \frac{(u_{\alpha,\beta+1}-u_{\alpha,\beta})^2}{h_l^2}\mathbf{dx}\\
&= \epsilon\frac{h_l}{h_d}\sum_{\alpha=0}^{2^d N}\sum_{\beta=1}^{2^l N}(u_{\alpha+1,\beta}-u_{\alpha,\beta})^2 + \frac{h_d}{h_l}\sum_{\alpha=1}^{2^d N}\sum_{\beta=0}^{2^l N}(u_{\alpha,\beta+1}-u_{\alpha,\beta})^2\\
&\le 2\epsilon\frac{h_l}{h_d}\sum_{\alpha=0}^{2^d N}\sum_{\beta=1}^{2^l N}(u^2_{\alpha+1,\beta}+u^2_{\alpha,\beta}) + 2\frac{h_d}{h_l}\sum_{\alpha=0}^{2^d N}\sum_{\beta=1}^{2^l N}(u^2_{\alpha,\beta+1}+u^2_{\alpha,\beta})\\
&= 4\epsilon\frac{h_l}{h_d}\sum_{\alpha=1}^{2^d N}\sum_{\beta=1}^{2^l N}u^2_{\alpha,\beta} + 4\frac{h_d}{h_l}\sum_{\alpha=1}^{2^d N}\sum_{\beta=1}^{2^l N}u^2_{\alpha,\beta}\\
&\le S_1(l)^{-1}|u_{h_l}|^2,
\end{aligned}$$

with $S_1(l) = h_d(4\epsilon + 4^{l-d+1})^{-1/2}$.

Secondly, we prove the left-hand side of (11.25). Due to Cauchy's inequality, we have

$$\begin{aligned}
|u_{h_l}(i,j)|^2 &= |\sum_{k=0}^{i-1}\nabla_{1h_d}u_{h_l}(k,j)h_d|^2\\
&\le \Big(\sum_{k=0}^{i-1}h_d\Big)\Big(\sum_{k=0}^{i-1}h_d|\nabla_{1h_d}u_{h_l}(k,j)|^2\Big)\\
&\le ih_d^2\sum_{k=0}^{2^d N}|\nabla_{1h_d}u_{h_l}(k,j)|^2,
\end{aligned}$$

similarly

$$|u_{h_l}(i,j)|^2 \le jh_l^2\sum_{k=0}^{2^l N}|\nabla_{2h_l}u_{h_l}(i,k)|^2.$$

Hence

$$|u_{h_l}(i,j)|^2 \le \frac{1}{1+\epsilon}\Big[\epsilon ih_d^2\sum_{k=0}^{2^d N}|\nabla_{1h_d}u_{h_l}(k,j)|^2 + jh_l^2\sum_{k=0}^{2^l N}|\nabla_{2h_l}u_{h_l}(i,k)|^2\Big]. \quad (11.26)$$

With the definition of $|\cdot|$ and (11.26), we have

$$\begin{aligned}
|u_{h_l}|^2 &= h_d h_l\sum_{i=0}^{2^d N}\sum_{j=0}^{2^l N}|u_{h_l}(i,j)|^2\\
&\le \frac{1}{1+\epsilon}\sum_{i=0}^{2^d N}\sum_{j=0}^{2^l N}h_d h_l\left[\epsilon ih_d^2\sum_{k=0}^{2^d N}|\nabla_{1h_d}u_{h_l}(k,j)|^2 + jh_l^2\sum_{k=0}^{2^l N}|\nabla_{2h_l}u_{h_l}(i,k)|^2\right]
\end{aligned}$$

$$\le \frac{1}{1+\epsilon}\left[\epsilon\sum_{i=0}^{2^dN} ih_d^2\sum_{j=0}^{2^lN}\sum_{k=0}^{2^dN}|\nabla_{1h_d}u_{h_l}(k,j)|^2h_dh_l\right.$$
$$\left.+\sum_{j=0}^{2^lN} jh_l^2\sum_{i=0}^{2^dN}\sum_{k=0}^{2^lN}|\nabla_{2h_l}u_{h_l}(i,k)|^2h_dh_l\right]$$
$$= \frac{1}{1+\epsilon}\left[\frac{2^dN(2^dN+1)}{2}h_d^2\epsilon\sum_{j=0}^{2^lN}\sum_{k=0}^{2^dN}|\nabla_{1h_d}u_{h_l}(k,j)|^2h_dh_l\right.$$
$$\left.+\frac{2^lN(2^lN+1)}{2}h_l^2\sum_{i=0}^{2^dN}\sum_{k=0}^{2^lN}|\nabla_{2h_l}u_{h_l}(i,k)|^2h_dh_l)\right]$$
$$\le \frac{1}{2(1+\epsilon)}\|u_{h_l}\|_{h_l}^2.$$

Thus, the inequality (11.25) holds true. □

Lemma 11.3. *For every function $y \in \mathcal{Y}^1$, we have*

$$S_2(h_1,h_d)|y|_\infty^2 \le |y|^2, \quad S_2(h_1,h_d) = h_1h_d, \tag{11.27}$$
$$\overline{S_1}(h_1,h_d)\|y\|_{h_d} \le |y|, \quad \overline{S}_1(h_1,h_d) = \frac{\sqrt{h_1h_d}}{\sqrt{4\epsilon 2^d+4}}, \tag{11.28}$$

where $|\cdot|_\infty$ is the maximum norm.

Proof. Let $K_{ij}^1 = [ih_d,(i+1)h_d)\times[jh_1,(j+1)h_1)$, $K_{ij}^d = [ih_d,(i+1)h_d)\times[jh_d,(j+1)h_d)$, $M_{ij}^1 = (ih_d, jh_1)$. Using the expression of y, we have

$$|y|^2 = \int_\Omega \sum_{i=1}^{2^dN}\sum_{j=1}^{N}\left(y(M_{ij}^1)\Psi_{h_1,M_{ij}^1}(\mathbf{x})\right)^2 d\mathbf{x}$$
$$= \sum_{i=1}^{2^dN}\sum_{j=1}^{N}\int_{K_{ij}^1}\sum_{i=1}^{2^dN}\sum_{j=1}^{N}\left(y(M_{ij}^1)\Psi_{h_1,M_{ij}^1}(\mathbf{x})\right)^2 d\mathbf{x}$$
$$= h_dh_1\sum_{i=1}^{2^dN}\sum_{j=1}^{N} y^2(M_{ij}^1) \ge h_1h_d|y|_\infty^2.$$

Now we prove the inequality (11.28). Using the inequality $(a-b)^2 \le 2(a^2+b^2)$ and the definitions of $\|\cdot\|_{h_d}$ and $|\cdot|$, it suffices to prove that

$$\|y\|_{h_d}^2 = \epsilon\int_\Omega \left(\frac{y(\mathbf{x}+h_de_1)-y(\mathbf{x})}{h_d}\right)^2 d\mathbf{x} + \int_\Omega\left(\frac{y(\mathbf{x}+h_de_2)-y(\mathbf{x})}{h_d}\right)^2 d\mathbf{x}$$
$$= \frac{\epsilon}{h_d^2}\sum_{i=1}^{2^dN}\sum_{j=1}^{N}\int_{K_{ij}^1}\left(y(\mathbf{x}+h_de_1)-y(\mathbf{x})\right)^2 d\mathbf{x}$$
$$+\frac{1}{h_d^2}\sum_{i=1}^{2^dN}\sum_{j=1}^{N}\int_{K_{ij}^1}\left(y(\mathbf{x}+h_de_2)-y(\mathbf{x})\right)^2 d\mathbf{x}$$

$$
\begin{aligned}
&= \frac{\epsilon}{h_d^2}\sum_{i=1}^{2^dN}\sum_{j=1}^{N} 2^d \int_{K_{ij}^d} \big(y((i+1)h_1, jh_1) - y(ih_1, jh_1)\big)^2 \mathrm{d}\mathbf{x} \\
&+ \frac{1}{h_d^2}\sum_{i=1}^{2^dN}\sum_{j=1}^{N} \int_{K_{ij}^d} \big(y(ih_1, (j+1)h_1) - y(ih_1, jh_1)\big)^2 \mathrm{d}\mathbf{x} \\
&= 2^d\epsilon \sum_{i=1}^{2^dN}\sum_{j=1}^{N} \big(y((i+1)h_1, jh_1) - y(ih_1, jh_1)\big)^2 \\
&+ \sum_{i=1}^{2^dN}\sum_{j=1}^{N} \big(y(ih_1, (j+1)h_1) - y(ih_1, jh_1)\big)^2 \\
&\le 2^d\epsilon \sum_{i=1}^{2^dN}\sum_{j=1}^{N} 2\big(y^2((i+1)h_1, jh_1) + y^2(ih_1, jh_1)\big) \\
&+ \sum_{i=1}^{2^dN}\sum_{j=1}^{N} 2\big(y^2(ih_1, (j+1)h_1) + y^2(ih_1, jh_1)\big) \\
&= (4\epsilon 2^d + 4) \sum_{i=1}^{2^dN}\sum_{j=1}^{N} \cdot y^2(ih_1, jh_1) \\
&\le \overline{S_1}(h_1, h_d)^{-1} |y|^2,
\end{aligned}
$$

where

$$
\overline{S_1}(h_1, h_d) = \frac{\sqrt{h_1 h_d}}{\sqrt{4\epsilon 2^d + 4}}. \qquad \square
$$

11.4.2 *Stability analysis*

Theorem 11.2 (Stability condition for Scheme I). *Assuming that $\tau \le \tau_0$ for some τ_0 fixed and*

$$
\frac{\tau}{h_d^2} \le \frac{3\cdot 2^d}{16\nu(1+2\cdot 2^d)}, \qquad \frac{\tau}{h_d^{2(q-1)}} \le \frac{2^{d(q-1)}}{4b_{2q-1}M_0^{q-1}}, \tag{11.29}
$$

with

$$
M_0 = |u_{h_d}^0|^2 + \frac{2}{\nu}(c_1 + \tau_0 c_2)|\Omega|,
$$

then we have the following estimate for any $n \ge 0$,

$$
|u_{h_d}^n|^2 = |y^n|^2 + |z^n|^2 \le M_0. \tag{11.30}
$$

Proof. In order to prove the stability of Scheme I, we just have to prove the stability of the following variational form:

$$
\begin{cases}
(\dfrac{y^{n+1}-y^n}{\tau}, \tilde{y}) + \nu((y^n + z^n, \tilde{y}))_{h_d} + (g(y^n), \tilde{y}) = 0, \quad \forall \tilde{y} \in Y^1, \\
(\dfrac{z^{n+1}-z^n}{\tau}, \tilde{z}) + \nu((y^n + z^n, \tilde{z}))_{h_d} = 0, \quad \forall \tilde{z} \in Z.
\end{cases} \tag{11.31}
$$

Let $\tilde{y} = 2\tau y^n$ and $\tilde{z} = 2\tau z^n$ in equation (11.31) and then add the two relations, we obtain with $2(a-b,b) = |a|^2 - |b|^2 - |a-b|^2$,

$$\Gamma_0 - |y^{n+1} - y^n|^2 - |z^{n+1} - z^n|^2 + 2\tau\nu\|y^n + z^n\|_{h_d}^2 + 2\tau(g(y^n), y^n) = 0,$$

where $\Gamma_0 = |y^{n+1}|^2 - |y^n|^2 + |z^{n+1}|^2 - |z^n|^2$. Form Lemma 11.1, we find

$$\begin{aligned} &\Gamma_0 - |y^{n+1} - y^n|^2 - |z^{n+1} - z^n|^2 \\ &+2\tau\nu\|y^n + z^n\|_{h_d}^2 + \tau b_{2q-1}\int_\Omega (y^n)^{2q}dx \le 2\tau c_1|\Omega|. \end{aligned} \tag{11.32}$$

Now, let $\tilde{y} = \tau(y^{n+1} - y^n)$ in the first equation of (11.31),

$$|y^{n+1} - y^n|^2 + \tau\nu((y^n + z^n, y^{n+1} - y^n))_{h_d} + \tau(g(y^n), y^{n+1} - y^n) = 0,$$

thanks to Cauchy-Schwarz inequality and Lemma 11.3,

$$\begin{aligned} |y^{n+1} - y^n|^2 &\le \tau\nu\|y^n + z^n\|_{h_d}\|y^{n+1} - y^n\|_{h_d} + \tau|g(y^{n+1})||y^{n+1} - y^n| \\ &\le \tau\nu\frac{1}{\overline{S_1}(h_1, h_d)}\|y^n + z^n\|_{h_d}|y^{n+1} - y^n| \\ &\quad + \tau^2|g(y^n)|^2 + \frac{1}{4}|y^{n+1} - y^n|^2 \\ &\le \frac{1}{2}|y^{n+1} - y^n|^2 + \tau^2|g(y^n)|^2 + \frac{\tau^2\nu^2}{\overline{S_1}(h_1, h_d)^2}\|y^n + z^n\|_{h_d}^2. \end{aligned}$$

Therefore,

$$|y^{n+1} - y^n|^2 \le 2\tau^2|g(y^n)|^2 + \frac{2\tau^2\nu^2}{\overline{S_1}(h_1, h_d)^2}\|y^n + z^n\|_{h_d}^2. \tag{11.33}$$

Now let $\tilde{z} = \tau(z_{n+1} - z^n)$ in the second equation of (11.31),

$$|z_{n+1} - z^n|^2 + \tau\nu((y^n + z^n, z^{n+1} - z^n))_{h_d} = 0.$$

Using Lemma 11.2 with $l = d$,

$$\begin{aligned} |z^{n+1} - z^n|^2 \le &\tau\nu\|y^n + z^n\|_{h_d}\|z^{n+1} - z^n\|_{h_d} \\ \le &\frac{\tau\nu}{S_1(d)}\|y^n + z^n\|_{h_d}^2|z^{n+1} - z^n|. \end{aligned}$$

Therefore

$$|z^{n+1} - z^n|^2 \le \frac{\tau^2\nu^2}{S_1(d)^2}\|y^n + z^n\|_{h_d}^2. \tag{11.34}$$

Combining (11.32), (11.33) and (11.34), we obtain the following inequality

$$\Gamma_0 + \Gamma_1(\epsilon)\|y^n + z^n\|_{h_d}^2 + \tau b_{2q-1}\int_\Omega (y^n)^{2q}dx \le 2\tau c_1|\Omega| + 2\tau^2|g(y^n)|^2,$$

where

$$\Gamma_1(\epsilon) = 2\tau\nu - \frac{4(\epsilon+1)\tau^2\nu^2}{h_d^2} - \frac{8\tau^2\nu^2(2^d\epsilon+1)}{2^d h_d^2}.$$

Due to the first condition of (11.29), Lemma 11.1 and Lemma 11.3, we obtain that

$$\tau^2|g(y^n)|^2 \leq \frac{2\tau^2 b_{2q-1}^2}{2^{d(q-1)}h_d^{2(q-1)}}|y^n|^{2q-2}\int_\Omega (y^n)^{2q}dx + \tau^2 c_2|\Omega|. \tag{11.35}$$

From $\|y^n + z^n\|_{h_d}^2 > 2|y^n + z^n|^2$, it follows that

$$\Gamma_0 + 2\Gamma_1(\epsilon)|y^n + z^n|^2 + \tau b_{2q-1}(1+2\Gamma_2)\int_\Omega (y^n)^{2q}dx \leq 2\tau c_1|\Omega| + 2\tau^2 c_2|\Omega|,$$

with

$$\Gamma_2 = -\frac{2\tau b_{2q-1}}{2^{d(q-1)}h_d^{2(q-1)}}|y^n|^{2q-2}, \quad \Gamma_3 = (2\tau c_1 + 2\tau^2 c_2)|\Omega|.$$

Since $0 < \epsilon \ll 1$, it suffices to prove that

$$\Gamma_0 + 2\Gamma_1(1)|y^n + z^n|^2 + \tau b_{2q-1}(1+2\Gamma_2)\int_\Omega (y^n)^{2q}dx \leq \Gamma_3.$$

We are now ready to prove Theorem 11.2 by induction:

- i=0 is obvious since $|y^0|^2 + |z^0|^2 \leq M_0$.
- Assuming (11.30) is correct up to i=n, we then have $|y^n|^2 + |z^n|^2 \leq M_0$.
- For i=n+1, using the second conditions of (11.29), we write

$$|y^{n+1}|^2 - |y^n|^2 + |z^{n+1}|^2 - |z^n|^2 + \tau\nu|y^n + z^n|^2 \leq \Gamma_3.$$

Note that $|y^n + z^n|^2 = |y^n|^2 + |z^n|^2$, therefore

$$|y^{n+1}|^2 + |z^{n+1}|^2 \leq (1-\tau\nu)(|y^n|^2 + |z^n|^2) + \Gamma_3.$$

That is

$$\begin{aligned}|y^{n+1}|^2 + |z^{n+1}|^2 \leq &(1-\tau\nu)^{n+1}(|y^0|^2 + |z^0|^2) + \Gamma_3\sum_{i=0}^{n}(1-\tau\nu)^i \\ \leq &(1-\tau\nu)^{n+1}(|y^0|^2 + |z^0|^2) + \frac{1}{\tau\nu}\Gamma_3 \\ = &(1-\tau\nu)^{n+1}(|y^0|^2 + |z^0|^2) + \frac{2}{\nu}(c_1 + \tau_0 c_2)|\Omega| \\ \leq &M_0.\end{aligned}$$

We complete the proof of the stability estimation (11.30). □

In order to study the stability of Scheme II, we firstly introduce the following lemma [Chen and Temam (1993c); Quarteroni and Valli (1994)].

Lemma 11.4 (Discrete Gronwall Lemma). *Let a^n, b^n be two nonnegative sequences satisfying*

$$\frac{a^{n+1} - a^n}{\tau} + \lambda a^{n+1} \leq b^n, \quad b^n \leq b, \quad \forall n \geq 0,$$

then, if $\tau > 0$ and $1 + \tau\lambda > 0$, we have

$$a^n \leq \frac{1}{(1+\tau\lambda)^n}a^0 + \frac{1+\tau\lambda}{\lambda}\left(1 - \frac{1}{(1+\tau\lambda)^{n+1}}\right)b, \qquad \forall n \geq 0.$$

Theorem 11.3 (Stability condition for Scheme II). *Assuming that $\tau \leq \tau_0$ for some τ_0 fixed and*

$$\frac{\tau}{h_d^{2(q-1)}} \leq \frac{2^{d(q-1)}}{2b_{2q-1}M_1^{q-1}}, \tag{11.36}$$

with

$$M_1 = |u_{h_d}^0|^2 + \frac{1+4\tau_0\nu}{4\nu}(2c_1 + \tau_0 c_2)|\Omega|,$$

then we have the following estimate:

$$|u_{h_d}^n|^2 = |y^n|^2 + |z^n|^2 \leq M_1, \quad \textit{for all} \ \ n \geq 0. \tag{11.37}$$

Proof. In order to prove the stability of Scheme II, we have to prove the stability of the following variational form:

$$\begin{cases} (\dfrac{y^{n+1}-y^n}{\tau}, \tilde{y}) + \nu((y^{n+1}+z^{n+1}, \tilde{y}))_{h_d} + (g(y^n), \tilde{y}) = 0, \quad \forall \tilde{y} \in \mathcal{Y}^1, \\ (\dfrac{z^{n+1}-z^n}{\tau}, \tilde{z}) + \nu((y^{n+1}+z^{n+1}, \tilde{z}))_{h_d} = 0, \quad \forall \tilde{z} \in \mathcal{Z}. \end{cases} \tag{11.38}$$

Now, we choose $\tilde{y} = 2\tau y^{n+1}$ and $\tilde{z} = 2\tau z^{n+1}$ in the equation (11.38) and then add them with $2(a-b,a) = |a|^2 - |b|^2 - |a-b|^2$,

$$\begin{aligned} &\Gamma_0 + |y^{n+1}-y^n|^2 + |z^{n+1}-z^n|^2 + 2\tau\nu\|y^{n+1}+z^{n+1}\|_{h_d}^2 \\ =& -2\tau(g(y^n, y^{n+1}-y^n) - 2\tau(g(y^n, y^n) \\ \leq& -2\tau(g(y^n), y^{n+1}-y^n) - \tau b_{2q-1}\int_\Omega (y^n)^{2q}dx + 2\tau c_1|\Omega|. \end{aligned}$$

Thus, we have

$$\begin{aligned} &\Gamma_0 + |y^{n+1}-y^n|^2 + |z^{n+1}-z^n|^2 + 2\tau\nu\|y^{n+1}+z^{n+1}\|_{h_d}^2 \\ &+ \tau b_{2q-1}\int_\Omega (y^n)^{2q}dx \leq -2\tau(g(y^n), y^{n+1}-y^n) + 2\tau c_1|\Omega|. \end{aligned} \tag{11.39}$$

Due to (11.35) and

$$\begin{aligned} -2\tau(g(y^n), y^{n+1}-y^n) &\leq 2\tau|g(y^n)||y^{n+1}-y^n| \\ &\leq |y^{n+1}-y^n|^2 + \tau^2|g(y^n)|^2, \end{aligned}$$

we obtain

$$\begin{aligned} &\Gamma_0 + |z^{n+1}-z^n|^2 + 2\tau\nu\|y^{n+1}+z^{n+1}\|_{h_d}^2 \\ &+\tau b_{2q-1}(1-\Gamma_2)\int_\Omega (y^n)^{2q} \leq 2\tau c_1|\Omega| + \tau^2 c_2|\Omega|. \end{aligned}$$

Thanks to Lemma 11.2 and condition (11.36), we get

$$\Gamma_0 + 4\tau\nu|y^{n+1}+z^{n+1}| \leq 2\tau c_1|\Omega| + \tau^2 c_2|\Omega|.$$

Due to $|y^{n+1}+z^{n+1}|^2 = |y^{n+1}|^2 + |z^{n+1}|^2$, the above inequality becomes

$$(1/\tau)\Gamma_0 + 4\nu(|y^{n+1}|^2 + |z^{n+1}|^2) \leq 2c_1|\Omega| + \tau c_2|\Omega|.$$

Now the discrete Gronwall lemma implies

$$|y^n|^2 + |z^n|^2 \leq \frac{|y^0|^2+|z^0|^2}{(1+4\tau\nu)^n} + \frac{1+4\tau\nu}{4\nu}\left(1 - \frac{1}{(1+4\tau\nu)^n}\right)(2c_1 + \tau_0 c_2)|\Omega|.$$

Therefore,

$$|y^n|^2 + |z^n|^2 \leq \frac{|y^0|^2+|z^0|^2}{(1+4\tau\nu)^n} + \frac{1+4\tau\nu}{4\nu}(2c_1 + \tau_0 c_2)|\Omega|.$$

The proof of stability estimate (11.37) is completed. □

Table 11.1 Condition number of the coefficient matrix

ϵ	0.1	0.01	0.001
WBIUs	23.622	21.859	21.682
WIUs	706.67	1247.5	1357.3

Table 11.2 CPU time of semi-implicit schemes

ϵ	0.1	0.01	0.001
WBIUs	115.33	97.391	83.359
WIUs	371.86	450.98	453.98

11.5 Numerical results

Considering the following equation

$$\begin{cases} \frac{\partial u}{\partial t} - (\epsilon\frac{\partial^2 u}{\partial x^2} + \frac{\partial^2 u}{\partial y^2}) + u^3 = f, & (x,y) \in \Omega, \quad t \in (0,1], \\ u(x,y,0) = (1-x)sin(\pi y)(e^{-x/\sqrt{\epsilon}} - 1), & (x,y) \in \Omega, \quad t = 0, \\ u(x,y,t) = 0, & (x,y) \in \partial\Omega, \quad t \in (0,1], \end{cases}$$

where $\Omega = (0,1)^2$ and

$$\begin{aligned} f =&(1-x)\sin(\pi y)e^t(e^{-x/\sqrt{\epsilon}} - 1) - (2\sqrt{\epsilon}\sin(\pi y)e^{t-x/\sqrt{\epsilon}} \\ &+ (1-x)\sin(\pi y)e^{t-x/\sqrt{\epsilon}}) + (1-x)\sin(\pi y)\pi^2 e^t(e^{-x/\sqrt{\epsilon}} - 1) \\ &+ ((1-x)\sin(\pi y)e^t(e^{-x/\sqrt{\epsilon}} - 1))^3. \end{aligned}$$

The exact solution is $(1-x)\sin(\pi y)e^t(e^{-x/\sqrt{\epsilon}} - 1)$.

Solving above equation with WIUs and WBIUs respectively, we find that when ϵ is small, the condition number of the WBIUs-type coefficient matrix is much smaller than WIUs-type's, see Table 11.1 and Fig. 11.1. From Tables 11.2 and 11.3, we can see that the CPU time with WBIUs is much less than WIUs. Figs. 11.2 and 11.3 give us the error images for the semi-implicit scheme and explicit scheme when $\epsilon = 0.1,\ 0.01$, $N = 1$ and $d = 4$, which tell us that the WBIUs-type schemes have the nearly same precision with WIUs, but they save lots of CPU time, so the WBIUs method is more efficient than WIUs method.

Table 11.3 CPU time of explicit scheme

ϵ	0.1	0.01	0.001
WBIUs	1193	1218	1197
WIUs	2363	2393	2338

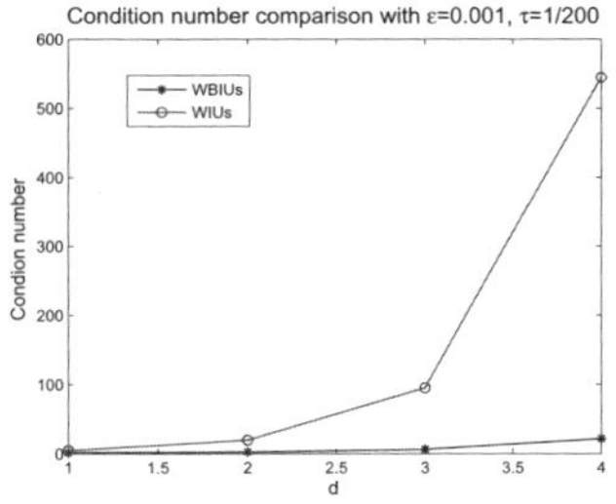

Fig. 11.1 The condition number comparison between WIUs and WBIUs.

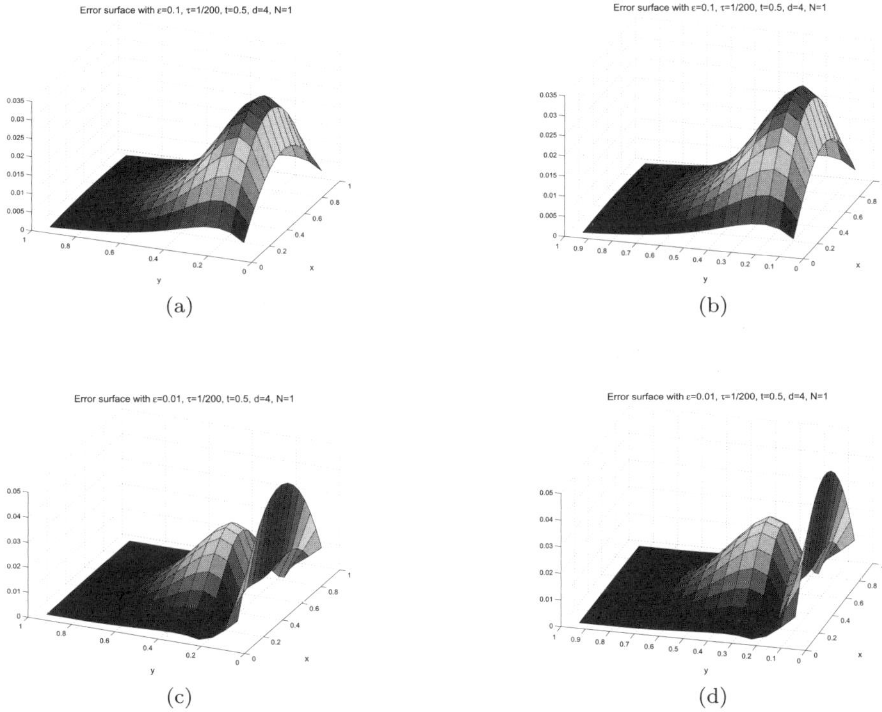

Fig. 11.2 The error comparison between WIUs and WBIUs (semi-implicit scheme).

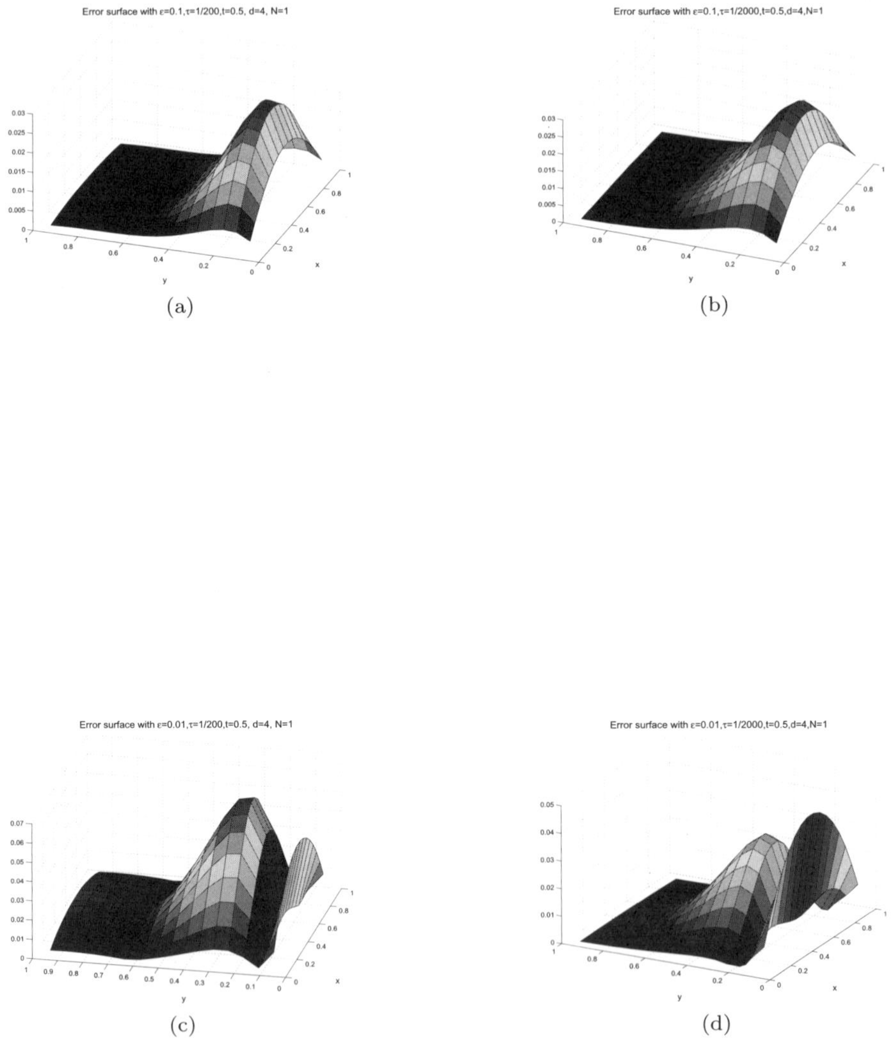

Fig. 11.3 The error comparison between WIUs and WBIUs (explicit scheme).

Bibliography

Chen, M. and Temam, R. (1993a). Incremental unknowns in finite differences: Condition number of the matrix, *SIAM J. Matrix Anal. Appl.* **14**, 2, pp. 432–455.

Chen, M. and Temam, R. (1993b). Nonlinear galerkin method in the finite difference case and wavelet-like incremental unknowns, *Numer. Math.* **64**, pp. 271–294.

Chen, M. and Temam, R. (1993c). Nonlinear Galerkin method with multilevel incremental unknowns, in *Contributions in numerical mathematics*, *World Sci. Ser. Appl. Anal.*, Vol. 2 (World Sci. Publ., River Edge, NJ), pp. 151–164.

Chen, Z.-Y., Wu, B. and Xu, Y. S. (2005). Multilevel augmentation methods for operator equations, *Numer. Math., J. Chin. Univ. (English Ser.)*, **14**, pp. 31-55.

Chen, Z. Y., Wu, B. and Xu, Y. S. (2006a). Multilevel augmentation methods for differential equations, *Adv. Comput. Math.* **24**, pp. 213-238.

Chen, Z. Y., Xu, Y. S. and Yang, H. Q. (2006b). A multilevel augmentation method for solving ill-posed operator equations, *Inverse Problems* **22**, 155-174.

Foias, C., Sell, G. and Temam, R. (1988). Inertial manifolds for nonlinear evolutionary equations, *J. Diff. Equ.* **73**, pp. 309–353.

Miranville, A. and Muresan, A. C. (2002). Block incremental unknowns for anisotropic elliptic equations, *Appl. Numer. Math.* **42**, 4, pp. 529–543.

Pouit, F. (1996). Stability study, error estimation, and condition number for semi-implicit schemes using incremental unknowns, *Numer. Meth. Part. Diff. Equ.* **12**, pp. 743–766.

Quarteroni, A. and Valli, A. (1994). *Numerical Approximation of Partial Differential Equations* (Springer-Verlag, Berlin).

Temam, R. (1990). Inertial manifolds and multigrid methods, *SIAM J. Math. Anal.* **21**, 1, pp. 154–178.

Wu, Y. J., Jia, X. X. and She, A. L. (2007). Semi-implicit schemes with multilevel wavelet-like incremental unknowns for solving reaction diffusion equation, *Hokk. Math. J.* **36**, pp. 711–728.

Wu, Y. J. and Yang, A. L. (2007). Incremental unknowns for the heat equation with time-dependent coefficients: semi-implicit θ-schemes and their stability, *J. Comput. Math.* **25**, 5, pp. 573–582.

Wu, Y. J., Yang, A. L. and Song, L. J. (2009). Wavelet-like block incremental unknowns for numerical computation of anisotropic parabolic equations, in *Proceedings of the 2009 WRI World Congress on Computer Science and Information Engineering* Volume 2, pp. 550–554.

Chapter 12

Models and dynamics of deterministically growing networks[1]

Weigang Sun[a,*], Jingyuan Zhang[a] and Guanrong Chen[b]
[a] *School of Science, Hangzhou Dianzi University, Hangzhou 310018, Zhejiang, PR China*
**wgsun999@yahoo.com.cn*
[b] *Department of Electronic Engineering, City University of Hong Kong, PR China*

On the basis of Koch networks constructed using Koch fractals, we propose a family of Koch networks with novel features including an initial state that is a complete graph with an arbitrary number of nodes as a generalization of a triangle. In the subsequent evolving steps, existing nodes create finite complete graphs. The analytical expressions for some topological and dynamical properties are obtained, including degree distribution, clustering coefficient, average path length, degree correlations, first passage time, average receiving time, mean return time, and mean first passage time. This family of Koch networks follows a power-law distribution, and has a large clustering coefficient, a small average path length, and degree correlations with rich dynamical properties. The mean return time of the new emerging nodes is identical and increases with some network parameter values. Moreover, the average sending time grows with the network order. In addition, the mean first passage time grows linearly with the number of nodes in the large limit of the network order and increases with the dimension of the polygon in the network.
Keywords: Random walk, Koch networks, average path length

[1]This work was supported by the National Natural Science Foundation of China (grant no.10872119), the Disciplinary Research of Hangzhou Dianzi University (grant nos. zx100204004-7 and KYF075610032), and the Hong Kong Research Grants Council under the GRF Grant CityU 1114/11E.

12.1 Introduction

The past decade has witnessed the emergence of a new and exciting research field of network science and engineering, which describes and characterizes a large number of natural and man-made inter-connected systems, such as the Internet, WWW, citation networks, power grids, airport networks, cellular and metabolic networks, and so on. Intensive and extensive research has shown that most of these networks exhibit small-world effects [Watts and Strogatz (1998)] and/or scale-free characteristics [Albert and Barabási (2002); Newman (2003); Dorogovtsev and Mendes (2002); Boccaletti *et al.* (2006); Costa *et al.* (2007)]. Recently, the main issues requiring further consideration about various complex networks include network models, functions, dynamics, and applications. Regarding network modeling and applications, studies in neuroscience have attracted specially high interest [Sporns and Zwi (2004); Bullmore and Sporns (2009)]. For example, brain functional networks extracted by functional magnetic resonance imaging (fMRI) are found to have scale-free and small-world characteristics [Eguíluz *et al.* (2005)], and the quantities of the networks constructed by electroencephalography (EEG) signals for patients in coma state are shown to be larger than those in the brain death state [Sun *et al.* (2011)].

With the discovery of small-world and scale-free [Barabási and Albert (1999)] network models, a lot of activities have been devoted to the study of various dynamical processes on complex networks. The goal is to understand the influences of network structures on dynamics. There exist many basic dynamical processes on complex networks, such as synchronization, epidemic spreading, robustness against attacks, community structural evolution, random walks, data routing, and so on. Especially, inner synchronization and its control on a network have been extensively investigated [Wang (2002); Arenas *et al.* (2008); Zhou *et al.* (2006); Wang, *et al.* (2008)] (see also many references cited therein). In addition, outer synchronization has been introduced in [Li *et al.* (2007)] for studying the synchronization between coupled networks. Recently, random walks on networks are widely investigated. A primary interest related to random walks is the first passage time (FPT) defined as the expected time for a walker to first reach a given destination node starting from a source node. This quantity plays an important role in many real applications, such as particles transport, neuron firing, disease spreading, and target searching [Condamin *et al.* (2007)]. Recently, there have been a growing number of studies on mean FPT (MPFT) in deterministic networks.

Through the studies on various network models, deterministically growing networks have attracted increasing attention because they provide precise analytical results for many network features such as degree distribution, clustering coefficient, average path length, degree correlation, random walk, spanning tree, among others. The first deterministic small-world and scale-free networks were proposed in [Comellas *et al.* (2000); Barabási *et al.* (2001)]. Later, models and dynamical processes of fractal networks, constructed by Koch fractals [Zhang and Sun (2010); Sun *et al.* (2011)], Sierpinski fractals [Zhang *et al.* (2009)], Vicsek fractals [Zhang *et al.* (2010)], have been widely studied. Although significant progress has been made in the field of network modeling, it is still a fundamental task and of current interest to put forward models and reproducing generic properties from different points of view.

Using Koch fractals as bases, Zhang *et al.* [Zhang *et al.* (2009)] proposed Koch networks and studied their topological properties. In [Zhang and Sun (2010)], we proposed generalized Koch networks, and transformed the initial state of the networks from a triangle to a polygon. Recently, a new type of Koch networks [Zhang *et al.* (2010)] has been introduced, in which existing nodes create a certain number of triangles in the subsequent steps.

Combined with the iterative algorithms proposed in the above mentioned studies [Zhang and Sun (2010); Zhang *et al.* (2010)], we further develop a family of Koch networks, which have some novel features such as an initial state that is a graph globally connected to a certain number of nodes. In the subsequent step, all the existing nodes produce some number of complete graphs. For this family of Koch networks, we obtain analytical results for the degree distribution, clustering coefficient, average path length, and degree correlations. In addition, we have investigated random walks on Koch networks, including first passage time, average receiving time (ART) and average sending time from a hub node to all other nodes (AST). Besides, for the network model described in [Zhang and Sun (2010)], we obtain the exact scaling of mean first passage time (MFPT), which shows that the MFPT grows linearly with the increasing order of the networks, and moreover the MFPT increases with the network parameter values.

12.2 A generation algorithm

Let $K_{m,r}(t)$ be the (m,r)-Koch network at step t. An iterative algorithm for creating the (m,r)-Koch networks is described as follows. Initially, $K_{m,r}(0)$ is a complete graph K_r. Each of the r nodes in the initial K_r then produces m groups of nodes. Each node group has $r-1$ nodes. These $r-1$ new nodes and their parent nodes are linked to form a new K_r. Thus, one obtains $K_{m,r}(1)$. Fig. 12.1 depicts $K_{m,r}(1)$ with $m=2$ and $r=4$. For $t \geq 1$, $K_{m,r}(t)$ is obtained from $K_{m,r}(t-1)$ by replacing each K_r in $K_{m,r}(t-1)$ with $K_{m,r}(1)$. The growing process is repeated until the network reaches a desired order.

Subsequently, the order and the size (number of edges) of $K_{m,r}(t)$ are calculated. To accomplish this, the total number of K_r existing at step t is first calculated. This number is denoted as $L_\diamond(t)$. Through the network construction, the quantity increases by a factor of $rm+1$, i.e., $L_\diamond(t) = (rm+1)L_\diamond(t-1)$. With the initial condition of $L_\diamond(0) = 1$, one obtains $L_\diamond(t) = (rm+1)^t$. Let $L_v(t)$ and $L_e(t)$ be the numbers of nodes and edges created at step t, respectively. Note that each K_r in

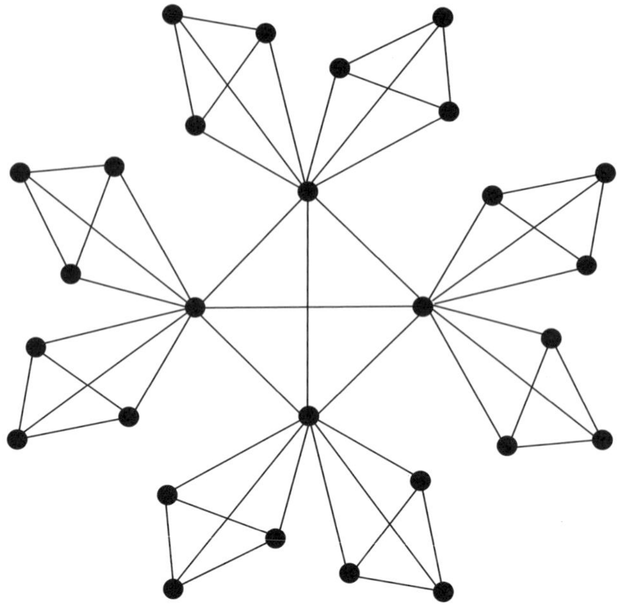

Fig. 12.1 $K_{m,r}(1)$ with $m = 2, r = 4$.

$K_{m,r}(t-1)$ will lead to an addition of $r \times (r-1)m$ new nodes and $r \times \frac{r(r-1)}{2}m$ new edges at step t. One can then easily obtains the following relationships:

$$L_v(t) = r(r-1)mL_\diamond(t-1) = r(r-1)m(rm+1)^{t-1},\ t > 0, \qquad (12.1)$$

$$L_e(t) = \frac{r^2(r-1)}{2}mL_\diamond(t-1) = \frac{r^2(r-1)}{2}m(rm+1)^{t-1},\ t > 0. \qquad (12.2)$$

The total numbers of vertices N_t and edges E_t at step t are respectively

$$N_t = \sum_{k=0}^{t} L_v(k) = (r-1)(rm+1)^t + 1$$

and

$$E_t = \sum_{k=0}^{t} L_e(k) = \frac{r(r-1)}{2}(rm+1)^t.$$

Thus, the average degree is

$$\langle k \rangle = \frac{2E_t}{N_t} = \frac{r(r-1)(rm+1)^t}{(r-1)(rm+1)^t + 1},$$

which is approximately r for a large t.

12.3 Structural properties

In this section, we study some relevant characteristics of the (m, r)-Koch network, focusing primarily on degree distribution, clustering coefficient, average path length, and degree correlations.

12.3.1 *Degree distribution*

Let $k_i(t)$ be the degree of node i at time t. When node i is added to the networks at step $t_i (t_i \geq 0)$, $k_i(t_i) = r - 1$. To obtain $k_i(t)$, the number of K_r involving node i at step t is first determined. This number is represented by $L_\diamond(i,t)$. These K_r will create new nodes connected to node i at step $t+1$. At step t_i, $L_\diamond(i, t_i) = 1$. From the network construction, one obtains $L_\diamond(i,t) = (m+1)L_\diamond(i,t-1)$, and thus $L_\diamond(i,t) = (m+1)^{t-t_i}$. The relationship between $k_i(t)$ and $L_\diamond(i,t)$ is

$$k_i(t) = (r-1)L_\diamond(i,t) = (r-1) \times (m+1)^{t-t_i}. \tag{12.3}$$

This way, the degree of node i at time t is explicitly computed. From Eq. (12.3), one obtain

$$k_i(t) = (m+1)k_i(t-1). \tag{12.4}$$

Equation (12.4) shows that the degree spectrum of the (m,r)-Koch network is discrete. Consequently, the cumulative degree distribution [Newman (2003)] is given by

$$P_{cum}(k) = \frac{1}{N_t}\sum_{\tau \leq t_i} L_v(\tau) = \frac{(r-1) \times (rm+1)^{t_i} + 1}{(r-1)(rm+1)^t + 1}. \tag{12.5}$$

Substituting $t_i = t - \frac{\ln(\frac{k}{r-1})}{\ln(m+1)}$ into Eq. (12.5) yields

$$P_{cum}(k) = \frac{(r-1) \times (rm+1)^t \times (\frac{k}{r-1})^{-\frac{\ln(rm+1)}{\ln(m+1)}} + 1}{(r-1) \times (rm+1)^t + 1}. \tag{12.6}$$

When t is sufficiently large, one obtains

$$P_{cum}(k) = (r-1)^{\frac{\ln(rm+1)}{\ln(m+1)}} \times k^{-\frac{\ln(rm+1)}{\ln(m+1)}}. \tag{12.7}$$

Thus, the degree distribution follows a power law with exponent $\gamma = \frac{\ln(rm+1)}{\ln(m+1)}$. When r satisfies the condition

$$m + 2 \leq r \leq m^2 + 3m + 3,$$

exponent γ belongs to the interval $[2, 3]$, which is the same range for the exponent of degree distribution in most real scale-free networks [Albert and Barabási (2002)].

12.3.2 *Clustering coefficient*

The clustering coefficient [Watts and Strogatz (1998)] of node i with degree k_i is given by $C_i = 2e_i/(k_i(k_i - 1))$, where e_i is the number of existing triangles attached to node i and $k_i(k_i - 1)/2$ is the total number of possible triangles, including node i. For a single node with degree k in $K_{m,r}(t)$, the clustering coefficient is denoted as $C(k)$. For node i with degree k, $(m+1)^{t-t_i}$ disjoint complete sub-graphs K_r are connected to this node. In each K_r, $(r-1)(r-2)/2$ triangles are attached to this node. Hence,

$$C(k) = \frac{r-2}{k-1},$$

which is approximately k^{-1} in the large limit of k.

After t-step growth, the average clustering coefficient is given by

$$C_t = \frac{1}{N_t} \sum_{l=0}^{t} \left[\frac{r-2}{G_l - 1} \times L_v(l) \right], \tag{12.8}$$

where the sum runs over all the nodes, and G_l is the degree of the nodes created at step l, which is given by Eq. (12.3). Equation (12.8) converges to a nonzero value C depending on parameters m and r. Figs. 12.2 and 12.3 show that the value of C_t increases with m(r) for fixed r(m). The clustering coefficient of the (m, r)-Koch network is higher than that of the Koch network [Zhang *et al.* (2009)].

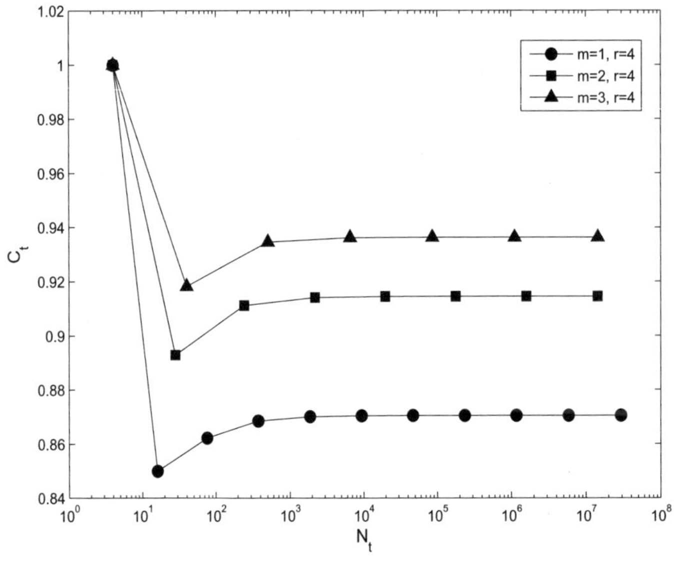

Fig. 12.2 Semilogarithmic graph of the clustering coefficient C_t versus network order N_t, $r = 4$.

12.3.3 *Average path length*

Let d_t denote the average path length of the (m, r)-Koch network. The average path length can be analytically computed by using the method in [Zhang *et al.* (2009)] because the (m, r)-Koch network is self-similar. We represent all the shortest path lengths of $K_{m,r}(t)$ as a matrix, in which entries d_{ij} are the shortest distance from nodes i to j. Hence,

$$d_t = \frac{D_t}{N_t(N_t - 1)/2}, \tag{12.9}$$

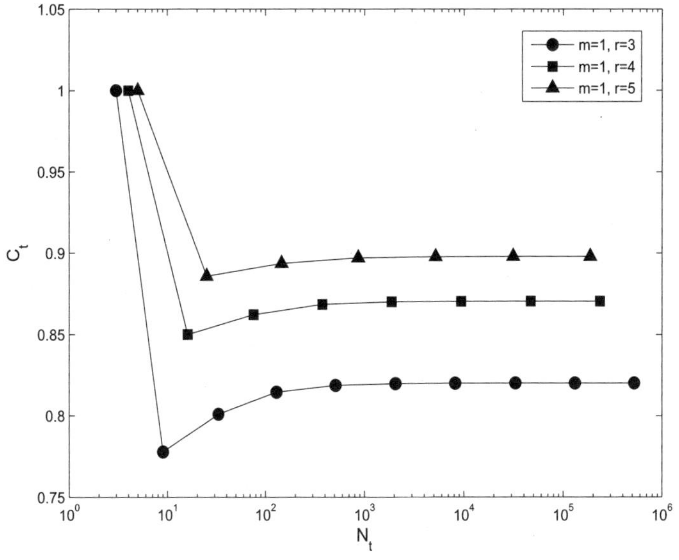

Fig. 12.3 Semilogarithmic graph of the clustering coefficient C_t versus network order N_t, $m = 1$.

where D_t is the total distance between all pairs of nodes, i.e.,

$$D_t = \sum_{\substack{i \in K_{m,r}(t),\, j \in K_{m,r}(t) \\ i \neq j}} d_{ij}. \tag{12.10}$$

Given that the (m, r)-Koch network is self-similar, one can make $rm + 1$ copies of $K_{m,r}(t)$ and join them at the hub nodes to obtain $K_{m,r}(t+1)$. Fig. 12.4 shows that $K_{m,r}(t+1)$ may be obtained by the juxtaposition of $rm+1$ copies of $K_{m,r}(t)$, which are labeled K_t^0, K_t^1, K_t^2, ..., K_t^{rm-1}, K_t^{rm}.

Then, sum D_{t+1} is written as

$$D_{t+1} = (rm + 1)D_t + \Delta_t, \tag{12.11}$$

where Δ_t is the sum of all the shortest paths whose endpoints are not in the same $K_{m,r}(t)$ branch. The solution of Eq. (12.11) is

$$D_t = (rm + 1)^{t-1} D_1 + \sum_{x=1}^{t-1} (rm + 1)^{t-1-x} \Delta_x. \tag{12.12}$$

The paths that contribute to Δ_t must all go through at least one of the r edge nodes where the different branches of $K_{m,r}(t)$ are connected. The analytical expression for Δ_t, called the length of crossing paths, is obtained as follows: $\Delta_t^{\alpha,\beta}$ is denoted as the sum of the lengths of all the shortest paths with endpoints in K_t^α and K_t^β. If K_t^α and K_t^β meet at an edge node, $\Delta_t^{\alpha,\beta}$ rules out the paths where either endpoint

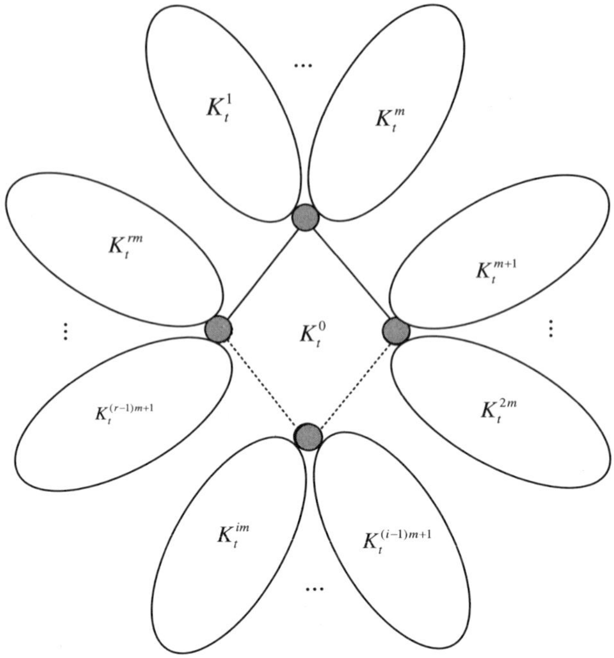

Fig. 12.4 $K_{m,r}(t+1)$ consisting of $K_t^0, K_t^1, \cdots, K_t^{rm}$.

is a shared edge node. If K_t^α and K_t^β do not meet, $\Delta_t^{\alpha,\beta}$ excludes the paths where either endpoint is any edge node. Total sum Δ_t is then given by

$$\begin{aligned}\Delta_t &= \sum_{0\le i<j\le rm} \Delta_t^{i,j}\\ &= \left[r \times \frac{m(m-1)}{2} + rm\right] \Delta_t^{0,1} + \frac{r(r-1)}{2} \times m^2 \times \Delta_t^{1,m+1}. \end{aligned} \tag{12.13}$$

To obtain $\Delta_t^{\alpha,\beta}$, define

$$s_t = \sum_{\substack{i \in K_{m,r}(t)\\ i \ne X}} d_{iX}.$$

Considering the self-similar network structure, one can easily obtain the following recursive relationship:

$$s_{t+1} = (m+1)s_t + (r-1)m \times (s_t + N_t - 1).$$

With regard to the initial condition $s_0 = r-1$, one obtains

$$s_t = [(r-1)^2 mt + (r-1)(rm+1)] \times (rm+1)^{t-1}. \tag{12.14}$$

Quantities $\Delta_t^{0,1}$ and $\Delta_t^{1,m+1}$ are then calculated. Using s_t, one obtains

$$\Delta_t^{0,1} = \sum_{\substack{i \in K_t^0,\, j \in K_t^1\\ i,j \ne X}} d_{ij}$$

$$= \sum_{\substack{i \in K_t^0, j \in K_t^1 \\ i, j \neq X}} (d_{iX} + d_{jX})$$
$$= (N_t - 1) \times \sum_{i \in K_t^0, i \neq X} d_{iX} + (N_t - 1) \times \sum_{j \in K_t^1, j \neq X} d_{jX}$$
$$= 2 \times (N_t - 1) \times \sum_{i \in K_t^0, i \neq X} d_{iX}. \tag{12.15}$$

and

$$\Delta_t^{1,m+1} = \sum_{\substack{i \in K_t^1, i \neq X \\ j \in K_t^{m+1}, j \neq Y}} d_{ij}$$
$$= \sum_{\substack{i \in K_t^1, i \neq X \\ j \in K_t^{m+1}, j \neq Y}} (d_{iX} + d_{XY} + d_{jX})$$
$$= 2 \times (N_t - 1) \times \sum_{i \in K_t^1, i \neq X} d_{iX} + (N_t - 1)^2, \tag{12.16}$$

where $d_{XY} = 1$. Substituting Eqs. (12.15) and (12.16) into Eq. (12.13) yields

$$\Delta_t = rm(m+1)(N_t - 1)s_t + \frac{r(r-1)}{2} m^2 \left[2(N_t - 1)s_t + (N_t - 1)^2\right]$$
$$= rm(N_t - 1)s_t(rm + 1) + \frac{r(r-1)}{2} \times m^2 (N_t - 1)^2. \tag{12.17}$$

Incorporating Eqs. (12.14) and (12.17) into Eq. (12.12) gives

$$D_t = (rm+1)^{t-1} D_1$$
$$+ rm(rm+1) \sum_{x=1}^{t-1} (rm+1)^{t-1-x} (N_x - 1) s_x$$
$$+ \frac{r(r-1)}{2} m^2 \sum_{x=1}^{t-1} (rm+1)^{t-1-x} (N_x - 1)^2$$
$$= (rm+1)^{t-1} D_1$$
$$+ (r-1)^3 rm^2 (rm+1) \sum_{x=1}^{t-1} (rm+1)^{t-2+x} x$$
$$+ (r-1)^2 rm (rm+1)^2 \sum_{x=1}^{t-1} (rm+1)^{t-2+x}$$
$$+ \frac{r(r-1)^3}{2} m^2 \sum_{x=1}^{t-1} (rm+1)^{t-1+x}$$
$$= (rm+1)^{t-1} D_1$$
$$+ \frac{1}{r} (r-1)^3 (rm+1)^t \times \{(rm+1)^{t-1}[rm(t-1) - 1] + 1\}$$

$$+(r-1)^2(rm+1)^{t+1}[(rm+1)^{t-1}-1]$$
$$+\frac{(r-1)^3}{2}m(rm+1)^t[(rm+1)^{t-1}-1], \qquad (12.18)$$

where

$$D_1 = \frac{r(r-1)}{2}(rm+1) + \left[r \times \frac{m(m-1)}{2} + rm\right] \times 2(r-1)^2$$
$$+\frac{r(r-1)}{2}m^2 \times 3(r-1)^2$$
$$= \frac{r(r-1)}{2}(rm+1) + rm(m+1)(r-1)^2 + \frac{3r(r-1)^3}{2} \times m^2.$$

The analytical result of d_t for the (m, r)-Koch network is easily obtained by using Eq. (12.9). When $t \to \infty$, $d_t \sim t$. Figs. 12.5 and 12.6 show the plotted average path length versus network order N_t with regard to parameter m or r.

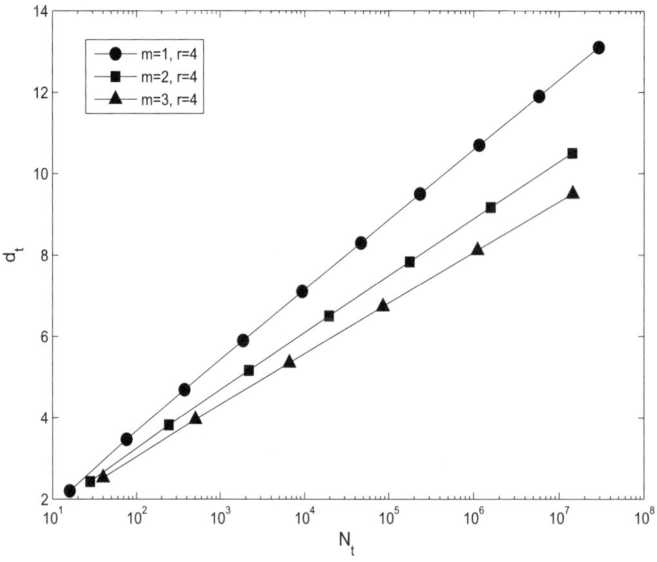

Fig. 12.5 Average path length d_t versus network order N_t on a semilogarithmic scale, $r = 4$.

12.3.4 *Degree correlations*

Degree correlations in networks have been an interesting research topic [Vázquez *et al.* (2002)] because they enable the study of the effects of some network structures. Let $k_{nn}(k)$ be the average degree of the nearest neighbors of nodes with degree k. When $k_{nn}(k)$ increases with k, the network is considered assortative. By contrast, if $k_{nn}(k)$ decreases with k, the network is disassortative. If correlations are absent,

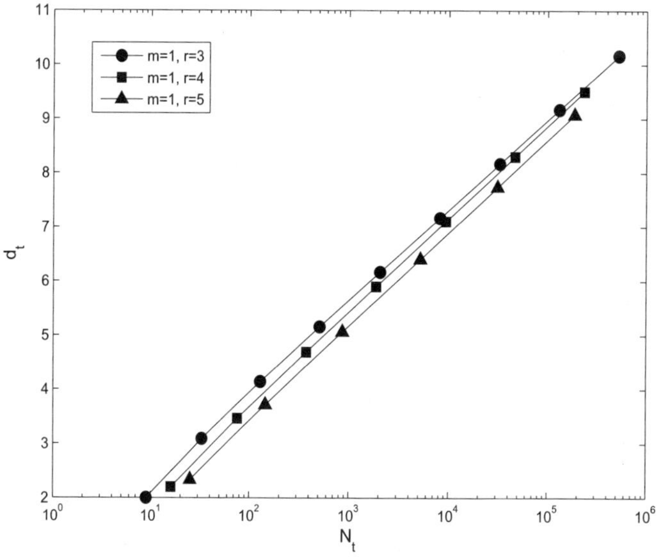

Fig. 12.6 Average path length d_t versus network order N_t on a semilogarithmic scale, $m = 1$.

$k_{nn}(k) = \text{const}$. Also, let $k(t_i, t)$ be the degree of a node at step t that was generated at step t_i. Using Eqs. (12.1) and (12.3), and the construction of the (m, r)-Koch network, one obtains the expressions of the degree correlations [Zhang *et al.* (2010)]

$$k_{nn}(k) = \frac{1}{L_v(t_i)k(t_i,t)}\left\{\sum_{t_i'=0}^{t_i-1} mL_v(t_i')k(t_i',t_i-1)k(t_i',t) + \sum_{t_i'=t_i+1}^{t} mL_v(t_i)k(t_i,t_i'-1)k(t_i',t)\right\} + r - 2. \tag{12.19}$$

To calculate Eq. (12.19), consider two cases: $m = r - 2$ and $m \neq r - 2$. When $m = r - 2$, one has

$$k_{nn}(k) = (r-2)(t-1) + 2r - 3.$$

In this case, the networks show absence of correlations independent of t. When $m \neq r - 2$, Eq. (12.19) is simplified to

$$k_{nn}(k) = \frac{rm+1}{m-r+2}\left[\frac{(m+1)^2}{rm+1}\right]^{t_i} - \frac{r^2-3r+m+3}{m-r+2} + \frac{m(r-1)}{m+1}(t-t_i). \tag{12.20}$$

Writing Eq. (12.20) in terms of k gives

$$k_{nn}(k) = \frac{rm+1}{m-r+2}\left[\frac{(m+1)^2}{rm+1}\right]^{t}\left(\frac{k}{r-1}\right)^{-\frac{\ln[\frac{(m+1)^2}{rm+1}]}{\ln(m+1)}}$$

$$-\frac{r^2 - 3r + m + 3}{m - r + 2} + \frac{m(r-1)}{m+1} \frac{\ln(\frac{k}{r-1})}{\ln(m+1)}. \tag{12.21}$$

Therefore, when $m > r - 2$, $k_{nn}(k)$ is approximately a power-law function of k with a negative exponent, indicating that the (m, r)-Koch network is disassortative. On the contrary, when $m < r - 2$, the network is assortative.

12.4 Random walks on Koch networks

In this section, we study random walks on this family of Koch networks, including first passage time (FPT), average return time (ART) and average sending time (AST) from one hub node to another node on the network.

12.4.1 *Evolutionary rule for first passage time*

Denote $F_{ij}(t)$ as the FPT of a walker moving from an arbitrary node i to an arbitrary node j. According to the network construction, the exact relation between $F_{ij}(t)$ and $F_{ij}(t+1)$ is then determined. Considering node i on network $K_{m,r}(t)$, Eq. (12.3) indicates that upon the growth of the networks from step t to $t+1$, the degree $k_i(t)$ of node i increases by m times, i.e., from $k_i(t)$ to $(m+1)k_i(t)$. Let A express the FPT from node i to any of its k_i old neighbors in $K_{m,r}(t)$ and B be the FPT that a walker walks from any of the mk_i new neighbors of node i generated at the $t+1$ iteration step to one of its k_i old neighbors existing in $K_{m,r}(t)$. Then,

$$\begin{aligned} A &= \frac{1}{m+1} \times 1 + \frac{m}{m+1}(1+B), \\ B &= \frac{1}{r-1}(1+A) + \frac{r-2}{r-1}(1+B), \end{aligned} \tag{12.22}$$

which leads to $A = mr + 1$. Therefore, the FPT from node i belonging to $K_{m,r}(t)$ to node j ($j \in K_{m,r}(t+1)$) increases mr times on average, namely,

$$F_{ij}(t+1) = (mr+1)F_{ij}(t), \tag{12.23}$$

where A is called the scaling efficiency, which is very important for the study of random walks on the generalized Koch network.

12.4.2 *Explicit expression for average return time*

Let $R_i(t)$ be the expected time that a walker starts from node i and returns to the starting node for the first time. Then,

$$R_i(t) = \frac{1}{k_i(t)} \sum_{j \in \Omega_i(t)} [1 + F_{ji}(t)], \tag{12.24}$$

where $\Omega_i(t)$ is the set of neighbors of node i in $K_{m,r}(t)$.

On the other hand, at time step $t+1$, for the network $K_{m,r}(t+1)$, among all the neighbors of node i, there are $\frac{m}{m+1}$ newly emerging nodes, and the rest are old neighbors. Consequently,

$$R_i(t+1) = \frac{m}{m+1} \times \tilde{R}_i(t+1) + \frac{1}{m+1} \times \frac{1}{k_i(t)} \sum_{j \in \Omega_i(t)} [1 + F_{ji}(t)]. \quad (12.25)$$

The first term on Eq. (12.25) shows that the walker returns to node i along the new neighbors. Because there are $\frac{m}{m+1}$ of new neighbors, such a process occurs with a probability of $\frac{m}{m+1}$ and it takes $\tilde{R}_i(t+1)$ time steps. The second term of Eq. (12.25) describes the case that the walker returns to node i through its old neighbors. Suppose the nodes $j_1, j_2, \cdots, j_{r-1}$ are new neighbors of node i in $K_{m,r}(t+1)$. Then

$$\begin{cases} \tilde{R}_i(t) = \frac{1}{r-1}\Big[1 + F_{j_1 i}(t)\Big] + \frac{1}{r-1}\Big[1 + F_{j_2 i}(t) + \cdots + \frac{1}{r-1}(1 + F_{j_{r-1} i}(t))\Big], \\ F_{j_k i}(t) = \frac{1}{r-1} \times 1 + \frac{1}{r-1} \times \sum_{m=1, m \neq k}^{r-1} \Big[1 + F_{j_m i}(t)\Big], k = 1, 2, \cdots, r-1. \end{cases} \quad (12.26)$$

Considering the new neighbors $j_1, j_2, \cdots, j_{r-1}$ are equivalent to node i, one has $F_{j_1 i} = F_{j_2 i} = \cdots = F_{j_{r-1} i}$. Solving Eqs. (12.26), gives $\tilde{R}_i(t) = r$. Hence,

$$\begin{aligned} R_i(t+1) &= \frac{mr}{m+1} + \frac{1}{(m+1)k_i(t)} \sum_{j \in \Omega_i(t)} [1 + F_{ji}(t)], \\ &= \frac{mr+1}{m+1} R_i(t). \end{aligned} \quad (12.27)$$

Next, we determine the mean return time (MRT) for newly generated nodes in $K_{m,r}(t)$. Let i_1 be a new neighbor of node i created at step t in $K_{m,r}(t-1)$. When node i_1 was generated, other $r-2$ neighbor nodes $i_2, i_3, \cdots, i_{r-1}$ were also produced, and they were connected to each other. Let M be the ART for a walker starting from node i without going through nodes $i_1, i_2, i_3, \cdots, i_{r-1}$ and returning back. Then,

$$R_{i_1}(t) = \frac{1}{r-1}[1 + F_{i i_1}(t)] + \frac{1}{r-1} \sum_{k=2}^{r-1} [1 + F_{i_k i_1}(t)], \quad (12.28)$$

$$F_{i_k i_1}(t) = \frac{1}{r-1} + \frac{1}{r-1} \sum_{m=2, m \neq k}^{r-1} [1 + F_{i_m i_1}(t)], \ k = 2, 3, \cdots, r-1, \quad (12.29)$$

$$F_{i i_1}(t) = \frac{1}{k_i(t)} + \frac{1}{k_i(t)} \sum_{k=2}^{r-1} [1 + F_{i_k i_1}(t)] + \frac{k_i(t) - r + 1}{k_i(t)} [M + F_{i i_1}(t)]. \quad (12.30)$$

To simplify Eqs. (12.28) and (12.30), the MRT of node i is written as

$$R_i(t) = \frac{1}{k_i(t)} \times r \times (r-1) + \frac{k_i(t) - r + 1}{k_i(t)} \times M. \tag{12.31}$$

The first part describes the process that the walker starts from node i and then goes through nodes $i_1, i_2, \cdots, i_{r-1}$ and finally returns to node i. The last part expresses the time that the walker returns without visiting nodes $i_1, i_2, \cdots, i_{r-1}$. Eliminating $F_{ii_1}(t), F_{i_2 i_1}(t), \cdots, F_{i_{r-1} i_1}(t)$, one obtains

$$R_{i_1}(t) = \frac{k_i(t)}{r-1} R_i(t). \tag{12.32}$$

Combining Eqs. (12.3) and (12.27) with $R_i(1) = r$, one obtains,

$$R_{i_1}(t) = r(mr+1)^t. \tag{12.33}$$

As shown in Figs. 12.7 and 12.8, $R_{i_1}(t)$ increases with m or r. It is noted that Eq. (12.33) does not depend on the degrees of the old nodes, which means that the newly created nodes i_1 have identical MRT.

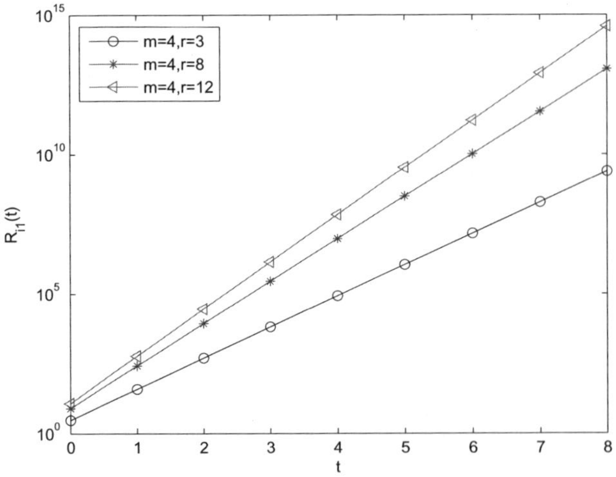

Fig. 12.7 Semilogarithmic graph of the MRT of the new emerging node i_1 versus step t, $m = 4$.

12.4.3 *Average sending time from a hub node to another node*

In this subsection, we study the average sending time (AST) from a hub node to another node on networks $K_{m,r}(t)$. Let $T_i(t)$ represent the FPT that a walker starting from the hub node and arriving at node i on network $K_{m,r}(t)$. The average of $T_i(t)$ over all target nodes in $K_{m,r}(t)$ is AST, denoted by $\langle T \rangle_t$. Let Δ_t be the set

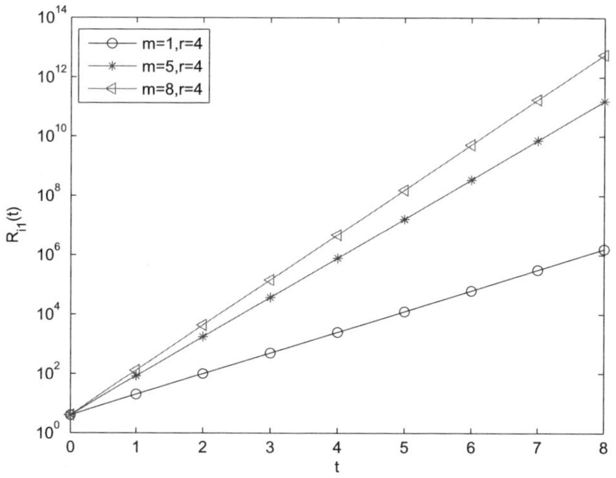

Fig. 12.8 Semilogarithmic graph of the MRT of new emerging node i_1 versus step t, $r = 4$.

of all the nodes existing in $K_{m,r}(t)$, and Δ'_t be the set of newly generated nodes at step t. Then, $\Delta_t = \Delta'_t + \Delta_{t-1}$ and $\langle T \rangle_t$ is given by

$$\langle T \rangle_t = \frac{1}{N_t - 1} T_{tot}(t), \tag{12.34}$$

where $T_{tot}(t) = \sum\limits_{i \in \Delta_t} T_i(t)$. Since all the nodes in $K_{m,r}(t)$ belong to either Δ'_t or Δ_{t-1}, $T_{tot}(t)$ reads as

$$T_{tot}(t) = \sum_{j \in \Delta'_t} T_j(t) + \sum_{j \in \Delta_{t-1}} T_j(t). \tag{12.35}$$

From Eq. (12.23), one has $T_j(t) = (mr + 1)T_j(t - 1)$, so Eq. (12.35) becomes

$$T_{tot}(t) = \sum_{j' \in \Delta'_t} T_{j'}(t) + (mr + 1)T_{tot}(t - 1). \tag{12.36}$$

Suppose that a walker starts from the hub node to visit node j_1, it first goes through node j which is an old neighbor of node j_1 existing in $K_{m,r}(t - 1)$. Then,

$$T_{j_1}(t) = T_j(t) + F_{jj_1}(t). \tag{12.37}$$

When node j_1 was born, the other $r - 2$ nodes $j_2, j_3, \cdots, j_{r-1}$ appear at the same time, and the r nodes are connected to node j, so that

$$R_{j_1}(t) = \frac{1}{r-1}[1 + F_{jj_1}(t)] + \frac{1}{r-1}\sum_{k=2}^{r-1}[1 + F_{j_k j_1}(t)],$$

$$F_{j_k j_1}(t) = \frac{1}{r-1} + \frac{1}{r-1}[1 + F_{jj_1}(t)] + \frac{1}{r-1}\sum_{m=2, m \neq k}^{r-1}[1 + F_{j_m j_1}(t)],$$

$k = 2, 3, \cdots, r-1.$

Solving the above equations, one obtains

$$F_{jj_1}(t) = \frac{r-1}{r-2} R_{j_1}(t) - r + 1. \tag{12.38}$$

Inserting Eq. (12.38) into Eq. (12.37) yields

$$T_{j_1}(t) = T_j(t) + \frac{r-1}{r-2} R_{j_1}(t) - r + 1. \tag{12.39}$$

Denote

$$T_{tot}^{(1)} = \sum_{j \in \Delta'_t} T_j(t) = \sum_{j \in \Delta'_t} \left[\frac{r-1}{r-2} R_j(t) - r + 1 \right]. \tag{12.40}$$

When node j in Δ_{t-1} was produced at step t_j, the number of K_r connected to node j at iteration $t-1$ is $L_\diamond(j, t-1) = (m+1)^{t-t_j-1}$, and each K_r leads to $m(r-1)$ new nodes at step t. Using Eqs. (12.33) and (12.40), $T_{tot}^{(1)}$ can be rewritten as

$$\begin{aligned} T_{tot}^{(1)} &= \sum_{j \in \Delta'_{t-1}} m(r-1) L_\diamond(j, t-1) T_j(t) \\ &\quad + (N_t - N_{t-1}) \left[\frac{r(r-1)}{r-2} (mr+1)^t - r + 1 \right] \\ &= \sum_{j \in \Delta'_{t-1}} m(r-1)(m+1)^{t-t_j-1} T_j(t) \\ &\quad + (N_t - N_{t-1}) \left[\frac{r(r-1)}{r-2} (mr+1)^t - r + 1 \right]. \end{aligned} \tag{12.41}$$

where the total number of vertices is $N_t = (r-1) \times (mr+1)^t + 1$.

Let $T_{sum} = \sum_{j \in \Delta'_{t-1}} m(r-1)(m+1)^{t-t_j-1} T_j(t)$. Then

$$T_{sum}(t) = (mr+1)(m+1) \sum_{j \in \Delta'_{t-1}} m(r-1)(m+1)^{t-t_j-2} T_j(t-1) \tag{12.42}$$

$$\begin{aligned} &+ m(r-1)(mr+1) \sum_{j \in \Delta'_{t-1}} T_j(t-1) \\ &= (mr+1)(m+1) T_{sum}(t-1) \\ &\quad + m(r-1)(mr+1) T_{tot}^{(1)}(t-1). \end{aligned} \tag{12.43}$$

Hence, Eq. (12.41) becomes

$$T_{tot}^{(1)}(t) = T_{sum}(t) + (N_t - N_{t-1}) \left[\frac{r(r-1)}{r-2} (mr+1)^t - r + 1 \right]. \tag{12.44}$$

For simplicity, fix the value of $r = 4$ to calculate $T_{sum}(t)$. Given initial conditions $T_{sum}(1) = 72m^2 + 18m$ and $T_{tot}^{(1)}(1) = 360m^2 + 54m$. Substituting Eq. (12.44) into Eq. (12.43) gives

$$T_{sum}(t) = (4m+1)(m+1) T_{sum}(t-1) + 3m(4m+1)$$

$$
\begin{aligned}
&\times\{T_{sum}(t-1)+(N_{t-1}-N_{t-2})[6(4m+1)^{t-1}-3]\}\\
&=(4m+1)^2T_{sum}(t-1)+3m(4m+1)(N_{t-1}-N_{t-2})[6(4m+1)^{t-1}-3]\\
&=(-144m^2-9m)(4m+1)^{2t-2}+216m^2t(4m+1)^{2t-2}+27m(4m+1)^{t-1}.
\end{aligned}
$$

and

$$
\begin{aligned}
T_{tot}^{(1)}(t)&=(-144m^2-9m)(4m+1)^{2t-2}+216m^2t(4m+1)^{2t-2}+27m(4m+1)^{t-1}\\
&\quad+12m(4m+1)^{t-1}[6(4m+1)^4-3]\\
&=(144m^2+216m^2t+63m)(4m+1)^{2t-2}-9m(4m+1)^{t-1}.
\end{aligned}
\tag{12.45}
$$

Inserting Eq. (12.45) into Eq. (12.36) gives

$$
T_{tot}(t)=\left(\frac{54m}{4m+1}t+\frac{9(16m-11)}{4(4m+1)}\right)(4m+1)^{2t}-\left(\frac{9m}{4m+1}t+\frac{64m-11}{4(4m+1)}\right)(4m+1)^t.
\tag{12.46}
$$

The analytical expression of AST is

$$
\begin{aligned}
\langle T\rangle_t=\frac{1}{3(4m+1)^t}&\left[\left(\frac{54m}{4m+1}t+\frac{9(16m-11)}{4(4m+1)}\right)(4m+1)^{2t}\right.\\
&\left.-\left(\frac{9m}{4m+1}t+\frac{64m-11}{4(4m+1)}\right)(4m+1)^t\right].
\end{aligned}
\tag{12.47}
$$

Next, express $\langle T\rangle_t$ as a function of the network order N_t. From the relation $(4m+1)^t=\frac{N_t-1}{3}$, one has $t=\frac{\ln(N_t-1)-\ln 3}{\ln(4m+1)}$. From this, one can rewrite Eq. (12.47) as

$$
\begin{aligned}
\langle T\rangle_t&=\frac{1}{4m+1}\left[\left(18mt+\frac{48m-33}{4}\right)(4m+1)^t-3mt-\frac{64m-11}{12}\right]\\
&=\frac{1}{4m+1}\left[\left(18m\frac{\ln(N_t-1)-\ln 3}{\ln(4m+1)}+\frac{48m-33}{4}\right)\frac{N_t-1}{3}\right.\\
&\quad\left.-3m\frac{\ln(N_t-1)-\ln 3}{\ln(4m+1)}-\frac{64m-11}{12}\right].
\end{aligned}
\tag{12.48}
$$

As $N_t\to\infty$, one has

$$
\langle T\rangle_t\sim\frac{6m(N_t-1)\ln(N_t-1)}{(4m+1)\ln(4m+1)},
\tag{12.49}
$$

which shows that the AST increases with the network order N_t at the rate of $N_t\ln N_t$. This asymptotic dependence of AST on the network order is in contrast with the linear scaling of receiving efficiency on network order [Zhang *et al.* (2009)]. The larger the value of m, the longer the AST. The details are summarized in Fig. 12.9.

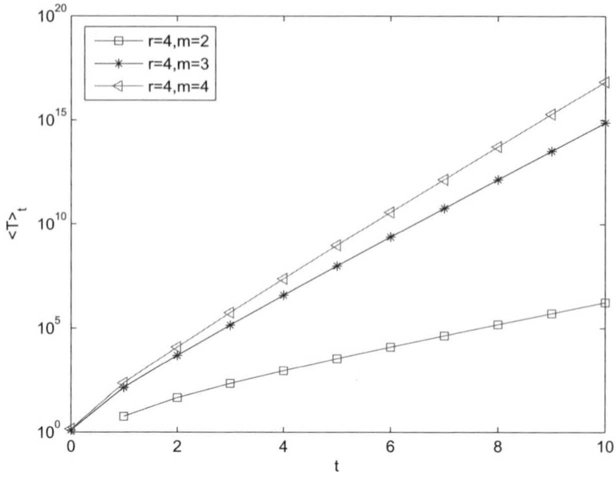

Fig. 12.9 Semilogarithmic graph of the AST with regard to step t, $r = 4$.

12.5 An exact solution for mean first passage time

In this section, we consider a special model of $K_{m,r}(t)$ with $m = 1$ and the initial state being a q-polygon [Zhang and Sun (2010)], denoted as $GK(t)$, see Fig. 12.10. We establish the MFPT to be absorbed by the trap on $GK(t)$, which dominates the evolution scaling of T_i^t at step t, where T_i^t is the first-passage time to be absorbed by the trap for a walker originating from node i on $GK(t)$.

According to the network construction, the total number V_t of nodes in $GK(t)$ is

$$V_t = (q-1)(q+1)^t + 1, \tag{12.50}$$

and the degree $k_i(t)$ of node i at step t, which was produced at step t_0 ($t_0 \geq 0$), is

$$k_i(t) = 2^{t+1-g_0} = 2k_i(t-1).$$

12.5.1 *First passage time at the first step*

It is obvious that $T_0^t = 0$, for all $t \geq 0$. Since $GK(0)$ is a q-polygon, the FPT $T_n^0 = y_n$ of node n at the initial state is

$$y_n = \frac{1}{2}(y_{n+1} + 1) + \frac{1}{2}(y_{n-1} + 1), \tag{12.51}$$

with initial conditions of $y_0 = 0$ and $y_q = 0$ for $1 \leq n \leq q-1$. The solution of Eq. (12.51) is

$$y_n = qn - n^2, \; 0 \leq n \leq q-1. \tag{12.52}$$

When $q = 3$, the values of T_1^0 and T_2^0 are 2, same as those obtained in [Zhang *et al.* (2009)].

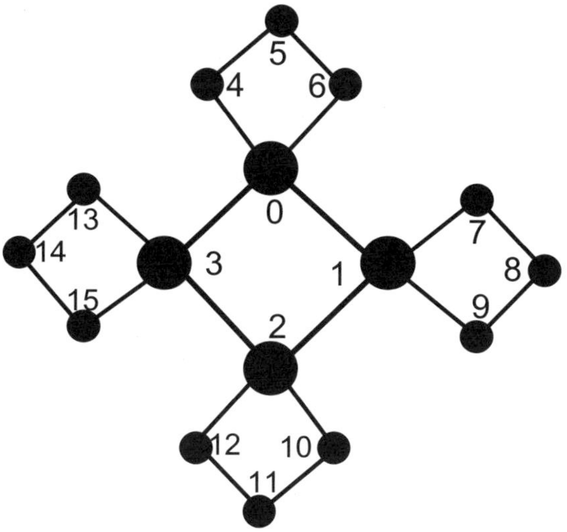

Fig. 12.10 $GK(t)$ at $t = 1$, the trap is located at node 0.

12.5.2 *Evolution scaling for the first passage time*

Consider an arbitrary node i in $GK(t)$. The mean transmit time for a random walker going from node i to any of its k_i old neighbors is denoted as X. Similarly, for going from any of its k_i new neighbors to one of the k_i old ones is Y, and for going from node A_n to one of the k_i old neighbors is a_n for $0 \leq n \leq (q-1)$, where nodes $A_0, A_1, \ldots, A_{q-1}$ form a q-polygon, node A_0 is node i, and nodes A_1 and A_{q-1} are two new neighbors of node i, as shown by Fig. 12.11.

Hence, one can establish the following underlying backward equations:

$$X = \frac{1}{2} + \frac{1}{2}(1+Y), \tag{12.53}$$

$$Y = \frac{1}{2}(1+X) + \frac{1}{2}(1+a_{q-2}), \tag{12.54}$$

$$a_n = \frac{1}{2}(a_{n+1}+1) + \frac{1}{2}(a_{n-1}+1), \tag{12.55}$$

with initial conditions of $a_1 = Y$ and $a_{q-1} = Y$ for $2 \leq n \leq q-2$. From Eq. (12.55), one obtains

$$\begin{aligned} a_{q-2} &= q(q-2) + \frac{Y + (q-1)(q-2-Y)}{2-q} - (q-2)^2 \\ &= Y + q - 3, \end{aligned} \tag{12.56}$$

Inserting Eq. (12.56) into Eq. (12.55) yields $X = q+1$. Therefore, the passage time from any node i to any node j increases by a factor of $q+1$, i.e.,

$$T_i^{t+1} = (q+1)T_i^t, \tag{12.57}$$

which is useful for deriving the formula for the MFPT below.

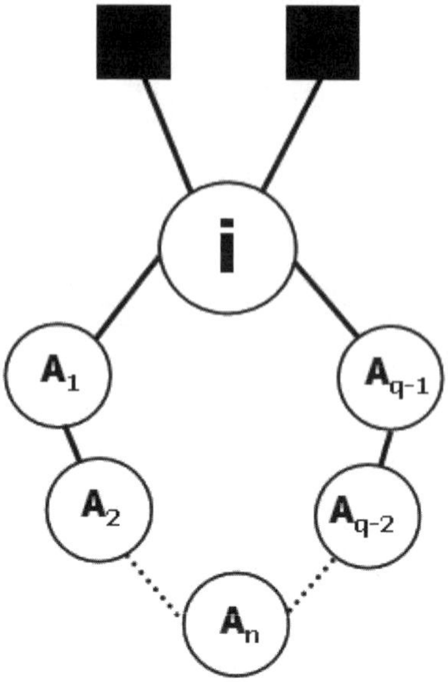

Fig. 12.11 Node $i \in GK(g)$ has k_i neighbor nodes(■) at step g and k_i new neighbor nodes(○) at step $g+1$. The nodes A_0(or node i), $A_1, \ldots, A_{q-1}$ form a q-polygon.

12.5.3 *Analytic formula for mean first passage time*

The set of nodes in $GK(t)$ is denoted as Ω_t, and the set of nodes created at step t as $\overline{\Omega}_t$. Obviously, $\Omega_t = \Omega_{t-1} \cup \overline{\Omega}_t$. Define, for $n \leq t$,

$$T^t_{n,sum} = \sum_{i \in \Omega_n} T^t_i, \tag{12.58}$$

and

$$\overline{T}^t_{n,sum} = \sum_{i \in \overline{\Omega}_n} T^t_i. \tag{12.59}$$

Then,

$$T^t_{t,sum} = T^t_{t-1,sum} + \overline{T}^t_{t,sum}. \tag{12.60}$$

The average of FPT $\langle F \rangle_t$ is

$$\langle F \rangle_t = \frac{1}{V_t - 1} T^t_{t,sum}, \tag{12.61}$$

where $T^t_0 = 0$. To this end, one first calculates $\overline{T}^t_{t,sum}$. At step $t = 1$, $\overline{\Omega}_1 = \{q, q+1, \ldots, 2q-2, 2q-1, \ldots, 3q-3, \ldots, q(q-1)+q-1\}$. Given nodes $0, q, q+1, \ldots, 2q-2$ form a q-polygon, the quantities for $T_0, T_q, T_{q+1}, \ldots, T_{2q-2}$ are exactly the q solutions of the following difference equations:

$$x_n = \frac{1}{2}(x_{n+1} + 1) + \frac{1}{2}(x_{n-1} + 1), \tag{12.62}$$

with initial conditions of $x_0 = T_0^1$ and $x_q = T_0^1$ for $0 \leq n \leq (q-1)$. From Eq. (12.62), one gets $x_n = qn - n^2 + T_0^1$ for $0 \leq n \leq q$. Thus,

$$\begin{aligned}(T_q + T_{q+1} + \ldots + T_{2q-2}) &= \sum_{i=1}^{q-1} x_i = \sum_{i=1}^{q-1}(qi - i^2 + T_0^1) \\ &= \frac{1}{6}q(q^2-1) + (q-1)T_0^1. \end{aligned} \quad (12.63)$$

Proceeding analogously, one obtains

$$\begin{aligned}&T_{i(q-1)+q-1+1} + T_{i(q-1)+q-1+2} + \ldots + T_{(i+1)(q-1)+q-1} \\ &\quad = \frac{1}{6}q(q^2-1) + (q-1)T_i^1, 1 \leq i \leq (q-1). \end{aligned} \quad (12.64)$$

It follows from the construction of $GK(1)$ that

$$\begin{aligned}\overline{T}^1_{1,sum} &= \sum_{i \in \overline{\Omega}_1} T_i^1 = (T_q + T_{q+1} + \ldots + T_{2q-2}) + (T_{2q-1} + \ldots + T_{3q-3}) \\ &\quad + \ldots + (T_{(q-1)(q-1)+q-1+1} + \ldots + T_{q(q-1)+q-1}) \\ &= \frac{1}{6}q^2(q^2-1) + (q-1)\sum_{i=0}^{q-1} T_i^1 \\ &= \frac{1}{6}q^2(q^2-1) + (q-1)\overline{T}^1_{0,sum}. \end{aligned} \quad (12.65)$$

At step $t = 2$, one has $\Omega_2 = \overline{\Omega}_0 \cup \overline{\Omega}_1 \cup \overline{\Omega}_2$. For node $i \in \overline{\Omega}_0$, its degree is $k_i(2) = 8$. There exists $k_i(2)/2 = 4$ q-polygons attached to this node, two of which are new q-polygons (a q-polygon is new if it is created in the present step). From Eqs. (12.63) and (12.64), one finds that each of new q-polygons attaching to node i contributes to the quantities $\frac{1}{6}q(q^2-1)$ and $(q-1)T_i^2$ for $i \in \overline{\Omega}_0$. Then, two new q-polygons attaching to node i contribute to the quantities $2 \times \frac{1}{6}q(q^2-1)$ and $2 \times (q-1)T_i^2$ for $i \in \overline{\Omega}_0$. For $i \in \overline{\Omega}_1$, it is the same case. Thus,

$$\begin{aligned}\overline{T}^2_{2,sum} &= (q-1) \times \sum_{i \in \overline{\Omega}_1} T_i^2 + 2(q-1) \times \sum_{i \in \overline{\Omega}_0} T_i^2 \\ &\quad + [(q+1)^2 - (q+1)] \times \frac{1}{6}(q-1)q(q+1) \\ &= (q-1)\overline{T}^2_{1,sum} + 2(q-1) \times \overline{T}^2_{0,sum} + \frac{1}{6}(q-1)q^2(q+1)^2. \end{aligned} \quad (12.66)$$

Furthermore,

$$\begin{aligned}\overline{T}^t_{t,sum} &= (q-1) \times \overline{T}^t_{t-1,sum} + 2(q-1) \times \overline{T}^t_{t-2,sum} \\ &\quad + 2^2(q-1) \times \overline{T}^g_{t-3,sum} + \ldots + 2^{t-1}(q-1) \times \overline{T}^t_{0,sum} \\ &\quad + \left[(q+1)^t - (q+1)^{t-1}\right] \times \frac{1}{6}q(q^2-1) \end{aligned} \quad (12.67)$$

and

$$\overline{T}^{g+1}_{t+1,sum} = (q-1)\overline{T}^{t+1}_{t,sum} + 2(q-1)\overline{T}^{t+1}_{t-1,sum}$$

$$+2^2(q-1)\times \overline{T}^{t+1}_{t-2,sum}+\ldots+2^t(q-1)\overline{T}^{t+1}_{0,sum} + \left[(q+1)^{t+1}-(q+1)^t\right]\frac{1}{6}q(q^2-1), \tag{12.68}$$

where $(q+1)^t-(q+1)^{t-1}$ and $(q+1)^{t+1}-(q+1)^t$ are the numbers of q-polygons generated at steps t and $t+1$, respectively. Using Eq. (12.68) to subtract Eq. (12.67) and then multiple the result by $2(q+1)$, and making use of Eq. (12.57), one obtains

$$\overline{T}^{t+1}_{t+1,sum}-2(q+1)\overline{T}^{t}_{t,sum} = (q-1)\overline{T}^{t+1}_{t,sum}+\frac{1}{6}q(q^2-1)\left[q(q+1)^t-2(q+1)\times q(q+1)^{t-1}\right], \tag{12.69}$$

which can be rewritten as

$$\overline{T}^{t+1}_{t+1,sum}=(q+1)^2\overline{T}^{t}_{t,sum}-\frac{1}{6}(q-1)q^2(q+1)^{t+1}. \tag{12.70}$$

Using $\overline{T}^{1}_{1,sum}=\frac{1}{6}q(q^2-1)(q^2+q-1)$, the solution of Eq. (12.70) can be written as

$$\overline{T}^{t}_{t,sum}=C(q+1)^{2t}+\frac{q(q-1)}{6}(q+1)^t, \tag{12.71}$$

where $C=\frac{q(q-1)(q^2+q-2)}{6(q+1)}$. Substituting Eq. (12.71) into Eq. (12.60) gives

$$\begin{aligned} T^{t}_{t,sum} &= T^{t}_{t-1,sum}+C(q+1)^{2t}+\frac{1}{6}q(q-1)(q+1)^t \\ &= (q+1)T^{t-1}_{t-1,sum}+C(q+1)^{2t}+\frac{1}{6}q(q-1)(q+1)^t. \end{aligned}$$

Using the initial condition of $T^{0}_{0,sum}=\frac{1}{6}q(q^2-1)$, one obtains

$$T^{t}_{t,sum}=\frac{(q-1)(q+1)^t}{6}[2+(q^2+q-2)(q+1)^t+qt]. \tag{12.72}$$

Plugging Eq. (12.72) into Eq. (12.61), we arrive at the accurate formula

$$\langle F\rangle_t=\frac{1}{N_t-1}T^{t}_{t,sum}=\frac{1}{3}+\frac{1}{6}(q^2+q-2)(q+1)^t+\frac{1}{6}qt, \tag{12.73}$$

for the average of the mean time to absorption by the trap located at node 0 on the generalized Koch networks $GK(t)$. When $m=3$, $\langle F\rangle_t=\frac{1}{3}+\frac{t}{2}+\frac{5}{3}\times 4^t$, which is the same as that shown in [Zhang *et al.* (2009)].

Next, we express the MPFT as a function of the network order, so as to obtain the scaling between these two quantities. From Eq. (12.50), one has $g=\log_{q+1}(\frac{V_t-1}{q-1})$. Hence, Eq. (12.73) is written as

$$\langle F\rangle_t=\frac{1}{3}+\frac{1}{6}(q+2)(V_t-1)+\frac{1}{6}q\log_{q+1}\left(\frac{V_t-1}{q-1}\right). \tag{12.74}$$

As $V_t\to\infty$, the mean trapping time $\langle F\rangle_t$ becomes

$$\langle F\rangle_t\approx\frac{1}{6}(q+2)V_t. \tag{12.75}$$

Thus, the mean trapping time grows linearly with the increasing order of the network and the parameter q. Numerical results are shown in Fig. 12.12.

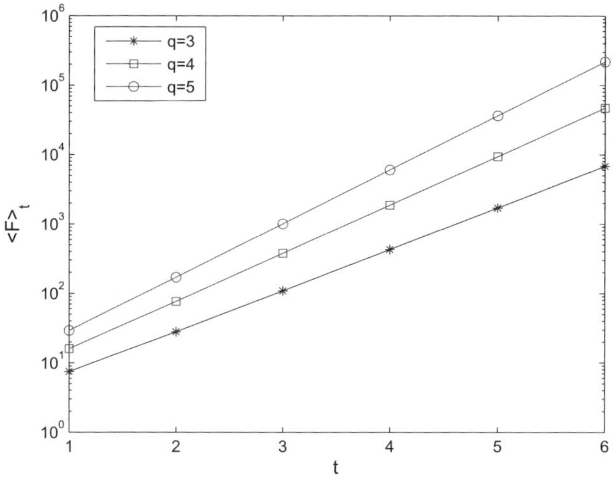

Fig. 12.12 Mean trapping time $\langle F \rangle_t$ versus step t on a semilogarithmic scale.

12.6 Conclusions

In this chapter, we have introduced a novel notion of evolving (m, r)-Koch networks based on the Koch network. The proposed network has an initial state that is a complete graph with r number of nodes. In each step of the network construction, every existing node produces m complete graphs. Explicit analytical solutions for the degree distribution, clustering coefficient, average path length, and degree correlations are obtained. The rigorous solutions show that the resultant networks follow a power-law distribution with exponent $\gamma \in [2, 3]$ under the condition of $m + 2 \leq r \leq m^2 + 3m + 3$. The clustering coefficient increases with m and r, whereas the average path length decreases. With regard to the properties of degree correlations, the resultant networks are either assortative, disassortative, or uncorrelated, depending on certain parameter values.

In addition, we have studied the random walks on the (m, r)-Koch network. We derived the scaling efficiency of the first passage time (FPT), concentrated on the average sending time (AST) for random walks from a hub node to all other nodes, and obtained an explicit formula for AST, which grows with network order, larger than the average return time (ART) from all other nodes to the hub. The rigorous solutions show that the mean return time (MRT) of the emerging nodes is identical. In addition, AST and ART both increase with parameters m and r.

Finally, we have investigated the random walk process on the generalized Koch network with a trap located at a hub node, mainly focusing on deriving scaling of the mean first passage time (MFPT). We have obtained the exact scaling efficiency exponents for the MFPT and shown that the MFPT behaves linearly with the number of network nodes and increases with the parameter m. Our findings are helpful to study random walks on various deterministically growing networks.

Bibliography

Albert, R. and Barabási, A. L. (2002). Statistical Mechanics of Complex Networks, *Rev. Mod. Phys.* **74**, 1, pp. 47–97.

Arenas, A., Díaz-Guilera, A., Kurths, J., Moreno, Y. and Zhou, C. S. (2008). Synchronization in complex networks, *Phys. Rep.* **469**, 3, pp. 93–154.

Barabási, A. L. and Albert, R. (1999). Emergence of scaling in random networks, *Science* **286**, 5439, pp. 509–512.

Barabási, A. L., Ravasz, E. and Vicsek, T. (2001). Deterministic scale-free networks, *Phys. A* **299**, 3-4, pp. 559–564.

Boccaletti, S., Latora, V., Moreno, Y., Chavez, M. and Hwang, D. U. (2006). Complex networks: structure and dynamics, *Phys. Rep.* **424**, 4-5, pp. 175–308.

Bullmore, E. and Sporns, O. (2009). Complex brain networks: graph theoretical analysis of structural and functional systems, *Nature Rev. Neurosci.* **10**, pp. 186–198.

Newman, M. E. J. (2003). The structure and function of complex networks, *SIAM Rev.* **45**, 2, pp. 167–256.

Comellas, F., Ozón, J. and Peters, J. G. (2000). Deterministic small-world communication networks, *Inf. Process. Lett.* **76**, 1-2, 83–90.

Condamin, S., Bénichou, O., Tejedor, V., Voituriez, R. and Klafter, J. (2007). First-passage times in complex scale-invariant media, *Nature* **450**, pp. 77–80.

Costa, L. da. F., Rodrigues, F. A., Travieso, G. and Boas, P. R. V. (2007). Characterization of complex networks: A survey of measurements, *Adv. Phys.* **56**, 1, pp. 167–242.

Dorogovtsev, S. N. and Mendes, J. F. F. (2002). Evolution of Networks, *Adv. Phys.* **51**, 4, pp. 1079–1187.

Eguíluz, V. M., Chialvo, D. R., Cecchi, G. A., Baliki, M. and Apkarian, A. V. (2005). Scale-free brain functional networks, *Phys. Rev. Lett.* **94**, 1, pp. 018102.

Li, C. P., Sun, W. G. and Kurths, J. (2007). Synchronization between two coupled complex networks, *Phys. Rev. E* **76**, 4, pp. 046204.

Sporns, O. and Zwi, J. D. (2004). The small world of the cerebral cortex, *Neuroinformatics* **2**, 2, pp. 145–162.

Sun, W. G., Cao, J. T. and Wang, R. B. (2011). Approach of complex networks for the determination of brain death, *Chin. Phys. Lett.* **28**, 6, pp. 068701.

Sun, W. G., Zhang, J. Y. and Wu, Y. Q. (2011). Novel evolving small-world scale-free Koch networks, *J. Stat. Mech.* **2011**, P03021.

Vázquez, A., Pastor-Satorras, R. and Vespignani, A. (2002). Large-scale topological and dynamical properties of the Internet, *Phys. Rev. E* **65**, 6, pp. 066130.

Wang, X. F. (2002). Complex networks: topology, dynamics and synchronization, *Int. J. Bifurcation Chaos* **12**, 5, pp. 885–916.

Wang, Q. Y., Duan, Z. S., Chen, G. R. and Feng, Z. S. (2008). Synchronization in a class of weighted complex networks with coupling delays, *Phys. A* **387**, 22, pp. 5616–5622.

Watts, D. J. and Strogatz, S. H. (1998). Collective dynamics of small-world networks, *Nature* **393**, 6684, pp. 440–442.

Zhang, Z. Z, Chen, L. C, Fang, L. J., Zhou, S. G., Zhang, Y. C. and Guan, J. H. (2009). The rigorous solution for the average distance of a Sierpinski network, *J. Stat. Mech.* **2**, 2, pp. 02034.

Zhang, Z. Z., Gao, S. Y., Chen, L. C., Zhou, S. G., Zhang, H. J. and Guan, J. H. (2010). Mapping Koch curves into scale-free small-world networks, *J. Phys. A: Math. Theor.* **43**, 39, pp. 395101.

Zhang, J. Y. and Sun, W. G. (2010). The structural properties of the generalized Koch network, *J. Stat. Mech.* **7**, 7, pp. 07011.

Zhang, Z. Z, Wu, B., Zhang, H. J, Zhou, S. G, Guan, J. H. and Wang, Z. G. (2010). Determining global mean-first-passage time of random walks on Vicsek fractals using eigenvalues of Laplacian matrices, *Phys. Rev. E* **81**, 3, pp. 031118.

Zhang, Z. Z., Zhou, S. G., Xie, W. L., Chen, L. C., Lin, Y. and Guan, J. H. (2009). Standard random walks and trapping on the Koch network with scale-free behavior and small-world effect, *Phys. Rev. E* **79**, 6, pp. 061113.

Zhou, C. S., Motter, A. E. and Kurths, J. (2006). Universality in the synchronization of weighted random networks, *Phys. Rev. Lett.* **96**, 3, pp. 034101.

Chapter 13

On different approaches to synchronization of spatiotemporal chaos in complex networks: a case study[1]

Yuan Chai and Li-Qun Chen*

Shanghai Institute of Applied Mathematics and Mechanics, Shanghai University, Shanghai 200072, PR China

**lqchen@straff.shu.edu.cn*

Three control schemes are presented in synchronization of spatiotemporal chaos in complex networks. The time-delay, the projection and the disturbances are accounted in the synchronization. The control laws are designed via the active sliding mode control, the backstepping control and the master stability function. The merits of each scheme are numerically demonstrated in synchronization of spatiotemporal Gray-Scott systems.

Keywords: Synchronization approaches, spatiotemporal chaos, complex networks

13.1 Introduction

During the past years, complex networks have attracted more and more attention in various fields such as physics, biology, chemistry and computer science [Strogatz (2001); Dorogovtesev and Mendes (2002); Boccaletti *et al.* (2006)]. Such chaotic systems in nature usually consist of a large number of interconnected dynamical units. Coupled biological and chemical systems, neural networks and the Internet are only a few of such examples. With the development of scientific research, the small-world was proposed by Watts and Strogatz [Watts and Strogatz (1998)] in 1998. Then, Newman and Watts (1999) modified it to generate another variant of the small-world model [Newman (1999)]. The scale-free network was introduced by Barabsi and Albert [Barabási and Albert (1999)] in 1999. Recently, the synchronization of complex network has attracted very wide attention among the scientists

[1]This work was supported by the National Outstanding Young Scientists Foundation of China (grant no. 10725209), the National Natural Science Foundation of China (grant nos. 90816001 and 10902064), Shanghai Subject Chief Scientist Project (grant no. 09XD1401700), Shanghai Leading Talent Program, Shanghai Leading Academic Discipline Project (grant no. S30106), the program for Changjiang Scholars and Innovative Research Team in University (grant no. IRT0844).

[Hwang *et al.* (2005); Sorrentino and Ott (2008); Silva *et al.* (2010); Ahn (2010); Gorochowski *et al.* (2010); Flunkert *et al.* (2010)]. Various types of approaches have been revealed to investigate chaos synchronization, such as adaptive control [Hassan and Aria (2009)], active control [Bhalekara and Daftardar-Gejji (2010)], impulsive control [Chen *et al.* (2011)], open-plus-closed-loop control [Chen (2001)], backstepping control, sliding mode control, master stability function control, etc. Especially, the last three approaches have become the most popular methods in chaos control [Njah *et al.* (2010); Bowong *et al.* (2004); Polyakov and Poznyak (2011); Roopaei *et al.* (2010); Fink *et al.* (2000)].

The backstepping control is mostly applied to the synchronization of two systems [Mascolo and Grassi (1997); Laoye *et al.* (2009); Sharma *et al.* (2009); Pishkenari *et al.* (2010); Yu and Li (2011)]. In aspect of backstepping control of complex networks, Chen et al. [Chen *et al.* (2010)] discussed the backstepping control for periodically time-varying systems using high-order neural network and Fourier series expansion. Lin et al. [Lin *et al.* (2010)] introduced the dynamic fuzzy neural networks modeling and adaptive backstepping tracking control of uncertain chaotic systems. Many colossal complex networks are composed of a chain structure, such as the railway networks, World Wide Web (WWW), food web, communication networks, biological neural networks, etc. Therefore, to research the synchronization of a chain network has values of definite magnitude. However, it is impossible to add controllers to a lot of nodes in the network to achieve synchronization. To reduce the number of controllers becomes a key point, very little results exist on this crucial aspect of network synchronization [Wang *et al.* (2002); Chen *et al.* (2007)].

Most of the above research on backstepping control is realized without considering time delays. However, the characteristic of time delays is very common in complex networks. Some of time delays can be ignorant, while some others can not be ignored. It is well known that time delay is ubiquitous when signals are communicated among neurons or secure communications [Li *et al.* (2004); Nana *et al.* (2009)], it is reasonable to require one system to synchronize the other system at a constant time delay [Huang *et al.* (2009)]. In the research area of lag synchronization, several results have been appeared in the literature [Du *et al.* (2010); Shahverdiev *et al.* (2002)].

Recently based on the advantages of the active control and the sliding mode control method, a new method called "active sliding mode control technique" is proposed to realize chaos synchronization, Haeri et al. [Haeri and Emadzadeh (2007)] investigated the synchronization of different chaotic systems using active sliding mode control. Zhang et al. [Zhang *et al.* (2004)] reported on the synchronization of chaotic systems with parametric uncertainty using active sliding mode control. Tavazoei et al. [Tavazoei and Haeri (2007)] discussed the determination of active sliding mode controller parameters in synchronizing different chaotic systems.

An interesting synchronization phenomenon has been created, called the modified projective synchronization [Li (2007)], in which the states of the multi-nodes synchronize up to a constant scaling matrix. The modified projective synchronization includes the complete synchronization, anti-synchronization and projective synchronization [Tang and Fang (2008)]. Therefore, modified projective synchronization is worth research because of its diversity. And some research results for the modified projective synchronization have been obtained in recent years [Park (2008); Wen (2011)]. Therefore, we propose modified projective lag synchronization. The complete synchronization, anti-synchronization and projective synchronization belong to the modified projective lag synchronization, which may be an appropriate technique to clearly indicate the synchronization of nature and artificial networks.

Most of the above studies are realized without any external perturbations. However, the noise disturbance is inevitable from a practical point of view. The chaotic systems are always in a changing environment and disturbed by some unknown factors from environment [Behzad *et al.* (2008)]. A small perturbation to spatiotemporal chaos will result in a drastic change in the chaotic behavior of the systems. Synchronization of chaos is unavoidably subject to external random disturbances. Therefore, investigation of synchronization for the chaotic systems by the impact of artificially adding random disturbances has become an important research topic [Li and Chang (2009)]. Cai et al. [Cai *et al.* (2010)] considered the modified projective synchronization of chaotic systems with disturbances via active sliding mode control. Synchronization of coupled chaotic systems is a widespread phenomenon occurring in physical and biological sciences, which have stimulated a great deal of theoretical work on synchronization in coupled networks. Increasingly, master stability functions (MSF) [Heagy *et al.* (1994); Pecora (1998); Barahona and Pecora (2002); Dhamala *et al.* (2004)] are fundamental to the study of synchronization in complex networks. Pecora and Carroll [Pecora and Carroll (1998)] synchronized coupled systems by the master stability functions. Huang et al. [Huang *et al.* (2009)] introduced the Generic behavior of master-stability functions in coupled nonlinear dynamical Systems.

For a network coupled by different chaotic systems, a necessary condition for synchronization to occur is that the maximum Lyapunov exponent at the corresponding normalized coupling parameters is negative. Sometimes a network may not be synchronized when a controller can not be added into the infrastructure of individual node. Thus several approaches have been proposed for the synchronization of complex networks without control [Chen and Duan (2008); Li and Lee (2007)], which are rather significant topics in both theoretical research and practical applications.

Most of the existing literature on synchronization usually assumes that the topological structure of all the nodes in a network is exactly identical. However, the assumption cannot be satisfied in real world. Furthermore, it has been found that the synchronization of non-identical chaotic systems, which is more universal than topological structure, plays an important role in realistic networks. Therefore, it is of great interest to investigate synchronization without controllers in networks consisting of nodes with non identical chaotic systems.

Temporal chaotic systems are selected for demonstration in most of papers, though a large number of natural systems are of spatiotemporal chaos [Lü and Meng (2011)]. Spatiotemporal chaos synchronization has greater potential application value to many fields [Chacón and Marcheggiani (2010)], such as secure communication, physics, auto-control, fluid, chemical, and biological systems. Chian et al. [Chian *et al.* (2010)] discussed the amplitude-phase synchronization at the onset of permanent spatiotemporal chaos. Therefore, the synchronization of spatiotemporal chaos has values of definite magnitude. However, synchronization of spatiotemporal chaos in chain networks via backstepping lag design, the generalized projective lag synchronization of spatiotemporal chaos with disturbances in star networks and the synchronization of spatiotemporal chaos with non-topological structure by MSF have been still less reported, the research in spatiotemporal chaos synchronization via above three methods have a vast unknown space to explore.

Yielding fruitful results puts an important step forward from the investigation of theory. Motivated by the above analysis, in this chapter, a backstepping approach is firstly proposed for the lag synchronization of chain networks between multi-spatiotemporal chaotic systems with topologically equivalent structures. The synchronization of multi-spatiotemporal chaotic systems is implemented by adding the control only at a terminal node, and controllers are designed via a corresponding update law. The control law is applied to spatiotemporal Gray-Scott systems. We secondly consider modified projective lag synchronization of identical spatiotemporal chaotic systems with disturbances in star networks. Based on the active sliding mode control technique, the sufficient conditions are given to assure the valid modified projective lag synchronization occurs. Finally, we formulate master stability functions for an arbitrary number of spatiotemporal nodes in a network. The master stability equation allows us to calculate the maximum Lyapunov exponent and determine the stability of the networks for any linear coupling scheme. Numerical results demonstrate the feasibility and the effectiveness of the proposed three approaches.

13.2 Design of the synchronization controller

The backstepping approach of complex networks is a multi-systematic control technique, which is characterized by a step-by-step procedure interlacing. At each step, a coordinate transformation and the design of a virtual control based on Lyapunov technique are required. Finally, the true terminal controller is obtained under the global stability. Here, our attention is focused on the spatiotemporal chaos systems. In this section, we give a terminal lag synchronization controller via backstepping design.

Suppose that a complex chain network consists of N nodes, and the state equations of the nodes are spatiotemporal chaos systems, in which the n-dimensional variable of node i at time t is $X_i(r,t)$. When the coupling action is not considered, each node can be described by the following equation:

$$\frac{\partial X_i(r,t)}{\partial t} = F(X_i(r,t)) \quad (i = 1, 2, ..., N), \tag{13.1}$$

where $X_i(r,t) = [X_{i1}(r,t), X_{i2}(r,t), \ldots, X_{in}(r,t)]^{\mathrm{T}} \in R^n$ are the state variables of node i, $F : R^n \to R^n$ is continuously differentiable, r and t are spatial and temporal variables.

Coupling on other nodes is added by using single state variables, and the state equation (13.1) of the i-th node is rewritten as

$$\frac{\partial X_i(r,t)}{\partial t} = F(X_i(r,t)) + \alpha \sum_{j=1}^{N} P_{ij} X_j(r, t - \tau_j) + \beta_i U(r,t). \tag{13.2}$$

α is the coupling strength between nodes of the network, and selected to be α =1 in the work, and P_{ij} being the matrix element of the inner coupling matrix, whose specific forms depend on the connection type of the network. Unidirectional chain-connection is discussed in the following parts. β_i is a terminal adding design, if there is a terminal adding, then β_i =1, otherwise β_i =0, $U(r,t)$ is a terminal synchronization controller.

The first error state variable of the chain network between the first node and the second node is defined as

$$e_1(r,t) = X_2(r, t - \tau_2) - X_1(r, t - \tau_1). \tag{13.3}$$

The first partial Lyapunov function is constructed as

$$V_1(r,t) = \frac{1}{2} e_1(r,t)^2. \tag{13.4}$$

The derivative of V_1 is

$$\frac{\partial V_1}{\partial t} = e_1(r,t) \frac{\partial e_1(r,t)}{\partial t} = -e_1(r,t)^2 + e_1(r,t)(e_1(r,t) + \frac{\partial e_1(r,t)}{\partial t}). \tag{13.5}$$

Letting

$$\begin{aligned} e_2(r,t) =& e_1(r,t) + \frac{\partial e_1(r,t)}{\partial t} \\ =& F(X_2(r, t - \tau_2)) - F(X_1(r, t - \tau_1)) \\ &+ (X_3(r, t - \tau_3) - X_2(r, t - \tau_2)). \end{aligned} \tag{13.6}$$

And introducing function W_2 is

$$W_2(r,t) = F(X_2(r, t - \tau_2)) - F(X_1(r, t - \tau_1)) + (X_3(r, t - \tau_3). \tag{13.7}$$

We have

$$e_2(r,t) = W_2(r,t) - X_2(r, t - \tau_2). \tag{13.8}$$

The second partial Lyapunov function is constructed as

$$V_2 = V_1 + \frac{1}{2} e_2(r,t)^2. \tag{13.9}$$

The derivative of V_2 is

$$\begin{aligned}\frac{\partial V_2}{\partial t} &= \frac{\partial V_1}{\partial t} + e_2(r,t)\frac{\partial e_2(r,t)}{\partial t} \\ &= -e_1(r,t)^2 - e_2(r,t)^2 + e_2(r,t)(e_1(r,t) + e_2(r,t) + \frac{\partial e_2(r,t)}{\partial t}).\end{aligned} \tag{13.10}$$

Letting

$$e_3(r,t) = e_1(r,t) + e_2(r,t) + \frac{\partial e_2(r,t)}{\partial t}. \tag{13.11}$$

And introducing function W_3 is

$$\begin{aligned}W_3(r,t) = e_1(r,t) + e_2(r,t) + \frac{\partial F(X_2(r,t-\tau_2))}{\partial t} - \frac{\partial F(X_1(r,t-\tau_1))}{\partial t} \\ + F(X_3(r,t-\tau_3)) + (X_4(r,t-\tau_4) - \frac{\partial X_2(r,t-\tau_2)}{\partial t}.\end{aligned} \tag{13.12}$$

We have

$$e_3(r,t) = W_3(r,t) - X_3(r,t-\tau_3). \tag{13.13}$$

$$\vdots$$

The $(i-1)$ partial Lyapunov function is constructed as

$$V_{i-1} = V_{i-2} + \frac{1}{2}e_{i-1}(r,t)^2. \tag{13.14}$$

The derivative of V_{i-1} is

$$\frac{\partial V_{i-1}}{\partial t} = \sum_{q=1}^{i-1} e_q(r,t)^2 + e_{i-1}(r,t)(e_{i-2}(r,t) + e_{i-1}(r,t) + \frac{\partial e_{i-1}(r,t)}{\partial t}). \tag{13.15}$$

Letting

$$e_i(r,t) = e_{i-2}(r,t) + e_{i-1}(r,t) + \frac{\partial e_{i-1}(r,t)}{\partial t}. \tag{13.16}$$

And introducing function W_i is

$$\begin{aligned}W_i(r,t) =& e_{i-2}(r,t) + e_{i-1}(r,t) + \sum_{l_1=1}^{i-3} \frac{\partial^{i-2-l_1} e_{l_1}(r,t)}{\partial t^{i-2-l_1}} + \sum_{l_2=2}^{i-2} \frac{\partial^{i-2-l_2} e_{l_2}(r,t)}{\partial t^{i-2-l_2}} \\ & - \frac{\partial^{i-2} F(X_1(r,t-\tau_1))}{\partial t^{i-2}} + \sum_{j_1=2}^{i} \frac{\partial^{i-j_1} F(X_{j_1}(r,t-\tau_{j_1}))}{\partial t^{i-j_1}} \\ & + X_{i+1}(r,t-\tau_{i+1}) \sum_{j_2=3}^{i-1} \frac{\partial^{i-j_2} X_{j_2}(r,t-\tau_{j_2})}{\partial t^{i-j_2}} - \frac{\partial^{i-2} X_2(r,t-\tau_2)}{\partial t^{i-2}}.\end{aligned} \tag{13.17}$$

We have

$$e_i(r,t) = W_i(r,t) - X_i(r,t-\tau_i). \tag{13.18}$$

$$\vdots$$

The $(N-2)$ partial Lyapunov function is constructed

$$V_{N-2} = V_{N-3} + \frac{1}{2}e_{N-2}(r,t)^2. \tag{13.19}$$

The derivative of V_{N-2} is

$$\frac{\partial V_{N-2}}{\partial t} = -\sum_{i=1}^{N-2} e_i(r,t)^2 + e_{N-2}(r,t)(e_{N-3}(r,t) + e_{N-2}(r,t) + \frac{\partial e_{N-2}(r,t)}{\partial t}). \tag{13.20}$$

Letting

$$e_{N-1}(r,t) = e_{N-3}(r,t) + e_{N-2}(r,t) + \frac{\partial e_{N-2}(r,t)}{\partial t}. \tag{13.21}$$

And introducing function W_{N-1} is

$$\begin{aligned} W_{N-1}(r,t) =& e_{N-3}(r,t) + e_{N-2}(r,t) + \sum_{l_1=1}^{N-4} \frac{\partial^{N-3-l_1} e_{l_1}(r,t)}{\partial t^{N-3-l_1}} \\ &+ \sum_{l_2=2}^{N-3} \frac{\partial^{N-2-l_2} e_{l_2}(r,t)}{\partial t^{N-2-l_2}} - \frac{\partial^{N-3} F(X_1(r,t-\tau_1))}{\partial t^{N-3}} \\ &+ \sum_{j_1=2}^{N-1} \frac{\partial^{N-1-j_1} F(X_{j_1}(r,t-\tau_{j_1}))}{\partial t^{N-1-j_1}} + X_N(r,t-\tau_N) \\ &- \sum_{j_2=3}^{N-2} \frac{\partial^{N-1-j_2} X_{j_2}(r,t-\tau_{j_2})}{\partial t^{N-1-j_2}} - \frac{\partial^{N-3} X_2(r,t-\tau_2)}{\partial t^{N-3}}. \end{aligned} \tag{13.22}$$

We have

$$e_{N-1}(r,t) = W_{N-1}(r,t) - X_{N-1}(r,t-\tau_{N-1}). \tag{13.23}$$

The $(N-1)$ partial Lyapunov function is constructed

$$V_{N-1} = V_{N-2} + \frac{1}{2}e_{N-1}(r,t)^2. \tag{13.24}$$

The derivative of V_{N-1} is

$$\frac{\partial V_{N-1}}{\partial t} = \sum_{i=1}^{N-1} e_i(r,t)^2 + e_{N-1}(r,t)(e_{N-2}(r,t) + e_{N-1}(r,t) + \frac{\partial e_{N-1}(r,t)}{\partial t}). \tag{13.25}$$

Letting

$$e_N(r,t) = e_{N-2}(r,t) + e_{N-1}(r,t) + \frac{\partial e_{N-1}(r,t)}{\partial t}. \tag{13.26}$$

Then

$$\frac{\partial V_{N-1}}{\partial t} = -\sum_{i=1}^{N-1} e_i(r,t)^2 + e_{N-1}(r,t)e_N(r,t). \tag{13.27}$$

A terminal node of controller $U(r,t)$ is designed in the chain network

$$\begin{aligned}
e_N(r,t) =& e_{N-2}(r,t) + e_{N-1}(r,t) + \frac{\partial e_{N-1}(r,t)}{\partial t} \\
=& e_{N-2}(r,t) + e_{N-1}(r,t) + \sum_{l_1=1}^{N-3} \frac{\partial^{N-2-l_1} e_{l_1}(r,t)}{\partial t^{N-2-l_1}} + \sum_{l_2=2}^{N-2} \frac{\partial^{N-1-l_2} e_{l_2}(r,t)}{\partial t^{N-1-l_2}} \\
& - \frac{\partial^{N-2} F(X_1(r,t-\tau_1))}{\partial t^{N-2}} + \sum_{j_1=2}^{N} \frac{\partial^{N-j_1} F(X_{j_1}(r,t-\tau_{j_1}))}{\partial t^{N-j_1}} + U(r,t) \\
& - \sum_{j_2=3}^{N-1} \frac{\partial^{N-j_2} X_{j_2}(r,t-\tau_{j_2})}{\partial t^{N-j_2}} - \frac{\partial^{N-2} X_2(r,t-\tau_2)}{\partial t^{N-2}}.
\end{aligned} \tag{13.28}$$

The structure of $U(r,t)$ is:

$$\begin{aligned}
U(r,t) = -\Bigg\{& e_{N-2}(r,t) + e_{N-1}(r,t) + \sum_{l_1=1}^{N-3} \frac{\partial^{N-2-l_1} e_{l_1}(r,t)}{\partial t^{N-2-l_1}} + \sum_{l_2=2}^{N-2} \frac{\partial^{N-1-l_2} e_{l_2}(r,t)}{\partial t^{N-1-l_2}} \\
& - \frac{\partial^{N-2} F(X_1(r,t-\tau_1))}{\partial t^{N-2}} + \sum_{j_1=2}^{N} \frac{\partial^{N-j_1} F(X_{j_1}(r,t-\tau_{j_1}))}{\partial t^{N-j_1}} \\
& - \sum_{j_2=3}^{N-1} \frac{\partial^{N-j_2} X_{j_2}(r,t-\tau_{j_2})}{\partial t^{N-j_2}} - \frac{\partial^{N-2} X_2(r,t-\tau_2)}{\partial t^{N-2}} \Bigg\}.
\end{aligned} \tag{13.29}$$

When the structure of synchronization controller $U(r,t)$ is taken as (13.29), we obtain:

$$\frac{\partial V_{N-1}}{\partial t} = -\sum_{i=1}^{N-1} e_i(r,t)^2. \tag{13.30}$$

According to Lyapunov stability theory, the synchronization of the chain network is then achieved when Eq. (13.30) is fulfilled.

13.3 Numerical results

In this section, we present numerical results to verify the analytical predictions obtained in the previous section, we use a terminal node controller to control the spatiotemporal chaos of the complex chain network. Numerical simulation results constitute excellent validations of our theoretical analysis.

The Gray-Scott system as a producer of spatiotemporal chaos is described as follows [Pearson (1993)]:

$$\begin{cases}
\dfrac{\partial X_1(r,t)}{\partial t} = -X_1(r,t)X_2(r,t)^2 + a(1 - X_1(r,t)) + d_1\nabla^2 X_1(r,t), \\
\dfrac{\partial X_2(r,t)}{\partial t} = X_1(r,t)X_2(r,t)^2 - (a+b)X_2(r,t) + d_2\nabla^2 X_2(r,t).
\end{cases} \tag{13.31}$$

where a and b are the system parameters. $X_1(r,t)$ and $X_2(r,t)$ are the state variables of the system, $X_1(r,t) \in [0,1]$ and $X_2(r,t) \in [0,1]$, d_1 and d_2 are diffusion coefficients. System size L is 2.5×2.5, and $r \in [0, L]$. The initial conditions are given as $X_1(r,0) = 1$, $X_2(r,0) = 1$, and time step $\triangle t = 1$, space step $\triangle l = 0.01$, diffusion coefficients $d_1 = 2 \times 10^{-5}$ and $d_2 = 2 \times 10^{-5}$, periodic boundary conditions are $X_1(0,t) = X_1(0,t) = 1$ and $X_2(0,t) = X_2(0,t) = 0$, and the spatiotemporal evolutions of state variables $X_1(r,t)$ and $X_2(r,t)$ are shown in Figs. 13.1–13.2, respectively.

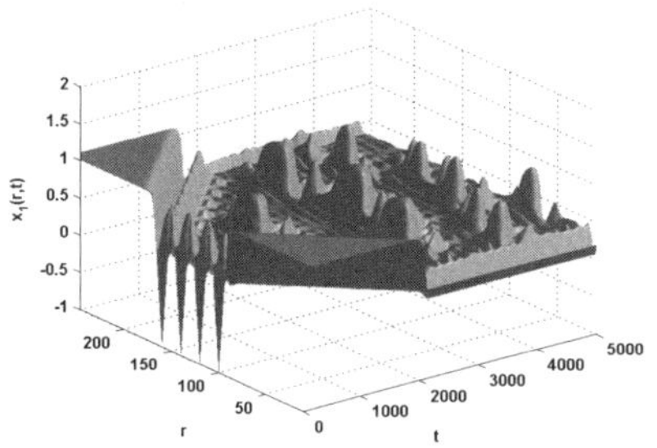

Fig. 13.1 The spatiotemporal evolution of $X_1(r,t)$.

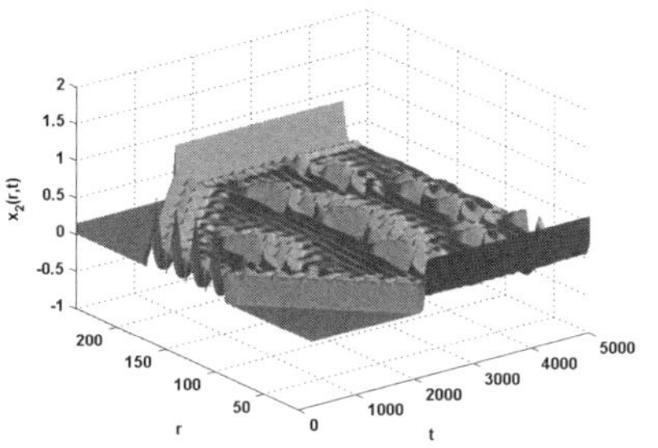

Fig. 13.2 The spatiotemporal evolution of $X_2(r,t)$.

The coupling matrix P is set as

$$P = \begin{bmatrix} -1 & 1 & 0 & \dots & 0 \\ 0 & -1 & 1 & \dots & 0 \\ \dots & \dots & \dots & \dots & \dots \\ 0 & 0 & \dots & -1 & 1 \\ 0 & 0 & \dots & 0 & 0 \end{bmatrix}. \tag{13.32}$$

In order to demonstrate briefly the effectiveness of the lag synchronization approach of a chain network with N nodes. Therefore, we first consider the chain network with three nodes. The state equation for each node is of Gray-Scott model described by Eq. (13.31), and state variables at three nodes are $(X_{11}(r,t), X_{12}(r,t))$, $(X_{21}(r,t), X_{22}(r,t))$ and $(X_{31}(r,t), X_{32}(r,t))$, respectively. Coupling is made by using the single state variable $X_i(r,t)$ in the Gray-Scott model according to Eq. (13.2)

$$\begin{cases} \dfrac{\partial X_{11}}{\partial t} = -X_{11}X_{12}^2 + a(1 - X_{11}) + d_1\nabla^2 X_{11} + (X_{21} - X_{11}), \\ \dfrac{\partial X_{12}}{\partial t} = X_{11}X_{12}^2 - (a+b)X_{12} + d_2\nabla^2 X_{12} + (X_{22} - X_{12}). \end{cases} \tag{13.33}$$

$$\begin{cases} \dfrac{\partial X_{21}}{\partial t} = -X_{21}X_{22}^2 + a(1 - X_{21}) + d_1\nabla^2 X_{21} + (X_{31} - X_{21}), \\ \dfrac{\partial X_{22}}{\partial t} = X_{21}X_{22}^2 - (a+b)X_{22} + d_2\nabla^2 X_{22} + (X_{32} - X_{22}). \end{cases} \tag{13.34}$$

$$\begin{cases} \dfrac{\partial X_{31}}{\partial t} = -X_{31}X_{32}^2 + a(1 - X_{31}) + d_1\nabla^2 X_{31} + U_1, \\ \dfrac{\partial X_{32}}{\partial t} = X_{31}X_{32}^2 - (a + b)X_{32} + d_2\nabla^2 X_{32} + U_2. \end{cases} \tag{13.35}$$

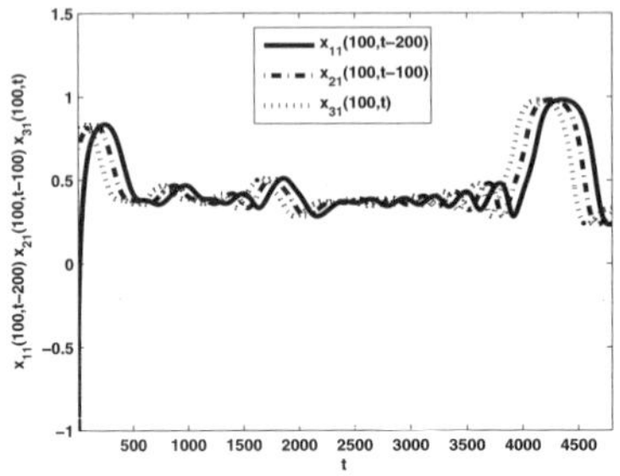

Fig. 13.3 The lag state trajectories of the X_{i1}.

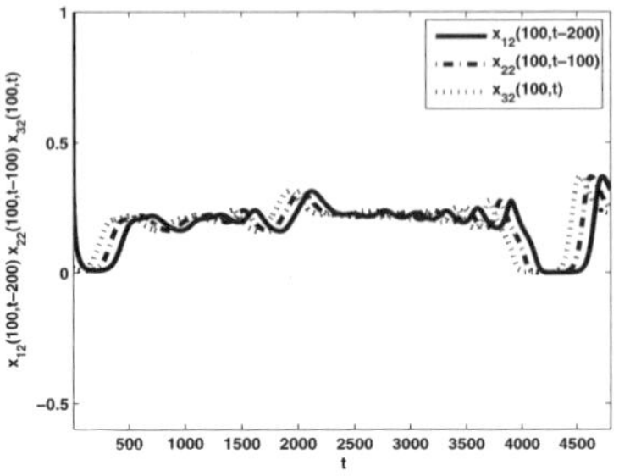

Fig. 13.4 The lag state trajectories of the X_{i2}.

Figs. 13.3–13.4 depicts the state trajectories of three chaotic systems evolve in the same directions with time delay $\tau_1 = 200$, $\tau_2 = 100$ and $\tau_3 = 0$ at step $r = 100$.

Iteration of the simulation is initiated at step $t = 2800$, and simulation results are shown in Figs. 13.5–13.8

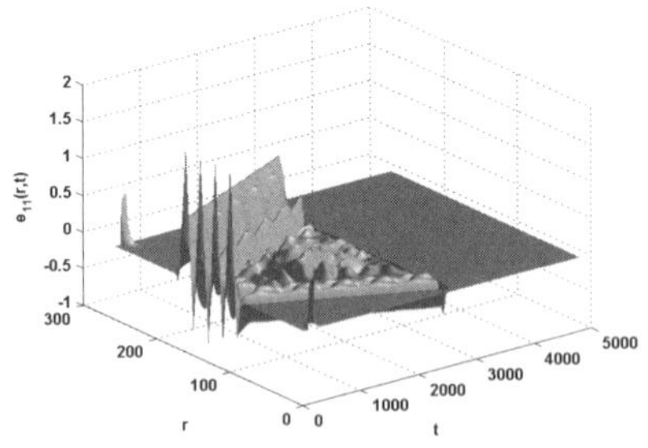

Fig. 13.5 The spatiotemporal evolution of $e_{11}(r, t)$.

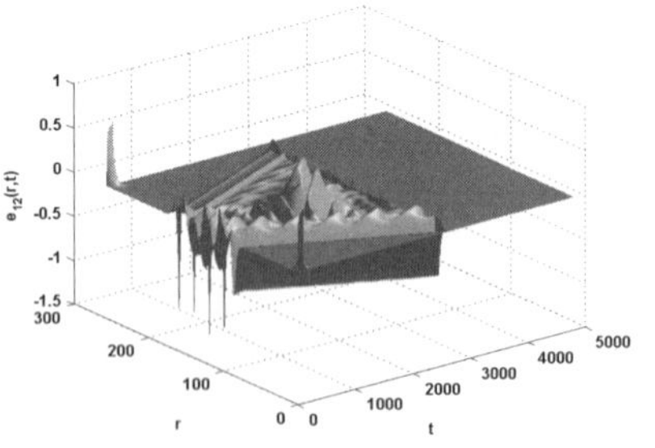

Fig. 13.6 The spatiotemporal evolution of $e_{12}(r, t)$.

It can be seen from Figs. 13.5–13.8 that all the error signals $e_1 = (e_{11}, e_{12})^{\mathrm{T}}$, $e_2 = (e_{21}, e_{22})^{\mathrm{T}}$ in the network approach to zero smoothly and rapidly in a short series of time once network coupling is added, which means that the chain network lag synchronization is realized.

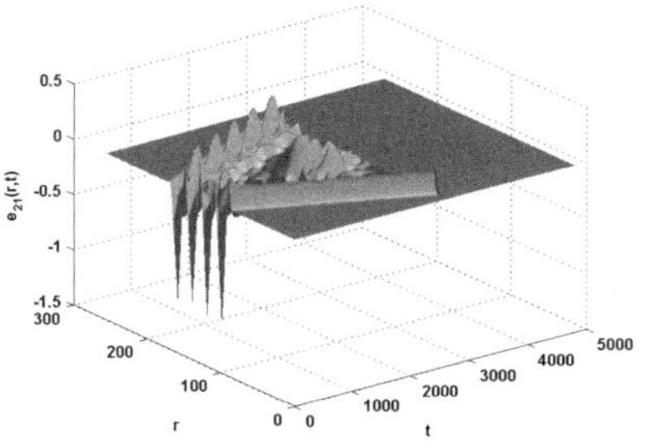

Fig. 13.7 The spatiotemporal evolution of $e_{21}(r,t)$.

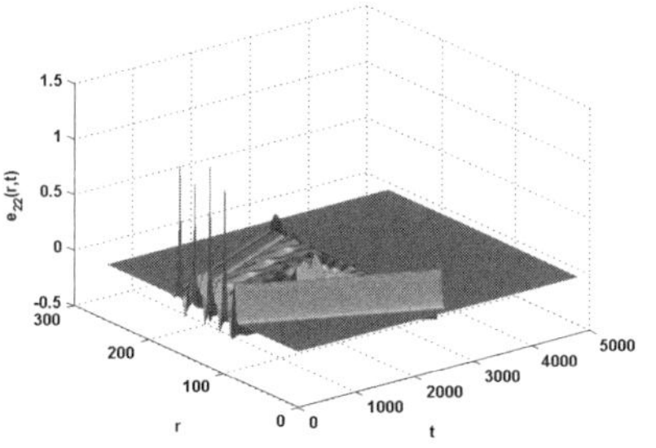

Fig. 13.8 The spatiotemporal evolution of $e_{22}(r,t)$.

13.4 Active sliding mode controller design

Suppose that a complex star network consists of N nodes, each node can be described by the following equation:

$$\frac{\partial X_i(r,t)}{\partial t} = AX_i(r,t) + F(X_i(r,t)), \quad i = 1, 2, ..., N. \tag{13.36}$$

N spatiotemporal chaotic systems (13.36) are taken as nodes to constitute a unidirectional star-network, where node i satisfies the following state equation:

$$\frac{\partial X_i(r,t)}{\partial t} = AX_i(r,t) + F(X_i(r,t)) + \sum_{j=1}^{N} g_{ij} X_j(r, t - \tau_j) + D_i(t) + U_i(r,t), \tag{13.37}$$

where $X_i(r,t) = [X_{i1}(r,t), X_{i2}(r,t), \ldots, X_{in}(r,t)]^{\mathrm{T}} \in R^n$ are the state variables of node i, and g_{ij} being the matrix element of the inner coupling matrix, whose specific forms depend on the connection type of the network, r and t are spatial and temporal variables, $A \in R^{n\times n}$ represents the constant matrices of the linear part and $F : R^n \to R^n$ is the nonlinear part of the system. Eq. (13.36), $D_i(t)$ are the disturbances. The $U_i(r,t) \in R^n$ is the controllers.

If there exists a constant matrix $P = \mathtt{diag}\{P_1, P_2, \ldots, P_{N-1}\}$ and P denotes a "scaling matrix" such that $\lim\limits_{t\to\infty} \|X_1(r, t-\tau_1) - PX_{i+1}(r, t-\tau_{i+1})\| = 0$, then we call the networks achieve the "modified projective lag synchronization", $\|D_1(r, t-\tau_1)\| < \delta_1$, $\|PD_{i+1}(r, t-\tau_{i+1})\| < \delta_2$, where δ_1, δ_2 are known constants.

Modified projective lag synchronization is the aggregation of complete synchronization, anti-synchronization and projective synchronization. All the dynamical states of modified projective lag synchronization are amplified or reduced synchronously. However, modified projective lag synchronization allows us to flex the scales of the different states independently because there are many scaling factors in a scaling matrix P.

Define the error state as $e_i(r,t) = X_1(r, t-\tau_1) - PX_{i+1}(r, t-\tau_{i+1})$, then the dynamics of the networks error can be expressed as

$$\begin{aligned}\frac{\partial e_i(r,t)}{\partial t} &= \frac{\partial X_1(r,t-\tau_1)}{\partial t} - P\frac{\partial X_{i+1}(r,t-\tau_{i+1})}{\partial t}\\ =&AX_1(r,t-\tau_1) + F(X_1(r,t-\tau_1)) + D_1(t-\tau_1) - P(AX_{i+1}(r,t-\tau_{i+1})\\ &+ F(X_{i+1}(r,t-\tau_{i+1})) + (X_1(r,t-\tau_1) - X_{i+1}(r,t-\tau_{i+1}))\\ &+ D_{i+1}(t-\tau_{i+1}) + U_{i+1}(r,t))\\ =&Ae_i(r,t) + F(X_1(r,t-\tau_1)) - PF(X_{i+1}(r,t-\tau_{i+1})) + D_1(t-\tau_1)\\ &- PD_{i+1}(t-\tau_{i+1}) - P(X_1(r,t-\tau_1) - X_{i+1}(r,t-\tau_{i+1})) - PU_{i+1}(r,t)\\ =&Ae_i(r,t) + F(r,t) + D(t) - P\triangle X(r,t) - PU_{i+1}(r,t),\end{aligned} \tag{13.38}$$

where $F(r,t) = F(X_1(r,t-\tau_1)) - PF(X_{i+1}(r,t-\tau_{i+1}))$, $\triangle X(r,t) = X_1(r,t-\tau_1) - X_{i+1}(r,t-\tau_{i+1})$ and $D(t) = D_1(t-\tau_1) - PD_{i+1}(t-\tau_{i+1})$. To synchronize the networks is to find controllers, such that the error system (13.38) is asymptotically stable, which implies the modified projective lag synchronization of the star network is realized, i.e.

$$\lim_{t\to\infty} \|e_i\| = \lim_{t\to\infty} \|X_1(r,t-\tau_1) - PX_{i+1}(r,t-\tau_{i+1})\| = 0. \tag{13.39}$$

In accordance to the active control design method, we choose the following input vector $U_i(r,t)$ to eliminate the nonlinear part of the error dynamics.

$$U_i(r,t) = P^{-1}F(r,t) - \triangle X(r,t) - P^{-1}H_i(r,t). \tag{13.40}$$

The error system (13.38) is rewritten as

$$\frac{\partial e_i(r,t)}{\partial t} = Ae_i(r,t) + H_i(r,t) + D(t). \tag{13.41}$$

Eq. (13.41) describes the error dynamics with a newly defined control input $H_i(r,t)$. In active sliding mode control, $H_i(r,t)$ is designed based on a sliding mode control law

$$H_i(r,t) = KW_i(r,t), \tag{13.42}$$

where $K = [k_1, k_2, \ldots, k_n]^{\mathrm{T}}$ is a constant gain vector and $W_i(r,t) \in R$ is the control input that satisfies

$$W_i(r,t) = \begin{cases} W_i^+(r,t) & S_i(e) \geq 0,\\ W_i^-(r,t) & S_i(e) \leq 0,\end{cases} \tag{13.43}$$

in which $S_i = S_i(e)$ is a switching surface which prescribes the desired dynamics. The resulting error dynamics is then

$$\frac{\partial e_i(r,t)}{\partial t} = Ae_i(r,t) + KW_i(r,t) + D(t). \tag{13.44}$$

The exact sliding mode controller will be designed according to the sliding mode control theory. The sliding surface can be defined as follows:

$$S_i(e) = Ce_i, \tag{13.45}$$

in which $C = [c_1, c_2, \ldots, c_n]$ is a constant vector.

$\dot{S}_i(e) = 0$ is a necessary condition for the state trajectory to stay on the switching surface $S_i(e) = 0$. Hence, when in sliding mode, the controlled system satisfies the following conditions:

$$\dot{S}_i(e) = Ce_i \quad and \quad S_i(e) = 0. \tag{13.46}$$

Based on (13.44)–(13.46), one can deduce the following result:

$$\dot{S}_i(e) = \frac{\partial S_i(e)}{\partial t}\dot{e}_i = C(Ae_i(r,t) + KW_i(r,t) + D(t)). \tag{13.47}$$

We consider the constant plus proportional rate reaching law in our study. Then, the reaching law is chosen as:

$$\dot{S}_i = -q\mathbf{sgn}(S_i) - rS_i, \tag{13.48}$$

where sgn(.) denotes the sign function. The gains $q > 0$ and $r > 0$ are determined such that the sliding condition is satisfied.

According to the Eqs. (13.44), (13.45) and (13.48), we have

$$W_i(r,t) = -(CK)^{-1}(CAe_i(r,t) + CD(t) + q\mathbf{sgn}(S_i) + rS_i). \tag{13.49}$$

The disturbances are unknown in the practical engineering applications. So the control input $W_i(r,t)$ is rewritten as follows:

$$W_i(r,t) = -(CK)^{-1}(CAe_i(r,t) + q\mathbf{sgn}(S_i) + rS_i). \tag{13.50}$$

The modified projective lag synchronization of the star network can be realized by using the control law $W_i(r,t)$ in Eq. (13.50), if the inequality $\|C\|(\delta_1 + \delta_2) < q$ is satisfied. That is to say, the error state trajectory converges to zero along the sliding mode surface.

Proof. Consider the following Lyapunov function:

$$V = \sum_{i=1}^{N} \frac{1}{2} S_i^2. \tag{13.51}$$

The time derivative of (13.51) is

$$\begin{aligned}
\dot{V} &= \sum_{i=1}^{N} S_i \dot{S}_i \\
&= \sum_{i=1}^{N} S_i C\{Ae_i(r,t) - K(CK)^{-1}[CAe_i(r,t) + q\mathbf{sgn}(S_i) + rS_i] + D(t)\} \\
&= \sum_{i=1}^{N} (-S_i q\mathbf{sgn}(S_i) - rS_i^2 + S_i CD_1(r, t-\tau_1) - S_i CPD_{i+1}(r, t-\tau_{i+1})) \\
&\leq \sum_{i=1}^{N} (-\|S_i\| q\mathbf{sgn}(S_i) - rS_i^2 + \|S_i C\|\|D_1(r,t-\tau_1)\| + \|S_i C\|\|PD_{i+1}(r,t-\tau_{i+1})\|) \\
&\leq \sum_{i=1}^{N} (-\|S_i\| q\mathbf{sgn}(S_i) - rS_i^2 + \|S_i C\|\delta_1 + \|S_i C\|\delta_2) \\
&= \sum_{i=1}^{N} (\|S_i\|(\|C\|(\delta_1 + \delta_2) - q) - rS_i^2).
\end{aligned} \tag{13.52}$$

Because the inequality $\|C\|(\delta_1+\delta_2)<q$ is hold, then the inequality $\dot{V}<0$ is satisfied. According to the Lyapunov stability theory, the error system is asymptotically stable.

13.5 Numerical results

In this section, we present numerical results to verify the analytical predictions obtained in the previous section. The Gray-Scott system as a producer of spatiotemporal chaos is demonstrated again. Unidirectional star-network is taken in the paper; the coupling matrix G is set as

$$G=\begin{bmatrix} 0 & 0 & 0 & \dots & 0 \\ 1 & -1 & 0 & \dots & 0 \\ \dots & \dots & \dots & \dots & \dots \\ 1 & 0 & \dots & -1 & 0 \\ 1 & 0 & \dots & 0 & -1 \end{bmatrix}. \tag{13.53}$$

In order to demonstrate briefly the effectiveness of the synchronization approach of a star network with N nodes. Therefore, we first consider the star-network with four nodes. The state equation for each node is also of Gray-Scott model described by Eq. (13.31), and state variables at three nodes are $(X_{11}(r,t),X_{12}(r,t))$, $(X_{21}(r,t),X_{22}(r,t))$, $(X_{31}(r,t),X_{32}(r,t))$,$(X_{41}(r,t),X_{42}(r,t))$ respectively. Coupling is made by using the single state variable $X_i(r,t)$ in the unidirectional star-network.

$$\begin{cases} \dfrac{\partial X_{11}}{\partial t}=-X_{11}X_{12}^2+a(1-X_{11})+d_1\nabla^2X_{11}+d_{11}, \\ \dfrac{\partial X_{12}}{\partial t}=X_{11}X_{12}^2-(a+b)X_{12}+d_2\nabla^2X_{12}+d_{12}. \end{cases}$$

$$\begin{cases} \dfrac{\partial X_{21}}{\partial t}=-X_{21}X_{22}^2+a(1-X_{21})+d_1\nabla^2X_{21}+(X_{11}-X_{21})+d_{21}+U_{21}, \\ \dfrac{\partial X_{22}}{\partial t}=X_{21}X_{22}^2-(a+b)X_{22}+d_2\nabla^2X_{22}+(X_{12}-X_{22})+d_{22}+U_{22}. \end{cases}$$

$$\begin{cases} \dfrac{\partial X_{31}}{\partial t}=-X_{31}X_{32}^2+a(1-X_{31})+d_1\nabla^2X_{31}+(X_{11}-X_{31})+d_{31}+U_{31}, \\ \dfrac{\partial X_{32}}{\partial t}=X_{31}X_{32}^2-(a+b)X_{32}+d_2\nabla^2X_{32}+(X_{12}-X_{32})+d_{32}+U_{32}. \end{cases}$$

$$\begin{cases} \dfrac{\partial X_{41}}{\partial t}=-X_{41}X_{42}^2+a(1-X_{41})+d_1\nabla^2X_{41}+(X_{11}-X_{41})+d_{41}+U_{41}, \\ \dfrac{\partial X_{42}}{\partial t}=X_{41}X_{42}^2-(a+b)X_{42}+d_2\nabla^2X_{42}+(X_{12}-X_{42})+d_{42}+U_{42}. \end{cases}$$

Here

$$U_i(r,t)=\begin{bmatrix} U_{i1}(r,t) \\ U_{i2}(r,t) \end{bmatrix}, \qquad A=\begin{bmatrix} -a & 0 \\ 0 & -(a+b) \end{bmatrix}. \tag{13.54}$$

$$F(X_i(r,t)) = \begin{bmatrix} -X_{i1}(r,t)X_{i2}^2(r,t) + a + d_1\nabla^2 X_{i1}(r,t) \\ X_{i1}(r,t)X_{i2}^2(r,t) + d_2\nabla^2 X_{i2}(r,t) \end{bmatrix}. \tag{13.55}$$

It is assume that the disturbances $D_i(t)$

$$D_i(t) = \begin{bmatrix} d_{i1} \\ d_{i2} \end{bmatrix} = \begin{bmatrix} (-1)^i 0.0001\texttt{cos}(20t) \\ (-1)^i 0.0001\texttt{cos}(10t) \end{bmatrix}, \tag{13.56}$$

where $P = \texttt{diag}\{2, -1, -\frac{1}{2}\}$ and $U_1(r,t) = 0$. The control parameters are chosen as $C = [11]$, $K = [11]^{\mathrm{T}}$, $r = 0.09$, $q = 0.0005$. The control input $w_i(r,t)$ is

$$\begin{aligned} W_i(r,t) &= \left[-0.062 - 0.009\right] e_i - 0.0005\texttt{sgn}(S_i) \\ &= \begin{cases} -0.062e_{i1} - 0.009e_{i2} - 0.0005 & S_i > 0, \\ -0.062e_{i1} - 0.009e_{i2} + 0.0005 & S_i < 0. \end{cases} \end{aligned} \tag{13.57}$$

Figs. 13.9–13.10 depict the two-dimensional state trajectories of four chaotic nodes evolve in the same directions with time delay $\tau_1 = 300$, $\tau_2 = 200$, $\tau_3 = 100$, $\tau_4 = 0$ at step $r = 100$. It can be seen from Figs. 13.14, 13.16, 13.20, 13.22, 13.26 and 13.28 that the state variables trajectories of the $X_{11}(r, t-\tau_1)$, $X_{12}(r, t-\tau_1)$ and $X_{i+11}(r, t-\tau_{i+1})$, $X_{i+12}(r, t-\tau_{i+1})$ are exactly the same after adding controllers, which implies the modified projective lag synchronization between the star-network is realized.

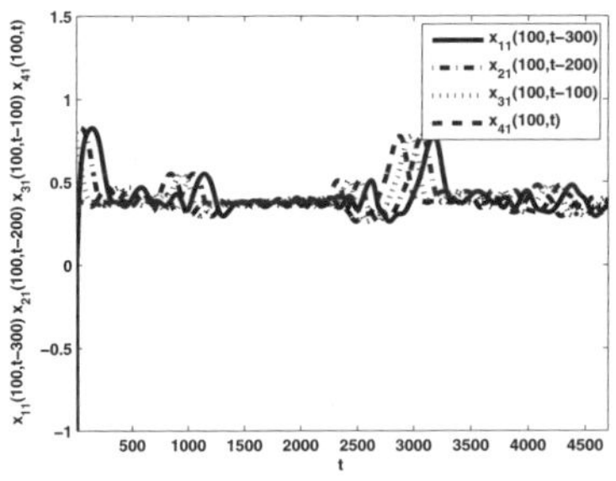

Fig. 13.9 The lag state trajectories of the X_{i1}.

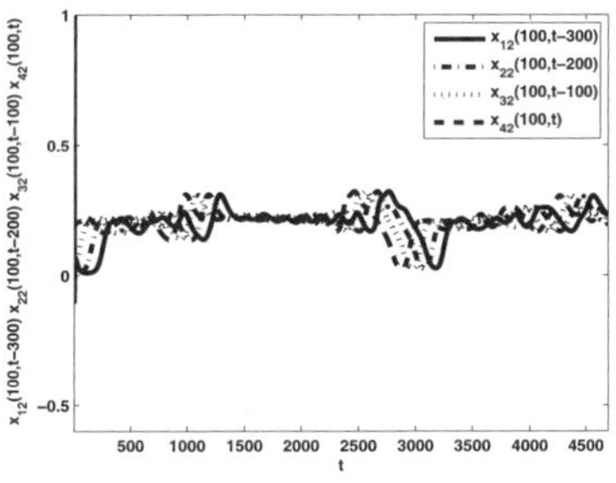

Fig. 13.10 The lag state trajectories of the X_{i2}.

The three-dimensional evolution of error signals are shown in Figs. 13.11 (13.12, 13.17, 13.18, 13.23, 13.24). It also can be seen that all the error signals $e_{i1}(r,t)$, $e_{i2}(r,t)$ approach to zero smoothly and rapidly in a short series of time once the design controller is added at step $t = 2800$, The simulation results demonstrate the star-network with disturbances can achieve the modified projective lag synchronization under the design controller.

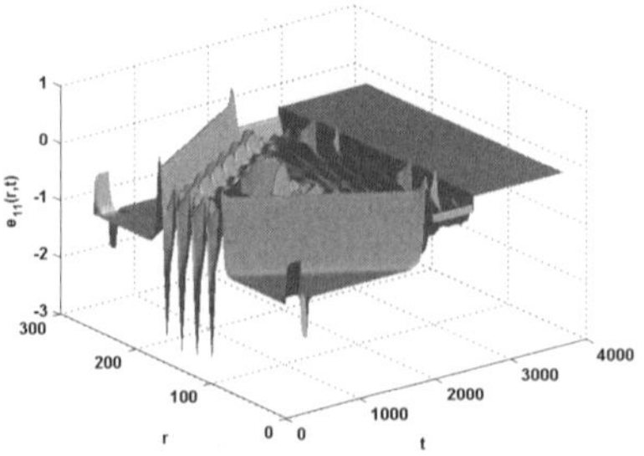

Fig. 13.11 The spatiotemporal evolution of $e_{11}(r, t)$.

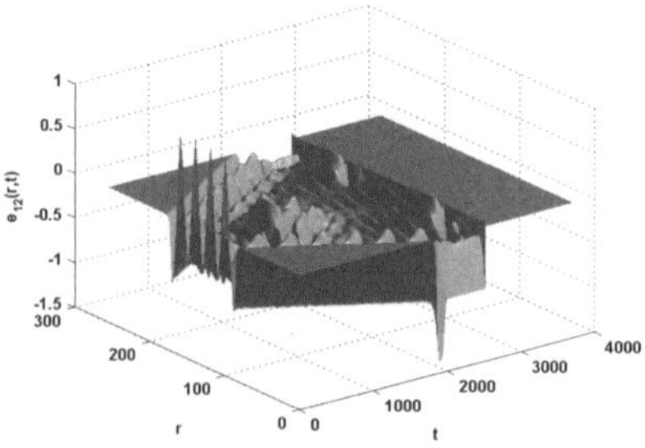

Fig. 13.12 The spatiotemporal evolution of $e_{12}(r, t)$.

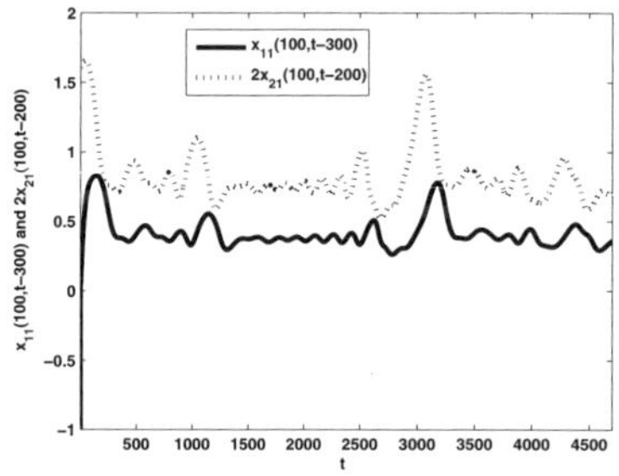

Fig. 13.13 The X_{11} and $2X_{21}$ without adding control.

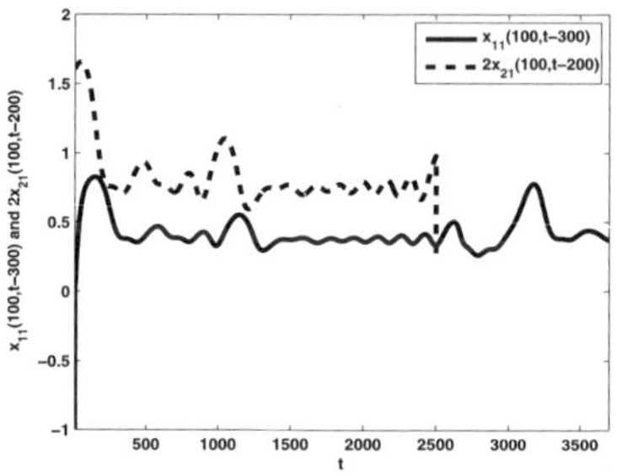

Fig. 13.14 The X_{11} and $2X_{21}$ with adding control.

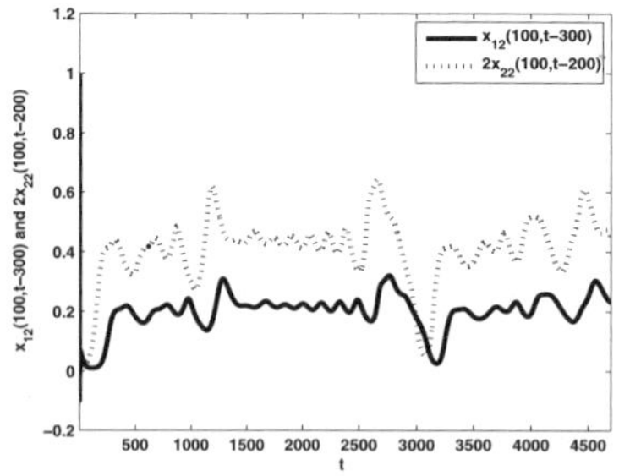

Fig. 13.15 The X_{12} and $2X_{22}$ without adding control.

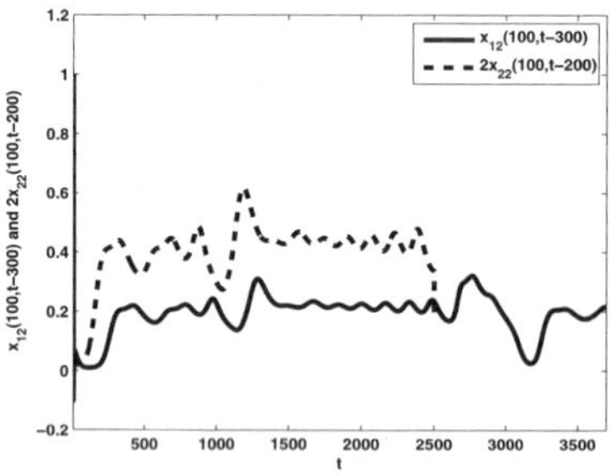

Fig. 13.16 The X_{12} and $2X_{22}$ with adding control.

13.6 Master stability functions

Given a complex dynamical system consisting of a number of non-identical coupled chaotic systems, an effective tool in the analyses of synchronization problems is the master stability function (MSF).

Suppose that a complex network consists of N nodes, and the state equations of the nodes are spatiotemporal chaos systems, in which the n-dimensional variable of

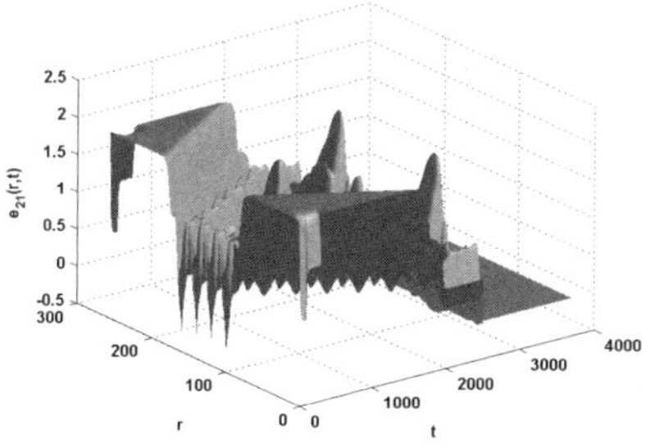

Fig. 13.17 The spatiotemporal evolution of $e_{21}(r,t)$.

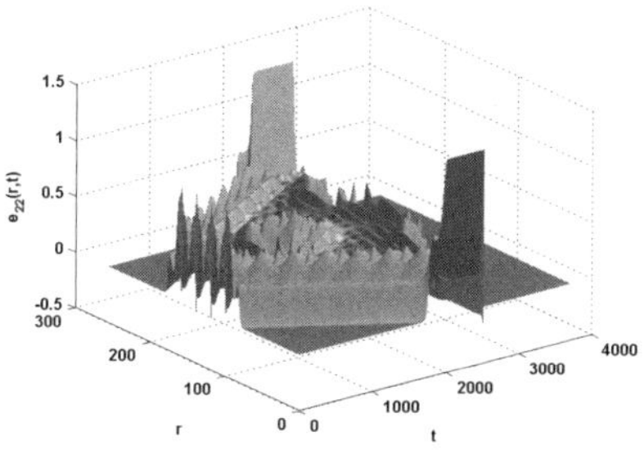

Fig. 13.18 The spatiotemporal evolution of $e_{22}(r,t)$.

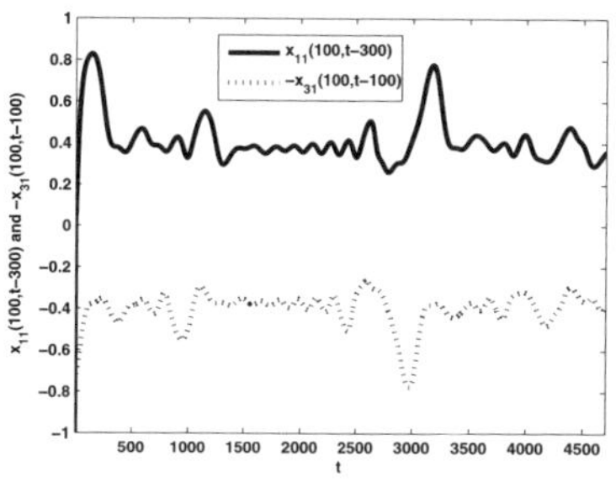

Fig. 13.19 The X_{11} and $-X_{31}$ without adding control.

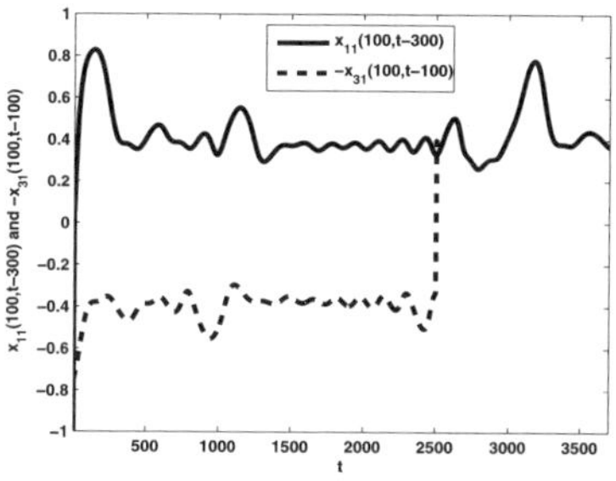

Fig. 13.20 The X_{11} and $-X_{31}$ with adding control.

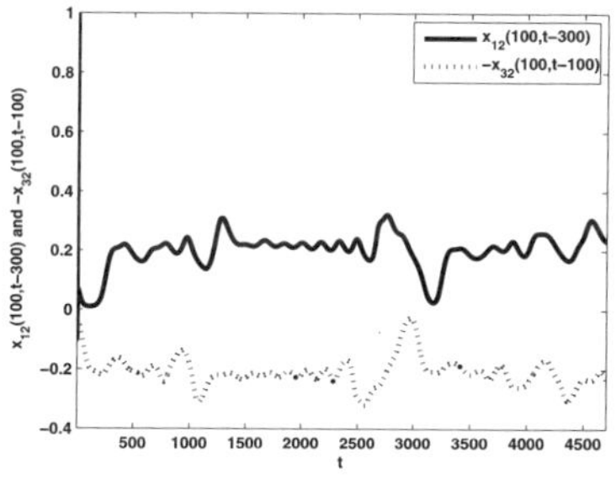

Fig. 13.21 The X_{12} and $-X_{32}$ without adding control.

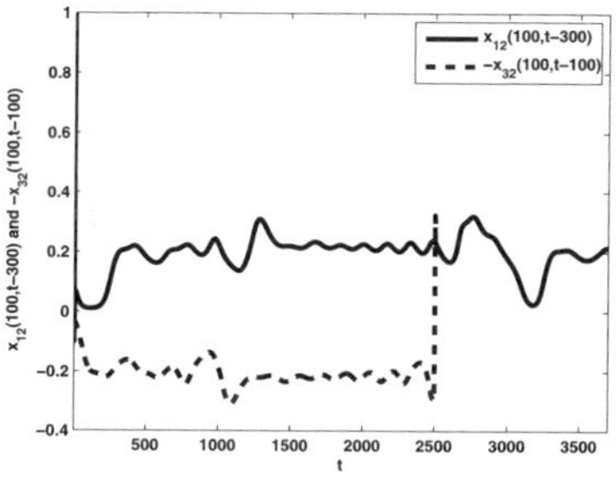

Fig. 13.22 The X_{12} and $-X_{32}$ with adding control.

node i at time t is $X_i(r,t)$. When the coupling action is not considered, each node can be described by the following equation:

$$\frac{\partial X_i(r,t)}{\partial t} = F_i(X_i(r,t)) \quad (i = 1, 2, ..., N), \tag{13.58}$$

where $X_i(r,t) = [X_{i1}(r,t), X_{i2}(r,t), \ldots, X_{in}(r,t)]^{\mathrm{T}} \in R^n$ are the state variables of node i, $F : R^n \to R^n$ is continuously differentiable, r and t are spatial and temporal variables.

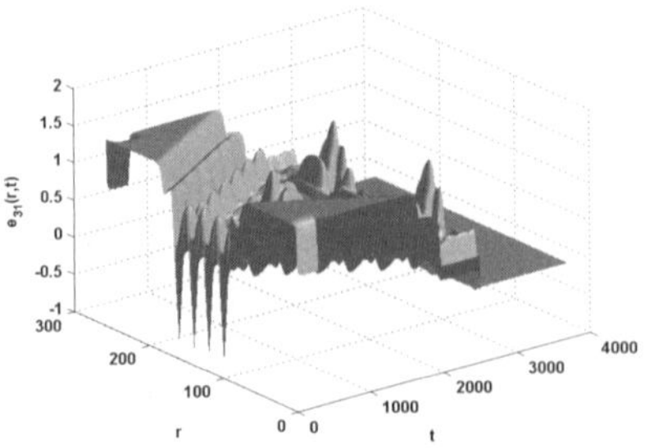

Fig. 13.23 The spatiotemporal evolution of $e_{31}(r, t)$.

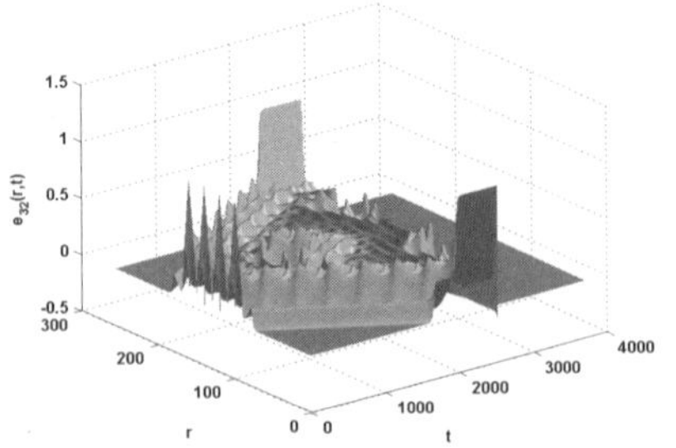

Fig. 13.24 The spatiotemporal evolution of $e_{32}(r, t)$.

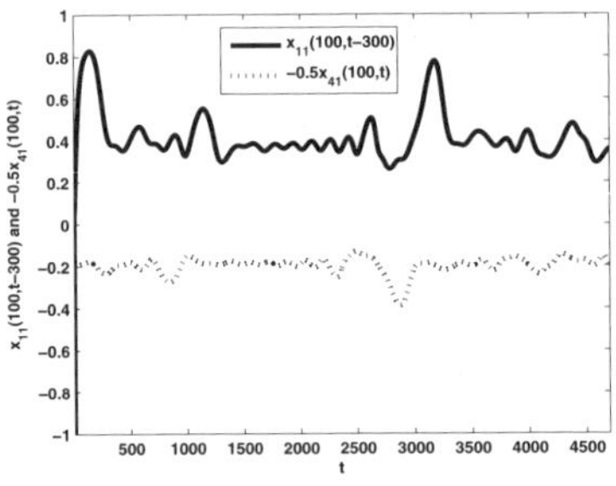

Fig. 13.25 The X_{11} and $-0.5X_{41}$ without adding control.

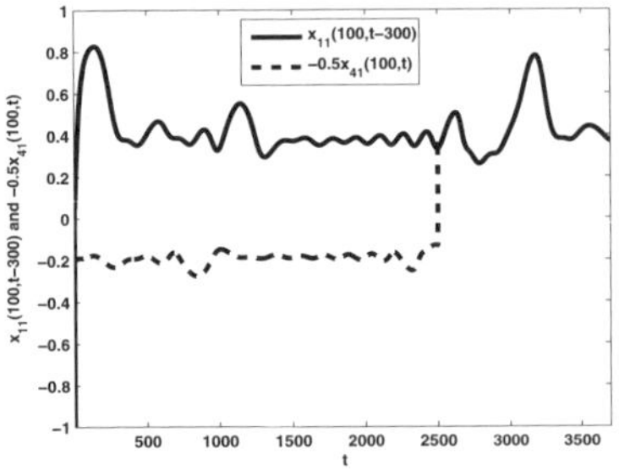

Fig. 13.26 The X_{11} and $-0.5X_{41}$ with adding control.

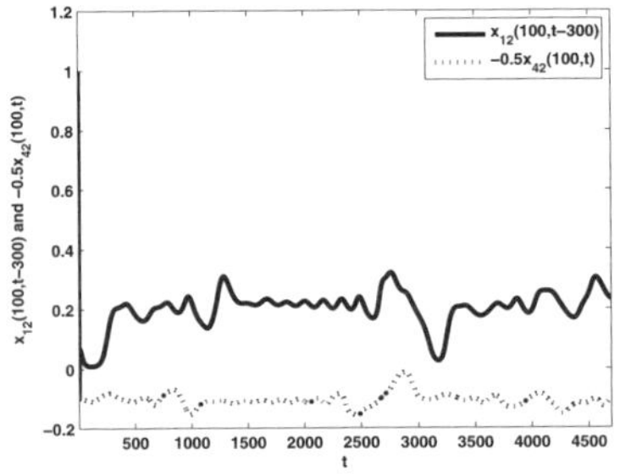

Fig. 13.27 The X_{11} and $-0.5X_{42}$ without adding control.

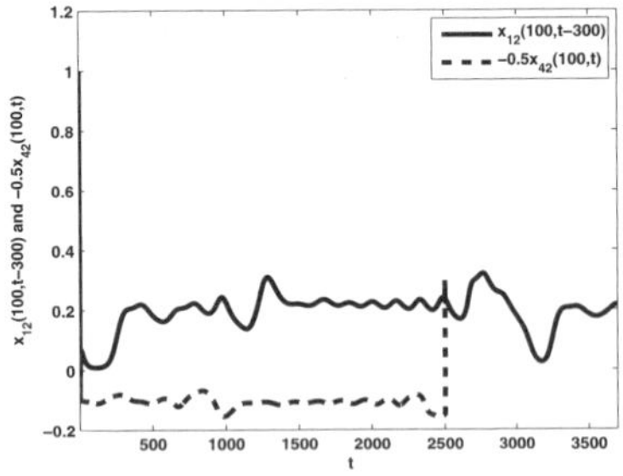

Fig. 13.28 The X_{11} and $-0.5X_{42}$ with adding control.

Coupling on other nodes is added by using single state variables, and the state equation (13.58) of the $i-th$ node is rewritten as

$$\frac{\partial X_i(r,t)}{\partial t} = F_i(X_i(r,t)) + c_i \sum_{j=1}^{N} a_{ij} H(X_j(r,t)), \tag{13.59}$$

in which $H(X_j(r,t))$ is a coupling function, c_i is a global coupling parameter, and A is a coupling matrix determined by the connection topology. The matrix A satisfies the condition $\sum_{j=1}^{N} a_{ij}$ for any i, where N is the network size, the synchronous state is $X_1 = X_2 = \ldots = X_N = S$. For the system described by Eq. (13.59), the variational equations $\chi_i(r,t) = X_i(r,t) - S(r,t)$ is

$$\frac{\partial \chi_i(r,t)}{\partial t} = DF_i(s(r,t))\chi_i(r,t) + c_i \sum_{j=1}^{N} a_{ij} DH(s(r,t))\chi_j(r,t). \tag{13.60}$$

Here $DF_i(s(r,t))$ and$DH(s(r,t))$ are the Jacobian matrices of the corresponding vector functions evaluated at $s(r,t)$. Let $\chi(r,t) = [\chi_1(r,t), \chi_2(r,t), \ldots \chi_N(r,t)]$, the Eq. (13.60) is rewritten as matrix equation

$$\frac{\partial \chi(r,t)}{\partial t} = DF_i(s(r,t))\chi(r,t) + CDH(s(r,t))\chi(r,t)A^{\mathrm{T}}, \tag{13.61}$$

where $A^T = S\Lambda S^{-1}$ is the Jordan decomposition of the matrix A and $\Lambda = \mathtt{diag}\{\lambda_l, \lambda_2, \ldots, \lambda_N\}$, the eigenvalues of the matrix is A is $\{\lambda_k\}_{k=1}^{N}$ and $\lambda_1 = 0$. Then, let $\chi(r,t)s(r,t) = [\delta_1(r,t), \delta_2(r,t), \ldots, \delta_N(r,t)]$, the Eq. (13.60) is rewritten as

$$\frac{\partial \delta_l(r,t)}{\partial t} = [DF_i(s(r,t)) + C\lambda_l DH(s(r,t))]\delta_l(r,t) \quad (l = 2, 3, ..., N). \tag{13.62}$$

The largest Lyapunov exponent determined from Eq. (13.62). If λ_{max} is negative, the synchronization of networks will succeed.

13.7 Numerical results

In this section, we present numerical results to verify the analytical predictions obtained in the previous section. The Fitzhugh-Nagumo model and the Panfilov model as a producer of spatiotemporal chaos are described as follows [Hildebrand *et al.* (1995); Panfilov and Hogeweg (1993); Sinha*et al.* (2001)]:

$$\begin{cases} \dfrac{\partial X_1}{\partial t} = \xi_1^{-1} X_1(1 - X_1)(X_1 - \dfrac{X_2 + b_1}{a_1}) + d\nabla^2 X_1, \\ \dfrac{\partial X_2}{\partial t} = f(X_1) - X_2, \end{cases} \tag{13.63}$$

in which

$$f(X_1) = \begin{cases} 0, \quad 0 \le X_1 \le \dfrac{1}{3}, \\ 1 - c_1 X_1 (X_1 - 1)^2, \quad \dfrac{1}{3} \le X_1 \le 1, \\ 1, \quad X_1 > 1, \end{cases} \tag{13.64}$$

where $X_1(r,t)$ and $X_2(r,t)$ are system state variables, a_1, b_1, c_1 and ξ_1 are system parameters, d is a diffusion coefficient. The system parameters are fixed at $a_1 = 0.84$, $b_1 = 0.07$, $c_1 = 6.75$, and the diffusion coefficient $d = 1$. For different values

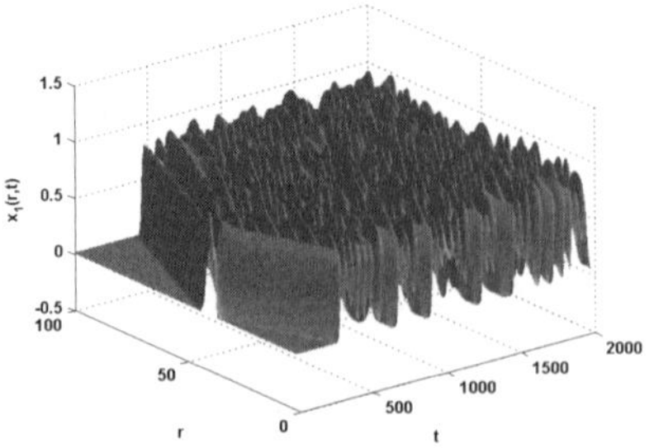

Fig. 13.29 The spatiotemporal evolution of $X_1(r,t)$.

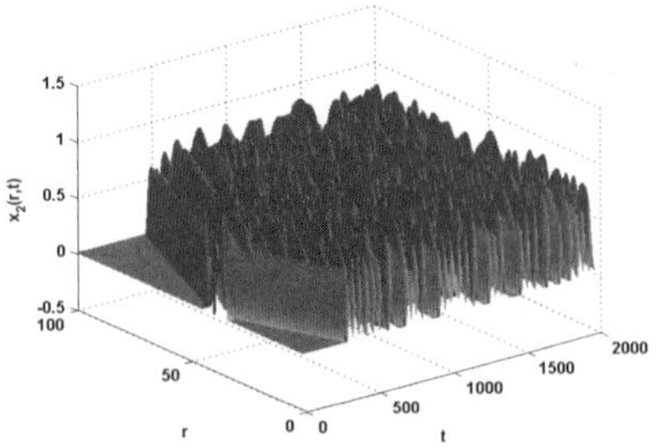

Fig. 13.30 The spatiotemporal evolution of $X_2(r,t)$.

of parameter ξ_1, the Fitzhugh-Nagumo model can present very complex and rich spatiotemporal dynamical behavior. See Figs. 13.29–13.30, in which the system size is 100×100, the space step is 1, the time step is 0.1 s.

The dynamical equations of the Panfilov model is

$$\begin{cases} \dfrac{\partial X_1}{\partial t} = \nabla^2 X_1 - f(X_1) - X_2, \\ \dfrac{\partial X_2}{\partial t} = \xi(k_1 X_1 - X_2), \end{cases} \tag{13.65}$$

in which

$$f(X_1) = \begin{cases} 20X_1 \quad X_1 < 0.837, \\ -3X_1 + 0.06 \quad 0.0026 \le X_1 \le 0.837, \\ 15(X_1 - 1) \quad X_1 > 0.837, \end{cases} \tag{13.66}$$

$$\xi = \begin{cases} \dfrac{1}{75} \quad X_1 < 0.837, \\ 1 \quad X_1 > 0.837, \\ 0.3 \quad X_1 < 0.0026, X_2 < 1.8, \end{cases} \tag{13.67}$$

where $X_1(r,t)$ and $X_2(r,t)$ are system state variables, k_1 are system parameters. The system parameters are fixed at $k_1 = 3$. The Panfilov model can also present very complex and rich spatiotemporal dynamical behavior. See Figs. 13.31–13.32 in which the system size is 100×100, the space step is 1, the time step is 0.1 s

In order to demonstrate briefly the effectiveness of the MSF synchronization approach of a chain network with non identical N nodes. Therefore, we first consider the chain network with three nodes. The state equation for each node is of the Fitzhugh-Nagumo model or the Panfilov model described by Eq. (13.63)–(13.66), and state variables at three nodes are $(X_{11}(r,t), X_{12}(r,t))$, $(X_{21}(r,t), X_{22}(r,t))$, $(X_{31}(r,t), X_{32}(r,t)$, respectively.

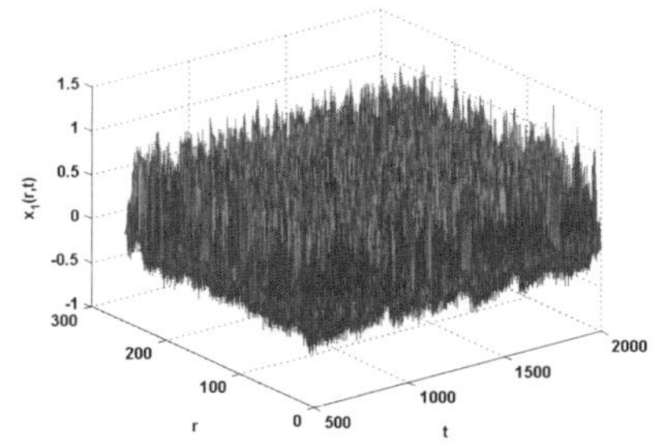

Fig. 13.31 The spatiotemporal evolution of $X_1(r,t)$.

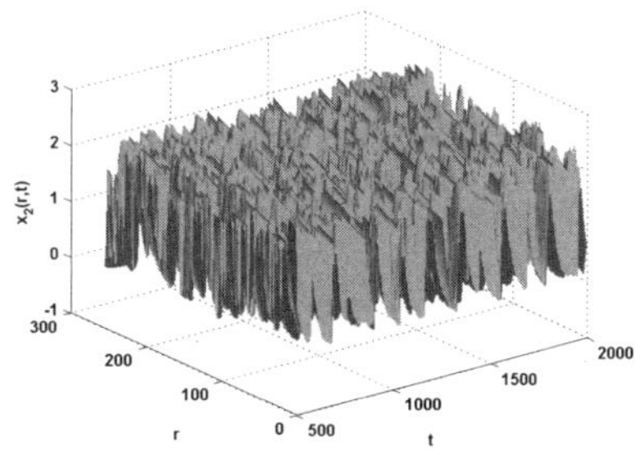

Fig. 13.32 The spatiotemporal evolution of $X_2(r,t)$.

The coupling matrix A is set as

$$A = \begin{bmatrix} 0 & 0 & 0 & \dots & 0 \\ 1 & -1 & 0 & \dots & 0 \\ \dots & \dots & \dots & \dots & \dots \\ 0 & \dots & 1 & -1 & 0 \\ 0 & 0 & \dots & 1 & -1 \end{bmatrix}. \tag{13.68}$$

Coupling is made by using the single state variable $X_i(r,t)$ between the Fitzhugh-Nagumo model and the Panfilov model.

$$\begin{cases} \dfrac{\partial X_{11}}{\partial t} = \xi_1^{-1} X_{11}(1 - X_{11})(X_{11} - \dfrac{X_{12} + b_1}{a_1}) + d\nabla^2 X_{11}, \\ \dfrac{\partial X_{12}}{\partial t} = f(X_{11}) - X_{12}. \end{cases} \tag{13.69}$$

$$\begin{cases} \dfrac{\partial X_{21}}{\partial t} = \nabla^2 X_{21} - f(X_{21}) - X_{22} + c_1(X_{11} - X_{21}), \\ \dfrac{\partial X_{22}}{\partial t} = \xi(k_1 X_{21} - X_{22}) + c_1(X_{12} - X_{22}). \end{cases} \tag{13.70}$$

$$\begin{cases} \dfrac{\partial X_{31}}{\partial t} = \xi_1^{-1} X_{31}(1 - X_{31})(X_{31} - \dfrac{X_{32} + b_1}{a_1}) + d\nabla^2 X_{31} + c_2(X_{21} - X_{31}), \\ \dfrac{\partial X_{32}}{\partial t} = f(X_{31}) - X_{32} + c_2(X_{22} - X_{32}). \end{cases} \tag{13.71}$$

From Figs. 13.33 we know, within the limits of the negative maximum Lyapunov exponent, coupling parameters c_1 and c_2 can be chose as 10 and 11 to synchronize network respectively.

Iteration of the simulation is initiated at step $t = 2800$, and simulation results are shown in Figs. 13.34–13.37.

It can be seen from Figs. 13.34–13.37 that all the error signals $e_1 = (e_{11}, e_{12})^{\mathrm{T}}$, $e_2 = (e_{21}, e_{22})^{\mathrm{T}}$ in the network approach to zero rapidly in a short series of time once network coupling is added, which means that the network synchronization is realized.

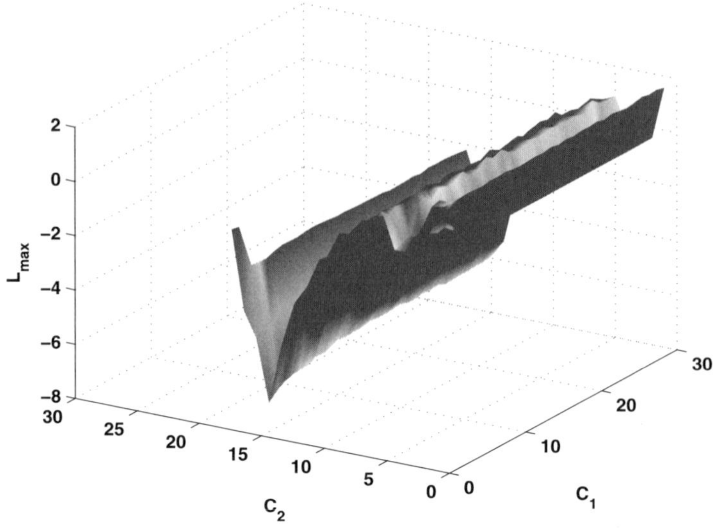

Fig. 13.33 The evolution of L_{max} with c_1 and c_2.

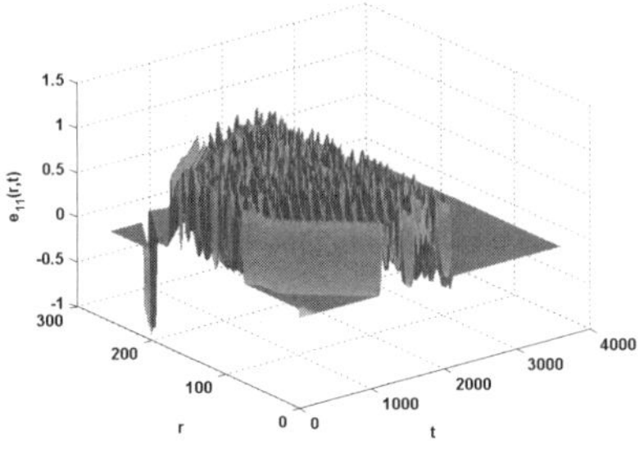

Fig. 13.34 The spatiotemporal evolution of $e_{11}(r,t)$.

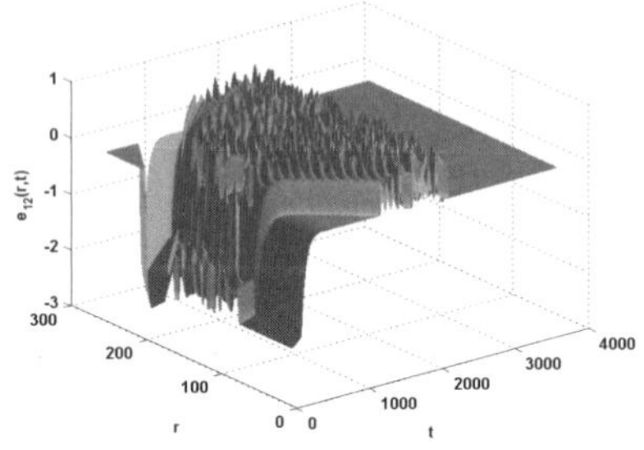

Fig. 13.35 The spatiotemporal evolution of $e_{12}(r,t)$.

13.8 Conclusion

A synchronization controller is designed to study spatiotemporal chaotic systems with identical structures and to realize synchronization between multi-Gray-Scott systems. The structure of terminal synchronization controller and the error equations of state variables between spatiotemporal chaotic systems are presented on the basis of Lyapunov stability theory. The active sliding mode controllers are designed to study modified projective lag synchronization of spatiotemporal chaos with disturbances and to realize synchronization between identical Gray-Scott systems. The structure of active sliding mode controllers are presented on the basis of combining the active control and sliding mode control method, the sufficient conditions

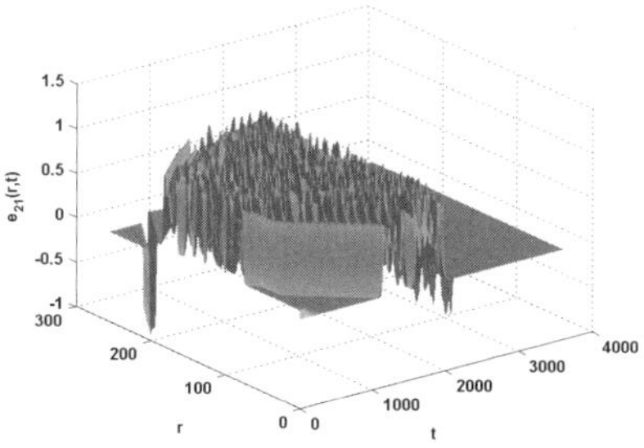

Fig. 13.36 The spatiotemporal evolution of $e_{21}(r,t)$.

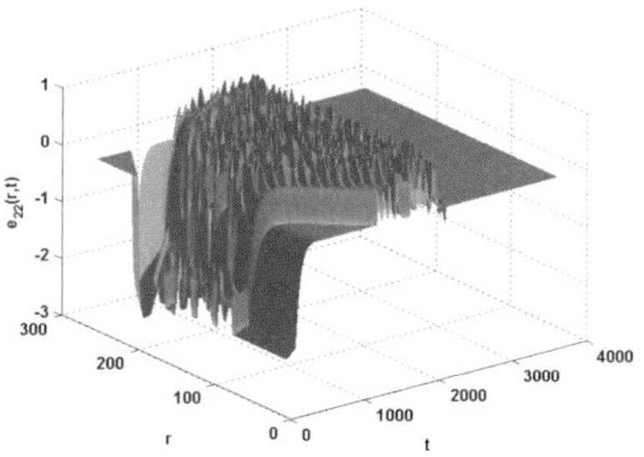

Fig. 13.37 The spatiotemporal evolution of $e_{22}(r,t)$.

of modified projective lag synchronization are obtained. Both the error signals approach zero smoothly and rapidly in a short series of time after adding controllers, and the phase maps are exactly the same, which means that the synchronization is achieved. The design of the terminal synchronization and the active sliding mode controllers are simple in principle. We also formulate master stability functions for an arbitrary number of spatiotemporal nodes in a network. The master stability equation allows us to calculate the maximum Lyapunov exponent and determine the stability of the networks for any linear coupling scheme. Within the limits of the negative maximum Lyapunov exponent, coupling parameters can be chosen to synchronize network arbitrarily. The sufficient conditions of non-topological structure of spatiotemporal chaos nodes in a network are obtained by the method of master stability function.

Bibliography

Ahn, C. K. (2010). Neural network $H\infty$ chaos synchronization, *Nonlinear Dyn.* **60**, pp. 295–302.

Barahona, M. and Pecora, L. M. (2002). Synchronization in small-world systems, *Phys. Rev. Lett.* **89**, pp. 054101.

Barabási, A. L. and Albert, R. (1999). Emergence of scaling in random networks, *Science* **286**, pp. 509–512.

Behzad, M., Salarieha, H. and Alastya, A. (2008). Chaos synchronization in noisy environment using nonlinear filtering and sliding mode control, *Chaos Solit. Fract.* **36**, pp. 12951304.

Boccaletti, S., Latora, V., Moreno, Y., Chavez, M. and Hwang, D. U. (2006). Complex networks: structure and dynamics, *Phys. Reports* **424**, pp. 175–308.

Bowong, S., Kakmeni, F. M. M. and Tchawoua, C. (2004). Controlled synchronization of chaotic systems with uncertainties via a sliding mode control design, *Phys. Rev. E* **70**, pp. 066217.

Bhalekara, S. and Daftardar-Gejji, V. (2010). Synchronization of different fractional order chaotic systems using active control, *Commun. Nonlinear Sci. Numer. Simulat.* **15**, pp. 3536–3546.

Cai, N., Jing, Y. W. and Zhang, S. Y. (2010). Modified projective synchronization of chaotic systems with disturbances via active sliding mode control, *Commun Nonlinear Sci. Numer. Simul.* **15**, pp. 161320.

Chen,L. Q. (2001). An open-plus-closed-loop control for discrete chaos and hyperchaos, *Phys. Lett. A* **281**, pp. 327–333.

Chen, T. P., Liu, X. W. and Lu, W. L. (2007). Pinning complex networks by a single controller, *IEEE Trans. Circ Syst.I* **54**, pp. 1317–1326.

Chen, G. and Duan, Z. (2008). Network synchronizability analysis: a graph-theoretic approach, *Chaos* **18**, pp. 037102.

Chen, W. S., Li, W. and Miao,Q. G. (2010). Backstepping control for periodically time-varying systems using high-order neural network and Fourier series expansion, *ISA Trans.* **49**, pp. 283–292.

Chen,J., Liu, H., Lu, J. A. and Zhang, Q. J. (2011). Projective and lag synchronization of a novel hyperchaotic system via impulsive control, *Commun. Nonlinear Sci. Numer. Simulat.* **16**, pp. 2033–2040.

Chacón, R. and Marcheggiani, L. (2010). Controlling spatiotemporal chaos in chains of dissipative Kapitza pendula, *Phys. Rev. E* **82**, pp. 016201.

Chian, A. C. L., Miranda, R. A., Rempel, E. L., Saiki, Y. and Yamada, M. (2010). Amplitude-phase synchronization at the onset of permanent spatiotemporal chaos,

Phys. Rev. Lett. **104**, pp. 254102.

Dhamala, M., Jirsa, V. K. and Ding, M. Z. (2004). Enhancement of neural synchrony by Time delay, *Phys. Rev. Lett.* **92**, pp. 074104.

Dorogovtesev, N. and Mendes, J. F. F. (2002). Evolution of networks, *Advances in Phys.* **51**, pp. 1079–1187.

Du, H. Y., Zeng, Q. S. and Lü, N. (2010). A general method for modified function projective lag synchronization in chaotic systems, *Phys. Lett. A* **374**, pp. 14936.

Fink, K. S., Johnson, G., Carroll,T., Mar, D. and Pecora, L. (2000). Three coupled oscillators as a universal probe of synchronization stability in coupled oscillator arrays, *Phys. Rev. E* **61**, pp. 5080–5090.

Flunkert, V., Yanchuk, S., Dahms, T. and Schll, E. (2010). Synchronizing distant nodes: A universal classification of networks, *Phys. Rev. Lett.* **105**, pp. 254101.

Gorochowski, T. E., Bernardo,M. D. and Grierson, C. S. (2010). Evolving enhanced topologies for the synchronization of dynamical complex networks, *Phys. Rev. E* **81**, pp. 056212.

Haeri, M. and Emadzadeh, A. A. (2007). Synchronizing different chaotic systems using active sliding mode control, *Chaos Solit. Fract.* **31**, pp. 11929.

Hassan, S. and Aria, A. (2009). Adaptive synchronization of two chaotic systems with stochastic unknown parameters, *Commun. Nonlinear Sci. Numer. Simulat.* **14**, pp. 508–519.

Heagy, J. F., Carroll, T. L. and Pecora, L. M. (1994). Synchronous chaos in coupled oscillator systems, *Phys. Rev. E* **50**, pp. 1874–1885.

Hildebrand, M., Eiswirth, M. and Carroll, M. (1995). Statistics of topological defects and spatiotemporal chaos in a Reaction-Diffusion system, *Phys. Rev. Lett.* **75**, pp. 1503–1506.

Huang, Y., Wang, Y. and Xiao, J. (2009). Generalized lag-synchronization of continuous chaotic system, *Chaos Solit. Fract.* **40**, pp. 76670.

Huang, L., Chen, Q. F., Lai,Y. C. and Pecora, L. M. (2009). Generic behavior of master-stability functions in coupled nonlinear dynamical systems, *Phys. Rev. E* **80**, pp. 036204.

Hwang, D., Chavez, M., Amann, A. and Boccaletti, S. (2001). Synchronization in complex networks with age ordering, *Phys. Rev. Lett.* **2005**, pp. 138701.

Laoye, J. A., Vincent, U. E. and Kareem, S. O. (2009). Chaos control of 4D chaotic systems using recursive backstepping nonlinear controller, *Chaos Solit. Fract.* **39**, pp. 356–362.

Li, C., Liao, X. and Wong, K. (2004). Chaotic lag-synchronization of coupled time-delayed systems and its applications in secure communication, *Phys. D* **194**, pp. 187202.

Li, G. H. (2007). Modified projective synchronization of chaotic system, *Chaos Solit. Fract.* **32** , pp. 178690.

Li, Z. and Lee, J. (2007). New eigenvalue based approach to synchronization in asymmetrically coupled networks, *Chaos* **17**, pp. 043117.

Li, W. L. and Chang, K. M. (2009). Robust synchronization of drive-response chaotic systems via adaptive sliding mode control, *Chaos Solit. Fract.* **39**, pp. 208692.

Lin, D., Wang, X. Y., Nian, F. Z. and Zhang, Y. L. (2010). Dynamic fuzzy neural networks modeling and adaptive backstepping tracking control of uncertain chaotic systems, *Neurocomputing* **73**, pp. 2873–2881.

Lü, L. and Meng, L. (2011). Parameter identification and synchronization of spatiotemporal chaos in uncertain complex network, *Nonlinear Dyn.* **66**, pp. 489–495.

Mascolo, S. and Grassi, G. (1997). Controlling chaos via backstepping design, *Phys. Rev. E* **56**, pp. 6166–6169.

Nana, B., Woafo, P. and Domngang, S. (2009). Chaotic synchronization with experimental application to secure communications, *Commun. Nonlinear Sci. Numer. Simulat.* **14**, pp. 226676.

Newman, M. E. J. and Watts, D. J. (1999). Renormalization group analysis of the small-world network model, *Phys. Lett. A* **263**, pp. 341–346.

Njah, A. N., Ojo, K. S., Adebayo, G. A. and Obawole, A. O. (2010). Generalized control and synchronization of chaos in RCL-shunted Josephson junction using backstepping design, *Phys. C* **470**, pp. 558–564.

Panfilov, A. and Hogeweg, P. (1993). Spiral breakup in a modified FitzHugh-Nagumo model, *Phys. Lett. A* **176**, pp. 295–299.

Park, J. H. (2008). Adaptive control for modified projective synchronization of a four-dimensional chaotic system with uncertain parameters, *J Comput. Appl. Math.* **213**, pp. 28893.

Pearson, J. E. (1993). Complex patterns in a simple system, *Science* **261**, pp. 189–192.

Pecora, L. M. (1998). Synchronization conditions and desynchronizing patterns in coupled limit-cycle and chaotic systems, *Phys. Rev. E* **58**, pp. 347–360.

Pecora, L. M. and Carroll, T. L. (1998). Master stability functions for synchronized coupled systems, *Phys. Rev. Lett.* **80**, pp. 2109–2112.

Pishkenari, H. N., Jalili, N., Mahboobi, S. H., Alasty, A. and Meghdari, A. (2010). Robust adaptive backstepping control of uncertain Lorenz system, *Chaos* **20**, pp. 023105.

Polyakov, A. and Poznyak, A. (2011). Invariant ellipsoid method for minimization of unmatched disturbances effects in sliding mode control, *Automatica* **47**, pp. 1450–1454.

Roopaei, M., Sahraei, B. R. and Lin, T. C. (2010). Adaptive sliding mode control in a novel class of chaotic systems, *Commun. Nonlinear Sci. Numer. Simulat.* **15**, pp. 4158–4170.

Shahverdiev, E. M., Sivaprakasam, S. and Shore, K. A. (2002). Lag-synchronization in time-delayed systems, *Phys. Lett. A* **292**, pp. 320–324.

Sharma, B. B. and Kar, I. N. (2009). Contraction theory based adaptive synchronization of chaotic systems, *Chaos Solit. Fract.* **41**, pp. 2437–2447.

Silva, J. A. L., Barrionuevo, J. A. and Giordani, F. T. (2010). Synchronism in population networks with non linear coupling, *Nonlinear Anal. Real World Appl.* **11**, pp. 1005–1016.

Sinha, S., Pande, A. and Pandit, R. (2001). Defibrillation via the Elimination of Spiral Turbulence in a Model for Ventricular Fibrillation, *Phys. Rev. Lett.* **86**, pp. 3678–3681.

Sorrentino, F. and Ott, E. (2008). Adaptive synchronization of dynamics on evolving complex networks, *Phys. Rev. Lett.* **100** , pp. 114101.

Strogatz, S. H. (2001). Exploring complex networks, *Nature* **410**, pp. 268–276.

Tang, Y. and Fang, J. A. (2008). General methods for modified projective synchronization of hyperchaotic systems with known or unknown parameters, *Phys Lett A* **372**, pp. 181626.

Tavazoei, M. S. and Haeri, M. (2007). Determination of active sliding mode controller parameters in synchronizing different chaotic systems, *Chaos Solit. Fract.* **32**, pp. 58391.

Wang, X. F. and Chen, G. R. (2002). Pinning control of scale-free dynamical networks, *Physica A* **310**, pp. 521–531.

Watts, D. J. and Strogatz, S. H. ((1998). Collective dynamics of 'small-world' networks, *Nature* **393**, pp. 440–442.

Wen, G. L. (2011). Designing Hopf limit circle to dynamical systems via modified projective synchronization, *Nonlinear Dyn.* **63**, pp. 38793.

Yu, Y. G. and Li, H. X. (2011). Adaptive hybrid projective synchronization of uncertain chaotic systems based on backstepping design, *Nonlinear Analysis: Real World Applications* **12**, pp. 388–393.

Zhang, H., Ma, X. K. and Liu, W. Z. (2004). Synchronization of chaotic systems with parametric uncertainty using active sliding mode control, *Chaos Solit. Fract.* **21**, pp. 124957.

Chapter 14

Chaotic dynamical systems on fractals and their applications to image encryption

Ruisong Ye[a,*], Yuru Zou[b] and Jian Lu[b]

[a] *Department of Mathematics, Shantou University, Shantou 515063, Guangdong, PR China*

[*] *rsye@stu.edu.cn*

[b] *College of Mathematics and Computational Science, Shenzhen University, Shenzhen 518060, Guangdong, PR China*

Thanks to the exceptionally good features of ergodicity, high sensitivity to initial conditions and control parameters, chaotic dynamical systems have good potential for image encryption. In this chapter, we will explore the applications of chaotic dynamical systems on fractals in digital image encryption. We will first introduce iterated function systems and the associated shift dynamical systems. The chaotic dynamics of the shift dynamical systems are then analyzed theoretically and numerically. A novel chaos-based image encryption scheme consisting of one permutation process and one diffusion process is finally proposed. In the permutation process, two shift dynamical systems are utilized to generate two chaotic orbits used to permute image pixel positions totally, while in the diffusion process, another two shift dynamical systems are employed to yield two pseudo-random gray value sequences for a two-way diffusion of pixel gray values. Experimental results show that the new image encryption scheme has satisfactory security thanks to its large key space and robust permutation-diffusion mechanism, which makes it a potential candidate for designing image encryption schemes.
Keywords: Chaotic dynamical system on fractal, image encryption, permutation, diffusion

14.1 Introduction

Chaos theory studies the behavior of dynamical systems that are highly sensitive to initial conditions; it attempts to explain the fact that complex and unpredictable results can and will occur in systems sensitive to their initial conditions. This sensitivity is popularly referred to as the butterfly effect. A common example would be fluttering butterfly wings in China could actually affect weather patterns in New York City. In other words, it is possible that a very small occurrence can produce unpredictable and sometimes drastic results by triggering a series of increasingly significant events. Stated differently, small differences in initial conditions (such as those due to rounding errors in numerical computation) can drastically change the long-term behavior of a system, and therefore rendering long-term prediction impossible in general. Many examples around us have been shown to possess chaotic properties, such as the dripping of a tap, the formation of clouds, the weather, the fibrillation of the human heart, and the movement of a simple pendulum under the influence of a number of magnets. The natural world has often had a chaotic way about it and the mathematical world has always been amazingly complex. Chaotic systems appear to be random, but they are deterministic, meaning that their future dynamics are fully determined by their initial conditions, with no random elements involved. These systems are very sensitive to initial conditions making the system fairly unpredictable but the systems have a sense of order and pattern [Gleick (1987); Stewart (1989); Devaney (2003); Peitgen (2004)]. Thanks to the rapid development computer technology, chaos theory has been evolved as a critical part of science, mathematics, art and the computing world. There is a strong link between chaos and fractal geometry, the geometry that describes the chaotic systems found in nature.

A fractal set is a rough or fragmented geometric shape that can be split into parts, each of which is (at least approximately) a reduced-size copy of the whole, a property called self-similarity. Roots of mathematical interest in fractals can be traced back to the late 19th century; however, the term "fractal" was coined by mathematician B. Mandelbrot in 1975 and was derived from the Latin fractus meaning "broken" or "fractured". A mathematical fractal is based on an equation that undergoes iteration, a form of feedback based on recursion. Fractal geometry is the geometry which is more suitable to describe the chaotic systems we find in nature. Most physical systems of nature and many human artifacts are not regular geometric shapes of the standard geometry derived from Euclid. Although Eulcidean geometry can easily define objects with smooth geometry, they are pretty useless if you want to model complex objects like clouds, grass or plants, since those possess infinitely non smooth, highly structured geometries. Fractal geometry offers almost unlimited ways of describing, measuring and predicting these natural phenomena. Many people are fascinated by the beautiful images termed fractals. Extending beyond the typical perception of mathematics as a body of complicated, boring formulas, fractal geometry mixes art with mathematics to demonstrate that equations are more than just a collection of numbers. What makes fractals even more interesting and applicable is that they are the best existing mathematical descriptions of many natural forms, such as coastlines, mountains, clouds, plants or parts of living organisms [Mandelbrot (1982); Barnsley and Demko (1985); Peitgen

(2004)].

Actually fractal geometry and chaos theory are strongly relative. The link between chaos and fractals needs to be investigated extensively and intensively. In this chapter, we will introduce iterated function system, or IFS for short, and explore the shift dynamical system associated with IFS. IFS is one of the common techniques to generate fractals [Hutchinson (1981); Demko et al. (1985); Barnsley (1993)]. IFS plays an important role in fractal geometry and has been found to have potential in different fields of physics, engineering, biology, geology, finance, etc. It also has been shown good performances in computer graphics and image processing, especially in the simulation of nature landscapes and digital image compression [Jacquin (1992); Fisher (1994)]. In this chapter, the chaotic dynamics of the shift dynamical systems associated with IFS are analyzed theoretically and numerically. For the first time, the derived shift dynamical systems associated with IFS are applied to constructing image encryption schemes, which may be regarded as a new application aspect of the dynamical systems on fractals.

With the rapid developments in digital image processing and network communication in the past decades, electronic publishing and wide-spread dissemination of digital multimedia data have been communicated over the Internet and wireless networks. Therefore it has become urgent to prevent them from leakages. Many applications, such as military image databases, confidential video conference, medical imaging system, online private photograph album, etc. require reliable, fast and robust secure system to store and transmit digital images. The requirements to fulfill the security needs of digital images have led to the development of effective image encryption algorithms. Digital images possess some intrinsic features, such as bulk data capacity, redundancy of data, strong correlation among adjacent pixels, being less sensitive as compared to the text data, etc. As a result, traditional encryption algorithms, such as DES, RSA [Schneier (1995)], are thereby not suitable for practical digital image encryption due to the weakness of low-level efficiency while encrypting images. Fortunately, chaos-based image encryption algorithms have shown their superior performance. Chaos consistently plays an active role in modern cryptography thanks to its ergodicity, unpredictability, pseudo-randomness, high sensitivity to initial conditions and control parameters, which are close to confusion and diffusion in cryptography. These properties make chaotic systems a potential choice for constructing cryptosystems [Matthews (1989); Fridrich (1998); Chen et al. (2004); Mao et al. (2004); Guan and Huang (2005); Lian et al. (2005); Wong et al. (2008); Patidar et al. (2009)].

Since Matthews [Matthews (1989)] firstly used a chaotic system to design a cryptographic algorithm in 1989, many chaos-based digital image encryption schemes have been proposed in the literature [Baptista (1998); Koca (2001); Guan and Huang (2005); Lian et al. (2005); Pareek at al. (2006); Tong and Cui (2008); Wong et al. (2008); Patidar et al. (2009)]. Chaotic maps can simulate random behavior in a quite impressive way. In particular, chaotic maps are easy to be implemented by microprocessors and personal computers. Therefore, chaotic cryptosystems generally have high speed with low cost, which makes them better candidates than many traditional ciphers for multimedia data encryption. Among the chaos-based encryption schemes, one-dimensional and two-dimensional chaotic systems, such as logistic map [Matthews (1989); Pareek at al. (2006)], skew tent map[Zhang and Liu (2011)], Arnold map [Chen et al. (2004); Ye (2011)], baker map [Fridrich (1998); Mao et al. (2004)], piecewise linear chaotic map [Masuda and Aihara (2002); Huang and Guan (2005); Liu and Wang (2011)], and standard map [Lian et al. (2005); Wong et al. (2008); Patidar et al. (2009)], were applied widely owing to the advantage of simple implementation. However, there are fundamental drawbacks in these chaotic systems, such as small key space, slow performance speed and weak security. As a matter of fact, some chaos-based image encryption algorithms have been broken recently [Li and Zheng (2002); Alvarez (2003, 2006); Li et al. (2009); Xiao et al. (2009); Liu and Qu (2010); Rhouma et al. (2010)].

To overcome the aforementioned drawbacks, in this essay, a novel chaos-based image encryption scheme with an efficient permutation-diffusion architecture is proposed. In both the permutation process and the diffusion process, shifted dynamical systems associated with IFS are utilized. First, the permutation process employs two chaotic shift dynamical systems associated with IFS to generate two chaotic orbits $\{x_k, k = 0, 1, \cdots\}$ of x_0 and $\{y_k, k = 0, 1, \cdots\}$ of y_0 with given control parameters $a_i(i = 1, \cdots, N_1 - 1)$, $b_j(j = 1, \cdots, N_2 - 1)$, where $a_i(i = 1, \cdots, N_1)$, $b_j(j = 1, \cdots, N_2 - 1)$ are the coefficients of the shift dynamical systems associated with IFS. $\{x_k, k = 1, \cdots, H\}$, $\{y_k, k = 1, \cdots, W\}$ (H and W are the width and the height of the processed image respectively) are then sorted to yield two index order sequences Ix, Iy applied to permuting the image pixel positions totally. To improve the diffusion effect, a two-way diffusion process is presented, where another two shift dynamical systems associated with IFS, whose control parameters are $c_i(i = 1, \cdots, N_3 - 1)$, $d_j(j = 1, \cdots, N_3 - 1)$, are utilized to generate two pseudo-random gray value sequences. The two sequences are then used to modify the pixel gray values sequentially. The yielded gray value sequences are not only sensitive to the control parameters and initial conditions of the considered shift dynamical systems associated with IFS, but also strongly depend on the plain-image processed, therefore the proposed scheme can resist statistical attack, differential attack, known-plaintext attack as well as chosen-plaintext attack. The proposed image encryption scheme possesses a huge key space, therefore efficiently frustrating brute-force attack. If N_1, N_2, N_3, N_4 are all set to be 6 for all the four shift dynamical systems associated with IFS, the key space will be 10^{384}. As a matter of fact, the key space will become 10^{16} times larger if each of $N_i(i = 1, \cdots, 4)$ increased by 1 for one shift dynamical system associated with IFS. The proposed image encryption scheme promotes the key space significantly.

The rest of the chapter is organized as follows. In Section 14.2, some related preliminaries on iterated function system and shift dynamical system associated with

IFS, like definitions and theorems, etc. are introduced. A special shift dynamical system associated with IFS is constructed and its chaotic properties are analyzed in Section 14.3. Section 14.4 proposes a novel image encryption scheme composed of one permutation process and one diffusion process based on shift dynamical systems associated with IFS. The security of the proposed scheme is evaluated via detailed analysis and experiments as well. Some conclusions are drawn in Section 14.5

14.2 Chaotic dynamical systems on fractals

14.2.1 *Iterated function systems*

Let (X, d) be a metric space and let $(H(X), h(d))$ denote the corresponding space of nonempty compact subsets from (X, d), with the Hausdorff metric $h(d)$. Barnsley defined an iterated function system in the following way [Barnsley (1993)]:

Definition 14.1. A (hyperbolic) iterated function system consists of a complete metric space (X, d) together with a finite set of contraction mappings $w_n : X \to X$, with respective contractivity factor s_n, for $n = 1, 2, \cdots, N$. The notation for the IFS just announced is $\{X; w_n, n = 1, 2, \cdots, N\}$ and its contractivity factor is $s = \max\{s_n, n = 1, 2, \cdots, N\}$.

The following theorem is extremely important and suggests an algorithm for computing the pre-fractals.

Theorem 14.1. *Let $\{X; w_1, w_2, \cdots, w_N\}$ be a hyperbolic iterated function system with contractivity factor s. We define $W(B) = \bigcup_{i=1}^{N} w_i(B), \forall B \in H(X)$. Then the following statements hold:*

a) $W : H(X) \to H(X)$ is a contraction mapping on the complete metric space $(H(X), h(d))$ with contractivity factor s.

b) Its unique fixed point $A \in H(X)$ obeys $A = W(A) = \bigcup_{i=1}^{N}(w_i(A))$ given by

$$A = \lim_{k \to \infty} W^k(B), \quad \forall B \in H(X).$$

The proof can be found in [Barnsley (1993)].

Definition 14.2. The fixed point $A \in H(X)$ as described in Theorem 14.1 is called the attractor of the IFS.

Barnsley presented two algorithms for computing fractals from iterated function systems. One is the Deterministic Algorithm, the other is the Random Iteration Algorithm. The Deterministic Algorithm is based on the idea of directly computing a sequence of sets $\{B_n = W^n(B_0)\}$ starting from an initial set B_0. As $n \to \infty$, the set $W^n(B_0)$ for any $B_0 \in H(X)$ will give us better and better approximations to the attractor A of the IFS. The Random Iteration Algorithm is founded in ergodic theory. The Deterministic Algorithm will usually take more time to produce a fractal pattern than the Random Iteration Algorithm.

The Deterministic Algorithm: Let $\{X; w_1, w_2, \cdots, w_N\}$ be a hyperbolic IFS. Choose any compact set $B_0 \in H(X)$. Then compute successively $\{B_n = W^n(B_0)\}$ according to

$$B_{n+1} = \bigcup_{i=1}^{N} w_i(B_n), \text{for } n = 0, 1, \cdots .$$

Then by Theorem 14.1, the constructed sequence $\{B_n, n = 0, 1, \cdots\} \subset H(x)$ will converge to the attractor of the IFS.

The Random Iteration Algorithm: Let $\{X; w_1, w_2, \cdots, w_N\}$ be a hyperbolic IFS, where probability $p_i > 0$ has been assigned to w_i for $i = 1, 2, \cdots, N$ and such that $\sum_{i=1}^{N} p_i = 1$. Choose $x_0 \in X$ and then choose recursively, independently,

$$x_n \in \{w_1(x_{n-1}), w_2(x_{n-1}), \cdots, w_N(x_{n-1})\}, \text{for } n = 1, 2, \cdots ,$$

where the probability of the event $x_n = w_i(x_{n-1})$ is p_i. Thus, construct a sequence $\{x_n : n = 0, 1, 2, \cdots\} \subset X$, which also converges to the attractor of the IFS.

14.2.2 *Chaotic dynamical systems on fractals*

We introduce the idea of a dynamical system and some of the associated terminologies related to dynamical systems on fractals of IFS.

Definition 14.3. A dynamical system is a transformation $f : X \to X$ on a metric space (X, d). It is denoted by $\{X; f\}$. The orbit of a point $x \in X$ is the sequence $\{f^n(x)\}_{n=0}^{\infty}$.

Dynamical systems are actually sources of deterministic chaos and fractals. The reasons for this deeply intertwined with IFS theory. In the sequel, we will introduce a special type of dynamical system, called the shift dynamical system associated with IFS. By studying the orbits of these systems, we will learn more about fractals. For example, we can see why the Random Iteration Algorithm successfully generates the images of attractors of IFS, and why chaos and fractals generally occur together.

Definition 14.4. Let $\{X; w_1, w_2, \cdots, w_N\}$ be a hyperbolic IFS. The code space associated with the IFS, (Σ, d_C) is defined to be the code space on N symbols $\{1, 2, \cdots, N\}$, with the metric d_C given by

$$d_C(\omega, \sigma) = \sum_{n=1}^{\infty} \frac{|\omega_n - \sigma_n|}{(N+1)^n} \quad \text{for all } \omega = \omega_1\omega_2\cdots, \sigma = \sigma_1\sigma_2\cdots \in \Sigma.$$

There exists a relationship between the code space associated with the IFS and the attractor of the IFS. In more details, we can construct a continuous transformation ϕ to realize the relationship. The following theorem states the construction detail of ϕ [Barnsley (1993)].

Theorem 14.2. *Let (X, d) be a complete metric space. Let $\{X; w_1, w_2, \cdots, w_N\}$ be a hyperbolic IFS. Let A denote the attractor of the IFS and let (Σ, d_C) denote the code space associated with IFS. For each $\sigma \in \Sigma, k \in \mathbb{N}$, and $x \in X$, let*

$$\phi(\sigma, k, x) = w_{\sigma_1} \circ w_{\sigma_2} \circ \cdots w_{\sigma_k}(x).$$

Then

$$\phi(\sigma) = \lim_{k \to \infty} \phi(\sigma, k, x)$$

exists, belonging to A, and is independent of $x \in X$. If K is a compact subset of X, then the convergence is uniform over $x \in K$. The function $\phi : \Sigma \to A$ is continuous and onto.

The following definition is about the classification of IFS.

Definition 14.5. Let $\{X; w_1, w_2, \cdots, w_N\}$ be a hyperbolic IFS with attractor A and associated code space (Σ, d_C). Let $\phi : \Sigma \to A$ be the continuous function from the code space onto the attractor of the IFS constructed in Theorem 14.2. An address of a point $a \in A$ is any member of the set

$$\phi^{-1}(a) = \{\omega \in \Sigma : \phi(\omega) = a\}.$$

This set is called the set of addresses of $a \in A$. The IFS is said to be totally disconnected if each point of its attractor possesses a unique address. The IFS is said to be just-touching if it is not totally disconnected yet its attractor contains an open set Θ such that

(i) $w_i(\Theta) \cap w_j(\Theta) = \emptyset, \forall i, j \in \{1, 2, \cdots, N\}$ with $i \neq j$;

(ii) $\displaystyle\bigcup_{i=1}^{N} w_i(\Theta) \subset \Theta$.

An IFS whose attractor obeys (i) and (ii) is said to obey the open set condition. The IFS is said to be overlapping if it is neither just-touching nor disconnected.

Theorem 14.3. *Let $\{X; w_1, w_2, \cdots, w_N\}$ be a hyperbolic IFS with attractor A. The IFS is totally disconnected if and only if*

$$w_i(A) \cap w_j(A) = \emptyset, \forall i, j \in \{1, 2, \cdots, N\} \text{ with } i \neq j.$$

We are ready to introduce the dynamical systems on fractals. The following lemma is needed.

Lemma 14.1. *Let $\{X; w_1, w_2, \cdots, w_N\}$ be a hyperbolic IFS with attractor A. Assume that the IFS is totally disconnected. Then for each $i \in \{1, 2, \cdots, N\}$, the transformation $w_i : A \to A$ is one-to-one.*

It implies from Lemma 14.1 that the following definition is well-defined.

Definition 14.6. Let $\{X; w_1, w_2, \cdots, w_N\}$ be a totally disconnected hyperbolic IFS with attractor A. The associated shift transformation on A is the transformation $S : A \to A$ defined by

$$S(a) = w_i^{-1}(a) \quad \text{for } a \in w_i(A), \tag{14.1}$$

where w_i is viewed as a transformation on A. The dynamical system $\{A; S\}$ is called the shift dynamical system associated with IFS.

There is a well-known dynamical system in the code space, namely the shift dynamical system $\{\Sigma; T\}$ with $T : \Sigma \to \Sigma$ defined by

$$T(\sigma_1\sigma_2\sigma_3\cdots) = \sigma_2\sigma_3\cdots, \forall\sigma = \sigma_1\sigma_2\sigma_3\cdots \in \Sigma.$$

The following theorem reports the relationship between the shift dynamical system $\{\Sigma; T\}$ and the shift dynamical system $\{A; S\}$ associated with IFS.

Theorem 14.4. *Let $\{X; w_1, w_2, \cdots, w_N\}$ be a totally disconnected hyperbolic IFS with attractor A. Let $\{A; S\}$ be the associated shift dynamical system. Let Σ be the associated code space of N symbols and let $\{\Sigma; T\}$ be the shift dynamical system on Σ. Then the two dynamical systems $\{A; S\}$ and $\{\Sigma; T\}$ are equivalent in the sense that*

$$\phi \circ T(\sigma) = S \circ \phi(\sigma), \forall\sigma \in \Sigma,$$

where the homeomorphism ϕ is defined in Theorem 14.2.

Theorem 14.4 is very useful for us to prove the dynamical properties of the shift dynamical system $\{A; S\}$ associated with a totally disconnected hyperbolic IFS. The reason lies in that $\{A; S\}$ is equivalent to the shift dynamical system $\{\Sigma; T\}$. As a matter of fact, this equivalent relationship means that the two systems have a number of properties in common; for example, $\{a_1, a_2, \cdots, a_p\}$ is a repulsive cycle of period p for S if, and only if, $\{\phi^{-1}(a_1), \phi^{-1}(a_2), \cdots, \phi^{-1}(a_p)\}$ is a repulsive cycle of period p for T, etc. A particularly important property that they share is that they are both chaotic dynamical systems. The following definitions are related to chaos.

Definition 14.7. Let (X, d) be a metric space. A subset $B \subset X$ is said to be dense in X if the closure of B equals X. A sequence $\{x_n\}_{n=0}^{\infty}$ of points in X is said to be dense in X if, for each point $a \in X$, there is a subsequence $\{x_{t_n}\}_{n=0}^{\infty}$ that converges to a. In particular an orbit $\{x_n\}_{n=0}^{\infty}$ of a dynamical system $\{X; f\}$ is said to be dense in X if the sequence is dense in X.

Definition 14.8. A dynamical system $\{X; f\}$ is transitive if, wherever U and V are open subsets of the metric space (X, d), there exists a finite integer n such that $U \cap f^n(V) \neq \emptyset$.

Definition 14.9. A dynamical system $\{X; f\}$ is said to be sensitive to initial conditions if there exists $\delta > 0$ such that, for any $x \in X$ and any ball $B(x, \epsilon)$ with radius $\epsilon > 0$, there is $y \in B(x, \epsilon)$ and an integer $n \geq 0$ satisfying $d(f^n(x), f^n(y)) > \delta$.

Definition 14.10. A dynamical system $\{X; f\}$ is chaotic if (a) it is transitive; (b) it is sensitive to initial conditions; (c) the set of periodic orbits of f is dense in X.

It is easy to show that the dynamical system $\{\Sigma; T\}$ is chaotic.

Theorem 14.5. *The dynamical system $\{\Sigma; T\}$ is chaotic.*

It follows from Theorem 14.4 and Theorem 14.5 that the shift dynamical system $\{A; S\}$ associated with a totally disconnected hyperbolic IFS is chaotic as well.

Theorem 14.6. *The shift dynamical system $\{A; S\}$ associated with a totally disconnected hyperbolic IFS consisting of two or more transformations is chaotic.*

This chaotic property makes the shift dynamical system $\{A; S\}$ play an important role in modern cryptography as we will show in the next section. We note that, in application, we pay more attentions to some hyperbolic IFS which are just-touching. The chaotic property is also valid for this type of IFS. We will present a special just-touching hyperbolic IFS which is acting on $[0, 1]$ and with the attractor $[0, 1]$. Although the associated shift dynamical system is simple and actually it is a piece-wise linear transformation acting on $[0, 1]$, it is very useful to construct image encryption scheme.

14.3 A special shift dynamical system associated with IFS

We consider the following IFS acting on the unit interval $X = [0, 1]$:

$$w_i(x) = p_i x + a_{i-1}, i = 1, \cdots, N, \tag{14.2}$$

where $0 = a_0 < a_1 < \cdots < a_{N-1} < a_N = 1$, and $p_i = a_i - a_{i-1}, i = 1, 2, \cdots, N$.

This IFS is hyperbolic with the contractivity $s = \max\{p_1, \cdots, p_N\} < 1$ and therefore there exists a unique attractor. It is easy to show that the attractor A is exactly the unit interval $[0, 1]$ in the sense that

$$A = \bigcup_{i=1}^{N} w_i(A).$$

The IFS is just-touching as $w_i(A)$ and $w_{i+1}(A)$ possess one intersection point a_i. We note that we can also well define the shift dynamical system with a minor revision to (14.1) even if the IFS is not totally disconnected. One simple way is to assign only one inverse map w_i^{-1} acting on the intersection region. In the just-touching case, the assignment of w_i^{-1} just happens at the intersection points $a_i (i = 1, 2, \cdots, N-1)$. As for the considered IFS (14.2), it is easy to define the shift dynamical system:

$$S(x) = \begin{cases} w_i^{-1}(x) & \text{for } x \in [a_{i-1}, a_i), i = 1, 2, \cdots, N-1, \\ w_N^{-1}(x) & \text{for } x \in [a_{N-1}, a_N]. \end{cases} \tag{14.3}$$

The revised shift dynamical system (14.3) is then deterministic. Indeed it is just the so called piece-wise linear map acting on $[0, 1]$. As $N = 2$ and $a_1 = 1/2$, the yielded shift dynamical system is the well-know Bernoulli shift map $B_0 : [0, 1] \to [0, 1]$ given by

$$B_0(x) = 2x \bmod 1 = \begin{cases} 2x, & \text{if } x \in [0, 1/2), \\ 2x - 1, & \text{if } x \in [1/2, 1]. \end{cases} \tag{14.4}$$

The Bernoulli shift map (14.4) yields a simple example for an essentially nonlinear stretch-and-cut mechanism, as it typically generates deterministic chaos. Such basic mechanisms are also encountered in more realistic dynamical systems. We may note that 'stretch and fold' or 'stretch, twist and fold' provide alternative mechanisms for generating chaotic behavior.

For general N and $a_i (i = 1, \cdots, N - 1)$, the yielded shift dynamical system becomes

$$S(x) = \begin{cases} (x - a_{i-1})/(a_i - a_{i-1}), \text{if } x \in [a_{i-1}, a_i), i = 1, \cdots, N - 1, \\ (x - a_{N-1})/(a_N - a_{N-1}), \text{if } x \in [a_{N-1}, a_N]. \end{cases} \tag{14.5}$$

The transformation (14.5) is not continuous at the $N - 1$ points $a_1, \cdots, a_{N-1}$ and linear on the interval $[a_{i-1}, a_i)(i = 1, \cdots, N)$ with slope $1/(a_i - a_{i-1}) > 1$. See Fig. 14.1 for the case of $N = 6$. We shall call map (14.5) a shift dynamical system associated with IFS hereafter.

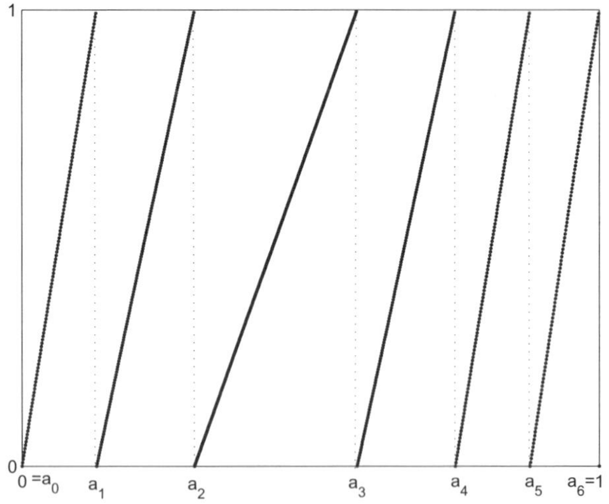

Fig. 14.1 The diagram of a shift dynamical system associated with IFS.

A typical orbit of x_0 derived from the shift dynamical system associated with IFS (14.5) is $\{x_k = S^k(x_0), k = 0, 1, \cdots\}$, which is shown in Fig. 14.2(a) for $a = [0\ 0.23\ 0.41\ 0.62\ 0.75\ 0.89\ 1.0], x_0 = 0.631$. Its waveform is quite irregular and indicates that the system is chaotic. The distribution of the points $\{x_k : k = 0, 1, \cdots, 20000\}$ of a typical orbit of length 20000 is represented by the histogram of Fig. 14.2(b). It can be seen that the points of the orbit spread more or less evenly over the unit interval in the course of time. This property will also be proved in the sequel. The shift dynamical system associated with IFS also possesses desirable auto-correlation and cross-correlation features. The iterated trajectory is used to calculate the correlation coefficients, which are shown in Figs. 14.2(c)-(d) respectively. The cross-correlation coefficients are calculated by the orbits of $x_0 = 0.631$ and $y_0 = 0.632$. The control parameter $a_1, \cdots, a_{N-1}$ and the initial condition x_0 can be regarded as cipher keys if the shift dynamical system associated with IFS is employed to design image encryption schemes.

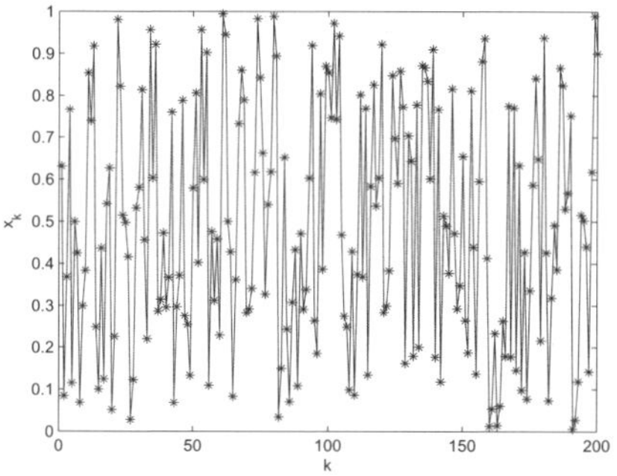

(a) The chaotic orbit of $x_0 = 0.631$

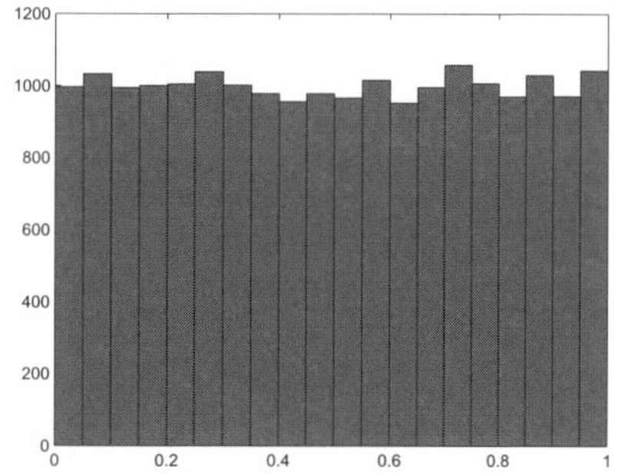

(b) Histogram of the points of a typical orbit of length 20000

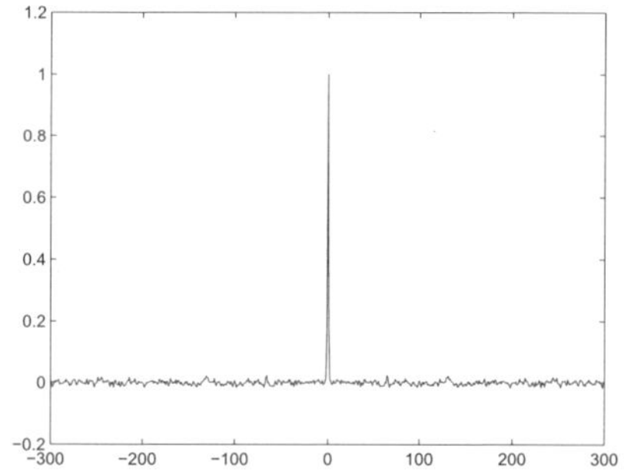

(c) The auto-correlation

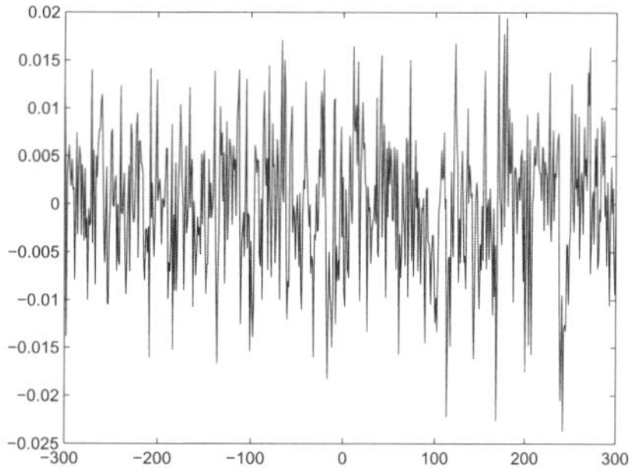

(d) The cross-correlation of two different orbits

Fig. 14.2 Orbits of the shift dynamical system associated with the IFS (14.2) with $a = [0\ 0.23\ 0.41\ 0.62\ 0.75\ 0.89\ 1.0]$.

Definition 14.11. A probability density function $\rho(x)$ on [0, 1] is invariant, if for each interval $[c, d] \subset [0, 1]$,

$$\int_c^d \rho(x)dx = \int_{f^{-1}([c,d])} \rho(x)dx, \tag{14.6}$$

where $f^{-1}([c, d]) = \{c \leq f(x) \leq d\}$.

As $p_i = a_i - a_{i-1}, i = 1, \cdots, N$, the unit interval is divided into N subinterval with length $p_i (i = 1, \cdots, N)$ such that $\sum_{i=1}^{N} p_i = 1$. It is not difficult to see that for the shift dynamical system (14.5), (14.6) amounts to the following equation

$$\begin{aligned}\rho(x) = p_1\rho(p_1x) + p_2\rho(a_1 + p_2x) + p_3\rho(a_2 + p_3x) + p_4\rho(a_3 + p_4x) + \cdots \\ +p_{N-1}\rho(a_{N-2} + p_{N-1}x) + p_N\rho(a_{N-1} + p_Nx).\end{aligned} \tag{14.7}$$

It is easy to see that $\rho(x) \equiv 1$ satisfies (14.7) and $\rho(x) \equiv 1$ is the unique solution to (14.7) according to the uniqueness of invariant probability density function [Hasler and Maistrenko (1997)]. It then follows from Birkhoff's ergodic theorem [Billingsley (1965)] that the the asymptotic distribution of the points of almost every trajectory is uniform. This fact has been illustrated by Fig. 14.2(b). The existence and unique value of the Lyapunov exponent also follows from the following theorem [Hasler and Maistrenko (1997)].

Theorem 14.7. *Suppose $f : [0, 1] \to [0, 1]$ is continuously differentiable except for a finite number of points. Let $\rho(x)$ be the unique invariant probability density of the function. Then for (Lebesgue-) almost all initial conditions, the Lyapunov exponent exists and has the value*

$$\lambda = \int_0^1 \ln|f'(x)|\rho(x)dx. \tag{14.8}$$

For the shift dynamical system (14.5) with the constant probability density $\rho(x) \equiv 1$, formula (14.8) becomes

$$\begin{aligned}\lambda &= \int_0^{a_1} \ln|\frac{1}{p_1}|dx + \int_{a_1}^{a_2} \ln|\frac{1}{p_2}|dx + \cdots + \int_{a_{N-1}}^{a_N} \ln|\frac{1}{p_N}|dx \\ &= -p_1 \ln p_1 - p_2 \ln p_2 - \cdots - p_N \ln p_N.\end{aligned}$$

It implies that λ is always larger than zero and the dynamical system is always chaotic. For $N = 6, a = [0\ 0.23\ 0.41\ 0.62\ 0.75\ 0.89\ 1.0]$, we obtain $p_1 = 0.23, p_2 = 0.18, p_3 = 0.21, p_4 = 0.13, p_5 = 0.14, p_6 = 0.11$, so $\lambda = 1.7577$. It is usually larger than the Lyapunov exponent $\ln 2 = 0.6931$ for the Bernoulli shift map (14.4). It implies that the shift dynamical system (14.5) is in a stronger sense chaotic, and therefore can perform better data mixing, which makes it a better choice for designing encryption schemes. We will propose an efficient image encryption scheme based on the chaotic shift dynamical system associated with IFS in Section 14.4.

14.4 The image encryption scheme based on the shift dynamical system associated with IFS

14.4.1 *Permutation process*

In this subsection, we propose a permutation process to confuse plain-image totally. Thanks to the chaotic nature of shift dynamical system associated with IFS on the unit interval $[0,1]$, one can easily get the chaotic orbit $\{x_k, k = 0, 1, \cdots\}$ of x_0 with given control parameters $a_i(i = 1, \cdots, N_1 - 1)$. We rearrange all the x_k values of the orbit to get a new sequence $\{\bar{x}_k, k = 0, 1, \cdots\}$ according to the order from small to large. As a result, we also get an index order number for every x_k and an index sequence Ix is generated. Similarly, another chaotic orbit $\{y_k, k = 0, 1, \cdots\}$ of y_0 with given control parameters $b_i(i = 1, \cdots, N_2 - 1)$ is generated and used to yield another index sequence Iy. The index order number sequences Ix, Iy can be applied to permuting the image pixel positions and therefore can confuse the image to get a shuffled image. The permutation process is stated as follows.

Step 1. Set the values of the control parameters $a_i(i = 1, \cdots, N_1 - 1)$, $b_j(j = 1, \cdots, N_2 - 1)$ and the initial conditions x_0, y_0. In the experiments, we set $a = [0\ 0.23\ 0.41\ 0.62\ 0.75\ 0.89\ 1.0], x_0 = 0.631$, $b = [0\ 0.17\ 0.33\ 0.5\ 0.68\ 0.83\ 1.0], y_0 = 0.728$.

Step 2. Iterate the shift dynamical systems (14.5) constructed by $a_i(i = 1, \cdots, N_1 - 1)$ and $b_j(j = 1, \cdots, N_2 - 1)$, respectively, to get the truncated orbit $\{x_k, k = 1, \cdots, H\}$ and $\{y_k, k = 1, \cdots, W\}$ where H, W are the height and the width of the processed image respectively.

Step 3. Sort $\{x_k, k = 1, \cdots, H\}$ to get one index order sequence $\{Ix(k), k = 1, \cdots, H\}$ and sort $\{y_k, k = 1, \cdots, W\}$ to get another index order sequence $\{Iy(k), k = 1, \cdots, W\}$.

Step 4. Permute the gray-scale value matrix of the processed plain-image P by Ix, Iy in the following way to get the shuffled image Q:

$$Q(i,j) = P(Ix(i), Iy(j)), i = 1, \cdots, H, j = 1, \cdots, W.$$

14.4.2 *Diffusion process*

It is necessary that a secure encryption algorithm should have a good mechanism of diffusion. On one hand, the diffusion processing can render the permutation process non-invertible, which therefore strengthen the security. On the other hand, the diffusion processing can significantly change the statistical properties of the plain-image by spreading the influence of each bit of the plain-image all over the cipher-image. Though the permutation process has changed the pixel positions of the plain-image, it can not change the histogram statistical property of the plain-image. The diffusion process will enhance the resistance to statistical attack and differential attack greatly, in which the histogram of the cipher-image is fairly uniform and is significantly different from that of the plain-image. The opponent can not find any useful clues between the plain-image and the cipher-image and so can not break the cryptosystem even after they spend a lot of time and effort. A good diffusion process should yield keystreams strongly related to plain-images. When encrypting different plain-images (even with the same cipher keys), the encryption scheme should generate different keystreams. The diffusion process is outlined as follows.

Step 1. Applying the permutation process to confuse the plain-image P and get a shuffled image Q. Set the values of the initial conditions z_0, w_0 and the control parameters $c_i(i = 1, \cdots, N_3 - 1), d_j(j = 1, \cdots, N_4 - 1)$, say $z_0 = 0.361, c = [0\ 0.15\ 0.31\ 0.57\ 0.78\ 0.89\ 1.0], w_0 = 0.633, d = [0\ 0.11\ 0.27\ 0.48\ 0.67\ 0.83\ 1.0]$, in the diffusion process.

Step 2. Let $i = 0$.

Step 3. Apply the following quantization formula to yield one 8-bit pseudo-random gray value $g(i)$:

$$g(i) = \text{floor}(L \times z_i),$$

where L is the color level (for a 256 gray-scale image, $L = 256$), the "floor" operation on x returns the largest value not greater than x.

Step 4. Compute the pixel gray value in the cipher-image by a two-point diffusion transmission:

$$C(i+1) = \phi(i+1) \oplus [(g(i) + C(i)) \text{mod } L],$$

where $\phi(i+1)$ is the gray value of the current operated pixel in the shuffled image which has been rearranged according to the order of row or column to a vector with length $H \times W$, $C(i)$ is the previous output cipher-pixel gray value. The diffusion process is well defined as the initial condition $C(0)$ is provided. $C(0)$ can be set to be part of the keys in the diffusion process or can just take the value of $g(0)$ for simplicity.

Step 5. Compute t by $t = 1 + [C(i+1) \text{mod } 2]$ to get the next z_{i+1} by iterating the shift dynamical system associated with IFS with control parameters $c_j(j = 1, \cdots, N_3 - 1)$ on z_i for t rounds, that is, $z_{i+1} = S^t(z_i)$. This is the crucial step to generate a keystream depending on the plain-image since t is related to $C(i+1)$, so is z_{i+1}. The encrypted image not only relates to the cipher keys, but also relates to the plain-image.

Step 6. Let $i = i + 1$ and return to Step 3 until i reaches $H \times W$.

The above diffusion process implies that it can not influence the pixels before the tampered pixel with a gray value change. As a remedy, we here add a reverse diffusion process as a supplement to the above diffusion process. Another shift dynamical system associated with IFS with control parameters $d_j(j = 1, \cdots, N_4 - 1)$ is used here.

Step 7. Iterate the following shift dynamical system associated with IFS to produce another pseudo-random gray value sequence

$$w_{k+1} = S(w_k),$$

$$\psi(k+1) = \text{floor}(L \times w_{k+1}), k = 0, 1, \cdots, H \times W - 1.$$

Step 8. Execute the reverse diffusion process:

$$D(i) = D(i+1) \oplus [(C(i) + \psi(i)) \text{mod } L], i = H \times W, \cdots, 2, 1,$$

where $D(i), i = 1, 2, \cdots, H \times W$ are the final encrypted vector consisting of the encrypted image pixel gray-scale values. The value of $D(H \times W + 1)$ should be provided to cipher out the sequence $D(i), i = 1, 2, \cdots, H \times W$. $D(H \times W + 1)$ can handled in the same way as $C(0)$.

The complete diffusion process is composed of Step 1 to Step 8. The permutation process and the diffusion process form the proposed image encryption scheme. The original plain-image Lena is encrypted and the resulted cipher-image is shown in Fig. 14.3(b).

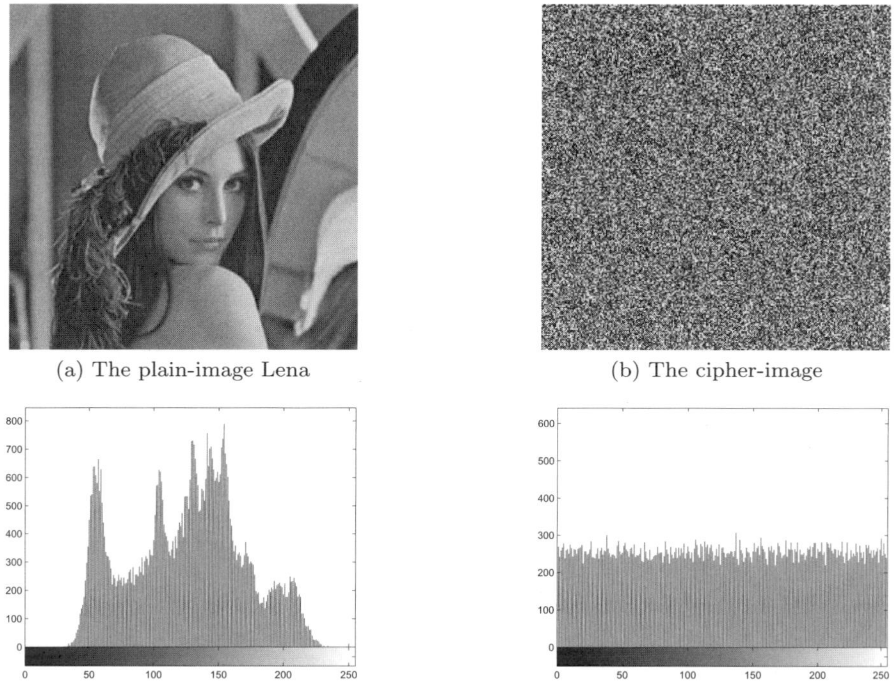

(a) The plain-image Lena

(b) The cipher-image

(c) Histogram of the plain-image

(d) Histogram of the cipher-image

Fig. 14.3 The encrypted results with $x_0 = 0.631, a = [0\ 0.23\ 0.41\ 0.62\ 0.75\ 0.89\ 1.0]$, $y_0 = 0.728, b = [0\ 0.17\ 0.33\ 0.5\ 0.68\ 0.83\ 1.0]$, $z_0 = 0.361, c = [0\ 0.15\ 0.31\ 0.57\ 0.78\ 0.89\ 1.0]$, $w_0 = 0.633, d = [0\ 0.11\ 0.27\ 0.48\ 0.67\ 0.83\ 1.0]$.

14.4.3 *Security analysis*

According to the basic principle of cryptology [Schneier (1995)], a good encryption scheme requires sensitivity to cipher keys, i.e., the cipher-text should have close correlation with the keys. An ideal encryption scheme should have a large key space to make brute-force attack infeasible; it should also well resist various kinds of attacks like statistical attack, differential attack, etc. In this section, some security analysis has been performed on the proposed image encryption scheme, including the most important ones like key space analysis, statistical analysis, and differential analysis. All the analysis shows that the proposed image encryption scheme is highly secure.

14.4.3.1 *Key space analysis*

A good image encryption scheme needs to contain sufficiently large key space for compensating the degradation dynamics in PC. It should be sensitive to cipher keys as well, and thus can effectively prevent invaders decrypting original data even after they invest large amounts of time and resources. The analysis results regarding the sensitivity and the key space are summarized as follows. Since the permutation process is irrelevant to the diffusion process, the key space consists of the cipher keys in both processes. In the permutation process, the control parameters $a_i(i = 1, \cdots, N_1 - 1)$, $b_j(j = 1, \cdots, N_2 - 1)$ and the initial conditions x_0, y_0 form the cipher keys. The cipher keys in the diffusion process consist of the initial conditions y_0, z_0, the control parameters $c_i(i = 1, \cdots, N_3 - 1), d_j(j = 1, \cdots, N_4 - 1)$ for two shift dynamical systems associated with IFS. The sensitive tests with respect to all cipher keys have been carried out. The sensitivity is generally measured by means of two criteria, namely, number of pixels change rate (NPCR) and unified average changing intensity (UACI) [Chen et al. (2004); Lian et al. (2005)]. They are defined as

$$\text{NPCR} = \frac{1}{W \times H} \sum_{i,j} D(i,j) \times 100\%, \tag{14.9}$$

$$\text{UACI} = \frac{1}{W \times H \times 255} \sum_{i,j} |C_1(i,j) - C_2(i,j)| \times 100\%, \tag{14.10}$$

where C_1, C_2 are the two cipher-images corresponding to two cipher keys with a minor change or two plain-images with only one pixel difference, D is a bipolar array with the same size as image C_1. $D(i,j)$ is determined as: if $C_1(i,j) = C_2(i,j)$, then $D(i,j) = 0$, otherwise $D(i,j) = 1$.

To verify the sensitivity of key parameter K, the original plain-image $I = (I(i,j))_{H \times W}$ is encrypted with $K = p, K = p - \Delta K$ and $K = p + \Delta K$ respectively while keeping the other key parameters unchanged. Here ΔK is the perturbing value. The corresponding encrypted images are denoted by I_1, I_2, I_3 respectively. The NPCR and UACI values are calculated for the cipher-image pairs (I_1, I_2) and (I_1, I_3). The greater the NPCR and the UACI, the more sensitive for the parameter K. Tables 14.1-14.2. show the results of the sensitivity test where the initial key values are set to be the following ($N_1 = N_2 = N_3 = N_4 = 6$):

$$\text{Permutation process}: \ x_0 = 0.631, a = [0\ 0.23\ 0.41\ 0.62\ 0.75\ 0.89\ 1.0],$$

$$\text{Diffusion process}: \begin{aligned} y_0 &= 0.728, b = [0\ 0.17\ 0.33\ 0.5\ 0.68\ 0.83\ 1.0], \\ z_0 &= 0.361, c = [0\ 0.15\ 0.31\ 0.57\ 0.78\ 0.89\ 1.0], \\ w_0 &= 0.633, d = [0\ 0.11\ 0.27\ 0.48\ 0.67\ 0.83\ 1.0]. \end{aligned}$$

All the variations ΔK for the considered parameters are set to be 10^{-16}.

We apply the proposed image encryption scheme one round with only perturbing one cipher key K with the corresponding variation value while fixing other parameters.

Table 14.1 Results regarding the sensitivity to cipher keys, part 1: calculated based on the cipher-images I_1, I_2.

K	x_0	a_1	a_2	a_3	a_4	a_5
NPCR(%)	99.60	99.56	99.57	99.56	99.63	99.62
UACI(%)	33.63	33.33	33.51	33.41	33.55	33.51

K	y_0	b_1	b_2	b_3	b_4	b_5
NPCR(%)	99.60	99.62	99.63	99.61	99.63	99.60
UACI(%)	33.39	33.38	33.70	33.49	33.57	33.48

K	z_0	c_1	c_2	c_3	c_4	c_5
NPCR(%)	99.64	99.54	99.61	99.62	99.63	99.60
UACI(%)	33.49	33.53	33.59	33.50	33.64	33.48

K	w_0	d_1	d_2	d_3	d_4	d_5
NPCR(%)	99.60	99.64	99.58	99.59	99.63	99.59
UACI(%)	33.39	33.55	33.55	33.39	33.47	33.51

The results in Tables 14.1-14.2 imply that the control parameters $a_i, b_i, c_i, d_i (i = 1, \cdots, 5)$ and the initial conditions x_0, y_0, z_0, w_0 are all strongly sensitive. It also implies from the results that the key space is more than 10^{384}, which is large enough to make brute-force attack infeasible. As a matter of fact, the key space will become 10^{16} times larger if $N-1$, the number of control parameters, increased by 1 for one shift dynamical system associated with IFS. The proposed image encryption scheme promotes the key space significantly.

The sensitivity tests can also be demonstrated visually, for example, see Figs. 14.4-14.5. In Fig. 14.4, the encrypted image with the key $c_2 = 0.31$ has 99.64% of difference from the encrypted image with the key $c_2 = 0.31 + 10^{-16}$; the image encrypted by the key $w_0 = 0.633$ has 99.62% of difference from the image encrypted by the key $w_0 = 0.633 + 10^{-16}$. Fig. 14.5 shows that the image encrypted by $z_0 = 0.361, d_4 = 0.67$ is not correctly decrypted by using the perturbed key $z_0 = 0.361 + 10^{-16}, d_4 = 0.67$ as well as $z_0 = 0.361, d_4 = 0.67 + 10^{-16}$.

Table 14.2 Results regarding the sensitivity to cipher keys, part 2: calculated based on the cipher-images I_1, I_3.

K	x_0	a_1	a_2	a_3	a_4	a_5
NPCR(%)	99.61	99.64	99.63	99.63	99.58	99.65
UACI(%)	33.44	33.57	33.38	33.44	33.38	33.40

K	y_0	b_1	b_2	b_3	b_4	b_5
NPCR(%)	99.60	99.62	99.58	99.59	99.58	99.62
UACI(%)	33.42	33.71	33.55	33.55	33.44	33.57

K	z_0	c_1	c_2	c_3	c_4	c_5
NPCR(%)	99.64	99.58	99.64	99.65	99.64	99.57
UACI(%)	33.40	33.54	33.42	33.45	33.42	33.51

K	w_0	d_1	d_2	d_3	d_4	d_5
NPCR(%)	99.59	99.62	99.64	99.60	99.58	99.60
UACI(%)	33.45	33.38	33.40	33.67	33.55	33.40

14.4.3.2 *Statistical analysis*

Shannon pointed out in his masterpiece [Shannon (1949)] the possibility to solve many kinds of ciphers by statistical analysis. Therefore, passing the statistical analysis on cipher-image is of crucial importance for a cryptosystem. Indeed, an ideal cryptosystem should be highly robust against any statistical attack. In order to prove the security of the proposed encryption scheme, the following statistical tests are performed.

(i) Histogram. Encrypt the plain-image Lena with one round, and then plot the histograms of plain-image and cipher-image as shown in Figs. 14.3(c)-(d), respectively. Fig. 14.3(d) shows that the histogram of the cipher-image is fairly uniform and significantly different from the histogram of the original image and hence it does not provide any useful information for the opponents to perform any effective statistical analysis attack on the encrypted image.

(ii) Correlation of adjacent pixels. To test the correlation between two adjacent pixels, the following performances are carried out. First, we select 6000 pairs of two adjacent pixels randomly from an image and then calculate the correlation coefficient of the selected pairs using the following formulae:

$$Cr = \frac{cov(x,y)}{\sqrt{D(x)}\sqrt{D(y)}}, \quad cov(x,y) = \frac{1}{T}\sum_{i=1}^{T}(x_i - E(x))(y_i - E(y)),$$

$$E(x) = \frac{1}{T}\sum_{i=1}^{T} x_i, \quad D(x) = \frac{1}{T}\sum_{i=1}^{T}(x_i - E(x))^2,$$

where x, y are the gray-scale values of two adjacent pixels in the image and T is

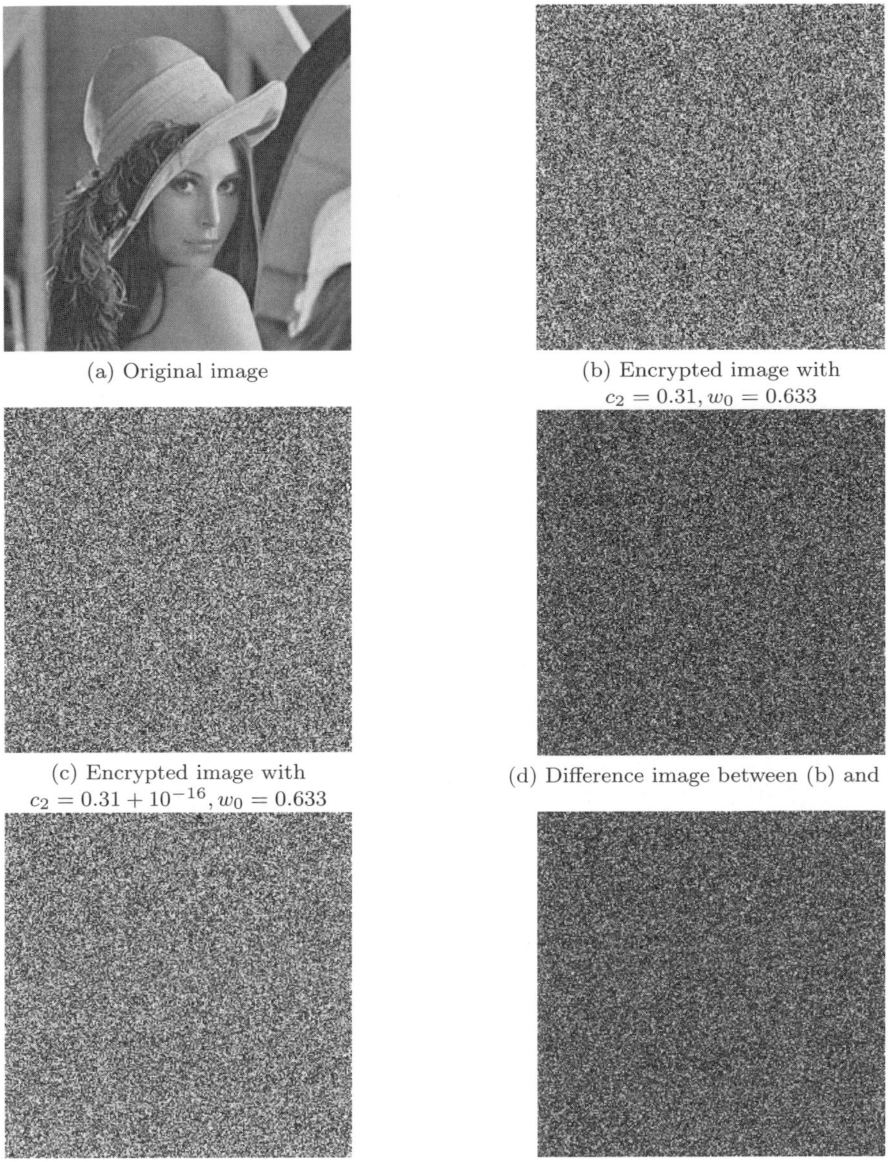

(a) Original image

(b) Encrypted image with $c_2 = 0.31, w_0 = 0.633$

(c) Encrypted image with $c_2 = 0.31 + 10^{-16}, w_0 = 0.633$

(d) Difference image between (b) and (c)

(e) Encrypted image with $c_2 = 0.31, w_0 = 0.633 + 10^{-16}$

(b) Difference image between (b) and (e)

Fig. 14.4 Key sensitive test: result 1.

the total pairs of pixels randomly selected from the image. The correlations of two adjacent pixels in the plain-image and in the cipher-image are shown in the Table 14.3.

The correlation distribution of two horizontally adjacent pixels in the plain-image and that in the cipher-image are shown in Fig. 14.6.

(a) Original image

(b) Encrypted image with $z_0 = 0.361, d_4 = 0.67$

(c) Decrypted image with $z_0 = 0.361 + 10^{-16}, d_4 = 0.67$

(d) Decrypted image with $z_0 = 0.361, d_4 = 0.67 + 10^{-16}$

Fig. 14.5 Key sensitive test: result 2.

Table 14.3 Correlation coefficients of two adjacent pixels in the plain-image and the cipher-image.

	plain-image	cipher-image
Horizontal	0.9435	0.0044
Vertical	0.9680	-0.0041
Diagonal	0.9157	-0.0028

(iii)Information entropy analysis. The entropy is the most outstanding feature of randomness. The entropy $H(m)$ of a message source m can be measured by

$$H(m) = -\sum_{i=0}^{L-1} p(m_i)\log(p(m_i)),$$

where L is the total number of symbols m, $p(m_i)$ represents the probability of occurrence of symbol m_i and log denotes the base 2 logarithm so that the entropy is expressed in bits. For a random source emitting 256 symbols, its entropy is $H(m) = 8$ bits. For the encrypted image of Lena, the corresponding entropy is

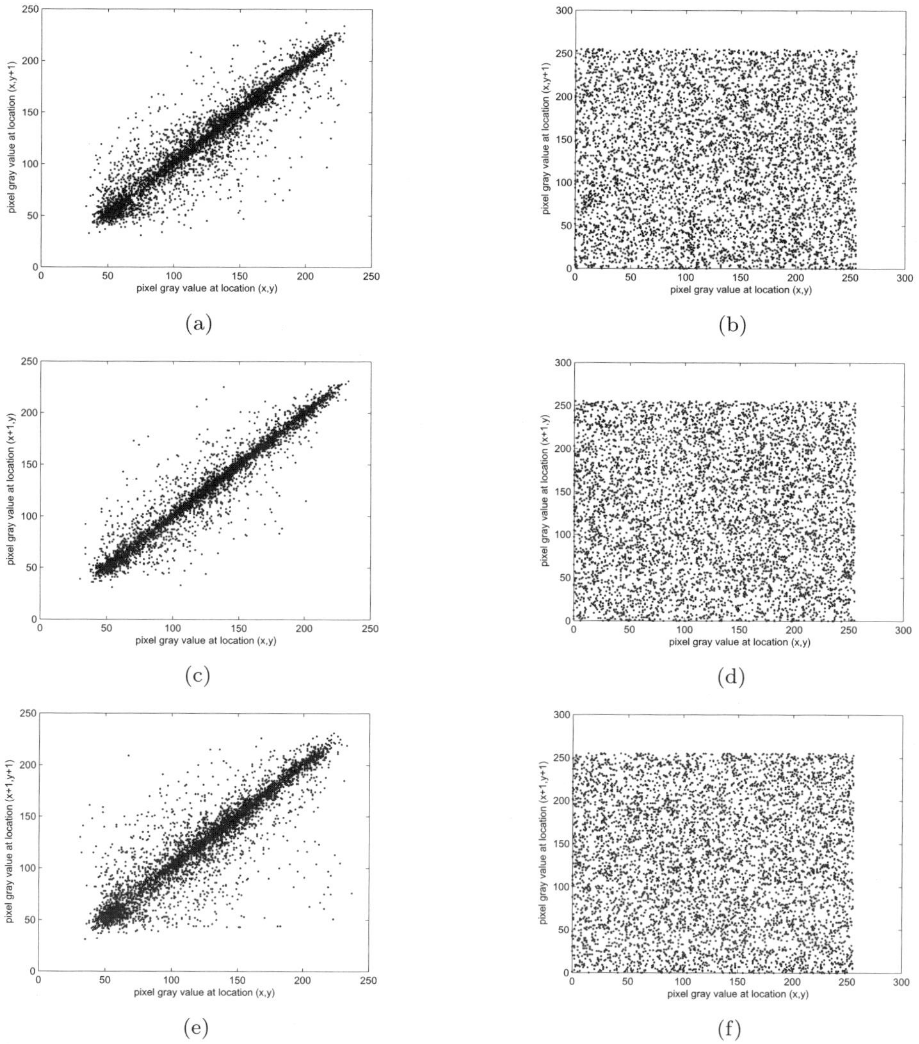

Fig. 14.6 Correlations of two adjacent pixels in the plain-image and in the cipher-image: (a), (c), (e) are for the plain-image; (b),(d),(f) are for the cipher-image.

7.9966bits. This means that the cipher-image is close to a random source and the proposed algorithm is secure against the entropy attack.

14.4.3.3 *Differential attack*

In general, attackers may make a slight change (e.g., modify only one pixel) of the plain-image to find out some meaningful relationships between the plain-image and the cipher-image. If one minor change in the plain-image will cause a significant change in the cipher-image, then the encryption scheme will resist the differential

Table 14.4 Results of NPCR and UACI tests of Lena.

Position	(154,5)	(34,253)	(16,246)	(213,129)	(82,96)
NPCR(%)	99.90	99.86	99.86	99.96	99.75
UACI(%)	46.49	45.39	27.56	48.17	39.04
Position	(184,28)	(100,129)	(11,237)	(196,35)	(15,142)
NPCR(%)	99.96	99.95	100	99.63	99.88
UACI(%)	26.07	48.20	50.05	32.97	11.36

attack efficiently. To test the influence of only one-pixel change in the plain-image over the whole cipher-image, two common measures NPCR and UACI, given by Eq. (14.9) and Eq. (14.10) respectively, are used [Chen et al. (2004)]. In this case, NPCR measures the percentage of different pixels numbers between the two cipher-images whose plain-images only have one-pixel difference; UACI measures the average intensity of differences between the two cipher-images. They indicate the sensitivity of the cipher-images to the minor change of plain-image. To resist difference attacks, the values of NPCR and UACI should be large enough. The test of the plain-image is Lena. We randomly select 10 pixels and change the gray values with a difference of 1, for example, we replace the gray value 79 of the pixel at position (154,5) by 80, and get the NPCR=99.90%, UACI=46.49%. The numerical results are shown in Table 14.4. The mean values of the ten NPCR and UACI values are 99.87% and 37.53% respectively. We observe from Table 14.4 that the two measure values are exceptionally good undergoing only one round of encryption.

14.4.3.4 *Resistance to known-plaintext and chosen-plaintext attacks*

In the diffusion process, a feedback from the cipher-image is employed to change the number of iterations of the shift dynamical system associated with IFS. In Step 3, $g(i)$ depends on the value of z_i which is related to the plain-image, implying that the keystream depends on the processed image. When different plain-images are encrypted, the corresponding keystreams are not the same. The attacker cannot obtain useful information by encrypting some special images since the resultant information is related to those chosen-images. Therefore, the attacks proposed in Refs. [Li et al. (2009); Xiao et al. (2009); Rhouma et al. (2010)] become ineffective on this new scheme. The proposed scheme can desirably resist known-plaintext attack and chosen-plaintext attack.

14.5 Conclusions

In this chapter, we explore the application of the shift dynamical system associated with IFS in digital image encryption. IFS and its associated shift dynamical systems are reviewed. The chaotic dynamics of the shift dynamical systems associated with IFS are then analyzed theoretically and numerically. An efficient image encryption scheme based on four shift dynamical systems associated with IFS is proposed. The proposed scheme can shuffle the plain-image efficiently in the permutation process. An effective two-way diffusion process is also presented to change the gray values of the whole image pixels. Security analysis like key space analysis, statistical attack analysis and differential attack analysis, etc. are performed numerically and visually. All the experimental results show that the proposed encryption scheme is secure thanks to its large key space, its highly sensitivity to the cipher keys and plain-images. The proposed encryption scheme is easy to manipulate and can be applied to any image with unequal width and height as well. All these satisfactory properties make the proposed scheme a potential candidate for encryption of multimedia data such as images, audios and even videos.

Bibliography

Alvarez, G. and Li, S. (2006). Breaking an encryption scheme based on chaotic baker map, *Physics Letters A* **352**, pp. 78–82.

Alvarez, G., Montoya, F., Romera, M. and Pastor, G. (2003). Cryptanalysis of a discrete chaotic cryptosystem using external key, *Physics Letters A* **319**, pp. 334–339.

Baptista, M. S. (1998). Cryptography with chaos, *Physics Letter A* **240**, pp. 50–54.

Barnsley, M. F. (1993). *Fractals Everywhere* (Academic Press).

Barnsley, M. F. and Demko, S. (1985). Iterated funvction systems and the galobal construction of fractals, *The Proceedings of the Royal Society of London* **A399**, pp. 243–275.

Billingsley, P. (1965). *Ergodic Theory and Information* (Wiley, New York).

Chen, G. R., Mao, Y. B. and Chui, C. K. (2004). A symmetric image encryption scheme based on 3D chaotic cat maps, *Chaos, Solitons & Fractals* **21**, pp. 749–761.

Demko, S., Hodges, L. and Naylor, B. (1985). Construction of fractal objects with iterated function systems, *SIGGRAPH* **3**, pp. 271–276.

Devaney, R. (2003). *An Introduction to Chaotic Dynamical Systems* Second Edition, (Westview Press).

Fisher, Y. (1994). *Fractal Image Compression-Theory and Application* (Springer-Verlag, New Yok).

Fridrich, J. (1998). Symmetric ciphers based on two-dimensional chaotic maps, *International Journal of Bifurcation and Chaos* **8**, pp. 1259–1284.

Gleick, J. (1987). *Chaos: Making a New Science* (Pengiun Books, New York).

Guan, Z. H., Huang, F. and Guan, W. (2005). Chaos-based image encryption algorithm, *Physics Letters A* **346**, pp. 153–157.

Hasler, M. and Maistrenko, Y. L. (1997). An introduction to the synchronization of chaotic systems: Coupled skew tent map, *IEEE Transactions on Circuits and Systems* **44**, pp. 856–866.

Huang, F. and Guan, Z. H. (2005). A modified method of a class of recently presented cryptosystems, *Chaos, Solitons and Fractals* **23**, pp. 1893–1899.

Hutchinson, J. E. (1981). Fractals and self similarity, *Indiana University Journal of Mathematics* **30**, pp. 713–747.

Jacquin, A. E. (1992). Image coding based on a fractal theory of iterated contractive image transformations, *IEEE Trans. Image Processing* **1**, pp. 18–30.

Kocarev, L. (2001). Chaos-based cryptography: a brief overview, *IEEE Circuits and Systems Magazine* **1**, pp. 6–21.

Li, C., Li, S., Chen, G. and Halang, W. A. (2009). Cryptanalysis of an image encryption scheme based on a compound chaotic sequence, *Image and Vision Computing* **27**, pp. 1035–1039.

Li, S. and Zheng, X. (2002). Cryptanalysis of a chaotic image encryption method, in: *Proc. IEEE Int. Symposium on Circuits and Systems* vol. II, pp. 708–711.

Lian, S., Sun, J. and Wang, Z. (2005). A block cipher based on a suitable use of the chaotic standard map, *Chaos, Solitons and Fractals* **26**, pp. 117–129.

Liu, J. M. and Qu, Q. (2010). Cryptanalysis of a substitution-diffusion based on cipher using chaotic standard and logistic map, in: *Third International Symposium on Information Processing* pp. 67–69.

Liu, H. and Wang, X. (2011). Color image encryption using spatial bit-level permutation and high-dimension chaotic system, *Optics Communications* **284**, pp. 3895–3903.

Mandelbrot, B. (1982). *The Fractal Geometry of Nature* (W. H. Freeman and Co., San Francisco).

Mao, Y. B., Chen, G. and Lian S. G. (2004). A novel fast image encryption scheme based on the 3D chaotic Baker map, *International Journal of Bifurcation and Chaos* **14(10)**, pp. 613–3624.

Masuda, N. and Aihara, K. (2002). Cryptosystems with discretized chaotic maps,*IEEE Trans. Circuits Syst.* I, 49, pp. 28–40.

Matthews, R. (1989). On the derivation of a chaotic encryption algorithm, *Cryptologia* 8, pp. 29–41.

Pareek, N. K., Patidar, V. and Sud, K. K. (2006). Image encryption using chaotic logistic map, *Image and Vision Computing* **24**, pp. 926–934

Patidar, V., Pareek, N. K. and Sud, K. K. (2009). A new substitution–diffusion based image cipher using chaotic standard and logistic maps, *Commun. Nonlinear Sci. Numer. Simulat.* **14**, pp. 3056–3075.

Peitgen, H. O., Jürgens, H. and Saupe, D. (2004). *Chaos and Fractals, New Frontiers of Science* Second Edition, (Springer-Verlag, New York).

Rhouma, R., Solak, E. and Belghith, S. (2010). Cryptanalysis of a new substitution-diffusion based image cipher, *Commun. Nonlinear Sci. Numer. Simulat.* **15**, pp. 1887–1892.

Schneier, B. (1995). *Cryptography: Theory and Practice* (CRC Press, Boca Raton).

Shannon, C. E. (1949). Communication theory of secrecy system, *Bell Syst. Tech. J.* **28**, pp. 656–715.

Stewart, I. (1989). *Does God play Dice? The Methematics of Chaos* (Blackwell, Cambridge MA).

Tong, X. and Cui, M. (2008). Image encryption with compound chaotic sequence cipher shifting dynamically, *Image and Vision Computing* **26**, pp. 843–850.

Wong, K. W., Kwok, B. and Law, W. S. (2008). A fast image encryption scheme based on chaotic standard map, *Physics Letters A* **372**, pp. 2645–2652.

Xiao, D., Liao, X. and Wei, P. (2009). Analysis and improvement of a chaos-based image encryption algorithm, *Chaos, Solitons and Fractals* **40**, pp. 2191–2199.

Ye, R. (2011). A novel chaos-based image encryption scheme with an efficient permutation-diffusion mechanism, *Optics Communications* **284**, pp. 5290–5298.

Zhang, G. J. and Liu, Q. (2011). A novel image encryption method based on total shuffling scheme, *Optics Communications* **284**, pp. 2775–2780.

Chapter 15

Planar crystallographic symmetric tiling patterns generated from invariant maps[1]

Ruisong Ye[a,*], Haiying Zhao[b] and Yuanlin Ma[c]

[a] *Department of Mathematics, Shantou University, Shantou 515063, Guangdong, PR China*

[*] *rsye@stu.edu.cn*

[b] *College of Computer Science and Technology, Xinjiang Normal University, Urumqi 830054, Xinjiang, PR China;*

School of Information Engineering, Beijing University of Science and Technology, Beijing 100083, PR China

[c] *Xinyang Vocational and Technical College, Xinyang 464000, Henan, PR China*

Thanks to the rapid development of computer technology, the visualization of chaotic attractors attracts extensive exploration, yielding the great potential for creating artistic patterns by computers. In this essay, we utilize the periodicity properties of trigonometric functions to construct invariant maps with respect to the seventeen planar crystallographic groups. We also apply the constructed invariant maps to iterating the points in plane and plotting the corresponding points with referred color based on the orbit property of the considered maps from a dynamical system's point of view. Colormaps together with orbit trap methods are applied to rendering fantastic colored images with local pseudo-3D effects. The proposed method is easy to implement and generalize, therefore providing a good approach for devising planar crystallographic symmetric tiling patterns.

Keywords: invariant map, planar crystallographic group, symmetry, tiling pattern

[1]The research is partially supported by the National Natural Science Foundation of China (grant no. 60863010), Pre-project of 973 Plan(grant no. 2010CB334709) and Natural Science Foundation of Xinjiang Province (grant no. 2010211a19).

15.1 Introduction

It has been known for several decades that iteration of maps can lead to chaotic attractors. One of the most famous attractors is the Lorez attractor arising from a weather model in which the lorenz attractor is three dimensional and possesses no apparent symmetries [Lorenz (1963)]. Another famous example is from logistic map which seems simple but could also show complex dynamics like chaotic attractors and fractal patterns [May (1976)]. Other nonlinear maps have been noted to have a great deal of inherent symmetry even while exhibiting chaotic behavior. When symmetries are present in equations, there are important relations between the possible symmetries of solutions and the group of symmetries of the equation [Pickover (1990); Field and Golubitsky (1995)]. In 1988, Chossat and Golubitsky observed that the iteration of a symmetric map could sometimes lead to a symmetric attractor [Chossat and Golubitsky (1988)]. As a matter of fact, the construction of various symmetric patterns has attracted the attention of researchers and artists. The Dutch artist M. C. Escher often used the simultaneous sense of symmetry and the bizarre to create different circle limits, square limits and planar tiling patterns with the symmetries of crystallographic groups by his paintbrush [Schattschneider (1990)]. Spurred by the rapid development of computer graphic techniques, there are extensive explorations on the visualization of dynamical systems' chaotic attractors, yielding the great potential for creating artistic patterns by dynamical systems. The dynamics of one system can be observed visually on the computer. Such graphic experiments not only reveal the characteristic of the system but also create exotic patterns. Since the 1990s, many articles have been published on the topic of how to construct different chaotic maps with symmetries and automatically generate artistic patterns. Among these proposed methods, the creation of aesthetic patterns by means of dynamical systems has been shown to have good performance. The generated patterns are simultaneously complex and artistic; the patterns with some kinds of symmetries are particularly striking. Field and Golubitsky investigated the generation of dihedral and crystallographic symmetric patterns in plane [Field and Golubitsky (1992)]. Carter et al. generated seven different kinds of chaotic attractors with frieze group symmetries and seventeen ones with crystallographic group symmetries in plane by taking advantage of trigonometric functions [Carter *et al.* (1998)]. Dumont et al. established a general method to create attractors with crystallographic symmetries in R^n [Dumont (2001)]. Lu et al. combined those functions constructed in [Carter *et al.* (1998)] with orbit trap rendering methods in the rectangular coordinates plane and yielded fascinating images with planar crystallographic symmetries [Lu *et al.* (2005)].

This essay focuses on the automatic generation of planar crystallographic symmetric tiling patterns. As a matter of fact, references [Field and Golubitsky (1992); Chung and Chan (1993); Carter *et al.* (1998); Lu *et al.* (2005)] are all dedicated to creating planar crystallographic symmetric tiling patterns by means of constructing dynamical systems equivariant with each crystallographic group. We here present another different approach to generate planar crystallographic symmetric tiling patterns by constructing nonlinear maps invariant with crystallographic groups. To the authors' knowledge, there are few papers applying invariant maps to generating planar crystallographic symmetric tiling patterns. We present a thorough investigation on the automatic generation of crystallographic symmetric tiling patterns from nonlinear maps $f(x, y)$ invariant with respect to group Γ in the sense that $f(\gamma(x, y)) = f(x, y)$ for any $\gamma \in \Gamma$. Different from those methods presented in [Field and Golubitsky (1992); Chung and Chan (1993); Carter *et al.* (1998)], we here intend to generate tiling patterns with the planar crystallographic symmetries from another viewpoint, that is, we don't concentrate on displaying the attractors of dynamical systems equivariant with each crystallographic group, however, we pay attention to just creating planar crystallographic symmetric tiling patterns by means of the properties of the considered maps invariant with respect to each crystallographic group. We apply invariant maps to acting on the points in the plane, as a result, those points in the plane which are planar crystallographic symmetric, namely, (x, y) and $\gamma(x, y), \gamma \in \Gamma$, will become the same point $f(x, y)$. If $f(x, y)$ is iterated ahead using any map $h(x, y)$, we will then get the same orbit $\{h(f(x, y)), h^2(f(x, y)), \cdots\}$ for the symmetric points (x, y) and $\gamma(x, y)$. In this sense, it is easier to generate symmetric patterns by invariant maps since $h(x, y)$ can be arbitrary maps without the restriction of invariance. In this essay, we only choose $h(x, y)$ to be equal to $f(x, y)$ for the sake of simplicity. If we produce tiling patterns by the properties of the orbit to every point in the plane, we will get a planar crystallographic symmetric tiling pattern. Therefore the patterns created here are completely different from those conventional ones in the aforementioned references. The method proposed here is more suitable to generate exotic symmetric patterns; it could be a novel method for the automatic generation of planar crystallographic symmetric tiling patterns by means of invariant maps. Motivated by the the construction of equivariant functions reported in [Carter *et al.* (1998)], we utilize the periodicity property of trigonometric functions to construct invariant maps with respect to one of the seventeen planar crystallographic groups. We also apply the constructed invariant maps to iterating the points in plane and plot the corresponding points with referred colors based on the orbit property of the considered maps from a dynamical system's point of view. The color rendering consists of colormaps and orbit trap methods [Carlson (1999); Ye (2002); Lu *et al.* (2005)], which derives fantastic colored images with local pseudo-3D effects. The proposed method is easy to implement and generalize, therefore providing a novel approach for devising planar crystallographic symmetric tiling patterns.

15.2 Planar crystallographic groups

In the classification of plane symmetry, there are 17 different crystallographic groups or wallpaper groups with the conventional names p1, p2, pm, pg, pmm, pmg, pgg, cm, cmm, p3, p3m1, p31m, p4, p4m, p4g, p6, p6m [Grunbaum and Shephard (1963); Coxeter and Moser (1980); Armstrong (1988)]. They are characterized by translations in two independent directions, giving rise to a lattice. These crystallographic groups exhibit one or a combination of the following symmetries: translation, reflection, glide reflection and rotations of order two, three, four, and six.

We will construct trigonometric maps invariant with each crystallographic group. Since Fourier series are often described in terms of functions that have period 2π, we are interested in such a kind of maps $H(x,y): R^2 \to R^2$ which are invariant with respect to $\sigma_1(x,y) = (x+2\pi, y)$ and $\sigma_2(x,y) = (x, y+2\pi)$. Here we have written pairs (x,y) in R^2 as a row instead of a column and we will use both notations freely for convenience. The following definition is the base to construct invariant maps, which are used to transform the points in plane to get the desired symmetric patterns.

Definition 15.1. Let $H(x,y): R^2 \to R^2$ be a map and $\gamma(x,y): \ R^2 \to R^2$ be an element of group Γ, $H(x,y)$ is said to be a map invariant with γ or a map possessing symmetry γ, if $H(x,y)$ is such that

$$H(\gamma(x,y)) = H(x,y). \tag{15.1}$$

If $H(x,y)$ is invariant with all the elements of group Γ, then $H(x,y)$ is said to be invariant with group Γ or possess symmetry Γ.

The simplest planar crystallographic group is p1 which is characterized by two independent translations and no other symmetry. It is well known that smooth functions periodic in both x and y directions can be represented as a two-variable Fourier series. It is natural to employ a double finite truncated Fourier series at both x and y directions to guarantee that our maps are periodic at both x and y directions with period 2π. For simplicity, we just truncate Fourier expansion at $2x$ and $2y$, thus we get a trigonometric map of the form

$$g_A(X) = \begin{pmatrix} g_1(X) \\ g_2(X) \end{pmatrix} = < 1, \cos(x), \cos(2x), \sin(x), \sin(2x) > \cdot A_{5\times 2\times 5} \cdot$$
$$< 1, \cos(y), \cos(2y), \sin(y), \sin(2y) > \bmod \begin{pmatrix} 2\pi \\ 2\pi \end{pmatrix}, \tag{15.2}$$

where $X = (x,y)$ and $A = (a_{ijk})$ is three-dimensional matrix, i,k stand for the corresponding terms of x, y, and $j = 0, 1$ denote the first term and second term of the map, for example, a_{101} represents the coefficient of the term $\cos(x)\cos(y)$ in $g_1(X)$, a_{011} is the coefficient of $\cos(y)$ in $g_2(X)$.

According to Definition 15.1, $g_A(X)$ must be invariant with $\sigma_1(x,y) = (x+2\pi, y)$ and $\sigma_2(x,y) = (x, y+2\pi)$ to assure that $g_A(X)$ is invariant with group p1, that is,

$$g_A(\delta(X)) = g_A(X) \tag{15.3}$$

must be satisfied as $\delta = \sigma_1, \sigma_2$, where $g_A(X)$ is of the form (15.2). It is straightforward to verify that equality (15.3) holds, implying that all general maps in the form (15.2) are invariant with p1.

We will use the invariant maps of the form (15.2) as our choice for the crystallographic groups except for the groups p3, p3m1, p31m, p6, p6m which use the hexagonal lattice. We deduce the conditions for the the invariant maps to possess the symmetries of crystallographic groups in the sequel.

15.2.1 *Groups p2, pm, pmm*

We now consider the crystallographic group p2 involving two-fold rotations. For the map $g_A(X)$ possessing the desired rotational symmetry, we require that it should be invariant with respect to $\delta(x, y) = (-x, -y)$, that is,

$$\begin{aligned} g_A(X) = g_A(\delta(X)) = g_A(-x, -y) & \\ &= <1, \cos(x), \cos(2x), \sin(-x), \sin(-2x)> \cdot A_{5\times2\times5} \cdot \\ &\quad <1, \cos(y), \cos(2y), \sin(-y), \sin(-2y)> \bmod \begin{pmatrix} 2\pi \\ 2\pi \end{pmatrix}. \end{aligned}$$

Expanding the above equality, we obtain two separate equalities

$$\begin{aligned} g_1(X) &= a_{000} + a_{100}\cos(x) + a_{200}\cos(2x) + a_{300}\sin(x) + a_{400}\sin(2x) + \cdots \\ = g_1(\delta(X)) &= a_{000} + a_{100}\cos(x) + a_{200}\cos(2x) - a_{300}\sin(x) - a_{400}\sin(2x) + \cdots. \end{aligned}$$

and

$$\begin{aligned} g_2(X) &= a_{010} + a_{110}\cos(x) + a_{210}\cos(2x) + a_{310}\sin(x) + a_{410}\sin(2x) + \cdots \\ = g_2(\delta(X)) &= a_{010} + a_{110}\cos(x) + a_{210}\cos(2x) - a_{310}\sin(x) - a_{410}\sin(2x) + \cdots. \end{aligned}$$

Comparing the coefficients of the corresponding terms at both sides of the above equations, we know that some coefficients of A must be zero, the others can be arbitrary:

$$\begin{aligned} a_{003} = a_{004} = a_{013} = 0, \quad a_{014} = a_{103} = a_{104} = 0, \\ a_{113} = a_{114} = a_{203} = 0, \quad a_{204} = a_{213} = a_{214} = 0, \\ a_{300} = a_{301} = a_{302} = 0, \quad a_{310} = a_{311} = a_{312} = 0, \\ a_{400} = a_{401} = a_{402} = 0, \quad a_{410} = a_{411} = a_{412} = 0, \end{aligned}$$

the other coefficients can be arbitrary real numbers. This means that 24 coefficients must be zero and the other 26 ones are unconstrained. These conditions allow us to construct the family of maps possessing symmetry p2 simply by creating a mask M_{p2} for the array A. We can now write our family of maps invariant with group p2 as $g_{A*M_{\mathrm{p2}}}(x, y)$, where M_{p2} is the array mask shown in Table 15.1. A mask consists of entries 0 and 1 where 0 means that the corresponding coefficient a_{ijk} is zero and 1 implies that the corresponding coefficient is arbitrary real number. Note that $A * M_{\mathrm{p2}}$ is an component-wise array multiplication of the matrices A and M_{p2}; we could produce the random matrix A and perform the computation $A * M_{\mathrm{p2}}$, which

Table 15.1 Masks for groups p2, pm, pmm, pg.

M_{p2}	M_{pm}	M_{pmm}	M_{pg}
1 1 1 0 0	1 1 1 0 0	1 1 1 0 0	1 1 1 0 0
1 1 1 0 0	1 1 1 0 0	1 1 1 0 0	1 1 1 0 0
1 1 1 0 0	1 1 1 0 0	1 1 1 0 0	0 0 0 1 1
1 1 1 0 0	1 1 1 0 0	1 1 1 0 0	0 0 0 1 1
1 1 1 0 0	1 1 1 0 0	1 1 1 0 0	1 1 1 0 0
1 1 1 0 0	1 1 1 0 0	1 1 1 0 0	1 1 1 0 0
0 0 0 1 1	1 1 1 0 0	0 0 0 0 0	0 0 0 1 1
0 0 0 1 1	1 1 1 0 0	0 0 0 0 0	0 0 0 1 1
0 0 0 1 1	1 1 1 0 0	0 0 0 0 0	1 1 1 0 0
0 0 0 1 1	1 1 1 0 0	0 0 0 0 0	1 1 1 0 0

forces some of the random entries to become zero, creating the desired symmetry p2.

The symmetry group pm is the most basic crystallographic group containing reflections. The lines of reflections are placed to make the axes of reflection parallel to one axis of the translation and perpendicular to the other axis of translation. A rectangular lattice is then created in the most general case. To generate a map $g_A(x, y)$ in the form (15.2) invariant with group pm, we require the map invariant with respect to $\delta_1(x, y) = (x, -y)$. Note that this yields a mirror reflection parallel to the horizontal translation vector. As we handle group p2, we could also get a mask M_{pm} for group pm shown in Table 15.1. The invariant maps for pm then could be written as the form $g_{A*M_{\mathrm{pm}}}(x, y)$.

The symmetry group pmm has perpendicular reflections. A map, which possesses symmetry pmm, should be invariant with respect to $\delta_1(x, y) = (x, -y)$ and $\delta_2(x, y) = (-x, y)$. We can make use of the existing mask for pm to create a map possessing the symmetry pmm. Since the mask for group pm forces invariance with respect to δ_1, we need only to design a mask which imposes invariance with respect to δ_2. We could derive the mask for δ_2 in a manner similar to get the mask for δ_1. Combining the two masks via component-wise array multiplication, we could get the mask for pmm as shown in Table 15.1.

15.2.2 *Groups pg, pmg, pgg, cm, cmm*

The groups discussed above are not related to glide reflections. Now we turn to discuss the conditions for maps invariant with respect to those crystallographic groups pg, pmg, pgg, cm, cmm containing glide reflections.

The most basic crystallographic group with glide reflections is pg. Note that the choice of a horizontal or vertical glide reflection axis is arbitrary. We take a horizontal glide reflection here for example. To guarantee a map invariant with respect to glide reflection $\delta_3(x, y) = (x + \pi, -y)$, we require that the map satisfies

$$g_A(\delta_3(X)) = g_A(x, y).$$

Expanding the above equality, yields

$$
\begin{aligned}
& < 1, \cos(x+\pi), \cos(2(x+\pi)), \sin(x+\pi), \sin(2(x+\pi)) > \cdot A_{5\times 2\times 5} \cdot \\
& < 1, \cos(y), \cos(2y), \sin(-y), \sin(-2y) > \bmod \begin{pmatrix} 2\pi \\ 2\pi \end{pmatrix} \\
= & < 1, \cos(x), \cos(2x), \sin(x), \sin(2x) > \cdot A_{5\times 2\times 5} \cdot \\
& < 1, \cos(y), \cos(2y), \sin(y), \sin(2y) > \bmod \begin{pmatrix} 2\pi \\ 2\pi \end{pmatrix}.
\end{aligned}
$$

We cane get the mask M_{pg} for group pg straightforwardly by comparing the corresponding terms at both sides of the above equality. We can now write our family of maps invariant with group pg as $g_{A*M_{\mathrm{pg}}}(x,y)$. The mask M_{pg} is shown in Table 15.1.

For the sake of convenience to describe the generation of the masks for other groups, we first derive the masks for $\delta_i, i = 1,2,3,4,5,6$. These masks are shown in Tables 15.2–15.3.

Table 15.2 Masks for $\delta_1, \delta_2, \delta_3$.

$\delta_1(x,y) = (x,-y)$	$\delta_2(x,y) = (-x,y)$	$\delta_3(x,y) = (x+\pi,-y)$
1 1 1 0 0	1 1 1 1 1	1 1 1 0 0
1 1 1 0 0	1 1 1 1 1	1 1 1 0 0
1 1 1 0 0	1 1 1 1 1	0 0 0 1 1
1 1 1 0 0	1 1 1 1 1	0 0 0 1 1
1 1 1 0 0	1 1 1 1 1	1 1 1 0 0
1 1 1 0 0	1 1 1 1 1	1 1 1 0 0
1 1 1 0 0	0 0 0 0 0	0 0 0 1 1
1 1 1 0 0	0 0 0 0 0	0 0 0 1 1
1 1 1 0 0	0 0 0 0 0	1 1 1 0 0
1 1 1 0 0	0 0 0 0 0	1 1 1 0 0

Table 15.3 Masks for δ_4, δ_5, δ_6.

$\delta_4(x,y) = (x+\pi,\pi-y)$	$\delta_5(x,y) = (-x,\pi+y)$	$\delta_6(x,y) = (\pi-x,\pi+y)$
1 0 1 1 0	1 0 1 0 1	1 0 1 0 1
1 0 1 1 0	1 0 1 0 1	1 0 1 0 1
0 1 0 0 1	1 0 1 0 1	0 1 0 1 0
0 1 0 0 1	1 0 1 0 1	0 1 0 1 0
1 0 1 1 0	1 0 1 0 1	1 0 1 0 1
1 0 1 1 0	1 0 1 0 1	1 0 1 0 1
0 1 0 0 1	0 1 0 1 0	1 0 1 0 1
0 1 0 0 1	0 1 0 1 0	1 0 1 0 1
1 0 1 1 0	0 1 0 1 0	0 1 0 1 0
1 0 1 1 0	0 1 0 1 0	0 1 0 1 0

The symmetry group pmg contains lines of reflection perpendicular to lines of glide reflection, that is, pmg contains δ_1, δ_5, where δ_5 is a glide reflection of length π

Table 15.4 Masks for pmg, pgg, cm, cmm.

M_{pmg}	M_{pgg}	M_{cm}	M_{cmm}
1 0 1 0 0	1 0 1 0 0	1 0 1 0 0	1 0 1 0 0
1 0 1 0 0	1 0 1 0 0	1 0 1 0 0	1 0 1 0 0
1 0 1 0 0	0 0 0 0 1	0 1 0 0 0	0 1 0 0 0
1 0 1 0 0	0 0 0 0 1	0 1 0 0 0	0 1 0 0 0
1 0 1 0 0	1 0 1 0 0	1 0 1 0 0	1 0 1 0 0
1 0 1 0 0	1 0 1 0 0	1 0 1 0 0	1 0 1 0 0
0 1 0 0 0	0 0 0 1 0	0 1 0 0 0	0 0 0 0 0
0 1 0 0 0	0 0 0 1 0	0 1 0 0 0	0 0 0 0 0
0 1 0 0 0	0 1 0 0 0	1 0 1 0 0	0 0 0 0 0
0 1 0 0 0	0 1 0 0 0	1 0 1 0 0	0 0 0 0 0

along the line $y = 0$. We multiply the corresponding masks for δ_1, δ_5 via component-wise array multiplication to get the mask M_{pmg} for group pmg, and the invariant maps take the form $g_{A*M_{\text{pmg}}}(x, y)$.

The symmetry group pgg has perpendicular glide reflections. It contains δ_3, δ_5, where δ_3 is a glide reflection of length π along the line $x = 0$. We multiply the corresponding masks for δ_3, δ_5 by component-wise array multiplication to get the mask M_{pgg} for group pgg, and the corresponding maps are of the form $g_{A*M_{\text{pgg}}}(x, y)$.

The symmetry group cm contains both reflections and glide reflections in parallel directions but no rotations. Therefore the maps invariant with group cm should be invariant with respect to two symmetries δ_1, δ_4, where δ_4 is a glide reflection of length π along the line $y = \pi/2$. Multiplying the two corresponding masks $M_{\delta_1}, M_{\delta_4}$ by component-wise array multiplication, we could get the mask M_{cm} and take the family of maps as $g_{A*M_{\text{cm}}}(x, y)$.

The group cmm not only contains perpendicular reflections, but also contains rotations of order two which do not lie on the reflections axes. Glide reflections also exist, resulting from combining perpendicular mirrors and half turns, so our map should be invariant with respect to four symmetries $\delta_1, \delta_2, \delta_4, \delta_6$. Similarly, we could get the masks M_{cmm} via component-wise array multiplication of the four masks and the desired map $g_{A*M_{\text{cmm}}}(x, y)$.

The masks for the groups discussed above are shown in Table 15.4.

15.2.3 *Groups p4, p4g, p4m*

In this subsection we focus on the groups p4, p4g, p4m, which all contain a four-fold rotation. The most basic one is p4. In order to possess this symmetry for a map, we require that the map should be invariant with respect to $\delta(x, y) = (-y, x)$, that is,

$$g_A(\delta(X)) = g_A(X).$$

Table 15.5 Masks for p4, p4m, p4g.

M_4	M_{p4m}	M_{p4g}
1 1 1 0 0	1 1 1 0 0	1 1 1 0 0
1 1 1 0 0	1 1 1 0 0	1 1 1 0 0
1 1 1 0 0	1 1 1 0 0	0 0 0 0 0
1 1 1 0 0	1 1 1 0 0	0 0 0 0 0
1 1 1 0 0	1 1 1 0 0	1 1 1 0 0
1 1 1 0 0	1 1 1 0 0	1 1 1 0 0
0 0 0 0 1	0 0 0 0 0	0 0 0 0 1
0 0 0 0 1	0 0 0 0 0	0 0 0 0 1
0 0 0 1 0	0 0 0 0 0	0 0 0 0 0
0 0 0 1 0	0 0 0 0 0	0 0 0 0 0

The above equality implies that the following conditions on the coefficients of the invariant map $g_A(X)$:

$$a_{001} = a_{100}, a_{002} = a_{200}, a_{102} = a_{201}, a_{304} = -a_{403},$$
$$a_{011} = a_{110}, a_{012} = a_{210}, a_{112} = a_{211}, a_{314} = -a_{413},$$

$a_{000}, a_{101}, a_{202}, a_{010}, a_{111}, a_{212}$ are arbitrary, and the other a_{ijk} are all zero. We can also write the array A satisfying the above conditions as the form M_{p4}, and the desired invariant maps take the form $g_{A*M_{\mathrm{p4}}}(x, y)$.

The symmetry group p4m contains the four-fold rotational symmetry p4 as well as perpendicular lines of reflection. The component-wise array multiplication of M_{pm} and M_{p4} gives rise to the mask for group p4m.

In a similar way to our extension of p4 to p4m, we extend p4 to p4g. Since p4g contains p4 and a glide reflection, we just need to multiply the mask M_{p4} and the mask M_{pg} by component-wise array multiplication to obtain the corresponding mask M_{p4g} and get the desired map invariant with group p4g. The masks for p4, p4m, p4g are given in Table 15.5.

15.2.4 *Groups p3, p3m1, p31m*

In this subsection we introduce the symmetry groups p3, p3m1, p31m. These groups contain three-fold rotation. Group p3 is the most basic one with third turns in addition to two independent translations. Since a square lattice does not map to itself by third turns, we need to use a different lattice for these groups. Indeed, a lattice with 120-degree angles between the independent directions is appropriate. We take $u_0 = 2\pi(1, 0), u_1 = 2\pi(-1/2, \sqrt{3}/2)$ as the generators to the lattice L. Connecting vertices in this lattice yields a regular tiling of the plane by equilateral triangles and blocks of hexagons; that is why we call it a hexagonal lattice. For convenience, we define a dual lattice L^* to L; the generators to L^* are $v_0 = (1, -1/\sqrt{3}), v_1 = (0, -2/\sqrt{3})$. It could be checked easily that for all

$i, j, u_i \cdot v_j = k \cdot 2\pi$ with integer k. We then get

$$\cos(v \cdot (X + u)) = \cos(v \cdot X), \quad \sin(v \cdot (X + u)) = \sin(v \cdot X), \tag{15.4}$$

where $u \cdot v$ represents the scalar product of two vectors, u, v are the elements from lattice L and L^* respectively, i. e., u, v take the following form

$$u = lu_0 + mu_1, v = nv_0 + pv_1, l, m, n, p \in \mathbb{Z}.$$

We define the following map of the form

$$F(X) = \sum_{j=0}^{2} (G_{v_j}(X) + H_{v_j}(X)) \mod L, \tag{15.5}$$

where

$$\begin{aligned}
&G_{v_j}(X) = \alpha_{v_j}(\cos(v_j \cdot X) + \cos(R(v_j) \cdot X) + \cos(R^2(v_j) \cdot X)),\\
&H_{v_j}(X) = \beta_{v_j}(\sin(v_j \cdot X) + \sin(R(v_j) \cdot X) + \sin(R^2(v_j) \cdot X)),\\
&\alpha_{v_j} = (b_{0j}\ b_{1j})^T, \quad \beta_{v_j} = (c_{0j}\ c_{1j})^T, 0 \le j \le 2,\\
&v_2 = v_0 + v_1, \quad X = (x, y).
\end{aligned}$$

$\alpha_{v_j}, \beta_{v_j}$ are random vectors. R is the transformation for three-fold rotation

$$R(X) = \begin{pmatrix} -1/2 & -\sqrt{3}/2 \\ \sqrt{3}/2 & -1/2 \end{pmatrix} \begin{pmatrix} x \\ y \end{pmatrix}.$$

It follows from (15.4) that if δ is any translational symmetry of L then the map (15.5) is invariant with respect to δ. We now turn to prove that $F(X)$ is invariant with respect to third turns about the origin. We first prove $G_v(R(X)) = G_v(X)$ as follows.

If R is a three-fold rotation transformation, then $R^3(\alpha_v) = \alpha_v$. It is also easy to show that $v \cdot R(X) = R^2(v) \cdot X$, $R(v) \cdot R(X) = v \cdot X$, $R^2(v) \cdot R(X) = R(v) \cdot X$, yields

$$\begin{aligned}
G_v(R(X)) &= \alpha_v(\cos(v \cdot R(X)) + \cos(R(v) \cdot R(X)) + \cos(R^2(v) \cdot R(X)))\\
&= \alpha_v(\cos(R^2(v) \cdot X) + \cos(v \cdot X) + \cos(R(v) \cdot X)) = G_v(X).
\end{aligned}$$

As far we prove $G_v(R(X)) = G_v(X)$. It is similar to show $H_v(X) = H_v(R(X))$. Therefore, we conclude that the map (15.5) possesses symmetry p3.

The symmetry group p3m1 contains symmetry p3 with the addition of mirrors going through each three-fold rotation. This can be accomplished by adding a mirror along y axis in our model for p3. Namely, we require that our map be invariant with respect to $\delta_2(x, y) = (-x, y)$. By expanding the invariant map for p3, we observe that $F(\delta_2(X)) \neq F(X)$. In order to construct invariant map to possessing symmetry p3m1, we make some changes on the map model for p3. Thanks to $v \cdot \delta_2(X) = \delta_2(v) \cdot X$, we define the map given by

$$\widehat{F}(X) = \sum_{j=0}^{2} (\widehat{G}_{v_j,\delta_2}(X) + \widehat{H}_{v_j,\delta_2}(X)) \mod L, \tag{15.6}$$

where

$$\begin{aligned}
\widehat{G}_{v_j,\delta_2}(X) &= \alpha_{v_j}(\cos(v_j \cdot X) + \cos(R(v_j) \cdot X) + \cos(R^2(v_j) \cdot X) \\
&\quad + \cos(\delta_2(v_j) \cdot X) + \cos(R(\delta_2(v_j)) \cdot X) + \cos(R^2(\delta_2(v_j)) \cdot X)), \\
\widehat{H}_{v_j,\delta_2}(X) &= \beta_{v_j}(\sin(v_j \cdot X) + \sin(R(v_j) \cdot X) + \sin(R^2(v_j) \cdot X) \\
&\quad + \sin(\delta_2(v_j) \cdot X) + \sin(R(\delta_2(v_j)) \cdot X) + \sin(R^2(\delta_2(v_j)) \cdot X)).
\end{aligned}$$

We can check that the map (15.6) possesses symmetry p3 by the same way to prove (15.5). We next verify that (15.6) is also invariant with respect to reflections about y axis. First we prove $\widehat{G}_{v,\delta_2}(\delta_2(X)) = \widehat{G}_{v,\delta_2}(X)$.

The following equalities are easy to verified.

$$\begin{aligned}
&v \cdot \delta_2(X) = \delta_2(v) \cdot X, \quad R(v) \cdot \delta_2(X) = R^2(\delta_2(v)) \cdot X, \\
&R^2(v) \cdot \delta_2(X) = R(\delta_2(v)) \cdot X, \quad \delta_2(v) \cdot \delta_2(X) = v \cdot X, \\
&R(\delta_2(v)) \cdot \delta_2(X) = R^2(v) \cdot X, \quad R^2(\delta_2(v)) \cdot \delta_2(X) = R(v) \cdot X.
\end{aligned}$$

Therefore

$$\begin{aligned}
&\widehat{G}_{v,\delta_2}(\delta_2(X)) = \alpha_v(\cos(v \cdot \delta_2(X)) + \cos(R(v) \cdot \delta_2(X)) + \cos(R^2(v) \cdot \delta_2(X)) \\
&\quad + \cos(\delta_2(v) \cdot \delta_2(X)) + \cos(R(\delta_2(v)) \cdot \delta_2(X)) + \cos(R^2(\delta_2(v)) \cdot \delta_2(X))) \\
&= \alpha_v(\cos(\delta_2(v) \cdot X) + \cos(R^2(\delta_2(v)) \cdot X) + \cos(R(\delta_2(v)) \cdot X) \\
&\quad + \cos(v \cdot X) + \cos(R^2(v) \cdot X) + \cos(R(v) \cdot X)) \\
&= \widehat{G}_{v,\delta_2}(X).
\end{aligned}$$

Similarly, it is straightforward to check $\widehat{H}_{v,\delta_2}(\delta_2(X)) = \widehat{H}_{v,\delta_2}(X)$. Therefore, the map (15.6) is invariant with group p3m1.

Group p31m contains $\delta_1(x, y) = (x, -y)$ and p3, so we can handle the group p31m by the same way as the group p3m1. We consider the map of the form

$$\widehat{F}(X) = \sum_{j=0}^{2} [\widehat{G}_{v_j,\delta_1}(X) + \widehat{H}_{v_j,\delta_1}(X)] \mod L, \tag{15.7}$$

where

$$\begin{aligned}
\widehat{G}_{v_j,\delta_1}(X) &= \alpha_{v_j}(\cos(v_j \cdot X) + \cos(R(v_j) \cdot X) + \cos(R^2(v_j) \cdot X) \\
&\quad + \cos(\delta_1(v_j) \cdot X) + \cos(R(\delta_1(v_j)) \cdot X) + \cos(R^2(\delta_1(v_j)) \cdot X)), \\
\widehat{H}_{v_j,\delta_1}(X) &= \beta_{v_j}(\sin(v_j \cdot X) + \sin(R(v_j) \cdot X) + \sin(R^2(v_j) \cdot X) \\
&\quad + \sin(\delta_1(v_j) \cdot X) + \sin(R(\delta_1(v_j)) \cdot X) + \sin(R^2(\delta_1(v_j)) \cdot X)).
\end{aligned}$$

It is easy to show that the map $\widehat{F}(X)$ defined in (15.7) is invariant with respect to group p31m.

15.2.5 Groups p6, p6m

The symmetry group p6 is generated by two independent translations and a sixth turn. Again we use a hexagonal lattice about the origin. Let S is the six-fold rotation transformation

$$S(X) = \begin{pmatrix} 1/2 & -\sqrt{3}/2 \\ \sqrt{3}/2 & 1/2 \end{pmatrix} \begin{pmatrix} x \\ y \end{pmatrix}.$$

We define the following map

$$F(X) = \sum_{j=0}^{2} [G_{v_j}(X) + H_{v_j}(X)] \mod L, \tag{15.8}$$

where

$$G_{v_j}(X) = \sum_{k=0}^{5} \alpha_{v_j} \cos(S^k(v_j) \cdot X),$$

$$H_{v_j}(X) = \sum_{k=0}^{5} \beta_{v_j} \sin(S^k(v_j) \cdot X).$$

We will make use the fact that $S^k(v) \cdot S(X) = S^{k-1}(v) \cdot X$ to show $F(S(X)) = F(X)$. First we show $G_v(S(X)) = G_v(X)$ as follows.

As $S^k(v) \cdot S(X) = S^{k-1}(v) \cdot X, S^6(\nu) = \nu$, we get $v \cdot S(X) = S^6(v) \cdot S(X) = S^5(v) \cdot X$, therefore

$$\begin{aligned} G_v(S(X)) &= \sum_{k=0}^{5} \alpha_v \cos(S^k(v) \cdot S(X)) \\ &= \alpha_v[\cos(v \cdot S(X)) + \cos(S(v) \cdot S(X)) + \cos(S^2(v) \cdot S(X)) \\ &\quad + \cos(S^3(v) \cdot S(X)) + \cos(S^4(v) \cdot S(X)) + \cos(S^5(v) \cdot S(X))] \\ &= \alpha_v[\cos(S^5(v) \cdot X) + \cos(v \cdot X) + \cos(S(v) \cdot X) + \cos(S^2(v) \cdot X) \\ &\quad + \cos(S^3(v) \cdot X) + \cos(S^4(v) \cdot X)] = G_v(X). \end{aligned}$$

It is similar to show $H_v(S(X)) = H_v(X)$. So $F(X)$ defined as (15.8) is invariant with group p6.

As for the group p6m, we consider the following map

$$\widehat{F}(X) = \sum_{j=0}^{2} [\widehat{G}_{v_j,\delta_2}(X) + \widehat{H}_{v_j,\delta_2}(X)] \mod L, \tag{15.9}$$

where

$$\widehat{G}_{v_j,\delta_2}(X) = \sum_{k=0}^{5} \alpha_{v_j} [\cos(S^k(v_j) \cdot X) + \cos(S^k(\delta_2(v_j)) \cdot X)],$$

$$\widehat{H}_{v_j,\delta_2}(X) = \sum_{k=0}^{5} \beta_{v_j} [\sin(S^k(v_j) \cdot X) + \sin(S^k(\delta_2(v_j)) \cdot X)].$$

We can show that $\widehat{F}(X)$ given by (15.9) is invariant with p6m in a similar way used to show that one map is invariant with p3m1.

15.3 Rendering method for planar crystallographic symmetric tiling patterns

In this section, we propose a method to create colored tiling patterns with the crystallographic symmetries from a dynamical system's point of view. Once we have constructed the invariant map $f(x, y)$, which possesses one crystallographic group symmetry Γ, we could use it to act on the points in plane, so that the symmetric points (x, y) and $\gamma(x, y), \forall\gamma \in \Gamma$ have the same function value $f(x, y)$.

If $f(x, y)$ is iterated ahead using any map $h(x, y)$, we get the same orbit $\{h(f(x, y)),\ h^2(f(x, y)), \cdots\}$ for the symmetric points (x, y) and $\gamma(x, y)$. In this essay, we choose $h(x, y)$ to be $f(x, y)$ for the sake of simplicity. We plot the point (x, y) in plane with a referred color according to the property of the orbit $\{h(f(x, y)), h^2(f(x, y)), \cdots\}$, which causes the colors for (x, y) and $\gamma(x, y), \forall\gamma \in \Gamma$ are the same since $f(\gamma(x, y)) = f(x, y)$. Consequently we could get one colored tiling pattern with crystallographic group symmetry Γ.

15.3.1 *Description of colormaps*

As for the color rendering, colormaps together with orbit trap methods are adopted to create colorful and striking tiling patterns. A colormap is an indexed mapping of colors for images. The index used here is denoted by ColorIndex. Two colormaps utilized to generate the tiling patterns presented here are shown in Tables 15.6–15.7. They consist of four hues and seven hues respectively. Usually, the RGB component values of one color vary linearly from the beginning to the end of each hue. In this essay, two new colormaps, based on the two colormaps in Tables 15.6–15.7, are generated using an exponential function [Lu *et al.* (2005)]. The two new colormaps are represented by the array variable ColorTable[Max_Color_Num], where the constant Max_Color_Num is a pre-defined maximum number of colors stored in ColorTable. The algorithm to yield the new colormaps is implemented as follows: For ColorIndex=0, 1, ..., Max_Color_Num, the value of ColorRatio is computed by the following formula

$$\text{ColorRatio} = \left(\frac{\text{ColorIndex}}{\text{ColorIndex} + 1}\right)^C, \tag{15.10}$$

where C is a constant, for example, $C = 17$ in the experiment. It implies ColorRatio $\in [0, 1)$, so we can determine the RGB values corresponding to the value of ColorIndex by linear interpolation of the ColorRatio to the nearest color ratios in the lookup array of Tables 15.6–15.7, and then store the RGB values in the new colormap ColorTable.

Table 15.6 Color table 1.

Hue	Color ratios Array	R	G	B
Black	0.0	0	0	0
Red	0.5	255	0	0
Yellow	0.9	255	255	0
Black	1.0	51	26	0

Table 15.7 Color table 1.

Hue	Color ratios Array	R	G	B
Black	0.0	0	0	0
Blue	0.6	0	0	255
Cyan	0.7	0	255	255
Aqua	0.75	153	230	128
Yellow	0.8	255	255	0
Red	0.9	255	0	0
Black	1.0	76	0	0

15.3.2 *Description of orbit trap methods*

In this subsection, we describe the orbit traps and introduce the rendering methods by orbit traps. Each orbit trap area as shown in Figs. 15.1–15.4 contains all color hues with the same distribution. Figs. 15.1–15.4 show the positions of the orbit traps which are circles here. If an orbit point falls within one of the circles, the distance from the orbit point to the center of the circle is computed. This distance is then divided by Ra, the circle radius, to obtain the value of TrapRatio which is always in [0,1). Consequently, ColorRatio can be set to be TrapRatio and used to compute ColorIndex by formula (15.10). Therefore, the pixel at the original point used for the iteration is plotted with the hue and intensity determined by the RGB values in the new colormap Colortable, as indexed by ColorIndex.

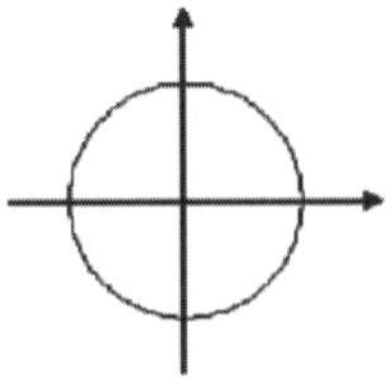

Fig. 15.1 Orbit trap 1 with Ra=1.5π.

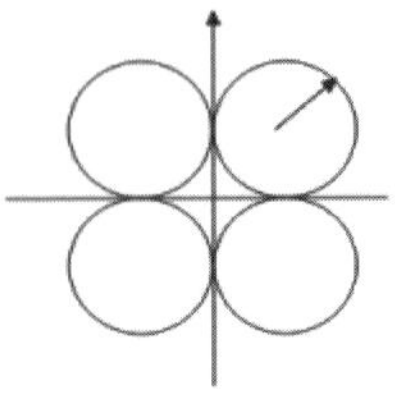

Fig. 15.2 Orbit trap 2 with Ra=3.5.

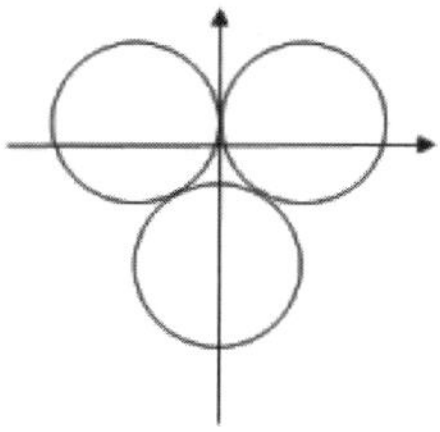

Fig. 15.3 Orbit trap 3 with Ra=$4\sqrt{3}$.

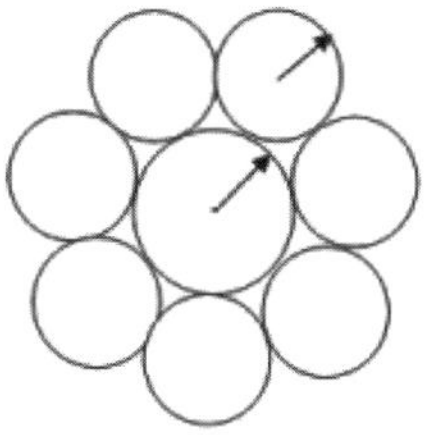

Fig. 15.4 Orbit trap 4 with Ra=0.8π.

15.3.3 *Description of the rendering scheme*

We present a rendering scheme to create planar crystallographic symmetric tiling patterns in this subsection. We assume that the region for displaying the tiling patterns on the screen consists of $N \times M$ pixels and the number of colors in the computer's setting is more than Max_Color_Num.

Step1. Set up a screen drawing region for the tiling patterns, for example, one rectangle with the top-left corner (0,0) and the bottom-right corner (N, M), and also set up the corresponding complex plane region W with the top-left corner $(xmin, ymin)$ and the bottom-right corner $(xmax, ymax)$. Let Itermax be the maximum iteration number.

For each point (nx, ny) in the display area, where $nx = 0, 1, \cdots, N-1$ and $ny = 0, 1, \cdots, M-1$, the following loop should be performed.

Step 2. Let $k = 0$ and

$$x[k] = xmin + nx \times (xmax - xmin)/N,$$

$$y[k] = ymin + ny \times (ymax - ymin)/M.$$

Step 3. Let the map invariant with one certain crystallographic group Γ to act on $(x[k], y[k])$, then we obtain the point $(x[k+1], y[k+1]) = f(x[k], y[k])$, and let $k = k+1$.

Step 4. Determine whether the point $(x[k], y[k])$ falls into the considered orbit trap.

(a) If $(x[k], y[k])$ falls into the trap, compute the distance d from the point $(x[k], y[k])$ to the center of the trap, and then divide d by Ra to obtain the value of TrapRatio, and then calculate the value of ColorIndex by formula (15.10), and go to Step 5.
(b) If k=Itermax, choose the RGB values used as the background color to be equal to ColorTable[Max_Color_Num], set Colorindex=Max_Color_Num and go to the Step 5.
(c) If $(x[k], y[k])$ is out of all traps, go to to Step 3.

Step 5. Plot the point (nx, ny) with the RGB values of ColorTable[ColorIndex], then turn to Step 1 for another loop.

By the above rendering scheme, we could easily generate all the 17 crystallographic symmetric tiling patterns, as shown in Figs. 15.5–15.20 corresponding to 16 crystallographic groups except for p1. We note that all the patterns always contains symmetry p1.

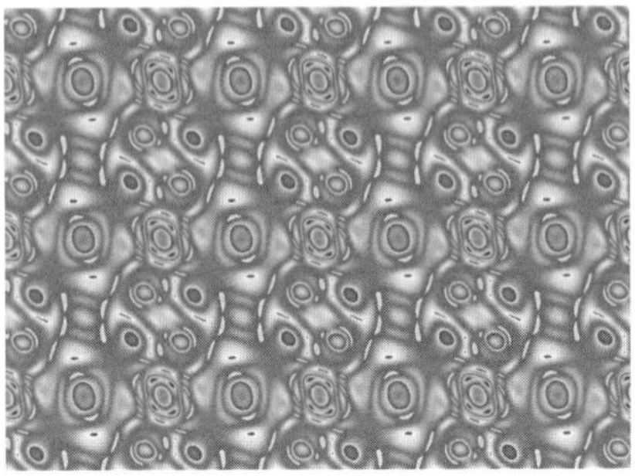

Fig. 15.5 p2 symmetric tiling with color table 1 and orbit trap 1.

Fig. 15.6 pm symmetric tiling with color table 1 and orbit trap 1.

Fig. 15.7 pmm symmetric tiling with color table 2 and orbit trap 1.

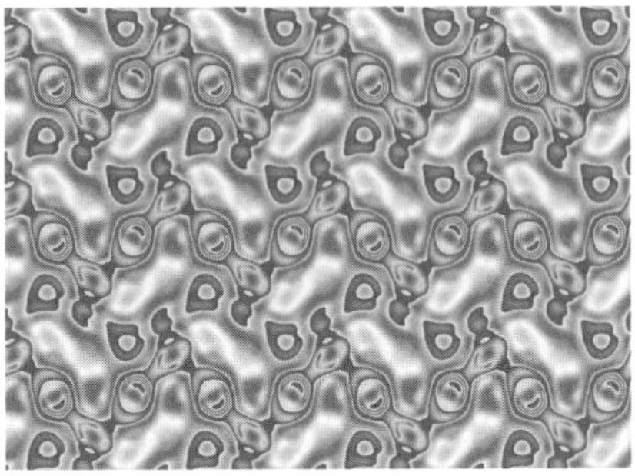

Fig. 15.8 pg symmetric tiling with color table 1 and orbit trap 1.

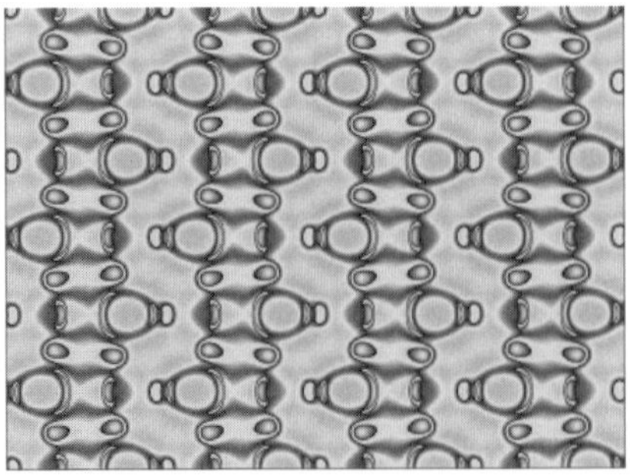

Fig. 15.9 pmg symmetric tiling with color table 2 and orbit trap 1.

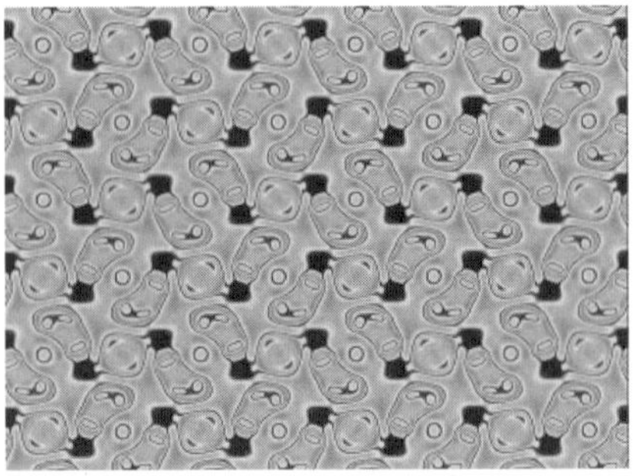

Fig. 15.10 pgg symmetric tiling with color table 2 and orbit trap 1.

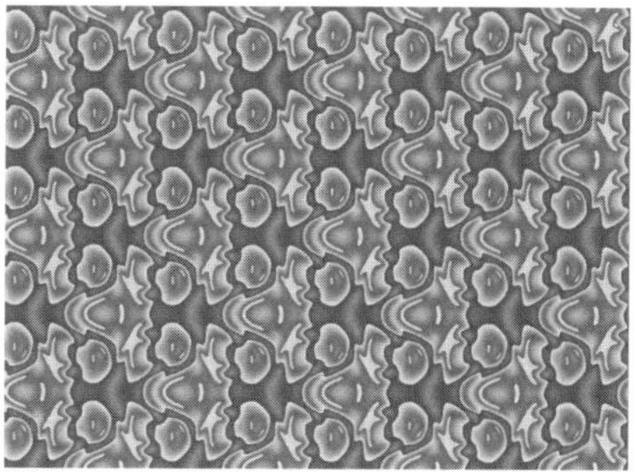

Fig. 15.11 cm symmetric tiling with color table 1 and orbit trap 1.

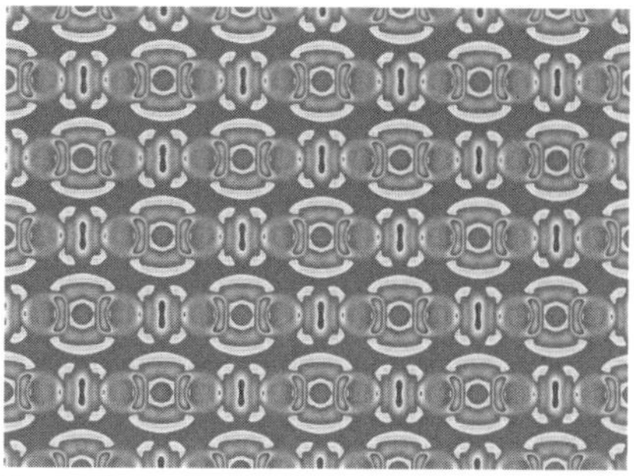

Fig. 15.12 cmm symmetric tiling with color table 1 and orbit trap 1.

Fig. 15.13 p4 symmetric tiling with color table 2 and orbit trap 2.

Fig. 15.14 p4m symmetric with color table 2 and orbit trap 2.

Fig. 15.15 p4g symmetric tiling with color table 2 and orbit trap 2.

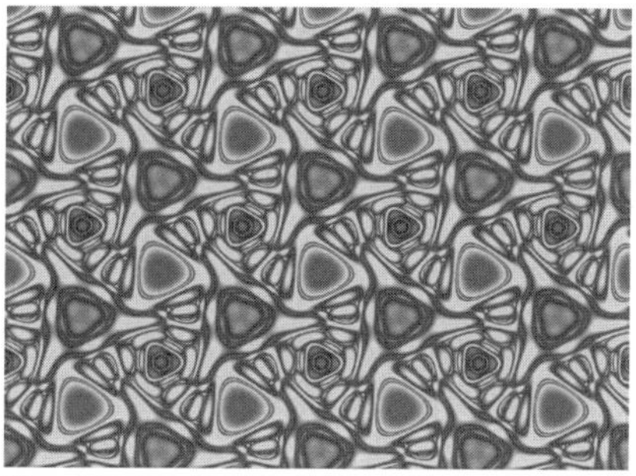

Fig. 15.16 p3 symmetric tiling with color table 2 and orbit trap 3.

15.4 Conclusions

Planar crystallographic symmetric tiling patterns are created by constructing suitable maps invariant with respect to each crystallographic group. Based on the constructed invariant maps acting on the points in plane, one could get symmetric

Fig. 15.17 p31m symmetric tiling with color table 2 and orbit trap 3.

Fig. 15.18 p3m1 symmetric tiling with color table 2 and orbit trap 3.

Fig. 15.19 p6 symmetric tiling with color table 2 and orbit trap 4.

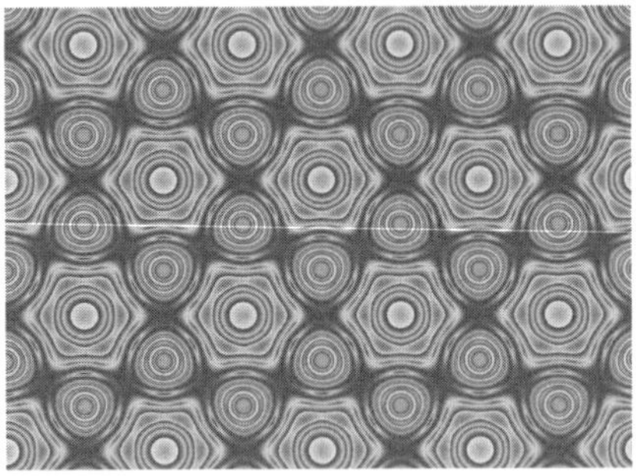

Fig. 15.20 p6m symmetric tiling with color table 2 and orbit trap 4.

tiling patterns since the invariant maps make those symmetric points in plane have the same orbit. Those symmetric points are plotted via a referred color according to the property of their orbits, resulting in a colored tiling patterns associated with the same symmetry. The invariant maps are truncated double Fourier series, so the construction of invariant maps are particularly easy. As a matter of fact, the proposed invariant maps are more suitable for designing crystallographic symmetric tiling patterns compared with the equivariant maps. Colormaps and orbit trap methods are together applied to creating tiling patterns with colorful and artistic appearance; the patterns are also often striking even if the Lyapunov exponent is not positive. It is known that the Lyapunov exponent of the dynamical system should be positive if one wants to get a chaotic attractor pattern with striking appearance. The method will therefore provide a good approach for devising planar crystallographic symmetric tiling patterns.

Bibliography

Armstrong, M. A. (1988). *Groups and Symmetry* (Springer-Verlag, New York).

Carlson, P. W. (1999). Two artistic orbit trap rendering methods for Newton M-set fractals, *Computers & Graphics* **23**, pp. 925–931.

Carter, N., Eagles, R., Grimes, S., Hahn, A. and Reiter, C. (1998). Chaotic attractors with discrete planar symmetries, *Chaos Solitons and Fractals* **9**, pp. 2031–2054.

Chossat, P. and Golubitsky, M.(1988). Symmetry-increasing bifurcation of chaotic attractors, *Physica D* **32**, pp. 423–436.

Chung, K. W. and Chan, H. S. Y. (1993). Symmetrical patterns from dynamics, *Computer Graphics Forum* **12**, pp. 33–40.

Coxeter, H. S. M. and Moser, W. O. J. (1980). *Generators and Relations for Discrete Groups*, 4th edition (Springer-Verlag, New York).

Dumont, J. P., Heiss, F. J., Jones, K. C., Reiter, C. A. and Vislocky L. M. (2001). *N*-dimensional chaotic attractors with crystallographic symmetry, *Chaos, Solitons and Fractals* **12**, pp. 761–784.

Field, M. and Golubitsky, M. (1992). *Symmetry in Chaos* (Oxford University Press, New York).

Field, M. and Golubitsky, M. (1995). Symmetric chaos: how and why, *Notices of the AMS* **42**, pp. 240–244.

Grunbaum, B. and Shephard, G. C. (1963). *Tilings and Patterns*, second edition (Cambridge University Press, Cambridge).

Lorenz, E. N. (1963). Deterministic nonperiodic flow, *Journal of Atmosphere Sciences* **20**, pp. 130–141.

Lu, J., Ye, Z., Zou, Y. and Ye, R. (2005). Orbit trap rendering methods for generating artistic images with crystallographic symmetries, *Computers & Graphics* **29**, pp. 787–794.

May, R. (1976). Simple mathematical models with very complicated dynamics, *Nature* **261**, pp. 459–467.

Pickover, C. A. (1990). *Computers, Pattern, Chaos and Beauty* (Alan Sutton Publishing, Stroud Glouscestershire).

Schattschneider, D. (1990). *Visions of symmetry: Notebooks, Periodic Drawings and Related Works of M. C. Escher* (Freeman, New York).

Ye, R. (2002). Another choice for orbit traps to generate artistic fractal images, *Computers & Graphics* **26**, pp. 629–633.

Chapter 16

Complex dynamics in a simple two-dimensional discrete system

Huiqing Huang[a] and Ruisong Ye[b,*]

[a]*School of Mathematics, Jiaying University, Meizhou 514015, Guangdong, PR China*

[b]*Department of Mathematics, Shantou University, Shantou 515063, Guangdong, PR China*

**rsye@stu.edu.cn*

In this essay, we investigate the complex dynamics of one simple two-dimensional noninvertible discrete system. The pitchfork bifurcation, flip bifurcation and Hopf bifurcation are investigated by means of center manifold theorem and bifurcation theory. Chaotic behavior in the sense of Li-Yorke's definition of chaos is proved as well. Numerical simulations including bifurcation diagram, computation of the maximum Lyapunov exponent and phase portraits are provided. The numerical results verify the theoretical analysis and display the interesting complex dynamics of the proposed dynamical system.

Keywords: invariant map, planar crystallographic group, symmetry, tiling pattern

16.1 Introduction

The study of complex dynamics in discrete systems has always been the focus of nonlinear science research [May (1976); Henon (1976); Paulo (2006); Richter (2007); Sun (2007); Gao (2007)]. Many classes of dynamical systems give rise to models in the form of maps in chaos control and synchronization, neural networks, secure communication and encryption, nonlinear electronics, etc. In the real world, many questions can use discrete systems to describe, even if those questions which are described by continuous dynamical systems, it also can be discrete [Wiggins (1990); Ott (1993)]. For example, by using Poincaré map's method of ordinary differential equation and the numerical calculus's format of differential equation, the continuous dynamical systems will become some difference equations, the research in this area therefore has great significance from both theoretical and practical viewpoints [Robinson (1994); Morris (2004)].

Among the study of nonlinear maps, two and higher dimensional noninvertible maps attract more attentions during the last decades. One reason of this situation comes from the fact that more and more mathematical models of dynamical process are related to such maps [Mira (1996)]. It is known that many nonlinear systems are dependent on control parameters which appear in the defining systems of equations. As the control parameters are changed, the qualitative structure of the orbits of the dynamical systems will change as well [Guckenheimer (1997)]. It is important to find when and where the dynamics will occur essential changes. It is then important to discover how the properties of orbits evolve as the control parameters of a dynamical system are changed [Ott (1993)]. It has been noted that there are complex dynamics even in simple nonlinear maps [May (1976)]. Among them, bifurcations and chaos are very important phenomena which needs intensive study. In this essay, we will mainly discuss the changing properties of one simple nonlinear map. we firstly analyze the stability of the fixed points, then analyze the bifurcations of the system using center manifold theorem and bifurcation theory [Wiggins (1990); Guckenheimer (1997)]. We observe that there exist pitchfork bifurcation, flip bifurcation and Hopf bifurcation in the considered map.

In 1975, Li and Yorke introduced the first mathematical definition of chaos and established a very simple criterion for one-dimensional discrete dynamical systems on the interval, i.e., "period three implies chaos" for brevity [Li (1975)]. Three years later, Marotto generalized this result to multi-dimensional discrete dynamical systems, showing that the existence of a snap-back repeller implies chaos in the sense of Li–Yorke [Marotto (1978)]. Yet, an error had been found in the condition of original Marotto theorem, see [Chen (1998); Lin (2002)] for example. Li and Chen investigated the issue further and presented an improved version of Morotto theorem in their work [Li (2003)], which is called Marotto–Li–Chen Theorem. We will apply this theorem here to proving rigorously that the two-dimensional discrete system considered in this work possesses chaotic phenomenon in the sense of Li–Yorke.

16.2 Fixed points and bifurcations

16.2.1 *The existence of fixed points*

Consider a two-dimensional map which takes the form

$$\begin{cases} x \mapsto ax + \frac{3}{2}y, \\ y \mapsto b(x - x^3), \end{cases} \tag{16.1}$$

where $b \neq 0$. The fixed points of map (16.1) satisfy the following equations

$$\begin{cases} x = ax + \frac{3}{2}y, \\ y = b(x - x^3). \end{cases}$$

By a simple calculation, it is easy to obtain the following proposition.

Proposition 16.1.
(i) If $1 + \frac{2a}{3b} - \frac{2}{3b} \leq 0$, *then map (16.1) has a unique fixed point* $O\,(0,0)$.
(ii) If $1 + \frac{2a}{3b} - \frac{2}{3b} > 0$, *then map (16.1) has three fixed points:* $O\,(0,0)\,, A\,(\alpha,\beta)$, $B\,(-\alpha,-\beta)$, *where* $\alpha = \sqrt{1 + \frac{2a}{3b} - \frac{2}{3b}}, \beta = \frac{2}{3}\,(1-a)\sqrt{1 + \frac{2a}{3b} - \frac{2}{3b}}$.

16.2.2 *The stability of fixed points and bifurcations*

The Jacobian matrix J of map (16.1) evaluated at the fixed point (x_0, y_0) is given by

$$J = \begin{pmatrix} a & \frac{3}{2} \\ b(1-3x_0^2) & 0 \end{pmatrix}.$$

The characteristic equation of the Jacobian matrix J can be written as

$$\lambda^2 - a\lambda - \frac{3}{2}b\left(1 - 3x_0^2\right) = 0. \tag{16.2}$$

A simple straightforward calculation can decide the linear stability of the fixed point $O\,(0,0)$ for map (16.1).

Theorem 16.1. *(i) As* $a^2 + 6b < 0$, *if* $-\frac{2}{3} < b < 0$, *then* $O\,(0,0)$ *is a stable focus; if* $b < -\frac{2}{3}$, *then* $O\,(0,0)$ *is an unstable focus.*
(ii) As $a^2 + 6b = 0$, *if* $|a| < 2$, *then* $O\,(0,0)$ *is a stable node; if* $|a| > 2$, *then* $O\,(0,0)$ *is an unstable node.*
(iii) As $a^2 + 6b > 0, 0 < a < 2$,

if $b < \frac{2}{3}(1-a)$, *then* $O\,(0,0)$ *is a stable node;*
if $\frac{2}{3}(1-a) < b < \frac{2}{3}(1+a)$, *then* $O\,(0,0)$ *is a saddle;*
if $b > \frac{2}{3}(1+a)$, *then* $O\,(0,0)$ *is an unstable node.*

(iv) As $a^2 + 6b > 0, -2 < a < 0$,

if $b < \frac{2}{3}(1+a)$, *then* $O\,(0,0)$ *is a stable node;*
if $\frac{2}{3}(1+a) < b < \frac{2}{3}(1-a)$, *then* $O\,(0,0)$ *is a saddle;*
if $b > \frac{2}{3}(1-a)$, *then* $O\,(0,0)$ *is an unstable node.*

(v) If $a^2 + 6b > 0, |a| > 2$, *then* $O\,(0,0)$ *is unstable.*

Proof. Because the characteristic equation associated with the linearization of map (16.1) at the fixed point $O\,(0,0)$ can be written as

$$\lambda^2 - a\lambda - \frac{3}{2}b = 0,$$

the eigenvalues of the characteristic equation are $\lambda_1, \lambda_2 = \frac{a \pm \sqrt{a^2+6b}}{2}$.

(i) As $a^2 + 6b < 0$, we get $\lambda_{1,2} = \frac{a \pm i\sqrt{-a^2-6b}}{2}$, so $|\lambda_1| = |\lambda_2| = \sqrt{-\frac{3}{2}b}$. Therefore, if $-\frac{2}{3} < b < 0$, $|\lambda_{1,2}| < 1$. Thus $O\,(0,0)$ is a stable focus; when $b < -\frac{2}{3}$, $|\lambda_{1,2}| > 1$, so $O\,(0,0)$ is an unstable focus.

(ii) As $a^2+6b=0$, we obtain $\lambda_{1,2}=\frac{a}{2}$, so $|\lambda_{1,2}|=\frac{|a|}{2}$. If $|a|<2$, $|\lambda_{1,2}|<1$, then $O\,(0,0)$ is a stable node; if $|a|>2$, $|\lambda_{1,2}|>1$, thus $O\,(0,0)$ is an unstable node.

(iii) As $a^2+6b>0$ and $0<a<2$, we have $\lambda_2=\frac{a-\sqrt{a^2+6b}}{2}<\lambda_1=\frac{a+\sqrt{a^2+6b}}{2}$. If $b<\frac{2}{3}(1-a)$, then $3b<2-2a$, so $a^2+6b<4-4a+a^2$ and $a+\sqrt{a^2+6b}<2$, therefore $0<\lambda_1<1$; meanwhile, $\lambda_2=\frac{a-\sqrt{a^2+6b}}{2}>\frac{a-(2-a)}{2}=a-1>-1$, so $O\,(0,0)$ is a stable node;

Similarly, we can prove that if $\frac{2}{3}(1-a)<b<\frac{2}{3}(1+a)$, then $\lambda_1>1, -1<\lambda_2<1$, thus, $O\,(0,0)$ is a saddle.

If $b>\frac{2}{3}(1+a)$, then $\lambda_1>1, \lambda_2<-1$, thus $O\,(0,0)$ is an unstable node.

(iv) It is similar to the proof of (iii).

(v) If $a^2+6b>0, |a|>2$, then $\lambda_1,\lambda_2=\frac{a\pm\sqrt{a^2+6b}}{2}$. At least one of the eigenvalues is greater than 1 in absolute value, and thus $O\,(0,0)$ is unstable. □

It follows from the above analysis that $O\,(0,0)$ loses its stability at $b=-\frac{2}{3},\frac{2}{3}(1-a),\frac{2}{3}(1+a)$. Now we consider the dynamical behavior in the vicinity of $O\,(0,0)$ as $b=-\frac{2}{3},\frac{2}{3}(1-a),\frac{2}{3}(1+a)$.

We first change b to $\bar{b}$ by $\bar{b}=b-\frac{2}{3}(1-a)$ and consider the parameter $\bar{b}$ as a new and dependent variable, then map (16.1) becomes

$$\begin{cases} x\mapsto ax+\frac{3}{2}y, \\ y\mapsto \frac{2}{3}(1-a)x+\bar{b}x-[\frac{2}{3}(1-a)x^3+\bar{b}x^3], \\ \bar{b}\mapsto \bar{b}. \end{cases} \tag{16.3}$$

Rewrite map (16.3) in the matrix form

$$\begin{pmatrix} x \\ y \\ \bar{b} \end{pmatrix} \mapsto \begin{pmatrix} a & \frac{3}{2} & 0 \\ \frac{2}{3}(1-a) & 0 & 0 \\ 0 & 0 & 1 \end{pmatrix} \begin{pmatrix} x \\ y \\ \bar{b} \end{pmatrix} + \begin{pmatrix} 0 \\ \bar{b}x-[\frac{2}{3}(1-a)x^3+\bar{b}x^3] \\ 0 \end{pmatrix}.$$

We construct an invertible matrix

$$T=\begin{pmatrix} \frac{3}{2(1-a)} & \frac{3}{2} & 0 \\ 1 & -1 & 0 \\ 0 & 0 & 1 \end{pmatrix},$$

and use the transform

$$\begin{pmatrix} x \\ y \\ \bar{b} \end{pmatrix} = T \begin{pmatrix} u \\ v \\ \mu \end{pmatrix}.$$

Then map (16.1) becomes

$$\begin{pmatrix} u \\ v \\ \mu \end{pmatrix} \mapsto \begin{pmatrix} 1 & 0 & 0 \\ 0 & a-1 & 0 \\ 0 & 0 & 1 \end{pmatrix} \begin{pmatrix} u \\ v \\ \mu \end{pmatrix} + \begin{pmatrix} f(u,v,\mu) \\ g(u,v,\mu) \\ 0 \end{pmatrix}, \tag{16.4}$$

where

$$f(u,v,\mu)=\frac{1-a}{2-a}\left(\frac{3u}{2(1-a)}+\frac{3v}{2}\right)\mu-\frac{1-a}{2-a}\left(\frac{2}{3}(1-a)+\mu\right)\left(\frac{3u}{2(1-a)}+\frac{3v}{2}\right)^3,$$

$$g(u,v,\mu)=\frac{1}{a-2}\left(\frac{3u}{2(1-a)}+\frac{3v}{2}\right)\mu-\frac{1}{a-2}\left(\frac{2}{3}(1-a)+\mu\right)\left(\frac{3u}{2(1-a)}+\frac{3v}{2}\right)^3.$$

The center manifold theorem implies that the stability of $(u,v)=(0,0)$ near $\mu=0$ can be determined by studying a one-parameter family of equations on a center manifold given by

$$W^c(0)=\{(u,v,\mu)\in R^3\,|v=h(u,\mu),h(0,0)=0,\,Dh(0,0)=0\}$$

for u and μ sufficiently small. We assume that the center manifold takes the form

$$h(u,\mu)=a_1u^2+a_2u\mu+a_3\mu^2+a_4u^3+O((|u|+|\mu|)^3).$$

The center manifold must satisfy

$$N(h(u,\mu))=h(u+f(u,h(u,\mu),\mu),\mu)-(a-1)h(u,\mu)-g(u,h(u,\mu),\mu)=0.$$

The map (16.4) restricted to $W^c(0)$ is given by

$$u\mapsto \bar{f}(u,\mu)=u+f(u,h(u,\mu),\mu).$$

By a direct calculation, the following equality is then obtained

$$\bar{f}(u,\mu)=u+\frac{3u\mu}{2(2-a)}-\frac{9u^3}{4(2-a)(1-a)}+O((|u|+|\mu|)^3).$$

Therefore $\bar{f}(0,0)=0$, $\frac{\partial\bar{f}}{\partial u}(0,0)=1$, $\frac{\partial\bar{f}}{\partial\mu}(0,0)=0$, $\frac{\partial^2\bar{f}}{\partial u\partial\mu}(0,0)=\frac{3}{2(2-a)}\neq 0(a\neq 2)$, $\frac{\partial^3\bar{f}}{\partial u^3}(0,0)=-\frac{27}{2(2-a)(1-a)}\neq 0(a\neq 1,2)$. It follows from bifurcation theory that the fixed point $(u,v,\mu)=(0,0,0)$ is a pitchfork bifurcation point for map (16.4).

Theorem 16.2. *If $a\neq 1,2$, then map (16.1) undergoes a pitchfork bifurcation at $O(0,0)$ for $b=\frac{2}{3}(1-a)$.*

Next we turn to consider the case of $b=\frac{2}{3}(1+a)$. Let $\bar{b}=b-\frac{2}{3}(1+a)$, then map(16.1) is changed to

$$\begin{cases} x\mapsto ax+\frac{3}{2}y,\\ y\mapsto \frac{2}{3}(1+a)x+\bar{b}x-[\frac{2}{3}(1+a)x^3+\bar{b}x^3],\\ \bar{b}\mapsto\bar{b}. \end{cases}\tag{16.5}$$

We construct an invertible matrix

$$T=\begin{pmatrix} -\frac{3}{2(1+a)} & \frac{3}{2} & 0\\ 1 & 1 & 0\\ 0 & 0 & 1\end{pmatrix},$$

and apply the transform

$$\begin{pmatrix} x\\ y\\ \bar{b}\end{pmatrix}=T\begin{pmatrix} u\\ v\\ \mu\end{pmatrix}.$$

Then map (16.5) can be written as

$$\begin{pmatrix} u \\ v \\ \mu \end{pmatrix} \mapsto \begin{pmatrix} -1 & 0 & 0 \\ 0 & a+1 & 0 \\ 0 & 0 & 1 \end{pmatrix} \begin{pmatrix} u \\ v \\ \mu \end{pmatrix} + \begin{pmatrix} f(u,v,\mu) \\ g(u,v,\mu) \\ 0 \end{pmatrix},$$

where

$$f(u,v,\mu) = \frac{1+a}{2+a}\left(\frac{3v}{2} - \frac{3u}{2(1+a)}\right)\mu - \frac{1+a}{2+a}\left(\frac{2}{3}(1+a)+\mu\right)\left(\frac{3v}{2} - \frac{3u}{2(1+a)}\right)^3,$$

$$g(u,v,\mu) = \frac{1}{a+2}\left(\frac{3v}{2} - \frac{3u}{2(1+a)}\right)\mu - \frac{1}{a+2}\left(\frac{2}{3}(1+a)+\mu\right)\left(\frac{3v}{2} - \frac{3u}{2(1+a)}\right)^3.$$

We once again apply the center manifold theorem to determine the bifurcation of the fixed point $(u,v) = (0,0)$ at $\mu = 0$. There exists a center manifold for map (16.5), which can be represented as

$$W^c(0) = \{(u,v,\mu) \in R^3 \,|v = h(u,\mu), h(0,0) = 0,\, Dh(0,0) = 0\}.$$

Assume $h(u,\mu) = b_1u^2 + b_2u\mu + b_3\mu^2 + b_4u^3 + O((|u| + |\mu|)^3)$, then the center manifold must satisfy

$$N(h(u,\mu)) = h(-u + f(u,h(u,\mu),\mu),\mu) - (a+1)h(u,\mu) - g(u,h(u,\mu),\mu) = 0.$$

By a direct calculation, yields

$$-ab_1u^2 + \left(\frac{3}{2(1+a)(2+a)} - (2+a)b_2\right)u\mu - ab_3\mu^2$$
$$-\left((2+a)b_4 + \frac{9}{4(2+a)(1+a)^2}\right)u^3 + O\left((|u| + |\mu|)^3\right) = 0.$$

Comparing the coefficients, we obtain

$$b_1 = 0, b_2 = \frac{3}{2(1+a)(2+a)^2}, b_3 = 0, b_4 = -\frac{9}{4(1+a)^2(2+a)^2}(a \neq 0,-1,-2).$$

Then the map (16.5) restricted to $W^c(0)$ is represented as

$$\hat{f} : u \mapsto -u + f(u,h(u,\mu),\mu).$$

A direct calculation leads to

$$\hat{f} : u \mapsto -u - \frac{3}{2(2+a)}u\mu + \frac{9}{a(1+a)(2+a)}u^3 + O\left((|u| + |\mu|)^3\right).$$

The second iteration of $\hat{f}$ is give by

$$\hat{f}^2 : u \mapsto u + \frac{3}{2+a}u\mu - \frac{9}{2(1+a)(2+a)}u^3 + O\left((|u| + |\mu|)^3\right).$$

Thus, $\hat{f}(0,0) = 0$, $\frac{\partial\hat{f}}{\partial u}(0,0) = -1$, $\frac{\partial\hat{f}}{\partial\mu}(0,0) = 0$, $\frac{\partial^2\hat{f}^2}{\partial u^2}(0,0) = 0$, $\frac{\partial^2\hat{f}^2}{\partial u\partial\mu}(0,0) = \frac{3}{2+a} \neq 0$, $\frac{\partial^3\hat{f}^2}{\partial u^3}(0,0) = -\frac{27}{(2+a)(1+a)} \neq 0 (a \neq -1,-2)$, the fixed point $(u,v,\mu) = (0,0,0)$ is a flip bifurcation point for map (16.5).

Theorem 16.3. *If $a \neq 0,-1,-2$, then map (16.1) undergoes a flip bifurcation at $(x,y,b) = (0,0,\frac{2}{3}(1+a))$.*

There exist Hopf bifurcation in the considered map as well. We need the Hopf bifurcation theorem.

Theorem 16.4 (Hopf bifurcation theorem [Guckenheimer (1997)]). *Let $f_\mu = f(x,\mu) : R^2 \to R^2$ be a one-parameter family of mappings which has a smooth family of fixed points $x(\mu)$ at which the eigenvalues are complex conjugates $\lambda(\mu), \bar{\lambda}(\mu)$. Assume*

(SH1) $|\lambda(\mu_0)| = 1$ *but* $\lambda^j(\mu_0) \neq 1$ *for* $j = 1, 2, 3, 4$.

(SH2) $\left.\frac{d(|\lambda(\mu)|)}{d\mu}\right|_{\mu=\mu_0} = d \neq 0$.

Then there is a smooth change of coordinates h so that the expression of $hf_\mu h^{-1}$ in polar coordinates has the form

$hf_\mu h^{-1}(r,\theta) = \big(r(1 + d(\mu - \mu_0) + \alpha r^2), \theta + c + \beta r^2\big)$ + *higher-order terms, where* $c = |\arg(\lambda)|$. *If, in addition*

(SH3) $\alpha \neq 0$.

Then there is a two-dimensional surface Σ (not necessarily infinitely differentiable) in $R^2 \times R$ having quadratic tangency with the plane $R^2 \times \{\mu_0\}$ which is invariant under f_μ. If $\Sigma \cap (R^2 \times \{\mu\})$ is larger than a point, then it is a simple closed curve. Assuming that the bifurcating system (restricted to the center manifold) is in the form

$$\begin{pmatrix} x \\ y \end{pmatrix} \mapsto \begin{pmatrix} \cos(c) & -\sin(c) \\ \sin(c) & \cos(c) \end{pmatrix} \begin{pmatrix} x \\ y \end{pmatrix} + \begin{pmatrix} f(x,y) \\ g(x,y) \end{pmatrix},$$

with eigenvalues $\lambda, \bar{\lambda} = e^{\pm ic}$, one obtains

$$\alpha = -Re\left[\frac{(1-2\lambda)\bar{\lambda}^2}{1-\lambda}\xi_{11}\xi_{20}\right] - \frac{1}{2}|\xi_{11}|^2 - |\xi_{02}|^2 + Re(\bar{\lambda}\xi_{21}),$$

where

$$\begin{aligned}
\xi_{20} &= \tfrac{1}{8}\left[(f_{xx} - f_{yy} + 2g_{xy}) + i(g_{xx} - g_{yy} - 2f_{xy})\right], \\
\xi_{11} &= \tfrac{1}{4}\left[(f_{xx} + f_{yy}) + i(g_{xx} + g_{yy})\right], \\
\xi_{02} &= \tfrac{1}{8}[(f_{xx} - f_{yy} - 2g_{xy}) + i(g_{xx} - g_{yy} + 2f_{xy})], \\
\xi_{21} &= \tfrac{1}{16}\left[(f_{xxx} + f_{xyy} + g_{xxy} + g_{yyy}) + i(g_{xxx} + g_{xyy} - f_{xxy} - f_{yyy})\right].
\end{aligned}$$

Using Theorem 16.4, we can deduce the following theorem on the Hopf bifurcation occurring in map (16.1).

Theorem 16.5. *If $|a| < 2, a \neq 0, 1$, then map* (16.1) *undergoes a Hopf bifurcation at the fixed point $O(0,0)$ for $b = -\frac{2}{3}$.*

Proof. Because the eigenvalues of the characteristic equation of the map (16.1) for the fixed point $O(0,0)$ are $\lambda, \bar{\lambda} = \frac{a \pm \sqrt{a^2+6b}}{2}$, the eigenvalues $\lambda, \bar{\lambda}$ are complex conjugate for $b < -\frac{a^2}{6}$. As $b = -\frac{2}{3}$, the eigenvalues at the fixed point $O(0,0)$ are complex conjugate with modulus 1.

It follows from the conditions $|a| < 2$ and $b = -\frac{2}{3}$ that $|\lambda| = \sqrt{-\frac{3}{2}b} = 1$ and $\left.\frac{d(|\lambda(b)|)}{db}\right|_{b=-\frac{2}{3}} = -\frac{3}{4} \neq 0$. In addition, when $a \neq 0, 1$, $\lambda^j(-\frac{2}{3}) \neq 1 (j = 1, 2, 3, 4)$. Let

$$T = \begin{pmatrix} -3\sqrt{4-a^2}/4 & -3a/4 \\ 0 & 1 \end{pmatrix}$$

and apply the following transform

$$\begin{pmatrix} x \\ y \end{pmatrix} = T \begin{pmatrix} u \\ v \end{pmatrix},$$

map (16.1) becomes

$$\begin{pmatrix} u \\ v \end{pmatrix} \mapsto \begin{pmatrix} a/2 & -\sqrt{4-a^2}/2 \\ \sqrt{4-a^2}/2 & a/2 \end{pmatrix} \begin{pmatrix} u \\ v \end{pmatrix} + \begin{pmatrix} f(u,v) \\ g(u,v) \end{pmatrix}, \tag{16.6}$$

where

$$f(u,v) = \frac{2a}{3\sqrt{4-a^2}} \left(\frac{3\sqrt{4-a^2}}{4} u + \frac{3a}{4} v \right)^3,$$

$$g(u,v) = -\frac{2}{3} \left(\frac{3\sqrt{4-a^2}}{4} u + \frac{3a}{4} v \right)^3.$$

Notice that (16.6) is exactly in the form on the manifold, the coefficient α in Theorem 16.4 is given by

$$\alpha = -Re\left[\frac{(1-2\lambda)\bar{\lambda}^2}{1-\lambda}\xi_{11}\xi_{20}\right] - \frac{1}{2}|\xi_{11}|^2 - |\xi_{02}|^2 + Re(\bar{\lambda}\xi_{21}),$$

where

$$\begin{aligned}
\xi_{20} &= \tfrac{1}{8}\left[(f_{uu} - f_{vv} + 2g_{uv}) + i(g_{uu} - g_{vv} - 2f_{uv})\right]\big|_{u=0,v=0} = 0, \\
\xi_{11} &= \tfrac{1}{4}\left[(f_{uu} + f_{vv}) + i(g_{uu} + g_{vv})\right]\big|_{u=0,v=0} = 0, \\
\xi_{02} &= \tfrac{1}{8}\left[(f_{uu} - f_{vv} - 2g_{uv}) + i(g_{uu} - g_{vv} + 2f_{uv})\right]\big|_{u=0,v=0} = 0,
\end{aligned}$$

and

$$\xi_{21} = \frac{1}{16}\left[(f_{uuu} + f_{uvv} + g_{uuv} + g_{vvv}) + i(g_{uuu} + g_{uvv} - f_{uuv} - f_{vvv})\right]\Big|_{u=0,v=0}$$

$$= -\frac{1}{16} \times \left(\frac{3}{2}\right)^3 \left(\frac{a^4}{2\sqrt{4-a^2}} + a^2\sqrt{4-a^2} + \frac{(4-a^2)\sqrt{4-a^2}}{2} \right) i.$$

All the above facts imply that $\alpha = Re\left(\bar{\lambda}\xi_{21}\right) = -\frac{27}{32} < 0$, by Theorem 16.4, we know that when $|a| < 2, a \neq 0, 1$, map (16.1) undergoes a Hopf bifurcation at the fixed point $O(0,0)$ for $b = -\frac{2}{3}$. □

So far, we discuss the the stability of the zero fixed point $O(0,0)$ and its corresponding bifurcation. We now turn to analyze the non-zero fixed points.

By (16.2), we know that when $1+\frac{2a}{3b}-\frac{2}{3b}>0$, there are two non-zero fixed points $A(\alpha,\beta), B(-\alpha,-\beta)$ in which $\alpha=\sqrt{1+\frac{2a}{3b}-\frac{2}{3b}}, \beta=\frac{2}{3}(1-a)\sqrt{1+\frac{2a}{3b}-\frac{2}{3b}}$. The characteristic equation associated with the linearization of map (16.1) at the two non-zero fixed points is given by

$$\lambda^2-a\lambda-3(1-a-b)=0.$$

The eigenvalues of the characteristic equation are $\lambda_{1,2}=\frac{a\pm\sqrt{a^2+12(1-a-b)}}{2}$. We discuss the following eight cases under the assumption of $1+\frac{2a}{3b}-\frac{2}{3b}>0$.

(i) If $b<0$, then $2-2a-3b>0$, and $\lambda_{1,2}=\frac{a\pm\sqrt{a^2+12(1-a-b)}}{2}=\frac{a\pm\sqrt{(a-2)^2+4(2-2a-3b)}}{2}$, so at least one of the eigenvalues is greater than unity in modulus and the fixed points $A(\alpha,\beta), B(-\alpha,-\beta)$ are always unstable.

(ii) If $a^2+12(1-a-b)>0, b>0, 0<a<2$, then $A(\alpha,\beta), B(-\alpha,-\beta)$ are stable nodes. It follows from the following facts. Owing to $1+\frac{2a}{3b}-\frac{2}{3b}>0, b>0$, one has $3b+2a-2>0$, that is, $1-a>3(1-a-b)$, or equivalently, $a^2+12(1-a-b)<(2-a)^2$, implying $a+\sqrt{a^2+12(1-a-b)}<2$. Therefore $0<\lambda_1<1$. A direct calculation shows that $-1<\lambda_2<\lambda_1<1$.

(iii) If $a^2+12(1-a-b)>0, b>0, |a|>2$, then $A(\alpha,\beta), B(-\alpha,-\beta)$ are unstable.

(iv) If $a^2+12(1-a-b)>0, b>\frac{2}{3}(1-2a), -2<a<0$, then $A(\alpha,\beta), B(-\alpha,-\beta)$ are stable nodes, which is deduced as follows. Due to $b>\frac{2}{3}(1-2a)$, we obtain $3b>2-4a$, $1+a>3(1-a-b)$, thus $(a+2)^2>a^2+12(1-a-b)$. A simple calculation implies $-1<\lambda_2<\lambda_1<1$.

(v) If $a^2+12(1-a-b)>0, \frac{2}{3}(1-a)<b<\frac{2}{3}(1-2a), -2<a<0$, then $A(\alpha,\beta), B(-\alpha,-\beta)$ are saddles. The proof for (iii) implies that if $b<\frac{2}{3}(1-2a)$, then $a-\sqrt{a^2+12(1-a-b)}<-2$, yielding $\lambda_2<-1$. The condition $\frac{2}{3}(1-a)<b$ results in $|\lambda_1|<1$.

(vi) If $a^2+12(1-a-b)>0, b<\frac{2}{3}(1-a), -2<a<0$, then $A(\alpha,\beta), B(-\alpha,-\beta)$ are unstable nodes.

(vii) If $a^2+12(1-a-b)<0, 0<b<\frac{4}{3}-a, a<\frac{4}{3}$, then $A(\alpha,\beta), B(-\alpha,-\beta)$ are stable foci. In fact, $a^2+12(1-a-b)<0$, $\lambda_{1,2}=\frac{a\pm i\sqrt{-a^2-12(1-a-b)}}{2}$, so $|\lambda|=\sqrt{-3(1-a-b)}$. $b<\frac{4}{3}-a$ implies $3b<4-3a$, therefore $-3(1-a-b)<1$ and then $|\lambda|<1$.

(viii) If $a^2+12(1-a-b)<0, b>\frac{4}{3}-a, a<\frac{4}{3}$, then $A(\alpha,\beta), B(-\alpha,-\beta)$ are unstable foci. The proof for (vii) implies that if $b>\frac{4}{3}-a$, $|\lambda|>1$. When $b=\frac{4}{3}-a$, $|\lambda|=1$, so we may expect a Hopf bifurcation to appear.

Theorem 16.6. *If $a<\frac{4}{3}, a^2+12(1-a-b)<0, 1+\frac{2a}{3b}-\frac{2}{3b}>0, a\neq 0,1$, then map (16.1) undergoes a Hopf bifurcation at the fixed point $A(\alpha,\beta), B(-\alpha,-\beta)$ for $b=\frac{4}{3}-a$.*

Proof. The eigenvalues of the characteristic equation of the map (16.1) for the fixed point $A(\alpha,\beta), B(-\alpha,-\beta)$ are $\lambda_{1,2} = \frac{a \pm i\sqrt{-a^2-12(1-a-b)}}{2}$, Therefore, when $b = \frac{4}{3} - a$, $|\lambda| = \sqrt{-3(1-a-b)} = 1$, and $\left.\frac{d(|\lambda(b)|)}{db}\right|_{b=\frac{4}{3}-a} = \frac{3}{2} \neq 0$, in addition when $a \neq 0, 1$, $\lambda^j(\frac{4}{3} - a) \neq 1 (j = 1, 2, 3, 4)$. By Theorem 16.4, when $a < \frac{4}{3}, a^2 + 12(1-a-b) < 0, 1 + \frac{2a}{3b} - \frac{2}{3b} > 0, a \neq 0, 1$, map (16.1) undergoes a Hopf bifurcation at the fixed point $A(\alpha,\beta), B(-\alpha,-\beta)$ for $b = \frac{4}{3} - a$. □

16.3 Existence of Marotto–Li–Chen chaos

In this section, we rigorously prove that map (16.1) possesses chaotic behavior in the sense of Li–York's definition. We first present Marotto–Li–Chen chaos theorem quoted from [Li (2003)].

Theorem 16.7. *(Marotto–Li–Chen Theorem). Consider the following n-dimensional discrete dynamical system:*

$$x_{k+1} = f(x_k), \quad x_k \in R^n, k = 0, 1, \cdots. \tag{16.7}$$

Suppose that $f(x)$ has a fixed point z. Assume also that

(i) $f(x)$ is continuously differential in $B_r(z)$ for some $r > 0$, where $B_r(z)$ is a closed ball in R^n of radius r centered at the point z.

(ii) all eigenvalues of $[f'(z)]^T f'(z)$ are greater than 1, in which $f'(z)$ denotes the Jacobian of $f(x)$ at z and superscript ^{T} denotes the transposition.

(iii) there exists a point $x^0 \in B'_r(z) = \{x : ||x - z|| \leq r' \leq r\}$ and all eigenvalues of $[f'(z)]^T f'(z)$ are larger than 1, with $x^0 \neq z$, such that $f^M(x^0) = z$ where $f^i(x^0) \in B_r(z), i = 0, 1, ..., M$, and the determinant $\left|[f^M(x)]'_{x=x^0}\right| \neq 0$, for some positive integer M.

Then, system (16.7) is chaotic in the sense of Li–Yorke, that is,

(a) There is a positive integer N such that for each integer $p \geq N$, $f(x)$ has a point x_p of period p, i.e., $f^p(x_p) = x_p$ but $x^k(x_p) \neq x_p$ for $1 \leq k < p$.

(b) There is an uncountable set S containing no periodic points of $f(x)$, such that

(b1) $f(S) \subset S$,

(b2) for every $x_s, y_s \in S$ with $x_s \neq y_s$,

$$\limsup_{k\to\infty} ||f^k(x_s) - f^k(y_s)|| > 0.$$

(b3) for every $x_s \in S$ and any periodic point y_p of $f(x)$,

$$\limsup_{k\to\infty} ||f^k(x_s) - f^k(y_p)|| > 0.$$

(c) There is an uncountable subset S_0 of S such that for every $x_{s_0}, y_{s_0} \in S_0$,

$$\liminf_{k\to\infty} ||f^k(x_{s_0}) - f^k(y_{s_0})|| = 0.$$

We now apply Theorem 16.7 to show that at $a = 1, b = -1.4$, the map (16.1) is chaotic in the sense of Li–Yorke. The map (16.1) becomes

$$\begin{cases} x \mapsto x + \frac{3}{2}y, \\ y \mapsto -1.4(x - x^3). \end{cases}$$

So the Jacobian matrix J of map (16.1) evaluated at the fixed point $O(0,0)$ is given by

$$J = \begin{pmatrix} 1 & 1.5 \\ -1.4 & 0 \end{pmatrix},$$

then $J'J = \begin{pmatrix} 2.96 & 1.5 \\ 1.5 & 2.25 \end{pmatrix}$. The characteristic equation of the matrix $J'J$ can be written as $\lambda^2 - 5.21\lambda + 4.41 = 0$, the eigenvalues of the characteristic equation are $\lambda_1 = 4.1464, \lambda_2 = 1.0636$. In order to find x^0, we assume (x', y') satisfies

$$\begin{cases} x' + \frac{3}{2}y' = 0, \\ -1.4(x' - x'^3) = 0, \end{cases}$$

the roots of the equation are $(x_1', y_1') = (1, -\frac{2}{3}), (x_2', y_2') = (-1, \frac{2}{3})$. Now we assume (x_0, y_0) satisfies

$$\begin{cases} x_0 + \frac{3}{2}y_0 = 1, \\ -1.4(x_0 - x_0^3) = -\frac{2}{3}, \end{cases}$$

the real root of this equation is $(x_0, y_0) = (-1.1841, 1.4561)$. Therefore, there exists $x^0 = (x_0, y_0) \neq (0,0)$ such that $f^2(x_0, y_0) = (0,0)$. Theorem 16.7 implies that when $a = 1, b = -1.4$, the map (16.1) is chaotic in the sense of Li-Yorke.

16.4 Numerical simulation results

In this section, some numerical simulation results are presented to verify the existence of bifurcations, chaotic behaviors in the considered map. The maximum Lyapunov exponents, phase portraits and bifurcation diagrams are calculated and depicted to show the complex dynamics of map (16.1). The phase portrait of map (16.1) for $a = 1$, $b = -1.4$ is given in Fig. 16.1. Let $a = 1$ and b range from -1.5 to 0.7, the maximum Lyapunov exponents are also calculated and plotted in Fig. 16.2. Fig. 16.3 is the bifurcation diagram, showing the output of x component with respect to the parameter b for fixed $a = 1$. From Figs. 16.2–16.3, one can observe that map (16.1) appears chaos for some regions of parameter b. All the numerical results presented here demonstrate that there exist complex dynamics even in simple two-dimensional noninvertible maps.

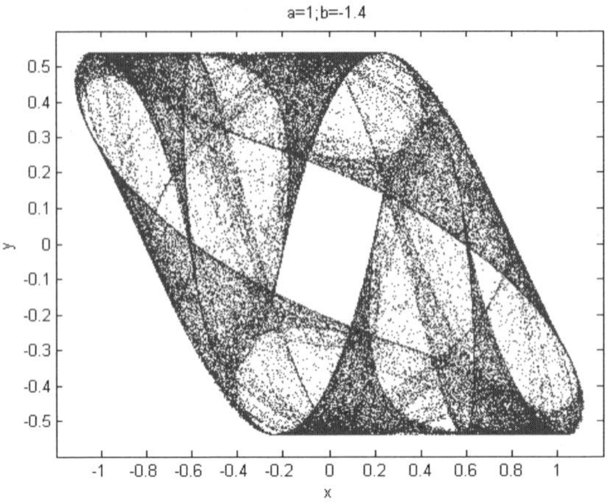

Fig. 16.1 Phase portrait of map (16.1) for $a = 1$ and $b = -1.4$.

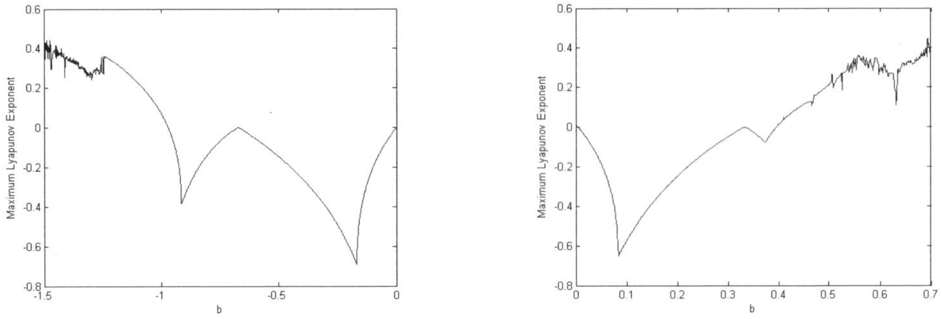

Fig. 16.2 The maximum Lyapunov exponent vs. the parameter b.

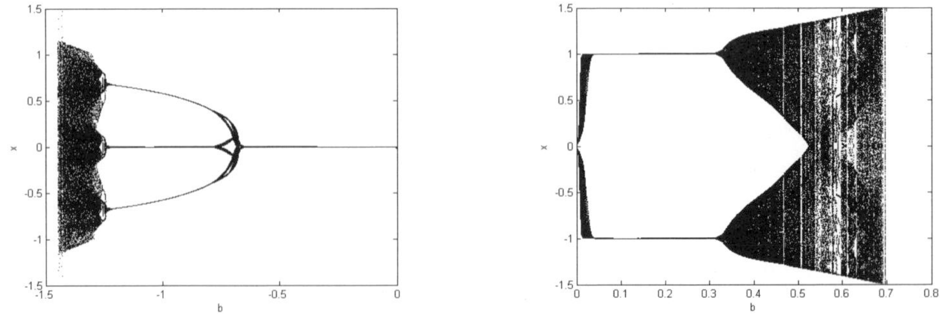

Fig. 16.3 Bifurcation diagram of x output vs. the parameter b.

Bibliography

Chen, G., Hsu, S. B. and Zhou, J. (1998). Snapback repellers as a cause of chaotic vibration of the wave equation with a van del Pol boundary condition and energy injection at the middle of the span, *J. Math. Phys.* **39**, pp. 6459–6489.

Gao, Y. (2009). Complex dynamics in a two-dimensional noninvertible map, *Chaos, Solitons and Fractals* **39**, pp. 1798–1810.

Guckenheimer, J. and Holmes, P. (1997). *Nonlinear Oscillations, Dynamical Systems and Bifurcations of Vector Fields* (New York: Springer-Verlag).

Henon, M. (1976). A two-dimensional mapping with a strange attractor, *Commun. Math. Phys.* **50**, pp. 69–77.

Li, T. Y. and Yorke, J. A. (1975). Periodic three implies chaos, *Am. Math. Monthly* **82**, pp. 985–992.

Li, C. P. and Chen, G. (2003). An improved version of the Marotto Theorem, *Chaos, Solitons and Fractals* **18**, pp. 69–77; Erratum: an improved version of the Marotto Theorem, *Chaos, Solitons and Fractals* **20**, pp. 855.

Lin, W., Ruan, J. and Zhao, W. R. (2002). On the mathematical clarification of the snap-back-repeller in high-dimensional system and chaos in a discrete neural network model, *Int. J. Bifurcation and Chaos* **12**, pp. 1129–1139.

Marotto, F. R. (1978). Snap-back repellers imply chaos in R^n, *J. Math. Anal. Appl.* **63**, pp. 199–223.

May, R. (1976). Simple mathematical models with very complicated dynamics, *Nature* **261**, pp. 459–467.

Mira, C., Gardini, L., Barugola, A. and Cathala, J. C. (1996). *Chaotic dynamics in two-dimensional noninvertible maps* (World Scientific, Singapore).

Morris, W. H., Stephen, S. and Robert, L. D. (2004). *Differential Equations, Dynamical Systems, and an Introduction to Chaos* (Elsevier, Singapore).

Ott, E. (1993). *Chaos in Dynamical Systems* (Cambridge University Press, Cambridge).

Paulo, C. Rech. (2008). Naimark-Sacker bifurcations in a delay quartic map, *Chaos, Solitons and Fractals* **37**, pp. 387–392.

Richter, H. (2008). On a family of maps with multiple chaotic attractors, *Chaos, Solitons and Fractals* **36**, pp. 559–571.

Robinson, C. (1994). *Dynamical Systems* (CRC, Boca Raton, FL).

Sun, H. and Cao, H. (2007). Bifurcation and chaos of a delayed ecological model, *Chaos, Solitons and Fractals* **33**, pp. 1383–1393.

Wiggins, S. (1990). *An Introduction to Applied Nonlinear Dynamics and Chaos* (Springer-Verlag, New York).

Chapter 17

Approximate periodic solutions of damped harmonic oscillators with delayed feedback[1]

Qian Guo

Department of Mathematics, Shanghai Normal University, Shanghai 200234, PR China

Scientific Computing Key Laboratory of Shanghai Universities, Shanghai 200234, PR China

qguo@shnu.edu.cn

In this chapter, taking the time delay as bifurcation parameter, we investigate the periodic dynamics of a harmonic oscillator with delayed feedback. First of all, we present concise conditions of Hopf bifurcation so as to be utilized conveniently. Thereafter, for damped harmonic oscillator, approximate analytical expression of periodic solution is derived by means of multiple-scales method. Meanwhile we discuss local stability of the trivial solution and periodic solution arising from trivial solution via Hopf bifurcation. Then we develop another analytic method, which is based on Lyapunov-Schmidt reduction method, to approximate periodic branch of a harmonic oscillator with delayed feedback near Hopf bifurcation point. We verify the accuracy of approximate expression by calculating the error, which illustrates effectiveness of our method. Furthermore, period-doubling route and chaos are explored by numerical experiment. Several period-doubling cascades are showed in detail so that one can intuitively know periodic dynamics of delay differential equation.

Keywords: delay differential equation, Lyapunov-Schmidt reduction, periodic solution

[1]The author would like to acknowledge the supports from Natural Science Foundation of China (grant no. 10901106), Natural Science Foundation of Shanghai (grant no. 09ZR1423200), Innovation Program of Shanghai Municipal Education Commission (grant no. 09YZ150), E-Institutes of Shanghai Municipal Education Commission (grant no. E03004), Shanghai Leading Academic Discipline Project (grant no. S30405).

17.1 Introduction

A harmonic oscillator with delayed feedback (HOwDF) is a kind of delay differential equation (DDE). Due to defining in an infinite dimensional space, a scalar DDE has more complex dynamics than a scalar ordinary differential equation (ODE). The dynamics of systems with time delay are attractive research object in a variety of scientific fields [Erneux (2009)]. As we know, a Hopf bifurcation point is important to analyze periodic solution. A periodic motion may arise via Hopf bifurcation of a stationary solution for a scalar DDE or DDE system. Furthermore, a periodic motion of DDE frequently transits into chaos. Hopf bifurcation analyses for some harmonic oscillators with delayed feedback have been studied by some authors [Campbell *et al.* (1995a,b); Deng *et al.* (2006); Guo and Li (2007); Xu *et al.* (2007); Song *et al.* (2008)]. Moreover, a double Hopf bifurcation (also called Hopf-Hopf bifurcation) occurs where two pure imaginary roots' curves intersect transversally, which has been studied when the feedback function is simple monotone negative [Campbell *et al.* (1995b)]. Recently, co-dimension two bifurcations of harmonic oscillators, such as Takens-Bogdanov bifurcation, were reported [Xu *et al.* (2007); Song *et al.* (2008); Peng *et al.* (2009)].

The approximate analytical formula is preferable to gain an insight into the periodic motion and their stability. There are different analytical methods usually used in finding an approximate analytical expression of a bifurcating periodic solution, such as *center manifold method* [Carr (1981)], *Lyapunov-Schmidt reduction method* (hereinafter called L-S method) [Chow and Hale (1982); Li and Chen (2001)], *perturbation method* [Nayfeh (1973); Li and Chen (2002)] (including multiple scales method, Poincaré-Lindstedt method, etc.). Another alternative is the treatment of Hopf bifurcation in the frequency domain thanks to Laplace transform, called graphical Hopf method and harmonic balance analysis (see [Moiola and Chen (1996)] and references therein).

The main idea of center manifold analysis and L-S method is to reduce the DDE system, which is infinite dimensional, to a much lower (often two) dimensional system by projecting the original dynamics onto the eigenvectors corresponding to purely imaginary eigenvalues. Center manifold reduction has been used to study periodic motion of some cases of harmonic oscillators [Campbell *et al.* (1995a,b); Xu *et al.* (2007); Song *et al.* (2008)]. We developed the L-S method for periodic solutions of DDE [Yang and Guo (2005); Guo and Li (2007)], which is effective in deriving approximate analytical expressions of periodic solutions. At the same time, we can attain the relationship between the bifurcation parameter and the period by means of L-S method. To the best of our knowledge, the first two methods depend on the *norm form* to analyze stability of periodic solution. About the harmonic balance method, as far as I know, it is with the aid of the *monodromy matrix* to analyze stability of periodic solution. It is helpful to analyze the stability of trivial solution and periodic solutions near the Hopf bifurcation points directly by multiple scales analysis [Wang and Hu (2003)]. Moreover, several examples have been used to show the differences between multiple scales method and center manifold reduction in [Nayfeh (2008)].

In this chapter, we study the periodic dynamics of

$$\ddot{y} + \beta \dot{y} + \alpha y = f(y_\tau), \tag{17.1}$$

where $\alpha, \beta \in \mathbb{R}$ are parameters, $y_\tau \triangleq y(s-\tau)$, $\tau \in \mathbb{R}^+$ represents time delay. $y(s)$ and $\dot{y}(s)$ represent the position and velocity at time s respectively, and f is a nonlinear smooth function which represents position feedback.

In what follows, we focus on two analytical approximate methods, Lyapunov-Schmidt reduction method and multiple time scales technique, for periodic solutions of a damped harmonic oscillator with delayed feedback. Meanwhile, we study period-doubling cascade numerically.

17.2 Hopf bifurcation analysis

The steady-state solution of (17.1), y^*, is solved from

$$\alpha y^* = f(y^*).$$

Taylor series expansion of $f(y_\tau)$ centered at $y = y^*$ is

$$\begin{aligned} f(y_\tau) = f(y^*) &+ \frac{df}{dy}\Big|_{y^*}\Big(y(s-\tau) - y^*\Big) \\ &+ \frac{1}{2}\frac{d^2 f}{dy^2}\Big|_{y^*}\Big(y(s-\tau) - y^*\Big)^2 + O\Big((y(s-\tau) - y^*)^3\Big). \end{aligned} \tag{17.2}$$

From (17.1) and (17.2), let $u(s) = y(s) - y^*$, we have

$$\frac{d^2u}{ds^2} + \beta\frac{du}{ds} + \alpha u = d_1 u(s-\tau) + d_2 u^2(s-\tau) + O\Big(u^3(s-\tau)\Big), \tag{17.3}$$

where $d_m = \frac{1}{m!}\frac{d^m f}{dy^m}|_{y^*}$, $m = 1, 2, \cdots$.

Taking the time delay as a bifurcation parameter, we introduce a new time variable t defined by $t = s/\tau$. (17.3) can be rewritten as

$$\frac{d^2x}{dt^2} + \tau\beta\frac{dx}{dt} + \tau^2\alpha x - \tau^2 d_1 x(t-1) = \tau^2 d_2 x^2(t-1) + O\Big(\tau^2 x^3(t-1)\Big), \tag{17.4}$$

where $x(t) \triangleq u(s)$, $x(t-1) \triangleq u(s-\tau)$.

DDE (17.4) has a trivial steady state solution and the characteristic equation for linearized equation of (17.4) is

$$\lambda^2 + \tau\beta\lambda + \tau^2\alpha - \tau^2 d_1 e^{-\lambda} = 0. \tag{17.5}$$

We set $\lambda(\tau) = \xi(\tau) \pm i\ \sigma(\tau)$, where τ is the bifurcation parameter and $\sigma(\tau) > 0$. By separating the real and imaginary parts, characteristic equation (17.5) yields

$$\begin{cases} \xi^2 - \sigma^2 + \tau\beta\xi + \tau^2\alpha - \tau^2 d_1 e^{-\xi}\cos\sigma = 0, & (17.6a) \\ 2\xi\sigma + \tau\beta\sigma + \tau^2 d_1 e^{-\xi}\sin\sigma = 0. & (17.6b) \end{cases}$$

Suppose when $\tau = \tau_0$, (17.5) has pure imaginary roots, that is, $\xi(\tau_0) = 0$ and $\sigma_1 \triangleq \sigma(\tau_0)$. Substituting τ with τ_0 in (17.6a) and (17.6b) yields

$$\begin{cases} \tau_0^2\alpha - \sigma_1^2 = \tau_0^2 d_1 \cos\sigma_1, & (17.7a) \\ -\tau_0\beta\sigma_1 = \tau_0^2 d_1 \sin\sigma_1, & (17.7b) \end{cases}$$

or

$$(\alpha - \sigma_0^2)^2 + \beta^2\sigma_0^2 = d_1^2, \tag{17.8}$$

where $\sigma_0 \triangleq \sigma_1/\tau_0$, τ_0 is a critical value of Hopf bifurcation.

Differentiating (17.5) with respect to τ gives

$$\left.\frac{d\xi}{d\tau}\right|_{\tau_0} = \frac{\sigma_1^2\tau_0[2\sigma_0^2 - (2\alpha - \beta^2)]}{\Delta_1}, \tag{17.9}$$

when $\Delta_1 \triangleq (\tau_0\beta + \tau_0^2\alpha - \sigma_1^2)^2 + (2\sigma_1 + \beta\sigma_1\tau_0)^2 \neq 0$.

Theorem 17.1. *When $d_1 \neq 0$, from (17.8) and (17.9), according to the Hopf bifurcation theorem [Hale and Verduyn Lunel (1993)], a Hopf bifurcation occurs under one of the following conditions:*

(1) $\beta \neq 0$ (damped oscillator case), $2\alpha \leq \beta^2$, $\alpha^2 < d_1^2$ and $\beta d_1 < 0$, here $\left.\frac{d\xi}{d\tau}\right|_{\tau_0} = \frac{\sigma_1^2\tau_0\Delta_2}{\Delta_1} \neq 0$, $\sigma_0 = \sigma_0^+ \triangleq \sqrt{\frac{2\alpha-\beta^2+\Delta_2}{2}}$, $\tau_0 = \tau_0^+ \triangleq \frac{1}{\sigma_0^+} arcsin(-\frac{\beta\sigma_0^+}{d_1})$;

(2) $\beta \neq 0$(damped oscillator case), $2\alpha \geq \beta^2$, $(\beta^2 - 2\alpha)^2 > 0 > 4(\alpha^2 - d_1^2)$ and $\beta d_1 < 0$, here $\left.\frac{d\xi}{d\tau}\right|_{\tau_0} = \frac{\sigma_1^2\tau_0\Delta_2}{\Delta_1} \neq 0$, $\sigma_0 = \sigma_0^+$, $\tau_0 = \tau_0^+$;

(3) $\beta \neq 0$(damped oscillator case), $2\alpha > \beta^2$, $(\beta^2 - 2\alpha)^2 > 4(\alpha^2 - d_1^2) > 0$ and $\beta d_1 < 0$, here $\left.\frac{d\xi}{d\tau}\right|_{\tau_0} = \pm\frac{\sigma_1^2\tau_0\Delta_2}{\Delta_1} \neq 0$, $\sigma_0 = \sigma_0^+$ or $\sigma_0 = \sigma_0^- \triangleq \sqrt{\frac{2\alpha-\beta^2-\Delta_2}{2}}$, $\tau_0 = \tau_0^+$ or τ_0^-, $\tau_0^- \triangleq \frac{1}{\sigma_0^-} arcsin(-\frac{\beta\sigma_0^-}{d_1})$;

(4) $\beta = 0$ (undamped oscillator case) and $\alpha > d_1$, here $\left.\frac{d\xi}{d\tau}\right|_{\tau_0} = \frac{-2d_1\sigma_1^2\tau_0}{\Delta_1} \neq 0$, $\sigma_0 = \sqrt{\alpha - d_1}$, $\tau_0 = 2k\pi/\sigma_0$, $k = 1, 2, 3, \cdots$;

(5) $\beta = 0$ (undamped oscillator case) and $\alpha > -d_1$, here $\left.\frac{d\xi}{d\tau}\right|_{\tau_0} = \frac{2d_1\sigma_1^2\tau_0}{\Delta_1} \neq 0$, $\sigma_0 = \sqrt{\alpha + d_1}$, $\tau_0 = (2k+1)\pi/\sigma_0$, $k = 0, 1, 2, 3, \cdots$,

where $\Delta_2 \triangleq \sqrt{(\beta^2 - 2\alpha)^2 - 4(\alpha^2 - d_1^2)}$. Moreover $(x, \tau) = (0, \tau_0)$ is a Hopf bifurcation point of (17.4) and a periodic solution bifurcates from the trivial steady-state solution.

Remark 17.1. It is easy to prove that $\Delta_1 \neq 0$ if $\Delta_2 \neq 0$ and $\Delta_2 \neq 0$ if one of the above conditions is satisfied.

Remark 17.2. Hopf-Hopf bifurcation occurs when case (3) is satisfied.

Remark 17.3. We derived the approximate analytical expression of periodic solution for *undamped* case by means of the L-S method [Guo and Li (2007)], so we focus on the *damped* case in this paper.

Remark 17.4. If $d_1 = 0$, then $\left.\frac{d\xi}{d\tau}\right|_{\tau_0} = 0$ so that Hopf bifurcation theorem can't be applied.

17.3 Lyapunov-Schmidt reduction approach for periodic solutions

17.3.1 *Preliminary: reformulation and projection operators*

Let $\tau = \tau_0 + \varepsilon, s = (1+\kappa)\hat{s}$, we reformulate Eq. (17.4) as follows

$$\begin{aligned}&(1+k)^2\frac{d^2u(s)}{ds^2} + (\tau_0+\varepsilon)(1+k)\beta\frac{du(s)}{ds} + \alpha(\tau_0+\varepsilon)^2u(s)\\ =&d_1(\tau_0+\varepsilon)^2u(s-(1+k)) + d_2(\tau_0+\varepsilon)^2\ u^2(s-(1+k))\\ &+O\Big(u^3(s-(1+k))\Big),\end{aligned} \tag{17.10}$$

where $u(s) \triangleq \hat{u}(\hat{s})$. Therefore, the T-period solution of Eq. (17.10) is equivalent to the $\frac{T}{1+\kappa}$-period solution of Eq. (17.4).

Denoting $u_{1+k} \triangleq u(s-1-k)$,

$$\begin{aligned}F(u,\varepsilon,k) \triangleq\ &(1+k)^2\frac{d^2u(s)}{ds^2} + (\tau_0+\varepsilon)(1+k)\beta\frac{du(s)}{ds} + \alpha(\tau_0+\varepsilon)^2u(s)\\ &-d_1(\tau_0+\varepsilon)^2u_{1+k} - d_2(\tau_0+\varepsilon)^2u_{1+k}^2 + O\Big(u_{1+k}^3\Big) = 0.\end{aligned} \tag{17.11}$$

Then, we define two spaces

$$\begin{aligned}C_p^0[0,\tfrac{2\pi}{\sigma}] &= \{u \in C^0(-\infty,\infty)|u(s) = u(s+\tfrac{2\pi}{\sigma})\},\\ C_p^2[0,\tfrac{2\pi}{\sigma}] &= \{u \in C^2(-\infty,\infty)|u(s) = u(s+\tfrac{2\pi}{\sigma})\}.\end{aligned}$$

$F(u,\varepsilon,k)$ is a mapping from $C_p^2[0,\frac{2\pi}{\sigma}] \times \mathbb{R} \times \mathbb{R}$ to $C_p^0[0,\frac{2\pi}{\sigma}]$. The linearized equation of Eq. (17.10) at Hopf bifurcation point is given by

$$L_0\varphi \triangleq \frac{d^2\varphi(s)}{ds^2} + \beta\tau_0\frac{d\varphi(s)}{ds} + \alpha\tau_0^2\varphi(s) - d_1\tau_0^2\varphi(s-1) = 0, \tag{17.12}$$

which leads the conjugate equation

$$L_0^*\psi \triangleq \frac{d^2\psi(s)}{ds^2} - \beta\tau_0\frac{d\psi(s)}{ds} + \alpha\tau_0^2\psi(s) - d_1\tau_0^2\psi(s+1) = 0. \tag{17.13}$$

Obviously,

$$\mathcal{N}(L_0) = \mathbf{span}\{\sin\sigma s, \cos\sigma s\}, \quad \mathcal{N}(L_0^*) = \mathbf{span}\{\sin\sigma s, \cos\sigma s\},$$

where $\mathcal{N}(L_0)$ and $\mathcal{N}(L_0^*)$ are the kernels of operators L_0 and L_0^* respectively. Therefore, we can make the following decomposition:

$$C_p^0[0, \frac{2\pi}{\sigma}] = \mathcal{N}(L_0^*) \oplus R(L_0), \quad C_p^2[0, \frac{2\pi}{\sigma}] = \mathcal{N}(L_0) \oplus M,$$

where $M = C_p^2[0, \frac{2\pi}{\sigma}] \cap \mathcal{N}(L_0)^{\perp}$.

We define projection operators $P : C_p^0[0, \frac{2\pi}{\sigma}] \to R(L_0), \quad Q : C_p^0[0, \frac{2\pi}{\sigma}] \to \mathcal{N}(L_0^*)$ as follows:

$$\begin{aligned} Qv &= < v, \ \sin\sigma s > \sin\sigma s + < v, \ \cos\sigma s > \cos\sigma s, \\ Pv &= Iv - Qv, \end{aligned}$$

where I is identity operator and the inner product is defined by

$$< v_1, \ v_2 > = \frac{\sigma}{\pi} \int_0^{\frac{2\pi}{\sigma}} v_1 v_2 \, ds.$$

Remark 17.5. $PL_0 = L_0$. □

Let

$$u(s) = (x \ \sin\sigma s + y \ \cos\sigma s) + w(s),$$

where coefficients $x, y \in \mathbb{R}$ and $w(s) \in M$.

Substituting $u(s)$ by $(x \ \sin\sigma s + y \ \cos\sigma s) + w(s)$ for Eq. (17.11), we get $\mathcal{G}(\omega(s); x, y, \varepsilon, k) = 0$. Thereafter, projecting it onto $R(L_0)$, we obtain

$$P\mathcal{G}(\omega; x, y, \varepsilon, k) = 0. \tag{17.14}$$

Moreover, we denote

$$g(x, y, \varepsilon, k) \triangleq \langle \mathcal{G}(\omega; x, y, \varepsilon, k), \sin\sigma s \rangle, \tag{17.15a}$$

$$h(x, y, \varepsilon, k) \triangleq \langle \mathcal{G}(\omega; x, y, \varepsilon, k), \cos\sigma s \rangle. \tag{17.15b}$$

The implicit function theorem guarantees that Eq.(17.14) has a unique solution

$$w(s) = w(s; x, y, \varepsilon, k) \tag{17.16}$$

near Hopf bifurcation point, which satisfying $w(s; 0, 0, 0, 0) \equiv 0$.

17.3.2 *Quadratic Taylor polynomial approximation*

To get the approximate expression of the periodic solution of Eq. (17.10), we explore the second-order Taylor series expansion for $w(s; x, y, \varepsilon, k)$ about $(x, y, \varepsilon, k) = (0, 0, 0, 0)$. $||u||$ is very small near the Hopf bifurcation point, without loss of generality, we choose second-order truncation equation as model, which is

$$\begin{aligned} F_2(u, \varepsilon, k) \triangleq \ & (1 + k)^2 \frac{d^2 u(s)}{ds^2} + (\tau_0 + \varepsilon)(1 + k)\beta \frac{du(s)}{ds} + \alpha(\tau_0 + \varepsilon)^2 u(s) \\ & - d_1(\tau_0 + \varepsilon)^2 u_{1+k} - d_2(\tau_0 + \varepsilon)^2 u_{1+k}^2 = 0. \end{aligned} \tag{17.17}$$

Thus, we need to know $\omega_x^0(s), \omega_y^0(s), \omega_\varepsilon^0(s), \omega_k^0(s)$, etc., where

$$\omega_x^0(s) = \frac{\partial \omega(s;0,0,0,0)}{\partial x}, \quad \omega_y^0(s) = \frac{\partial \omega(s;0,0,0,0)}{\partial y},$$

$$\omega_\varepsilon^0(s) = \frac{\partial \omega(s;0,0,0,0)}{\partial \varepsilon}, \quad \omega_k^0(s) = \frac{\partial \omega(s;0,0,0,0)}{\partial k}, \cdots.$$

First of all, we have

Lemma 17.1. *If $v(s) \in M$ solves*

$$\frac{d^2 v(s)}{ds^2} + \beta\tau_0 \frac{dv(s)}{ds} + \alpha\tau_0^2 v(s) - d_1\tau_0^2 v(s-1) = 0, \tag{17.18}$$

then

$$v(s) \equiv 0.$$

Proof. Since $v(s)$ is the solution of Eq. (17.18), we have $v(s) = c_1 \sin\sigma s + c_2 \cos\sigma s$. On the other hand, $v(s) \in M$ means that

$$\langle v(s), \sin\sigma s\rangle = 0, \quad \langle v(s), \cos\sigma s\rangle = 0,$$

which imply $c_1 = c_2 = 0$. □

Differentiating Eq. (17.14) with respect to x at $(x, y, \varepsilon, \kappa) = (0, 0, 0, 0)$, with *Remark 17.5*, we get

$$\frac{d^2 \omega_x^0(s)}{ds^2} + \beta\tau_0 \frac{d\omega_x^0(s)}{ds} + \alpha\tau_0^2 \omega_x^0(s) - d_1\tau_0^2 \omega_x^0(s-1) = 0,$$

that is, $\omega_x^0(s)$ satisfies Eq. (17.18).

Noticing that $\omega_x^0(s) \in M$, by Lemma 17.1,

$$\omega_x^0(s) \equiv 0,$$

thereafter, differentiating Eqs. (17.15a) and (17.15b) with respect to x at $(x, y, \varepsilon, \kappa) = (0, 0, 0, 0)$, we have

$$\begin{aligned} g_x^0 &= -\sigma^2 + \tau_0^2\alpha - d_1\tau_0^2 \cos\sigma, \\ h_x^0 &= \tau_0\beta\sigma + d_1\tau_0^2 \sin\sigma. \end{aligned}$$

Moreover, we get the following results in the same way

$$\begin{aligned} &\omega_y^0(s) \equiv 0, \quad \omega_\varepsilon^0(s) \equiv 0, \quad \omega_\kappa^0(s) \equiv 0, \quad g_\varepsilon^0 = 0, \quad g_\kappa^0 = 0, \quad h_\varepsilon^0 = 0, \quad h_\kappa^0 = 0, \\ &g_y^0 = -\tau_0\beta\sigma - d_1\tau_0^2 \sin\sigma, \quad h_y^0 = -\sigma^2 + \tau_0^2\alpha - d_1\tau_0^2 \cos\sigma. \end{aligned}$$

Differentiating Eq. (17.14) with respect to x twice at $(x, y, \varepsilon, k) = (0, 0, 0, 0)$, it yields that $\omega_{xx}^0(s)$ satisfies the following equation

$$\frac{d^2 \omega_{xx}^0(s)}{ds^2} + \tau_0\beta \frac{d\omega_{xx}^0(s)}{ds} + \alpha\tau_0^2 \omega_{xx}^0(s) - d_1\tau_0^2 \omega_{xx}^0(s-1) - 2d_2\tau_0^2 P(\sin^2\sigma(s-1)) = 0. \tag{17.19}$$

We notice that $P(\sin^2\sigma(s-1)) = \sin^2\sigma(s-1)$ and $\omega^0_{xx}(s) \in M$, setting

$$\omega^0_{xx}(s) = c_{11}\ \sin 2\sigma s + c_{12}\ \cos 2\sigma s + c_{13}, \tag{17.20}$$

substituting (17.20) into Eq. (17.19) leads to that

$$\begin{aligned}
&(4\sigma^2 - \alpha\tau_0^2 + d_1\tau_0^2\cos 2\sigma)\ c_{11} + (2\tau_0\beta\sigma + d_1\tau_0^2\sin 2\sigma)\ c_{12} = -d_2\tau_0^2\sin 2\sigma,\\
&(2\tau_0\beta\sigma + d_1\tau_0^2\sin 2\sigma)\ c_{11} - (4\sigma^2 - \alpha\tau_0^2 + d_1\tau_0^2\cos 2\sigma)\ c_{12} = d_2\tau_0^2\cos 2\sigma,\\
&\alpha\tau_0^2\ c_{13} - d_1\tau_0^2\ c_{13} - d_2\tau_0^2 = 0.
\end{aligned}$$

Solving this system of linear equations, we have

$$\begin{aligned}
c_{11} &= \frac{d_2\tau_0^2(4\sigma^2\sin 2\sigma - \alpha\tau_0^2\sin 2\sigma - 2\tau_0\beta\sigma\cos 2\sigma)}{(4\sigma^2-\alpha\tau_0^2+d_1\tau_0^2\cos 2\sigma)^2+(2\tau_0\beta\sigma+d_1\tau_0^2\sin 2\sigma)^2},\\
c_{12} &= \frac{d_2\tau_0^2(2\tau_0\beta\sigma\sin 2\sigma+4n^2\cos 2\sigma-\alpha\tau_0^2\cos 2\sigma+d_1\tau_0^2)}{(4\sigma^2-\alpha\tau_0^2+d_1\tau_0^2\cos 2\sigma)^2+(2\tau_0\beta\sigma+d_1\tau_0^2\sin 2\sigma)^2},\\
c_{13} &= \frac{d_2}{\alpha-d_1}.
\end{aligned}$$

Differentiating Eqs. (17.15a) and (17.15b) with respect to x twice at trivial solution, we have $g^0_{xx} = 0,\ \ h^0_{xx} = 0$.

The others second-order derivatives can be treated in the same way, for simplicity, we summarize the derivatives which may *not vanish* as follows:

$$\begin{aligned}
\omega^0_{xy}(s) &= -c_{12}\ \sin 2\sigma s + c_{11}\ \cos 2\sigma s;\\
\omega^0_{yy}(s) &= -c_{11}\ \sin 2\sigma s - c_{12}\ \cos 2\sigma s + c_{13};\\
g^0_{x\varepsilon} &= 2\tau_0\alpha - 2d_1\tau_0\sin\sigma,\quad h^0_{x\varepsilon} = \beta\sigma + 2d_1\tau_0\sin\sigma;\\
g^0_{y\varepsilon} &= -(\beta\sigma + 2d_1\tau_0\sin\sigma),\quad h^0_{y\varepsilon} = 2\tau_0\alpha - 2d_1\tau_0\sin\sigma;\\
g^0_{xk} &= -2\sigma^2 + d_1\tau_0^2\sigma\sin\sigma,\quad h^0_{xk} = \tau_0\beta\sigma + d_1\tau_0^2\sigma\cos\sigma;\\
g^0_{yk} &= -(\tau_0\beta\sigma + d_1\tau_0^2\sigma\cos\sigma),\quad h^0_{xk} = -2\sigma^2 + d_1\tau_0^2\sigma\sin\sigma.
\end{aligned}$$

Therefore, near Hopf bifurcation point, the quadratic Taylor polynomial approximation for $\omega(s; x, y, \varepsilon, \kappa)$ is

$$\omega(s) \approx \omega_2(s) \triangleq \gamma_{11}\ \frac{x^2}{2} + \gamma_{12}\ xy + \gamma_{22}\ \frac{y^2}{2},$$

and

$$u(s) \approx u_2(s) \triangleq x\sin\sigma s + y\cos\sigma s + \omega_2(s), \tag{17.21}$$

where $\gamma_{11} = c_{11}\sin 2\sigma s + c_{12}\cos 2\sigma s + c_{13}$, $\gamma_{12} = -c_{12}\ \sin 2\sigma s + c_{11}\ \cos 2\sigma s$, $\gamma_{22} = -c_{11}\ \sin 2\sigma s - c_{12}\ \cos 2\sigma s + c_{13}$.

17.3.3 *Bifurcation equations*

In summary, we have the following approximate expressions:

$$g(x, y, k, \varepsilon) \approx g_2(x, y, k, \varepsilon) \triangleq [\Gamma_1 + \Gamma_2\varepsilon + \Gamma_3\kappa]x + [\Gamma_4 + \Gamma_5\varepsilon + \Gamma_6\kappa]y,$$

$$h(x, y, k, \varepsilon) \approx h_2(x, y, k, \varepsilon) \triangleq [\Gamma_4 + \Gamma_5\varepsilon + \Gamma_6\kappa](-x) + [\Gamma_1 + \Gamma_2\varepsilon + \Gamma_3\kappa]y,$$

where $\Gamma_1 = -\sigma^2 + \tau_0^2\alpha - d_1\tau_0^2\cos\sigma$, $\Gamma_2 = 2\tau_0\alpha - 2d_1\tau_0\sin\sigma$, $\Gamma_3 = -2\sigma^2 + d_1\tau_0^2\sigma\sin\sigma$, $\Gamma_4 = -\tau_0\beta\sigma - d_1\tau_0^2\sin\sigma$, $\Gamma_5 = -(\beta\sigma + 2d_1\tau_0\sin\sigma)$, $\Gamma_6 = -(\tau_0\beta\sigma + d_1\tau_0^2\sigma\cos\sigma)$.

Because $\mathcal{G}(\omega(s); x, y, \varepsilon, k) = 0$, the approximate bifurcation equations are

$$g_2(x, y, \varepsilon, k) = 0, \tag{17.22a}$$

$$h_2(x, y, \varepsilon, k) = 0. \tag{17.22b}$$

Moreover, we have

$$\varepsilon \approx \varepsilon_2 = \frac{\Gamma_1\Gamma_6 - \Gamma_3\Gamma_4}{\Gamma_3\Gamma_5 - \Gamma_2\Gamma_6}, \tag{17.23a}$$

$$\kappa \approx \kappa_2 = \frac{\Gamma_1\Gamma_5 - \Gamma_2\Gamma_4}{\Gamma_2\Gamma_6 - \Gamma_3\Gamma_5}. \tag{17.23b}$$

Remark 17.6. Here, we don't go into detail about three order Taylor polynomial approximation, but give the main results about the approximate bifurcation equations.

$$\begin{aligned} g(x, y, k, \varepsilon) &\approx g_3(x, y, k, \varepsilon) \\ &\triangleq [\Gamma_1 + \Gamma_2\varepsilon + \Gamma_3\kappa + \Gamma_7(x^2 + y^2)]x + [\Gamma_4 + \Gamma_5\varepsilon + \Gamma_6\kappa + \Gamma_8(x^2 + y^2)]y, \\ h(x, y, k, \varepsilon) &\approx h_3(x, y, k, \varepsilon) \\ &\triangleq [\Gamma_4 + \Gamma_5\varepsilon + \Gamma_6\kappa + \Gamma_8(x^2 + y^2)](-x) \\ &\quad + [\Gamma_1 + \Gamma_2\varepsilon + \Gamma_3\kappa + \Gamma_7(x^2 + y^2)]y, \end{aligned}$$

where $\Gamma_7 = -\frac{1}{2}d_2\tau_0^2(c_{11}\sin\sigma - c_{12}\cos\sigma + 2c_{13}\cos\sigma)$, $\Gamma_8 = \frac{1}{2}d_2\tau_0^2(c_{11}\cos\sigma + c_{12}\sin\sigma - 2c_{13}\sin\sigma)$, then the relations between parameters and periodic solution are

$$\varepsilon \approx \varepsilon_3 = \frac{\Gamma_1\Gamma_6 - \Gamma_3\Gamma_4 + (\Gamma_6\Gamma_7 - \Gamma_3\Gamma_8)(x^2 + y^2)}{\Gamma_3\Gamma_5 - \Gamma_2\Gamma_6}, \tag{17.24a}$$

$$\kappa \approx \kappa_3 = \frac{\Gamma_1\Gamma_5 - \Gamma_2\Gamma_4 + (\Gamma_5\Gamma_7 - \Gamma_2\Gamma_8)(x^2 + y^2)}{\Gamma_2\Gamma_6 - \Gamma_3\Gamma_5}. \tag{17.24b}$$

□

17.3.4 *Accuracy of approximation*

If $(\tilde{u}, \tilde{\varepsilon}, \tilde{\kappa})$ is the exact solution of Eq. (17.17), we denote

$$error_1 \triangleq |F_2(u_2(s), \varepsilon_2, \kappa_2) - F_2(\tilde{u}, \tilde{\varepsilon}, \tilde{\kappa})| = |F_2(u_2(s), \varepsilon_2, \kappa_2) - 0|,$$

$$error_2 \triangleq |F_3(u_2(s), \varepsilon_2, \kappa_2) - F_3(\bar{u}, \bar{\varepsilon}, \bar{\kappa})| = |F_3(u_2(s), \varepsilon_2, \kappa_2) - 0|,$$

where

$$\begin{aligned} F_3(u, \varepsilon, k) \triangleq\ & (1+k)^2\frac{d^2u(s)}{ds^2} + (\tau_0 + \varepsilon)(1+k)\beta\frac{du(s)}{ds} + \alpha(\tau_0 + \varepsilon)^2u(s) \\ & - d_1(\tau_0 + \varepsilon)^2u_{1+k} - d_2(\tau_0 + \varepsilon)^2u_{1+k}^2 - d_3(\tau_0 + \varepsilon)^2u_{1+k}^3, \end{aligned}$$

and $(\bar{u}, \bar{\varepsilon}, \bar{\kappa})$ is the exact solution of $F_3(u, \varepsilon, k) = 0$.

Here, for illustrating accuracy by the $error_1$ and $error_2$, we summarize them in Tables 17.1–17.6. Therefore, we can see that (u, ε, k) is well approximated by $(u_2(s), \varepsilon_2, \kappa_2)$. Especially, Table 17.5 and Table 17.6 show that our method may be also effective in Hopf-Hopf case described in *Remark 17.2*.

Table 17.1 Errors of approximate periodic solution. $\alpha = 1, \beta = 1.5$, $d_1 = -5, d_2 = 0.2, d_3 = 2, x = 0.01, y = 0.01$.

s	1	2	3	4	5	6
$error_1$	$1.07E{-6}$	$1.07E{-6}$	$1.07E{-6}$	$1.07E{-6}$	$1.07E{-6}$	$1.07E{-6}$
$error_2$	$1.29E{-6}$	$1.67E{-6}$	$1.38E{-6}$	$1.08E{-6}$	$1.00E{-6}$	$0.57E{-6}$

Table 17.2 Errors of approximate periodic solution. $\alpha = 1, \beta = 1.5$, $d_1 = -5, d_2 = 0.2, d_3 = 2, x = 0.1, y = 0.1$.

s	1	2	3	4	5	6
$error_1$	$1.08E{-4}$	$1.08E{-4}$	$1.07E{-4}$	$1.07E{-4}$	$1.06E{-4}$	$1.06E{-4}$
$error_2$	$3.24E{-4}$	$7.11E{-4}$	$4.17E{-4}$	$1.13E{-4}$	$0.39E{-4}$	$3.92E{-4}$

Table 17.3 Errors of approximate periodic solution. $\alpha = 1.3, \beta = 1.5$, $d_1 = -5, d_2 = 0.2, d_3 = 2, x = 0.01, y = 0.01$.

s	1	2	3	4	5	6
$error_1$	$1.08E{-6}$	$1.08E{-6}$	$1.08E{-6}$	$1.08E{-6}$	$1.08E{-6}$	$1.08E{-6}$
$error_2$	$1.30E{-6}$	$1.69E{-6}$	$1.37E{-6}$	$1.08E{-6}$	$0.98E{-6}$	$0.52E{-6}$

Table 17.4 Errors of approximate periodic solution. $\alpha = 1.3, \beta = 1.5$, $d_1 = -5, d_2 = 0.2, d_3 = 2, x = 0.1, y = 0.1$.

s	1	2	3	4	5	6
$error_1$	$1.09E{-4}$	$1.09E{-4}$	$1.08E{-4}$	$1.08E{-4}$	$1.07E{-4}$	$1.07E{-4}$
$error_2$	$3.27E{-4}$	$7.21E{-4}$	$3.92E{-4}$	$1.10E{-4}$	$4.21E{-4}$	$4.48E{-4}$

17.4 Multiple scales analysis for periodic solutions

In this section, we derive the approximate periodic solution of the damped case of (17.4) by the multiple scales method. First of all, we denote the nonlinear terms' coefficients of (17.4) as follows: $d_l = \epsilon^{m_l} c_l$, $l \geq 2$ and $l \in \mathbb{Z}^+$, where c_l is a bounded quantity and $0 < \epsilon \ll 1$, $m_l \in \mathbb{Z}^+$.

Assume that $\tau = \tau_0 + \epsilon\delta$, $\delta \in \mathbb{R}$ is a bounded quantity, then the solution of

Table 17.5 Errors of approximate periodic solution. $\alpha = 4, \beta = -1.8$, $d_1 = 3.5, d_2 = 0.2, d_3 = 2, x = 0.01, y = 0.01$.

s	1	2	3	4	5	6
$error_1$	$6.05E{-6}$	$6.06E{-6}$	$5.97E{-6}$	$5.94E{-6}$	$6.03E{-6}$	$6.07E{-6}$
$error_2$	$7.26E{-6}$	$7.52E{-6}$	$5.22E{-6}$	$3.98E{-6}$	$6.44E{-6}$	$8.57E{-6}$

Table 17.6 Errors of approximate periodic solution. $\alpha = 4, \beta = -1.8$, $d_1 = 3.5, d_2 = 0.2, d_3 = 1, x = 0.1, y = 0.1$.

s	1	2	3	4	5	6
$error_1$	$6.46E{-4}$	$6.55E{-4}$	$5.63E{-4}$	$5.40E{-4}$	$6.31E{-4}$	$6.68E{-4}$
$error_2$	$12.8E{-4}$	$14.3E{-4}$	$2.10E{-4}$	$3.96E{-4}$	$8.51E{-4}$	$19.8E{-4}$

(17.4) is approximated by the following expansion in three time scales:

$$\begin{aligned} x(t;\epsilon) =& X_0(T_0,T_1,T_2) + \epsilon X_1(T_0,T_1,T_2) \\ &+ \epsilon^2 X_2(T_0,T_1,T_2) + O(\epsilon^3), \end{aligned} \tag{17.25}$$

where $T_r = \epsilon^r t,\ r = 0,1,2.$

The first and second order derivatives of $x(t)$, and the delay term $x(t-1)$ are expanded as

$$\frac{dx}{dt} = \frac{\partial X_0}{\partial T_0} + \epsilon\Big[\frac{\partial X_0}{\partial T_1} + \frac{\partial X_1}{\partial T_0}\Big] + \epsilon^2\Big[\frac{\partial X_0}{\partial T_2} + \frac{\partial X_1}{\partial T_1} + \frac{\partial X_2}{\partial T_0}\Big] + O(\epsilon^3), \tag{17.26}$$

$$\begin{aligned} \frac{d^2x}{dt^2} =& \frac{\partial^2 X_0}{\partial T_0^2} + \epsilon\Big[2\frac{\partial^2 X_0}{\partial T_1 \partial T_0} + \frac{\partial^2 X_1}{\partial T_0^2}\Big] \\ &+\epsilon^2\Big[\frac{\partial^2 X_0}{\partial T_1^2} + \frac{\partial^2 X_2}{\partial T_0^2} + 2\frac{\partial^2 X_1}{\partial T_1 \partial T_0} + 2\frac{\partial^2 X_0}{\partial T_2 \partial T_0}\Big] + O(\epsilon^3), \end{aligned} \tag{17.27}$$

$$\begin{aligned} x(t-1) =& X_0(T_0-1,T_1,T_2) + \epsilon\Big[X_1(T_0-1,T_1,T_2) \\ &-\frac{\partial X_0}{\partial T_1}(T_0-1,T_1,T_2)\Big] + \epsilon^2\Big[X_2(T_0-1,T_1,T_2) \\ &-\frac{\partial X_1}{\partial T_1}(T_0-1,T_1,T_2) - \frac{\partial X_0}{\partial T_2}(T_0-1,T_1,T_2) \\ &+\frac{1}{2}\frac{\partial^2 X_0}{\partial T_1^2}(T_0-1,T_1,T_2)\Big] + O(\epsilon^3). \end{aligned} \tag{17.28}$$

Without loss of generality, we assume $m_2 = 1$, $m_3 = 2$, $m_l > 2$ for $l > 3$.

Substituting (17.25)-(17.28) into (17.4) and equating coefficients of equal powers of ϵ, from the terms of first three orders, we obtain

$$\frac{\partial^2 X_0}{\partial T_0^2} + \beta\tau_0\frac{\partial X_0}{\partial T_0} + \alpha\tau_0^2 X_0 - d_1\tau_0^2 X_0(T_0-1) = 0, \tag{17.29}$$

$$\begin{aligned} &\frac{\partial^2 X_1}{\partial T_0^2} + 2\frac{\partial^2 X_0}{\partial T_1\partial T_0} + \beta\tau_0\Big[\frac{\partial X_0}{\partial T_1} + \frac{\partial X_1}{\partial T_0}\Big] + \beta\delta\frac{\partial X_0}{\partial T_0} \\ &+\alpha\tau_0^2 X_1 + 2\alpha\tau_0\delta X_0 - d_1\tau_0^2\Big[X_1(T_0-1) - \frac{\partial X_0}{\partial T_1}(T_0-1)\Big] \\ &-2d_1\tau_0\delta X_0(T_0-1) - c_2\tau_0^2 X_0^2(T_0-1) = 0, \end{aligned} \tag{17.30}$$

$$\begin{aligned} &\frac{\partial^2 X_2}{\partial T_0^2} + \frac{\partial^2 X_0}{\partial T_1^2} + 2\frac{\partial^2 X_1}{\partial T_1\partial T_0} + 2\frac{\partial^2 X_0}{\partial T_2\partial T_0} + \\ &\beta\tau_0\Big[\frac{\partial X_0}{\partial T_2} + \frac{\partial X_1}{\partial T_1} + \frac{\partial X_2}{\partial T_0}\Big] + \beta\delta\Big[\frac{\partial X_0}{\partial T_1} + \frac{\partial X_1}{\partial T_0}\Big] \end{aligned}$$

$$
\begin{aligned}
&+\alpha\tau_0^2X_2+2\alpha\tau_0\delta X_1+\alpha\delta^2X_0-d_1\tau_0^2\Big[X_2(T_0-1)\\
&-\frac{\partial X_1}{\partial T_1}(T_0-1)-\frac{\partial X_0}{\partial T_2}(T_0-1)+\frac{1}{2}\frac{\partial^2 X_0}{\partial T_1^2}(T_0-1)\Big]\\
&-2d_1\tau_0\delta\Big[X_1(T_0-1)-\frac{\partial X_0}{\partial T_1}(T_0-1)\Big]-d_1\delta^2X_0(T_0-1)\\
&-c_2\tau_0^2\Big\{2X_0(T_0-1)\Big[X_1(T_0-1)-\frac{\partial X_0}{\partial T_1}(T_0-1)\Big]\Big\}\\
&-2c_2\tau_0\delta X_0^2(T_0-1)-c_3\tau_0^2X_0^3(T_0-1)=0,
\end{aligned}
\tag{17.31}
$$

where

$$
\frac{\partial^2 X_r}{\partial T_i\partial T_j}\triangleq\frac{\partial^2 X_r}{\partial T_i\partial T_j}(T_0,T_1,T_2),\quad \frac{\partial X_r}{\partial T_i}\triangleq\frac{\partial X_r}{\partial T_i}(T_0,T_1,T_2),
$$

$$
\frac{\partial X_r}{\partial T_i}(T_0-1)\triangleq\frac{\partial X_r}{\partial T_i}(T_0-1,T_1,T_2),\quad X_r\triangleq X_r(T_0,T_1,T_2),
$$

$$
X_r(T_0-1)\triangleq X_r(T_0-1,T_1,T_2),\quad r,i,j=0,1,2.
$$

(17.29) has the following fundamental harmonic solution:

$$
X_0(T_0,T_1,T_2)=A(T_1,T_2)e^{i\sigma_1T_0}+\bar{A}(T_1,T_2)e^{-i\sigma_1T_0},
\tag{17.32}
$$

and let

$$
A(T_1,T_2)=R(T_1,T_2)e^{i\varphi(T_1,T_2)}.
\tag{17.33}
$$

Substituting (17.32) into (17.30) yields

$$
\begin{aligned}
&\frac{\partial^2 X_1}{\partial T_0^2}+\beta\tau_0\frac{\partial X_1}{\partial T_0}+\alpha\tau_0^2X_1-d_1\tau_0^2X_1(T_0-1)\\
=\;&\Big[(-2i\sigma_1-\beta\tau_0-d_1\tau_0^2e^{-i\sigma_1})\frac{\partial A}{\partial T_1}\\
&+(-i\beta\delta\sigma_1-2\alpha\delta\tau_0+2d_1\tau_0\delta e^{-i\sigma_1})A\Big]e^{i\sigma_1T_0}\\
&+c_2\tau_0^2(A^2e^{-2i\sigma_1}e^{2i\sigma_1T_0}+R^2)+cc,
\end{aligned}
\tag{17.34}
$$

where 'cc' represents the complex conjugate of preceding terms of this symbol.

To remove the non-periodic (also called secular) terms, we eliminate the coefficient of $e^{i\sigma_1T_0}$ of the right-hand side of (17.34) (see [Nayfeh (1973)], Pages 23-26). So, we obtain

$$
(-2i\sigma_1-\beta\tau_0-d_1\tau_0^2e^{-i\sigma_1})\frac{\partial A}{\partial T_1}-(i\beta\delta\sigma_1+2\alpha\delta\tau_0-2d_1\tau_0\delta e^{-i\sigma_1})A=0.
\tag{17.35}
$$

With (17.33), after separating real and imaginary parts of (17.35), we have

$$
\begin{cases}
\dfrac{\partial R}{\partial T_1}=\Big[\dfrac{\tau_0\delta\sigma_1^2}{H_1^2+H_2^2}(2\sigma_0^2-2\alpha+\beta^2)\Big]R, & (17.36\text{a})\\[2ex]
\dfrac{\partial \varphi}{\partial T_1}=\dfrac{\delta\sigma_1H_3}{H_1^2+H_2^2}, & (17.36\text{b})
\end{cases}
$$

where $H_1 = -\beta\tau_0 - \tau_0^2 d_1 \cos\sigma_1$, $H_2 = 2\sigma_1 - \tau_0^2 d_1 \sin\sigma_1$, $H_3 = 4\tau_0\sigma_0^2 + \beta\sigma_1^2 + \beta^2\tau_0 + \alpha\beta\tau_0^2$.

From (17.34) and (17.35), we have

$$\begin{aligned} &\frac{\partial^2 X_1}{\partial T_0^2} + \beta\tau_0 \frac{\partial X_1}{\partial T_0} + \alpha\tau_0^2 X_1 - d_1\tau_0^2 X_1(T_0 - 1) \\ = \ & c_2\tau_0^2(A^2 e^{-2i\sigma_1} e^{2i\sigma_1 T_0} + R^2) + cc, \end{aligned} \tag{17.37}$$

where 'cc' represents the complex conjugate of preceding terms.

The fundamental harmonic solution of (17.37) is

$$X_1(T_0, T_1, T_2) = C(T_1, T_2) + D(T_1, T_2)e^{2i\sigma_1 T_0} + cc, \tag{17.38}$$

where 'cc' represents $\bar{C}(T_1, T_2) + \bar{D}(T_1, T_2)e^{-2i\sigma_1 T_0}$.

Substituting (17.38) into (17.37), we have

$$\begin{cases} C = \dfrac{c_2 R^2}{\alpha - d_1}, & (17.39a) \\ D = \dfrac{c_2\tau_0^2 R^2 e^{2i(\varphi - \sigma_1)}}{-4\sigma_1^2 + 2i\beta\tau_0\sigma_1 + \alpha\tau_0^2 - d_1\tau_0^2 e^{-2i\sigma_1}}, & (17.39b) \end{cases}$$

when $\alpha \neq d_1$.

Substituting (17.32) and (17.38) into (17.31) and eliminating secular term, let

$$\begin{aligned} &(-2i\sigma_1 - \beta\tau_0 - d_1\tau_0^2 e^{-i\sigma_1})\frac{\partial A}{\partial T_2} \\ = \ & (1 - \frac{1}{2}d_1\tau_0^2 e^{-i\sigma_1})\frac{\partial^2 A}{\partial T_1^2} + (\beta\delta + 2d_1\tau_0\delta e^{-i\sigma_1})\frac{\partial A}{\partial T_1} \\ &+(\alpha\delta^2 - d_1\delta^2 e^{-i\sigma_1})A - 2c_2\tau_0^2(2CA + D\bar{A})e^{-i\sigma_1} \\ &-3c_3\tau_0^2 A^2\bar{A}e^{-i\sigma_1}. \end{aligned} \tag{17.40}$$

To differentiate both sides of (17.33) with respect to T_1, T_2, we have

$$\frac{\partial A}{\partial T_n} = (\frac{\partial R}{\partial T_n} + iR\frac{\partial \varphi}{\partial T_n})e^{i\varphi}, \quad n = 1, 2, \tag{17.41}$$

$$\frac{\partial^2 A}{\partial T_1^2} = \left[\frac{\partial^2 R}{\partial T_1^2} - R(\frac{\partial\varphi}{\partial T_1})^2 + 2i\frac{\partial R}{\partial T_1}\frac{\partial\varphi}{\partial T_1}\right]e^{i\varphi}. \tag{17.42}$$

Substituting (17.33), (17.39a), (17.39b), (17.41) and (17.42) into (17.40), we obtain

$$\frac{\partial R}{\partial T_2} = \frac{1}{H_1^2 + H_2^2}\Big[(H_1P_1 - H_2Q_1)\delta^2 R + (H_1P_2 - H_2Q_2)R^3\Big], \tag{17.43}$$

where

$$\begin{aligned} P_1 =& \frac{1}{(H_1^2 + H_2^2)^2}\Big\{\Big(1 - \frac{1}{2}\alpha\tau_0^2 + \frac{1}{2}\sigma_1^2\Big)\Big[\tau_0^6\sigma_0^4(2\sigma_0^2 - 2\alpha + \beta^2)^2 - \sigma_1^2(4\tau_0\sigma_0^2 + \beta\sigma_1^2 \\ &+ \beta^2\tau_0 + \alpha\beta\tau_0^2)^2\Big] + \frac{1}{2}\beta\tau_0^2\sigma_0(16\sigma_0^7\tau_0^5 - 16\sigma_1^5\alpha + 12\sigma_1^5\beta^2 + 4\sigma_0^7\tau_0^6\beta + 2\sigma_0^5\tau_0^6\beta^3 \\ &- 4\sigma_0^3\tau_0^5\beta^2\alpha + 2\sigma_0^3\tau_0^5\beta^4 - 4\sigma_0^3\tau_0^6\alpha^2\beta + 2\sigma_0^3\tau_0^6\alpha\beta^3)\Big\} + \frac{1}{H_1^2 + H_2^2}\Big[(\beta + 2\alpha\tau_0 \\ &- 2\tau_0\sigma_0^2)(2\tau_0^3\sigma_0^4 - 2\tau_0^3\sigma_0^2\alpha + \tau_0^3\sigma_0^2\beta^2) - 2\beta\sigma_1(4\tau_0^2\sigma_0^3 + \sigma_1^3\beta + \tau_0^2\sigma_0\beta^2\sigma_0\tau_0^3\alpha\beta)\Big] \\ &+ \sigma_0^2, \end{aligned}$$

$$
\begin{aligned}
Q_1 = &\frac{1}{(H_1^2+H_2^2)^2}\Big\{-\frac{1}{2}\beta\tau_0^2\sigma_0\Big[\tau_0^6\sigma_0^4(2\sigma_0^2-2\alpha+\beta^2)^2-\sigma_1^2(4\tau_0\sigma_0^2+\beta\sigma_1^2+\beta^2\tau_0\\
&+\alpha\beta\tau_0^2)^2\Big]+(1-\frac{1}{2}\alpha\tau_0^2+\frac{1}{2}\sigma_1^2)(16\sigma_0^7\tau_0^5-16\sigma_1^5\alpha+12\sigma_1^5\beta^2+4\sigma_0^7\tau_0^6\beta+2\sigma_0^5\tau_0^6\beta^3\\
&-4\sigma_0^3\tau_0^5\beta^2\alpha+2\sigma_0^3\tau_0^5\beta^4-4\sigma_0^3\tau_0^6\alpha^2\beta+2\sigma_0^3\tau_0^6\alpha\beta^3)\Big\}+\frac{1}{H_1^2+H_2^2}\Big[2\beta\sigma_1(2\tau_0^3\sigma_0^4\\
&-2\tau_0^3\sigma_0^2\alpha+\tau_0^3\sigma_0^2\beta^2)+(\beta+2\alpha\tau_0-2\tau_0\sigma_0^2)(4\tau_0^2\sigma_0^3+\sigma_1^3\beta+\tau_0^2\sigma_0\beta^2+\sigma_0\tau_0^3\alpha\beta)\Big]\\
&+\beta\sigma_0,
\end{aligned}
$$

$$
\begin{aligned}
P_2 = &-\frac{4c_2^2\tau_0^2(\alpha-\sigma_0^2)}{d_1(\alpha-d_1)}-\frac{2c_2^2\tau_0^2}{9\sigma_0^4+\beta^2\sigma_0^2}\Big(\frac{4\alpha^4}{d_1^3}-\frac{28\sigma_0^2\alpha^3}{d_1^3}+\frac{60\sigma_0^4\alpha^2}{d_1^3}-\frac{3\alpha^2}{d_1}-\frac{52\alpha\sigma_0^6}{d_1^3}\\
&+\frac{15\alpha\sigma_0^2}{d_1}+\frac{16\sigma_0^8}{d_1^3}-\frac{12\sigma_0^4}{d_1}-\frac{8\sigma_0^4\beta^4}{d_1^3}+\frac{6\sigma_0^2\beta^2}{d_1}-\alpha+\sigma_0^2\Big)-\frac{3c_3\tau_0^2\alpha}{d_1}+\frac{3c_3\sigma_1^2}{d_1},
\end{aligned}
$$

$$
\begin{aligned}
Q_2 = &-\frac{4c_2^2\tau_0^2\beta\sigma_0}{d_1(\alpha-d_1)}-\frac{2c_2^2\tau_0^2}{9\sigma_0^4+\beta^2\sigma_0^2}\Big(\frac{16\sigma_0^5\beta^3}{d_1^3}+\frac{8\sigma_0^7\beta}{d_1^3}-\frac{18\sigma_0^3\beta}{d_1}+\frac{24\sigma_0^3\beta\alpha^2}{d_1^3}-\frac{8\sigma_0\beta\alpha^3}{d_1^3}\\
&+\frac{9\sigma_0\beta\alpha}{d_1}-\frac{24\sigma_0^5\beta\alpha}{d_1^3}-\frac{4\alpha\sigma_0^3\beta^3}{d_1^3}-\beta\sigma_0\Big)-\frac{3c_3\tau_0^2\beta\sigma_0}{d_1}.
\end{aligned}
$$

Thereafter, with (17.36a) and (17.43), the dynamic equation of amplitude R gives

$$
\begin{aligned}
\frac{dR}{dt} &= \epsilon\frac{dR}{dT_1}+\epsilon^2\frac{dR}{dT_2}=\Big[\frac{\tau_0\epsilon\delta\sigma_1^2}{H_1^2+H_2^2}(2\sigma_0^2-2\alpha+\beta^2)\Big]R\\
&+\frac{1}{H_1^2+H_2^2}\Big[(H_1P_1-H_2Q_1)\epsilon^2\delta^2R+(H_1P_2-H_2Q_2)\epsilon^2R^3\Big].
\end{aligned}
\tag{17.44}
$$

A periodic solution is a nontrivial steady-state solution of (17.4), that is, $R\neq 0$ and $dR/dt=0$. The equilibrium $\hat{R}$ can be solved as

$$
\hat{R}=\sqrt{\frac{\tau_0\sigma_1^2(2\sigma_0^2-2\alpha+\beta^2)\epsilon\delta+(H_1P_1-H_2Q_1)(\epsilon\delta)^2}{-(H_1P_2-H_2Q_2)\epsilon^2}},
\tag{17.45}
$$

when the term in the radical sign is greater than 0.

Thus, we can obtain the approximate expression of the bifurcating periodic solution

$$
\hat{x}(t;\epsilon)=\epsilon\frac{c_2\hat{R}^2}{\alpha-d_1}+\hat{R}e^{i(\sigma_1+\frac{\delta\sigma_1H_3}{H_1^2+H_2^2}\epsilon)t}+\epsilon\hat{D}e^{2i\sigma_1t}+cc+O(\epsilon^2),
$$

where

$$
\hat{D}=\frac{c_2\tau_0^2\hat{R}^2e^{2i(\varphi-\sigma_1)}}{-4\sigma_1^2+2i\beta\tau_0\sigma_1+\alpha\tau_0^2-d_1\tau_0^2e^{-2i\sigma_1}},
$$

and 'cc' represents the complex conjugate of preceding terms.

The stability of $\hat{x}(t;\epsilon)$ can be determined by the sign of the derivative at $\hat{R}$ of the right-hand side of (17.44) with respect to R, which is as follows

$$\begin{aligned}\lambda_p &= \Big[\frac{\tau_0\epsilon\delta\sigma_1^2}{H_1^2+H_2^2}(2\sigma_0^2-2\alpha+\beta^2)\Big]+\\ &\quad \frac{1}{H_1^2+H_2^2}\Big[(H_1P_1-H_2Q_1)\epsilon^2\delta^2+3(H_1P_2-H_2Q_2)\epsilon^2\hat{R}^2\Big]\\ &= \frac{1}{H_1^2+H_2^2}\Big[-2\tau_0\epsilon\delta\sigma_1^2(2\sigma_0^2-2\alpha+\beta^2)-2(H_1P_1-H_2Q_1)\epsilon^2\delta^2\Big]\\ &= \frac{-2}{H_1^2+H_2^2}\Big[\tau_0\sigma_1^2(2\sigma_0^2-2\alpha+\beta^2)+(H_1P_1-H_2Q_1)(\tau-\tau_0)\Big](\tau-\tau_0),\end{aligned} \tag{17.46}$$

when $\hat{R}$ exists.

The corresponding derivative at the trivial solution is

$$\begin{aligned}\lambda_0 &= \Big[\frac{\tau_0\epsilon\delta\sigma_1^2}{H_1^2+H_2^2}(2\sigma_0^2-2\alpha+\beta^2)\Big]+\frac{1}{H_1^2+H_2^2}\Big[(H_1P_1-H_2Q_1)\epsilon^2\delta^2\Big]\\ &= \frac{1}{H_1^2+H_2^2}\Big[\tau_0\sigma_1^2(2\sigma_0^2-2\alpha+\beta^2)+(H_1P_1-H_2Q_1)(\tau-\tau_0)\Big](\tau-\tau_0).\end{aligned} \tag{17.47}$$

Then we have the following theorem.

Theorem 17.2. *According to* (17.47), *near the Hopf bifurcation point, for critical time delay* $\tau_0=\tau_0^+$, *if* $\tau<\tau_0$ *the trivial solution of* (17.4) *is stable and the trivial solution of* (17.4) *is* not *stable if* $\tau>\tau_0$.

Proof. Near the Hopf bifurcation point, we can find $r>0$ such that

$$|\tau_0\sigma_1^2(2\sigma_0^2-2\alpha+\beta^2)|>|H_1P_1-H_2Q_1|\,|\tau-\tau_0|,$$

when $|\tau-\tau_0|\le r$.

Then the sign of λ_0 is the same as sign of $(2\sigma_0^2-2\alpha+\beta^2)(\tau-\tau_0)$ because $\tau_0\sigma_1^2/(H_1^2+H_2^2)>0$. When $\tau_0=\tau_0^+$, $(2\sigma_0^2-2\alpha+\beta^2)(\tau-\tau_0)=\sqrt{\Delta_2}(\tau-\tau_0^+)$. Therefore, trivial solution is unstable when $\tau>\tau_0$ and trivial solution is stable when $\tau<\tau_0$. □

According to (17.46) and (17.47), if $\lambda_p<0$ ($\lambda_p>0$) then $\lambda_0>0$ ($\lambda_0<0$), and therefore the trivial solution is unstable when periodic solution is stable and vice versa. According to the damped cases in Theorem 17.1, using formula (17.46), we show some examples of stability of periodic solutions in Table 17.7.

17.5 Simulation of period-doubling cascade

In this section, we numerically investigate periodic dynamics by using program 'dde23' [Shampine and Thompson (2001)] with high accuracy setting in *MATLAB*.

We explore the location of period-doubling bifurcation points by adjusting time delay τ manually. Table 17.8 shows an example with strong damping for period-doubling sequence when parameters satisfy case (2) of Theorem 17.1. Furthermore, another example with weak damping for case (3) of Theorem 17.1 is in Table 17.9.

Table 17.7 Some examples of stability of periodic solution near τ_0^+.

Parameter Value					Sign of	Stability of	Phase diagram	τ_0^+
α	β	d_1	d_2	d_3	λ_p	periodic solution		
1	1.5	-5	0.2	0.01	-	stable	Fig. 17.1a	0.33
1	-1.5	5	0.2	0.01	+	unstable	/	0.33
1.3	1.5	-5	0.2	0.01	-	stable	Fig. 17.1b	0.33
1.3	-1.5	5	0.2	0.01	+	unstable	/	0.33
4	1.8	-3.9	0.2	0.01	-	stable	Fig. 17.1c	0.66
4	-1.8	3.5	0.2	0.01	-	stable	Fig. 17.1d	0.77

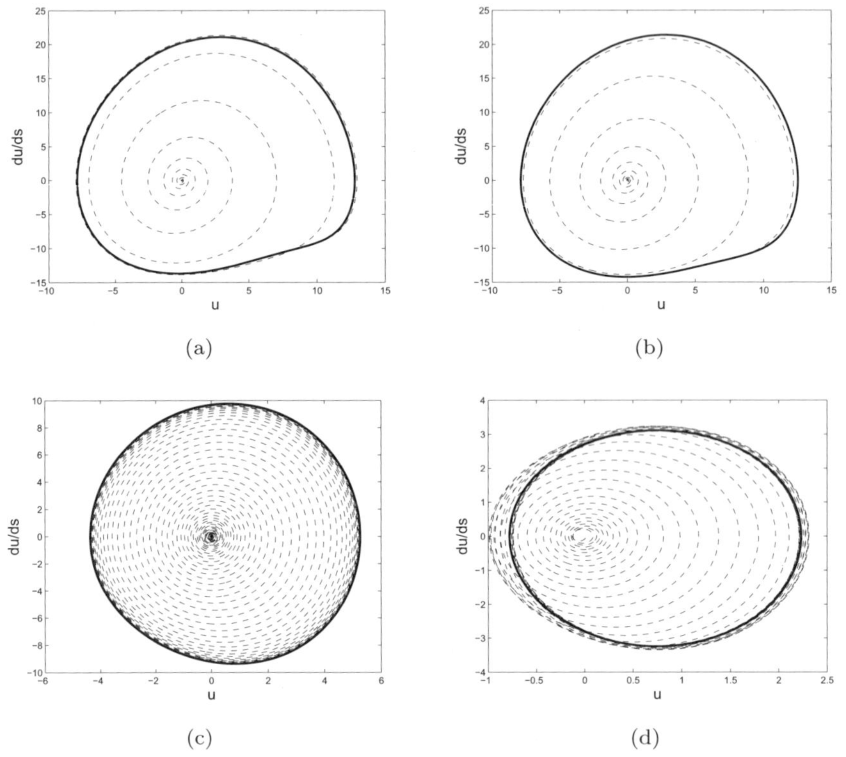

Fig. 17.1 Phase diagrams for periodic solutions near Hopf bifurcation point, solid curves represent stable limit cycles of $d^2u/ds^2 + \beta\, du/ds + \alpha u = d_1 u(s-\tau) + d_2 u^2(s-\tau) + d_3 u^3(s-\tau)$ with initial condition $(u(s), \dot{u}(s)) = (1, 0)$. (a) $\alpha = 1, \beta = 1.5, d_1 = -5, d_2 = 0.2, d_3 = 0.01, \tau = 0.5$ $(\tau > \tau_0^+ = 0.33)$. (b) $\alpha = 1.3, \beta = 1.5, d_1 = -5, d_2 = 0.2, d_3 = 0.01, \tau = 0.5$ $(\tau > \tau_0^+ = 0.33)$. (c) $\alpha = 4, \beta = 1.8, d_1 = -3.9, d_2 = 0.2, d_3 = 0.01, \tau = 0.8$ $(\tau > \tau_0^+ = 0.66)$. (d) $\alpha = 4, \beta = -1.8, d_1 = 3.5, d_2 = 0.2, d_3 = 0.01, \tau = 0.8$ $(\tau > \tau_0^+ = 0.77)$.

The *Feigenbaum constant*, 4.6692..., is the limiting ratio of each bifurcation interval to the next. Feigenbaum originally related this number to the period-doubling bifurcations in the logistic map, but also showed it to hold for all one-dimensional maps with a single quadratic maximum [Feigenbaum (1978)]. In the

Table 17.8 Period-doubling bifurcation points, parameters are as follows $\alpha = 4, \beta = 1, d_1 = -5, d_2 = 0.2, d_3 = 0.0$.

Period	Critical value of τ	Ratio	Phase diagram
1	0.227		Fig. 17.2a
2	0.580	(0.580-0.227)/(0.643-0.580)≈ 5.6032	Fig. 17.2b
4	0.643	(0.643-0.580)/(0.6562-0.643) ≈ 4.7727	Fig. 17.2c
8	0.6562	(0.6562-0.643)/(0.659-0.6562)≈ 4.7143	Fig. 17.2d
16	0.659		Fig. 17.2e

Table 17.9 Period-doubling bifurcation points, parameters are as follows $\alpha = 4, \beta = 0.01, d_1 = -2, d_2 = 0.2, d_3 = 0.01$.

Period	Critical value of τ	Ratio	Phase diagram
1	0.02		Fig. 17.3a
2	0.5485	(0.5485-0.02)/(0.587-0.5485) ≈ 13.7273	Fig. 17.3b
4	0.587	(0.587-0.5485)/(0.596-0.587) ≈ 4.2778	Fig. 17.3c
8	0.596	(0.596-0.587)/(0.5979-0.596)≈ 4.7368	Fig. 17.3d
16	0.5979		Fig. 17.3e

present paper, we aim to test the possibility of existence of the constant in such an infinite dimensional dynamical system. We calculate the ratios of the adjacent intervals between the period-doubling bifurcation points (see Tables 17.8 and 17.9). The existence of the universal constant in delayed system might be reasonable, because a periodic solution of delay differential equation corresponds to a fixed point of Poincaré map [Hale and Verduyn Lunel (1993)]. The ratios in Tables 17.8 and 17.9 seem to be an approximate Feigenbaum constant, so this parameter maybe provide a useful tool to numerically determine the value of the universal constant in dynamics of delay differential equation.

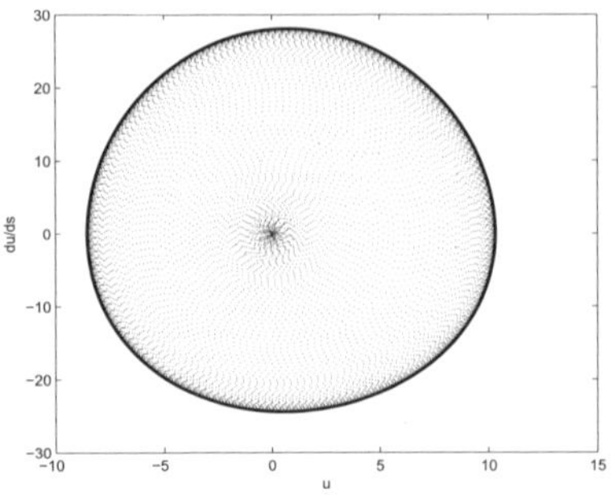

(a)

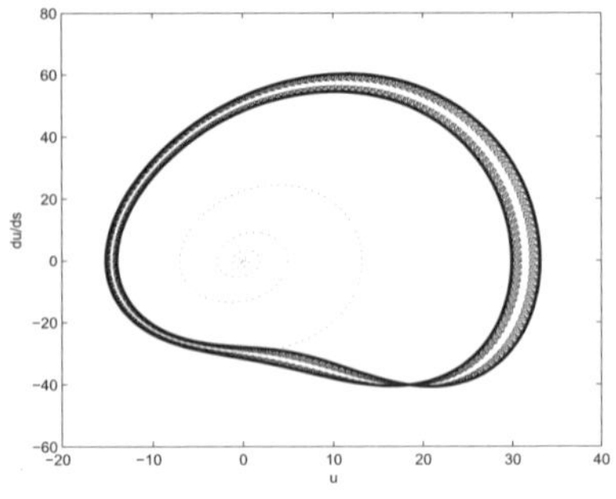

(b)

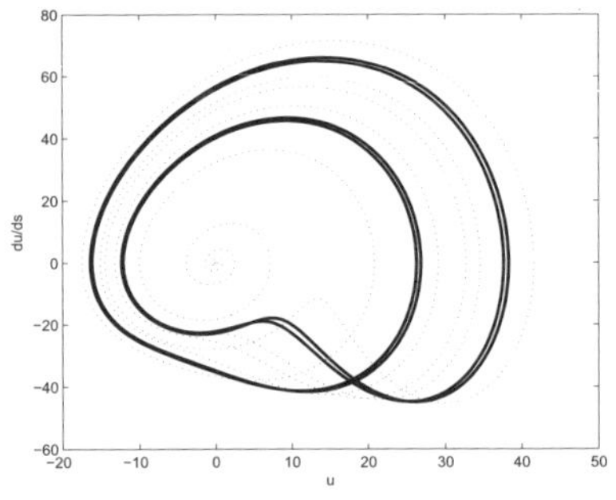

(c)

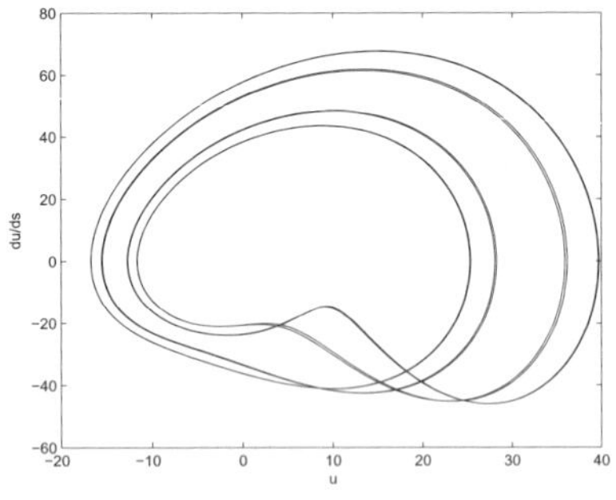

(d)

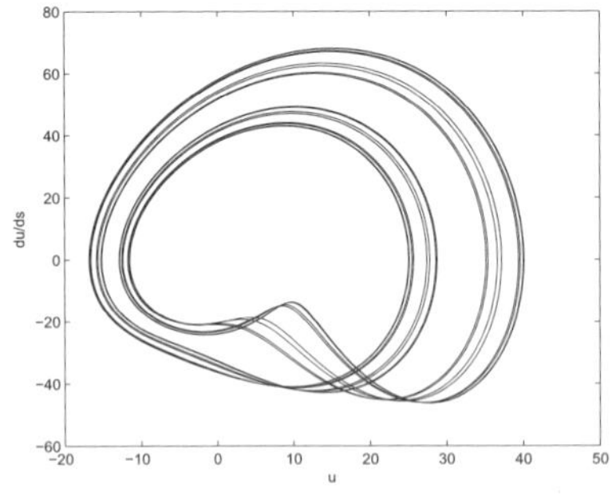

(e)

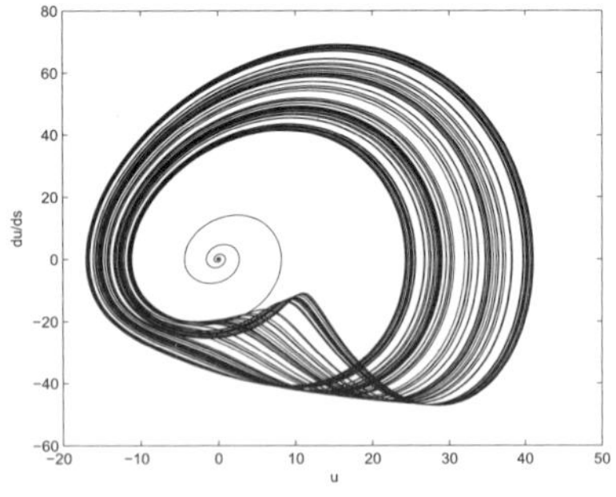

(f)

Fig. 17.2 Phase diagrams for periodic solutions and chaos of $d^2u/ds^2 + \beta\, du/ds + \alpha u = d_1 u(s - \tau) + d_2 u^2(s-\tau) + d_3 u^3(s-\tau)$ with initial condition $(u(s), \dot{u}(s)) = (1, 0)$, where $\alpha = 4, \beta = 1, d_1 = -5, d_2 = 0.2, d_3 = 0.0$. (a) Period-1 solution, $\tau = 0.227$. (b) Period-2 solution, $\tau = 0.58$. (c) Period-4 solution, $\tau = 0.643$. (d) Period-8 solution, $\tau = 0.6562$. (e) Period-16 solution, $\tau = 0.659$. (f) Chaos attractor, $\tau = 0.675$.

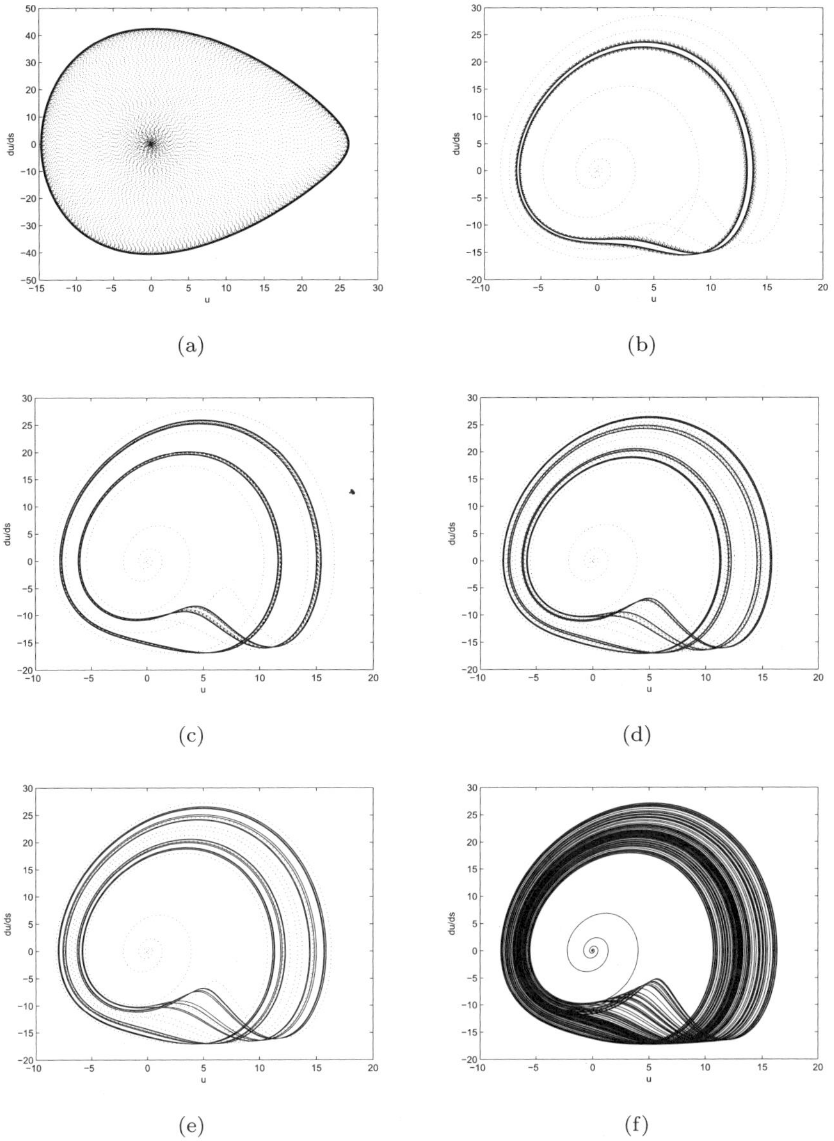

Fig. 17.3 Phase diagrams for periodic solutions and chaos of $d^2u/ds^2 + \beta\, du/ds + \alpha u = d_1 u(s - \tau) + d_2 u^2(s - \tau) + d_3 u^3(s - \tau)$ with initial condition $(u(s), \dot{u}(s)) = (1, 0)$, where $\alpha = 4, \beta = 0.01, d_1 = -2, d_2 = 0.2, d_3 = 0.01$. Solid curves represent stable periodic solutions. (a) Period-1 solution, $\tau = 0.02$. (b) Period-2 solution, $\tau = 0.55$. (c) Period-4 solution, $\tau = 0.588$. (d) Period-8 solution, $\tau = 0.597$. (e) Period-16 solution, $\tau = 0.5984$. (f) Chaos attractor, $\tau = 0.61$.

Bibliography

Campbell, S. A., Bélair, J., Ohira, T., and Milton, J. (1995a). Complex dynamics and multistability in a damped harmonic oscillator with delayed negative feedback, *Chaos* **5**, pp. 640–645.

Campbell, S. A., Bélair, J., Ohira, T. and Milton, J. (1995b). Limit cycles, tori and complex dynamics in a second order differential equation with delayed negative feedback, *J. Dyn. Differ. Equ.* **7**, pp. 213–236.

Carr, J. (1981). *Applications of Centre Manifold Theory* (Springer-Verlag, New York).

Chow, S. N. andHale, J. K. (1982). *Methods of Bifurcation Theory* (Springer-Verlag, New York).

Deng, W. H., Wu, Y. J. and Li, C. P. (2006). Stability analysis of differential equations with time-dependent delay, *Int. J. Bifurc. Chaos* **16**, pp. 465–472.

Erneux, T. (2009). *Applied Delay Differential Equations* (Springer-Verlag, New York).

Feigenbaum, M. J. (1978). Quantitative universality for a class of nonlinear transformations, *J. Statist. Phys.* **19**, pp. 25–52.

Guo, Q. and Li, C. P. (2007). Hopf bifurcation of a delayed differential equation, *Int. J. Bifurc. Chaos* **17**, pp. 1367–1374.

Hale, J. K. and Verduyn Lunel, S. M. (1993). *Introduction to Functional Differential Equations* (Springer-Verlag, New York).

Li, C. P. and Chen, G. (2001). Bifurcations of one dimensional reaction diffusion equations, *Int. J. Bifurc. Chaos* **11**, pp. 1295–1306.

Li, C. P. and Chen, G. (2002). Bifurcation from an equilibrium of the steady state Kuramoto-Sivashinsky equation in two spatial dimensions, *Int. J. Bifurc. Chaos* **12**, pp. 103–144.

Moiola, J. L. and Chen, G. (1996). *Hopf bifurcation analysis-A frequency domain approach* (World Scientific, Singapore).

A.H. Nayfeh (1973). *Perturbation Methods* (Wiley-Interscience, New York).

Nayfeh, A. H. (2008). Order reduction of retarded nonlinear systems-the method of multiple scales versus center-manifold reduction, *Nonlinear Dyn.* **51**, pp. 483–500.

Peng, G. J., Jiang Y. L. and Li, C. P. (2009). Bifurcations of a Holling-type II predator-prey system with constant rate harvesting, *Int. J. Bifurc. Chaos* **19**, pp. 2499–2514.

Shampine, L. F. and Thompson, S. (2001). Solving DDEs in MATLAB, *Appl. Numer. Math.* **37**, pp. 441-458.

Song, Y. L., Zhang, T. H. and Tadé, M. O. (2008). Stability and multiple bifurcations of a damped harmonic oscillator with delayed feedback near zero eigenvalue singularity, *Chaos* **18**, 043113.

Wang ,H. L. and Hu, H. Y. (2003). Remarks on the perturbation methods in solving the second-order delay differential equations, *Nonlinear Dyn.* **33**, pp. 379–398.

Xu, X., Hu, H.Y. and Wang, H. L. (2007). Stability, bifurcation and chaos of a delayed oscillator with negative damping and delayed feedback control, *Nonlinear Dyn.* **49**, pp. 117–129.

Yang, Z. H. and Guo, Q. (2005). Bifurcation analysis of delayed logistic equation, *Appl. Math. Comput.* **167**, pp. 454–476.

Chapter 18

The numerical methods in option pricing problem[1]

Xiong Bo

School of Statistics and Mathematics, Zhongnan University of Economics and Law, Wuhan 430072, Hubei, PR China

xiongb@163.com

Option is the most important finance derivative security. The numerical computation of option value is very important in the fields of finance engineering and real options. Option pricing problem plays a key role in option theory. Black-Scholes first gave an option price formula to calculate European option. However, for American option, no analytic formula and exact solution can be obtained. So the study of numerical methods for various options is very significant. In this chapter we study finite difference methods and binomial tree methods for pricing option. Numerical examples show the efficiency and convergence of our algorithm.

Key words: Option pricing problem, Binomial tree methods, Finite difference methods

18.1 Introduction

Option is a contract which can give the holder right to buy a certain amount of financial assets according to the predetermined price in a certain time, besides the holder do not have to undertake the obligation of buying. The stated financial assets in option is underlying assets, the agreed price of the underlying assets is exercise price. According to the business type, the option can be divided into call option and put option. Call option is a contract which means you can buy a certain financial assets at exercise price on specified date. Put option is a contract which means you can sell a certain financial assets at exercise price on specified date. According to the exercise time, the option can be divided into European options and American options. European option is an option that may only be exercised on

[1]The present work was supported by the Youth Foundation of Zhongnan University of Economics and Law (grant no. 31541111206).

expiration. American option is an option that may be exercised on any trading day on or before expiry. Option is the most important derivatives, which can reward the buyer good results from bad results, and can also make great loss for the seller. Option is not free, and option price is the only variable in contract, subjected to the market changes. Option price is the core problem in option trading, for it directly influences the profit and loss of the contracting parties. As for the holder, he can exercise or not during the option period, otherwise for the seller, he has potential obligation to keep an agreement in case that the option holder intends to exercise. There is a certain value in option, for the option endows the holder a certain right but the holder does not have to undertake the obligation. The value can not be reflected directly in the market, which brings the problem of valuation of options.

After Portfolio Theory and Capital Asset Pricing Model, the Option Pricing Theory becomes an important theory in financial application from which we can gain the Nobel Prize. In 1973, Fisher Black and Myron Scholes published "the pricing of options liabilities", and presented the famous Option Pricing Theory. The ideological sprout of option thought can retrospect to Code of Hammurabi in 1800 BC. The rudiment of option trading appeared in the business between ancient Greek and ancient Phoenician Country in 1200 BC. The acknowledged originator of Option Pricing Theory is French mathematician Louis Bachelier, whose doctoral dissertation is Theories de la speculation, in which the trend of share price was first given strict mathematical description. It is acknowledged as the milestone of finance. He assumed that a stock price process is no drift and variance per unit time with a pure standard Brown motion, and obtained the call option price

$$P(X,T) = X\Phi\Big(\frac{X-K}{\sigma\sqrt{T}}\Big) - K\Phi\Big(\frac{X-K}{\sigma\sqrt{T}}\Big) + \sigma\sqrt{T}\phi\Big(\frac{X-K}{\sigma\sqrt{T}}\Big), \tag{18.1}$$

where $P(X,T)$ means price of option when the stock price is X at time T. X means the stock price, K means the option exercise price, $\Phi(\cdot)$ means the standard normal distribution function, $\phi(\cdot)$ means the standard normal distribution density function. Now it appears that the model is slightly less than the following two aspects: Firstly, the Absolute Brown Motion allows the stock price is negative, the average expected price change is zero, this assumption is divorced from reality. Secondly, the model does not consider the time value of money. But Bachelier's work did not cause the financial community's attention until the 1950s, the option pricing theory was developed rapidly, during the progress of option, pricing theory is used mainly in the application of econometric models. The call option prices of Kassouf's work [Gemmill (1992)] is as follows

$$P = K[(X/K)^\gamma + 1]^{1/\gamma} - 1, \qquad 1 \leq \gamma + \infty, \tag{18.2}$$

where K, X are the strike price and stock price respectively, γ is a parameter required by the expiration time, stock returns and other variables to estimate.

In 1961, Sprekle [Merton (1973)] developed the option pricing model, he got the call option value formula:

$$P(X,T) = Xe^{\alpha T}\Phi\Big(\frac{\ln(X/K) + (\alpha + \sigma^2/2)T}{\sigma\sqrt{T}}\Big) - (1-\pi)K\Phi\Big(\frac{\ln(X/K) + (\alpha - \sigma^2/2)T}{\sigma\sqrt{T}}\Big). \tag{18.3}$$

In 1964, Boness considered the time value of money, and concluded the model:

$$P(X,T) = X\Phi\Big(\frac{\ln(X/K) + (\alpha + \sigma^2/2)T}{\sigma\sqrt{T}}\Big) - e^{-\alpha T}K\Phi\Big(\frac{\ln(X/K) + (\alpha - \sigma^2/2)T}{\sigma\sqrt{T}}\Big). \tag{18.4}$$

In the same year, Sharpe and Lintner [Broadie and Detemple (2004)] also made a call option pricing model, but their results depended on investors' risk aversion. In 1965, Samuelson [Black and Scholes (1973)] gave the formula for European call options

$$P(X,T) = e^{(\alpha-\beta)T}X\Phi\Big(\frac{\ln(X/K) + (\alpha + \sigma^2/2)T}{\sigma\sqrt{T}}\Big) - e^{-\beta T}K\Phi\Big(\frac{\ln(X/K) + (\alpha - \sigma^2/2)T}{\sigma\sqrt{T}}\Big). \tag{18.5}$$

Obviously, it is Boness who proposed pricing model when $\alpha = \beta$. Boness and Samuelson's call option formula is basically close to the Black-Scholes option pricing formula. But these formulas contain one or more parameters, and these parameters are decided by investors to the risk or the return rate of stock preference decisions, so it directly effects the use of formula.

18.2 Black–Scholes option pricing theory assumptions

Modern option pricing theory revolution occurred in 1973, the U.S. financial experts Fisher Black and Myron Scholes used the hedging strategies to introduce the continuous trading Black-Scholes pricing model. This model is based on an efficient market and the stock price which follows five assumptions. The innovation of this model is that it does not depend on investor preferences, which leads all investors to the same risk-neutral world with a risk-free interest rate as the rate of investment returns. In 1997, because of the contribution of the Black-Scholes formula and the resulting series of option pricing theory, M. Scholes (F. Black deceased) won the Nobel Prize in Economics. Black-Scholes option pricing theory [Black and Scholes (1973)] is based on the following assumptions.

(1)The underlying asset price follows geometric Brown motion:

$$dX = \mu X dT + \sigma X dW,$$

where μ is the expected return rate of the underlying asset, and σ is the price volatility. Both of them are constants. And dW is a Wiener process, $dW = \varepsilon\sqrt{dT}$,

and ε is subject to the standard normal distribution;
(2) Risk-free rate r is a constant and it is the same for all maturities;
(3) do not pay transaction cost and tax;
(4) there are no risk-free arbitrage opportunities;
(5) allow the use of all assets derived from short selling.

Black-Scholes option pricing model lies in the design of a hedging portfolio strategy, making the options market investment risk be zero, this is the idea of option pricing formula for modeling high-level summary. Black-Scholes established the following partial differential equation of $P(x,t)$ which is the underlying asset value of call option price at the time of t with price x.

$$\begin{cases} \frac{\partial P(x,t)}{\partial t} - rP(x,t) + rx\frac{\partial P(x,t)}{\partial x} + \frac{1}{2}\sigma^2 x^2 \frac{\partial^2 P(x,t)}{\partial x^2} = 0, \\ P(x,T) = \max\{0, x-K\}, \quad x > 0, \end{cases}$$

where T is the option expiration date, r means the risk-free interest rate, σ^2 subjects to the rate of change in the underlying asset return variance, representing the volatility of the underlying asset price, K means the option's strike price. It is Easy to fined the equation elimination of the expected rate of return μ, so the equation is independent investors' preferences. By solving the above equation, one can get the pricing formula of European call option.

$$P(x,t) = x\Phi(d_1) - K\Phi(d_2)e^{-r(T-t)},$$

where $\Phi(\cdot)$ is the standard normal distribution function,

$$d_1 = \frac{\ln(x/K) + (r + \sigma^2/2)(T-t)}{\sigma\sqrt{T-t}}, \qquad d_2 = \frac{\ln(x/K) + (r - \sigma^2/2)(T-t)}{\sigma\sqrt{T-t}},$$

and European put option pricing formula:

$$P(x,t) = -x\Phi(-d_1) - K\Phi(-d_2)e^{-r(T-t)}.$$

For standard European option, we can directly use the Black-Scholes pricing formula. However, for the American put option or pay dividends of American call options and profit and European option with complex loss function, there is no analytical formula, and one can not find the exact solution, so numerical methods become very important pricing methods.

18.3 Binomial tree methods

Cox, Ross and Robinstein stated Binomial tree methods [Ahn and Song (2007)], which divide the available period of option into several small enough time interval. Assume that in each interval Δt, the price of Underlying Asset moves from x to two new values xu, xd at the rate of p and $1-p$, namely rises up from x to xu, drop down from x to xd. If it is for a long interval, such assumption is not proper, however, when interval is very short, the assumption that assets price changes just into two directions is acceptable. In fact, Binomial tree model simulates continuous assets movement by a large number of discrete small amplitude.

Suppose that current time is $t = 0$, underlying asset's price is x. Then at time Δt, underlying asset's price is xu or xd; at $2\Delta t$, the price is xu^2, xud, xd^2, and so on. At $i\Delta t$, underlying asset's price has $i+1$ possibilities, $xu^j d^{i-j}$ ($j = 0, 1, \ldots, i$). F_{ij} ($j = 0, 1, \ldots, i$) denotes the options at $i\Delta t$. The price of call options is $\max\{x - K, 0\}$; the price of put option is $\max\{K - x, 0\}$. We can conclude the terminal options' price just from the expiring date of it. Totally speaking, in a high risky environment, the options' price at $i\Delta t$ can get from the expected value of options at the moment of $(i+1)\Delta t$ which proceeds by risk-free rate of interest during Δt.

$$F_{ij} = e^{-r\Delta t}[pF_{i+1,j+1} + (1-p)F_{i+1,j}], \qquad j = 0, 1, \ldots, i, \quad 0 \le i \le n-1.$$

For European put options, option at the moment 0 is $F = F_{00}$; For American put options, it is

$$F_{ij} = \max_{0 \le j \le i, 0 \le i < n} \{K - xu^i d^{i-j}, e^{-r\Delta t}[pF_{i+1,j+1} + (1-p)F_{i+1,j}]\}.$$

As we compute it by using backward method at $i\Delta t$ moment, the option at T not only indicates the affect of the value of option by executing this possibility at time $i\Delta t$, but also indicates that executing this possibility ahead during the following time. In the formula, there are three parameters p, u, d which need to be made properly. Since the fluctuation ratio of Underlying Asset follows Gaussian distributions, by Risk-Neutral Valuation, one can conclude $p = \frac{s-d}{u-d}$, $u = e^{\sigma\sqrt{\Delta t}}$, $d = e^{-\sigma\sqrt{\Delta t}}$, $s = e^{r\Delta t}$.

Assume that the underlying price x after time Δt later, with probability p and $1-p$, can get the possible values of p and $1-p$ to get the possible values of xu and xd ($u > 1$, $d < 1$). According to the risk-neutral assumption, at the time scale Δt, the subject of price expectation is $xe^{r\Delta t}$, so

$$xe^{r\Delta t} = pxu + (1-p)xd.$$

The underlying price model is

$$\Delta x = x\mu\Delta t + x\sigma\sqrt{\varepsilon\Delta t}, \tag{18.6}$$

where ε is normally distributed random value. Let $Y = x + \Delta x$ indicate the underlying price changes at Δt time period, the EY meets

$$E(Y^2) - E^2(Y) = x^2\sigma^2\Delta t. \tag{18.7}$$

Combining (18.6) and (18.7) leads to

$$\sigma^2\Delta t = pu^2 + (1-p)d^2 - e^{2r\Delta t}. \tag{18.8}$$

Let $s = e^{r\Delta t}$, then (18.6) and (18.8) can be written as

$$s = pu + (1-p)d, \tag{18.9}$$

$$\sigma^2 \Delta t = pu^2 + (1-p)d^2 - s^2. \tag{18.10}$$

Letting $d = 1/u$, and combining (18.9) and (18.10), we have

$$p = \frac{s-d}{u-d}, \quad u = e^{\sigma\sqrt{\Delta t}}, \quad d = e^{-\sigma\sqrt{\Delta t}}, \quad s = e^{r\Delta t}. \tag{18.11}$$

But there are obvious flaws in the parameters selection. First, setting up $d = 1/u$ is not realistic. This assumption will cause inaccurate results. Second, for smaller volatility σ, one may get the probability greater than 1 or negative probability by the formula (18.11). This parameter selection does not make sense. Therefore we must reset parameters.

Theorem 18.1. *Based on risk assumptions, parameters p, u, and d satisfy the formula*

$$p = 1/2, \quad u = s + \sigma\sqrt{\Delta t}, \quad d = s - \sigma\sqrt{\Delta t}, \quad s = e^{t\Delta t}. \tag{18.12}$$

Proof. According to (18.9), one obtains $p = s - d/u - d$. Inserting p into (18.10) yields

$$\sigma^2 \Delta t = p(u^2 - d^2) - s^2 + d^2 = (s-d)(u-s).$$

The formulae for calculating the parameters are then obtained

$$p = (s-d)/(u-d), \quad d = s - \frac{\sigma^2 \Delta t}{u-s}, \quad s = e^{r\Delta t}. \tag{18.13}$$

Assume the current target price is x, in the risk-neutral world, after the time Δt, the expected value of x is $E(x + \Delta x) = xr^{r\Delta t}$, $E(u) = e^{r\Delta t}$. According to (18.6),

$$\frac{x + \Delta x}{x} = 1 + \mu\Delta t + \sigma\sqrt{\varepsilon\Delta t}, \quad \sigma^2(u) = \sigma^2\Delta t.$$

So a reasonable value of u should be $u = E(u) \pm \sigma(u)$. Since $0 \leq p \leq 1$, (18.13) means $u \geq E(u)$, we take $u = E(u) + \sigma(u)$ according to (18.13). We get the calculations of parameters p, u, d. □

This result is reasonable, $p = 1/2$ means x by the same probability rise and fall, which is in line with the actual situation; and this calculation does not yield a probability greater than 1 or a negative probability.

18.4 Finite difference method

The basic idea is to use finite-difference method [Zhao *et al.* (2007)] to approximate the derivative of partial differential equations, using the numerical method to solve the differential equation satisfied by derivative assets in order to estimate the price of them. The differential equations are transferred to a series of differential equations, from the expiration time of derivative assets pour back to the initial time. The iterative method is utilized to solve these differential equations. As previously mentioned, the put option must satisfy the following equation:

$$\begin{cases} \frac{\partial P(x,t)}{\partial t} - rP(x,t) + rx\frac{\partial P(x,t)}{\partial x} + \frac{1}{2}\sigma^2 x^2 \frac{\partial^2 P(x,t)}{\partial x^2} = 0, & (t,x) \in [0,T] \times (0,+\infty), \\ P(x,T) = \max\{0, K-x\}, & x > 0, \end{cases} \tag{18.14}$$

1) Meshing

Insert N points: $\Delta t, 2\Delta t, \cdots, n\Delta t$, with $\Delta t = \frac{T}{N}$. They will be divided into $N+1$ small time periods by equally spaced intervals. Let X represent the price which can reach a high enough price of the underlying asset. When the price reaches, American put option is actually no value.

Delimit $\Delta x = \frac{x_{\max}}{M}$, so there are $M+1$ prices $0, \Delta x, 2\Delta x, \cdots x_{\max}$. There are $(M+1)(N+1)$ nodes. Each node (i,j) corresponds to the time $i\Delta t$ of the underlying asset price $j\Delta x$. Let P_{ij} represent the option price at node (i,j).

2) Difference

Use central difference to approximate $\frac{\partial P}{\partial x}$, so $\frac{\partial P}{\partial x} \approx \frac{P_{i,j+1}-P_{i,j-1}}{2\Delta x}$; use forward difference to approximate $\frac{\partial P}{\partial t}$, so $\frac{\partial P}{\partial t} \approx \frac{P_{i+1,j}-P_{i,j}}{\Delta t}$; use second-order central difference to approximate $\frac{\partial^2 P}{\partial x^2}$, then $\frac{\partial^2 P}{\partial x^2} \approx \frac{P_{i,j+1}-2P_{i,j}+P_{i,j-1}}{(\Delta x)^2}$. Substituting the approximation into the differential equation and taking into account $x = j\Delta x$ yield

$$\frac{P_{i+1,j}-P_{ij}}{\Delta t} + rj\Delta x\frac{P_{i,j+1}-P_{i,j-1}}{2\Delta x} + \frac{1}{2}\sigma^2 j^2 \Delta x^2 \frac{P_{i,j+1}-2P_{ij}+P_{i,j-1}}{(\Delta x)^2} = rP_{ij}.$$

Sorted out

$$a_j P_{i,j-1} + b_j P_{i,j} + c_j P_{i,j+1} - P_{i+1,j} = 0 \quad (i = 0,1,\cdots,N-1,\ j = 0,1,\cdots,M-1),$$

where $a_j = -\frac{1}{2}(-rj+\sigma^2 j^2)\Delta t,\ b_j = 1 + (\sigma^2 j^2 + r)\Delta t,\ c_j = -\frac{1}{2}(rj+\sigma^2 j^2)\Delta t.$

3) Boundary conditions

(1)The price of T moment of American put option is $\max\{K - x_T, 0\}$, so

$$P = \max\{K - j\Delta x, 0\}, \quad j = 0,1,\cdots,M.$$

(2)When the mark price is 0, the price of American put option is K,

$$P_{i,0} = K \quad i = 0,1,\cdots,N.$$

(3)When the mark price is infinite, and the price of American put option is 0,

$$P_{i,M} = 0 \quad i = 0,1,\cdots,N.$$

The three conditions above give $x = 0, x = x_{\max}, t = T$, the option value boundaries of the chart.

4) Calculation

At time $i\Delta t$, $i = 0,1,...,N-1$, according to

$$a_j P_{i,j-1} + b_j P_{i,j} + c_j P_{i,j+1} - P_{i+1,j} = 0 \quad (i = 0,1,\cdots,N-1,\ j = 0,1,\cdots,M-1),$$

we get $M-1$ formulae

$$a_j P_{N-1,j-1} + b_j P_{N-1,j} + c_j P_{N-1,j+1} - P_{N,j} = 0 \quad (j = 1,\cdots,M-1).$$

Because of boundary conditions, we suppose $P_{N-1,0} = K$, $P_{N-1,M} = 0$.

Solve the tridiagonal equation, we can get $P_{N-1,j}$, $j = 1, 2, \cdots, M-1$. Compare $P_{N-1,j}$ and $K - j\Delta x$, if $P_{N-1,j} < K - j\Delta x$, we ought to exercise the option ahead of time $(N-1)\Delta t$ and let $P_{N-1,j} = K - j\Delta x$. Following this method, we can obtain P_{0j}, $j = 1, 2, \cdots, M-1$. Finally we get the corresponding option value according to the market price.

5) Numerical examples

Consider five-month American put option of a non-dividend paying stocks. The stock price x is 50 dollars, exercise price K is 50 dollars, $r = 0.1, \sigma = 0.4$, expiration date is $T = \frac{5}{12}$. The option's expiration date is divided into N parts, $\Delta t = \frac{T}{N}$. Let $M = 20$, $N = 10$, $\Delta x = 5$, then $\Delta t = 0.04167$, $x_{\max} = M\Delta x = 100$. The results are shown in the following table.

j	Stock	Expiration time (month)										
	price	5	4.5	4	3.3	3	2.5	2	1.5	1	0.5	0
20	100	0.00	0.00	0.00	0.00	0.00	0.00	0.00	0.00	0.00	0.00	0.00
19	95	0.02	0.02	0.01	0.01	0.00	0.00	0.00	0.00	0.00	0.00	0.00
18	90	0.05	0.04	0.03	0.02	0.01	0.00	0.00	0.00	0.00	0.00	0.00
17	85	0.00	0.00	0.00	0.00	0.00	0.00	0.00	0.00	0.00	0.00	0.00
16	80	0.16	0.12	0.09	0.07	0.04	0.03	0.02	0.01	0.00	0.00	0.00
15	75	0.27	0.22	0.17	0.13	0.09	0.06	0.03	0.02	0.01	0.00	0.00
14	70	0.47	0.39	0.32	0.25	0.18	0.13	0.08	0.04	0.02	0.00	0.00
13	65	0.82	0.71	0.60	0.49	0.38	0.28	0.19	0.11	0.06	0.02	0.00
12	60	1.42	1.27	1.11	0.95	0.78	0.62	0.45	0.30	0.16	0.05	0.00
11	55	2.43	2.24	2.05	1.83	1.61	1.36	1.09	0.81	0.51	0.22	0.00
10	50	4.07	3.88	3.67	3.45	3.19	2.91	2.57	2.17	1.66	0.99	0.00
9	45	6.58	6.44	6.29	6.13	5.96	5.77	5.57	5.36	5.17	5.02	5.00
8	40	10.15	10.10	10.05	10.00	10.00	10.00	10.00	10.00	10.00	10.00	10.00
7	35	15.00	15.00	15.00	15.00	15.00	15.00	15.00	15.00	15.00	15.00	15.00
6	30	20.00	20.00	20.00	20.00	20.00	20.00	20.00	20.00	20.00	20.00	20.00
5	25	25.00	25.00	25.00	25.00	25.00	25.00	25.00	25.00	25.00	25.00	25.00
4	20	30.00	30.00	30.00	30.00	30.00	30.00	30.00	30.00	30.00	30.00	30.00
3	15	35.00	35.00	35.00	35.00	35.00	35.00	35.00	35.00	35.00	35.00	35.00
2	10	40.00	40.00	40.00	40.00	40.00	40.00	40.00	40.00	40.00	40.00	40.00
1	5	45.00	45.00	45.00	45.00	45.00	45.00	45.00	45.00	45.00	45.00	45.00
0	0	50.00	50.00	50.00	50.00	50.00	50.00	50.00	50.00	50.00	50.00	50.00

Both the binary tree method and the finite difference method use the future value of options to trace the initial value of the option. So they can be used to calculate the American option. However, the biggest three shortcomings of both methods are computationally intensive, low computational efficiency and difficult to calculate options depending on the history of the state variable path complications. Binary tree method is more suitable for the calculation of a small amount. When faced with a large number of option value calculations, finite difference methods have advantages. With the deepening of option pricing theory, scholars have made a lot of option pricing methods. For example, Monte Carlo simulation method is a kind of European-style derivative asset valuation methods, its advantages rely on the solvability of the underlying asset expected returns and volatility function in more complex situations. The computing time linearly increases with the number of variables increasing. So it is effective in computation. But Monte Carlo simulation method has also its flaws. It is not suitable for calculating American options. In the case of in-complete financial market without completed replication strategies, the traditional option pricing methods and binary, finite difference method are no longer applicable. Other scholars presented ε-arbitrage pricing method and interval pricing method. See, e.g., in [Henderson (2007)]. One may notice that binary tree method is the special case of ε-arbitrage pricing method. The result obtained by ε-arbitrage pricing method must be in the interval obtained by interval pricing method. These two methods are both applicable to the complete financial market and the in-complete financial market. Various types of option pricing methods are ongoing studied and developed. The development of options is endless in theory. The option is complex and is used widely in practice. Therefore, in-depth study and exploration of option pricing methods always play important roles in theory and practice.

Bibliography

Ahn, J. and Song, M. (2007). Convergence of the trinomial tree method for pricing European/American options, *Appl. Math. Comput.* **189**, pp. 575–582.

Black, F. and Scholes, M. (1973). The pricing of options and corporate liabilities, *Journal of Political Economy* **81**, pp. 637–654.

Broadie, M. and Detemple, J. (2004). Option pricing:valuation models and applications, *Management Science* **59**, pp. 1145–1177.

Gemmill, G. (1992). *Option Pricing* (McGraw Hill, New York), pp. 203–245.

Henderson, V. (2007). Valuing the option to invest in an incomplete market *Spring-Verlag* **1**, pp. 103–128.

Merton, R. (1973). Theory of rational option pricing, *Bell Journal of Ecnocomics and Management Science* **4**, pp. 141–183.

Zhao, J., Davison, M. and Corless, R. M. (2007). Coppact finite difference method for American option pricing, *J. Comput. Appl. Math.* **206**, pp. 306–321.

Chapter 19

Synchronization and its control between two coupled networks

Yongqing Wu* and Minghai Lü

Basic Teaching Department, Liaoning Technical University, Huludao 125105, Liaoning, PR China

**wuyongqing1001@yahoo.com.cn*

With the discovery of the small-world and scale-free networks, the researchers in physics, biology, mathematics and computer science dedicate to the study of complex networks. The reason for this increased attention is that complex networks can describe and characterize many natural complex systems. Among the issues of complex networks, synchronization and its control inside a network has attracted considerable interest. Recently, outer synchronization was introduced to study the dynamics between two coupled networks. This chapter mainly makes a presentation about the outer synchronization and its control between two networks with (or without) time delays. Firstly, anti-synchronization between two networks with two special interactions is presented. In addition, anti-synchronization for the two networks with delayed coupling is investigated via the pinning control technique. Finally, generalized synchronization between two coupled complex networks is studied. Numerical examples are given to show the efficiency of the obtained theoretical results.

Keywords: complex network, generalized synchronization, anti-synchronization

19.1 Introduction

Generally speaking, a complex network is defined as a large set of nodes and the connections among them, in which the nodes and connections can have different meanings in different situations. Many real world systems can be described by complex networks, such as the Internet, the World Wide Web (WWW), food webs, neural networks and social networks, etc. Consequently, complex networks have received rapidly increasing attention in the variety of fields including biology, physics, mathematics, engineering and so on [Watts and Strogatz (1998); Barabási and Albert (1999); Albert and Barabási (2002); Newman (2003)].

In particular, interesting attention has been focused on the synchronization phenomenon of complex networks, in which each node is regarded as a dynamical element[Boccaletti *et al.* (2006); Duan *et al.* (2007); Arenas *et al.* (2008)]. The synchronization problems for small-world and scale-free networks have been studied in [Wang and Chen (2002a)] and [Wang and Chen (2002b)], respectively. And various control schemes have been applied to the above mentioned problems, such as adaptive control [Huang (2006); Chen and Zhou (2006); Zhou *et al.* (2006); Sorrentino and Ott (2008)], impulsive control [Zhou *et al.* (2007); Zhang *et al.* (2007)], pinning control [Wang and Chen (2002c); Li *et al.* (2004); Chen *et al.* (2007); Yu *et al.* (2009)], etc.

The above mentioned work focussed on the synchronization inside a network, which was called "inner synchronization" as in stated [Li *et al.* (2007)]. In 2007, Li *et al.* pioneered in studying "outer synchronization", which refers to the synchronization phenomenon between two or more coupled networks regardless of "inner synchronization", in Ref. [Li *et al.* (2007)]. An important example of outer synchronization is the spread of infectious diseases between different communities. In [Li *et al.* (2009)], outer synchronization between two discrete-time networks is theoretically and numerically studied. Therefore, how one realizes the synchronization between different networks is a very interesting and challenging work. Later on through the adaptive controllers, synchronization between two networks are studied in [Tang *et al.* (2008); Li *et al.* (2008); Chen *et al.* (2009); Zheng *et al.* (2009); Sun *et al.* (2009); Wang *et al.* (2009)], which could deal with more complicated cases, such as different node dynamics, nonidentical topological structures, or time-varying delays. On the other hand, generalized synchronization [Wu *et al.* (2009); Shang *et al.* (2009); Wu *et al.* (2012)] and anti-synchronization [Wu *et al.* (2011)] of complex networks have also been investigated.

In this chapter, outer synchronization and its control between two networks are investigated. The main contents can be summarized as follows. In Section 19.2, anti-synchronization between two coupled networks with nonlinear signal's connection and the inter-network actions is presented. In Section 19.3, a sufficient condition for the anti-synchronization between two general complex dynamical networks with non-delayed and delayed coupling is obtained based on the pinning controllers. In Section 19.4, generalized synchronization between two coupled complex networks is studied, where the node vectors in different networks are not the same, and the numbers of nodes of both networks are not necessarily equal. A sufficient criterion for generalized synchronization of two coupled networks is established based on the auxiliary system method. Numerical examples are also included which coincide with the theoretical analysis.

19.2 Anti-synchronization between two coupled networks with nonlinear signal's connection and the inter-network actions

The coupled equations of two networks can generally be expressed as follows:

$$\dot{x}_i(t) = f(x_i(t)) + \sum_{j=1}^{N} a_{ij}\Gamma x_j(t) + \mathbb{C}_{\mathrm{YX}}(Y, X), \quad i = 1, 2, \cdots, N, \tag{19.1}$$

$$\dot{y}_i(t) = g(y_i(t)) + \sum_{j=1}^{N} b_{ij}\Gamma y_j(t) + \mathbb{C}_{\mathrm{XY}}(X, Y), \quad i = 1, 2, \cdots, N, \tag{19.2}$$

where x_i, $y_i \in R^n$, $i = 1, 2, \cdots, N$, $f(\cdot), g(\cdot) : R^n \to R^n$ are continuously differentiable functions which determine the dynamical behavior of the nodes in networks $\mathbb{X}$ and $\mathbb{Y}$ respectively. $\Gamma \in R^{n\times n}$ is a constant 0-1 matrix linking coupled variables. For simplicity, one often assumes that $\Gamma = \mathrm{diag}(r_1, r_2, \cdots, r_n) \geq 0$ is a diagonal matrix. $A = (a_{ij})_{N\times N}$ and $B = (b_{ij})_{N\times N}$ represent the coupling configurations of both networks, whose entries a_{ij} and b_{ij} are defined as follows: if there is a connection between node i and node j $(j \neq i)$, then set $a_{ij} = 1$ and $b_{ij} = 1$, otherwise $a_{ij} = 0$, $b_{ij} = 0\,(j \neq i)$; the matrices A and B can be symmetric or asymmetric, each line sum of A and B is equal to zero. $\mathbb{C}_{\mathrm{XY}}(X, Y)\,(\mathbb{C}_{\mathrm{YX}}(Y, X))$ represents the interaction from network $\mathbb{X}(\mathbb{Y})$ to network $\mathbb{Y}(\mathbb{X})$. There are lots of active forms between two networks, for instance, communicated by signals, special nodes or bidirectional actions. This section focuses on the nonlinear signal's connection and bidirectional actions [Wu *et al.* (2011)].

Assumption 19.1. Assume that the functions $f(x), g(x)$ are Lipschitz continuous, i.e., there exist positive constants L_1, L_2, satisfying

$$\|f(y) - f(x)\| \leq L_1\|y - x\|, \quad \|g(y) - g(x)\| \leq L_2\|y - x\|,$$

where y and x are time-varying vectors.

Assumption 19.2. The functions $f(x), g(x)$ are odd functions of x, i.e., $f(-x) = -f(x), g(-x) = -g(x)$ for any $x \in R^n$.

Lemma 19.1 (Barbalat Lemma). *If function $\phi(t)$ is uniformly continuous, and $\lim\limits_{t\to\infty} \int_0^t |\phi(\tau)|\mathrm{d}\tau$ is bounded, then $\phi(t) \to 0$ when $t \to \infty$.*

Definition 19.1. If $\lim\limits_{t\to+\infty} \|x_i(t) + y_i(t)\| = 0$, $i = 1, 2, \cdots, N$, we call that network $\mathbb{X}$ and network $\mathbb{Y}$ achieve the anti-synchronization.

19.2.1 *Two coupled networks with nonlinear signals*

Choosing $\mathbb{C}_{\mathrm{XY}}(X, Y) = (1 - \theta)[f(y_i(t)) + g(x_i(t))]$ and $\mathbb{C}_{\mathrm{YX}}(Y, X) = \theta[f(y_i(t)) + g(x_i(t))]$, where $0 \leq \theta \leq 1$ represents the nonlinear coupling parameter between these two networks. If there doesn't exist such an active form, which makes anti-synchronization happen, the control strategies are often applied. Then Eqs. (19.1)-(19.2) read as

$$\dot{x}_i(t) = f(x_i(t)) + \sum_{j=1}^{N} a_{ij}\Gamma x_j(t) + \theta[f(y_i(t)) + g(x_i(t))], \tag{19.3}$$

$$\dot{y}_i(t) = g(y_i(t)) + \sum_{j=1}^{N} b_{ij}\Gamma y_j(t) + (1-\theta)[f(y_i(t)) + g(x_i(t))] + U_i, \tag{19.4}$$

where U_i is the controller for node i to be designed.

The main result is summarized in the following theorem.

Theorem 19.1. *Suppose that Assumptions* 19.1 *and* 19.2 *hold. If anti-synchronization is achieved under the nonlinear connected action, let* $U_i = 0$*; otherwise, the network* (19.3)–(19.4) *can realize anti-synchronization with the following adaptive control scheme:*

$$U_i = \sum_{j=1}^{N} \hat{a}_{ij}\Gamma x_j(t) + \sum_{j=1}^{N} \hat{b}_{ij}\Gamma y_j(t) - E_i e_i, \; i = 1, 2, \cdots, N, \tag{19.5}$$

where $e_i = x_i + y_i$, $\dot{E}_i = \varepsilon_i \|e_i\|^2$, ε_i *are arbitrary positive constants, and* $\dot{\hat{a}}_{ij} = -e_i^T \Gamma x_j, \dot{\hat{b}}_{ij} = -e_i^T \Gamma y_j$.

19.2.2 *Two coupled networks with reciprocity*

In this subsection, choosing $\mathbb{C}_{XY}(X,Y) = \sum_{j=1,j\neq i}^{N} d_{ij}\Gamma[x_j(t) + y_i(t)]$ and $\mathbb{C}_{YX}(Y,X) = \sum_{j=1,j\neq i}^{N} c_{ij}\Gamma[y_j(t) + x_i(t)]$, the network equations are written as:

$$\dot{x}_i(t) = f(x_i(t)) + \sum_{j=1}^{N} a_{ij}\Gamma x_j(t) + \sum_{j=1,j\neq i}^{N} c_{ij}\Gamma[y_j(t) + x_i(t)], \tag{19.6}$$

$$\dot{y}_i(t) = g(y_i(t)) + \sum_{j=1}^{N} b_{ij}\Gamma y_j(t) + \sum_{j=1,j\neq i}^{N} d_{ij}\Gamma[x_j(t) + y_i(t)] + U_i, \tag{19.7}$$

where C is an $N \times N$ dimensional coupling matrix, whose entries (c_{ij}) represent the intensity of the direct interaction from i in network $\mathbb{X}$ to j in network $\mathbb{Y}$. Similarly the entries of (d_{ij}) have the same meaning.

Next, an anti-synchronous theorem for network model (19.6)-(19.7) is derived.

Theorem 19.2. *Suppose that Assumptions* 19.1 *and* 19.2 *hold. If the coupling matrices* A, B, C, D *make network* (19.6)–(19.7) *anti-synchronize, then take* $U_i = 0$, *or else*

$$U_i = f(y_i(t)) + g(x_i(t)) + \sum_{j=1}^{N} \tilde{p}_{ij}\Gamma y_j + \sum_{j=1,j\neq i}^{N} \tilde{q}_{ij}\Gamma(x_j + y_i) - H_i e_i, \tag{19.8}$$

where $\dot{\tilde{p}}_{ij} = -e_i^T\Gamma y_j$, $\dot{\tilde{q}}_{ij} = -e_i^T\Gamma(x_j + y_i), \dot{H}_i = \mu_i\|e_i\|^2$, $\mu_i > 0$, *then anti-synchronization between networks* (19.6) *and* (19.7) *can be realized under the controllers* U_i, $i = 1, 2, \cdots, N$.

19.2.3 *Numerical examples*

Choosing the Chua's systems as the network node dynamics. The Chua' system is described by [Chua *et al.* (1993)],

$$\begin{aligned}
\dot{x}_{i1} &= \alpha(x_{i2} - x_{i1} - \phi(x_{i1})), \\
\dot{x}_{i2} &= x_{i1} - x_{i2} + x_{i3}, \\
\dot{x}_{i3} &= -\beta x_{i2},
\end{aligned}$$

where

$$\phi(x_{i1}) = bx_{i1} + \frac{1}{2}(a - b)[|x_{i1} + 1| - |x_{i1} - 1|]$$

and a, b, α, β are parameters. For simplicity, we choose $f(\cdot)$, $g(\cdot)$ as Chua's systems with different parameters. Network $\mathbb{X}$: $a = -1.27$, $b = -0.68$, $\alpha = 10.00$, $\beta = 14.87$; while network $\mathbb{Y}$: $a = -1.39$, $b = -0.75$, $\alpha = 10.00$, $\beta = 18.60$. The inner-coupling matrix Γ is an identity matrix. In the numerical simulation, the initial values of state vectors X_0, Y_0 and control $E_i(0), H_i(0)$ are randomly chosen in $(0, 1)$. Let

$$\|e(t)\| = \sqrt{\sum_{i=1}^{N}\sum_{j=1}^{3}[(y_{ij}(t) + x_{ij}(t))^2]}$$

be the 2-norm of the total anti-synchronization errors at time t, for $t \in (0, +\infty)$.

Example 18.1. Consider the network (19.3)-(19.4). The network size N is taken as 10 and configuration matrices of networks $\mathbb{X}$ and $\mathbb{Y}$ are given as follows,

$$A = \begin{pmatrix}
-3 & 0 & 1 & 0 & 0 & 1 & 0 & 0 & 1 & 0 \\
0 & -4 & 1 & 0 & 0 & 1 & 0 & 1 & 0 & 1 \\
1 & 0 & -3 & 0 & 0 & 0 & 1 & 0 & 1 & 0 \\
0 & 1 & 0 & -4 & 1 & 0 & 1 & 0 & 0 & 1 \\
1 & 0 & 0 & 0 & -3 & 0 & 1 & 1 & 0 & 0 \\
1 & 1 & 0 & 1 & 0 & -5 & 0 & 1 & 1 & 0 \\
1 & 0 & 1 & 0 & 1 & 0 & -4 & 0 & 0 & 1 \\
0 & 1 & 0 & 1 & 0 & 1 & 0 & -4 & 0 & 1 \\
0 & 1 & 0 & 1 & 0 & 0 & 1 & 0 & -3 & 0 \\
1 & 0 & 1 & 0 & 0 & 1 & 1 & 0 & 0 & -4
\end{pmatrix}, \tag{19.9}$$

$$B = \begin{pmatrix} -2 & 0 & 1 & 0 & 0 & 0 & 1 & 0 & 0 & 0 \\ 0 & -4 & 1 & 0 & 1 & 0 & 1 & 0 & 0 & 1 \\ 0 & 0 & -4 & 1 & 0 & 1 & 1 & 0 & 0 & 1 \\ 0 & 1 & 0 & -3 & 1 & 0 & 0 & 0 & 1 & 0 \\ 0 & 1 & 1 & 0 & -4 & 0 & 0 & 1 & 1 & 0 \\ 0 & 1 & 0 & 0 & 1 & -5 & 0 & 1 & 1 & 1 \\ 1 & 0 & 0 & 0 & 1 & 0 & -3 & 1 & 0 & 0 \\ 1 & 0 & 1 & 1 & 1 & 0 & 0 & -5 & 1 & 0 \\ 0 & 0 & 1 & 0 & 1 & 0 & 0 & 1 & -4 & 1 \\ 0 & 0 & 0 & 1 & 0 & 0 & 1 & 1 & 0 & -3 \end{pmatrix}. \tag{19.10}$$

If the controllers $U_i = 0, i = 1, 2, \cdots, 10$, through the numerical simulation, we find that anti-synchronization doesn't happen for arbitrary value of θ in $[0, 1]$. Now we take the value of θ is 0.5, and use the adaptive controllers proposed in Theorem 19.1 to make these two networks anti-synchronize. Fig. 19.1 plots the anti-synchronization error for $\varepsilon_i = 2$, $i = 1, \cdots, 10$.

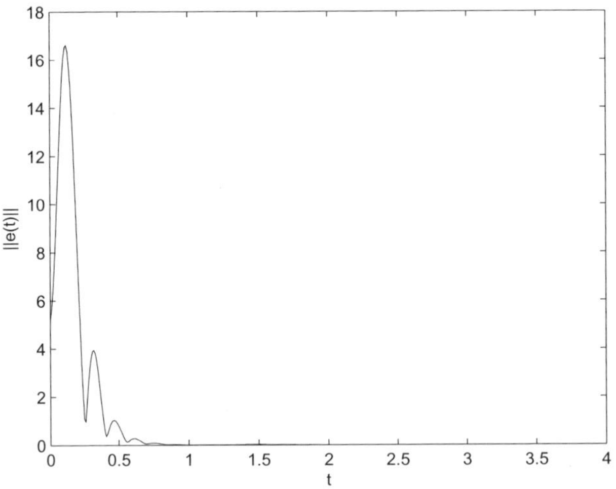

Fig. 19.1 Anti-synchronization error between networks (19.3) and (19.4) with $\varepsilon_i = 2$, $i = 1, \cdots, 10$.

Example 18.2. Consider the network model (19.6)-(19.7). Let configuration matrices A and B be (19.9) and (19.10) respectively. Here, C and D are 10×10 matrices with random entries, chosen from a uniform distribution on the interval $(0, 0.4)$. Anti-synchronization also doesn't exist between networks (19.6)–(19.7) for $U_i = 0$, and the adaptive controllers in Theorem 19.2 are applied. The anti-synchronization between networks (19.6) and (19.7) is shown in Fig. 19.2 with $\mu_i = 2$, $i = 1, \cdots, 10$.

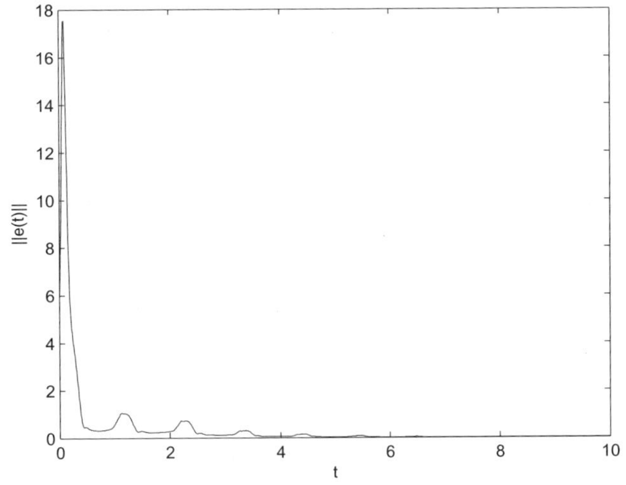

Fig. 19.2 Anti-synchronization error between networks (19.6) and (19.7) with $\mu_i = 2, i = 1, \cdots, 10$.

19.3 Pinning anti-synchronization between two general complex dynamical networks

Consider the following drive-response networks

$$\dot{x}_i(t) = f(x_i(t), t) + c_1 \sum_{j=1}^{N} a_{ij}\Gamma x_j(t) + c_2 \sum_{j=1}^{N} b_{ij}\Gamma x_j(t - \tau(t)), \tag{19.11}$$

$$\dot{y}_i(t) = f(y_i(t), t) + c_1 \sum_{j=1}^{N} a_{ij}\Gamma y_j(t) + c_2 \sum_{j=1}^{N} b_{ij}\Gamma y_j(t - \tau(t)) + u_i, \tag{19.12}$$

for $i = 1, 2, \ldots, N$. Here $x_i, y_i \in \mathbb{R}^n$ are respectively the state vector of the ith node in drive network $\mathbb{X}$ and response network $\mathbb{Y}$, $f : \mathbb{R}^n \times \mathbb{R}^+ \to \mathbb{R}^n$ is a continuously differentiable nonlinear vector-valued function, $c_1, c_2 > 0$ are the coupling strength, Γ is a inner-coupling matrix between nodes, $\tau(t)$ is the time-varying delay, and $A = (a_{ij})_{N\times N}, B = (b_{ij})_{N\times N}$ are respectively the coupling matrices representing the topological structure of the networks $\mathbb{X}$ and $\mathbb{Y}$ for non-delayed configuration and delayed one. The entries a_{ij} (b_{ij}) are defined as follows: a_{ij} $(b_{ij}) > 0$ if there is a connection between node i and node j $(i \neq j)$; otherwise a_{ij} $(b_{ij}) = 0$ $(i \neq j)$, and the diagonal entries $a_{ii} = -\sum_{j=1,j\neq i}^{N} a_{ij}, b_{ii} = -\sum_{j=1,j\neq i}^{N} b_{ij}$, $i = 1, 2, \ldots, N$. u_i is a controller to be designed.

To investigate the anti-synchronization between networks $\mathbb{X}$ and $\mathbb{Y}$, the pinning strategy will be introduced. Without loss of generality, assuming that the first $l\,(1 \leq l \leq N)$ nodes of network $\mathbb{Y}$ are selected. The pinning adaptive feedback controllers u_i can be described by

$$\begin{cases} u_i = -c_1 d_i(t)\Gamma e_i(t), \quad \dot{d}_i(t) = q_i e_i^T(t)\Gamma e_i(t), \quad i = 1, 2, \ldots, l, \\ u_i = 0, \quad i = l+1, \ldots, N, \end{cases} \tag{19.13}$$

where q_i are positive constants.

Suppose that the matrix A is symmetric and irreducible. Let $A = \begin{pmatrix} A_1 & A_3 \\ A_3^T & A_2 \end{pmatrix}$ where $A_1 \in \mathbb{R}^{l\times l}$, $A_2 \in \mathbb{R}^{(N-l)\times(N-l)}$, $A_3 \in \mathbb{R}^{l\times(N-l)}$. According to (19.11), (19.12) and (19.13), we obtain the following error dynamical network

$$\begin{cases} \dot{e}_i(t) = f(y_i(t), t) + f(x_i(t), t) + c_1 \sum_{j=1}^{N} a_{ij}\Gamma e_j(t) \\ \qquad + c_2 \sum_{j=1}^{N} b_{ij}\Gamma e_j(t - \tau(t)) - c_1 d_i(t)\Gamma e_i(t), \quad i = 1, 2, \ldots, l, \\ \dot{d}_i(t) = q_i e_i^T(t)\Gamma e_i(t), \quad i = 1, 2, \ldots, l, \\ \dot{e}_i(t) = f(y_i(t), t) + f(x_i(t), t) + c_1 \sum_{j=1}^{N} a_{ij}\Gamma e_j(t) \\ \qquad + c_2 \sum_{j=1}^{N} b_{ij}\Gamma e_j(t - \tau(t)), \quad i = l+1, \ldots, N. \end{cases} \tag{19.14}$$

19.3.1 *Pinning anti-synchronization criterion*

In order to derive the anti-synchronization criterion, some useful assumptions and lemmas are presented as follows.

Assumption 19.3. For any $x = (x_1, x_2, \ldots, x_n)^T \in \mathbb{R}^n$, $y = (y_1, y_2, \ldots, y_n)^T \in \mathbb{R}^n$, there exists a positive constant L_1 such that

$$(y - x)^T (f(y, t) - f(x, t)) \leq L_1 (y - x)^T \Gamma (y - x),$$

where Γ is a positive definite matrix. Here x and y are time-varying vectors.

Assumption 19.4. $f(x, t)$ is an odd function of x, i.e., $f(-x, t) = -f(x, t)$ for arbitrary $x \in \mathbb{R}^n$.

Assumption 19.5. $\tau(t)$ is a differential function with $0 \leq \dot{\tau}(t) \leq \varepsilon < 1$. Clearly, constant time delay is a special case of this assumption.

Lemma 19.2 (Schur complement). *The following linear matrix inequality (LMI)*

$$\begin{pmatrix} \mathcal{A}(x) & \mathcal{B}(x) \\ (\mathcal{B}(x))^T & \mathcal{C}(x) \end{pmatrix} > 0,$$

where $\mathcal{A}(x) = (\mathcal{A}(x))^T$, $\mathcal{C}(x) = (\mathcal{C}(x))^T$, *is equivalent to one of the following conditions:*
(*a*) $\mathcal{A}(x) > 0$ *and* $\mathcal{C}(x) - (\mathcal{B}(x))^T(\mathcal{A}(x))^{-1}\mathcal{B}(x) > 0$;
(*b*) $\mathcal{C}(x) > 0$ *and* $\mathcal{A}(x) - \mathcal{B}(x)(\mathcal{C}(x))^{-1}(\mathcal{B}(x))^T > 0$.

Here, the pinning adaptive anti-synchronization between the drive network $\mathbb{X}$ and the response network $\mathbb{Y}$ is investigated, and the main result is summarized in the following theorem.

Theorem 19.3. *Suppose that Assumptions* 19.3, 19.4 *and* 19.5 *hold,* Γ *is a positive definite matrix and* A *is symmetric and irreducible. If*

$$\lambda_{max}(A_2) < -L/c_1, \tag{19.15}$$

where $P = (B^T B)\otimes\Gamma^T$, $k = \lambda_{max}(P)/\lambda_{min}(I_N\otimes\Gamma)$, $L = L_1 + (c_2 k)/(2(1-\varepsilon)) + c_2/2$, *then the networks* $\mathbb{X}$ *and* $\mathbb{Y}$ *can realize anti-synchronization.*

19.3.2 *Numerical simulations*

In the numerical simulations throughout this subsection, the coupling matrices A and B obey the scale-free distribution of the BA network model [Barabási and Albert (1999)] with $m_0 = m = 8$, $N = 100$ and the small-world model [Watts and Strogatz (1998)] with the link probability $p = 0.1$, $m = 4$, $N = 100$, respectively, the initial values are randomly chosen in the interval (0, 1) and the quantity

$$E(t) = \max\{\|x_i(t) + y_i(t)\| : i = 1, 2, \ldots, N\}, \text{ for } t \in [0, +\infty)$$

is used to measure the quality of the anti-synchronization process.

The dynamics at every node of both networks $\mathbb{X}$ and $\mathbb{Y}$ are described by Chua's system [Chua *et al.* (1993)],

$$\dot{x}(t) = f(x(t), t) = \begin{cases} \alpha(x_2(t) - x_1(t) - \phi(x_1(t))), \\ x_1(t) - x_2(t) + x_3(t), \\ -\beta x_2(t), \end{cases} \tag{19.16}$$

where $x(t) = (x_1(t), x_2(t), x_3(t))^T$, $\phi(x_1(t)) = bx_1(t) + \frac{1}{2}(a-b)(|x_1(t)+1| - |x_1(t)-1|)$ and $a = -1.27$, $b = -0.68$, $\alpha = 10.00$, $\beta = 14.87$. For simplicity, setting $\Gamma = \mathrm{diag}(2, 2, 2)$, $c_1 = 30$, $c_2 = 1$, $\tau(t) = 1$, then one gets $k = 72.5671$, $\varepsilon = 0$. It is easy to see that $f(x(t), t)$ is an odd function of $x(t)$. Thus, one has

$$\begin{aligned}
&(y(t) - x(t))^T(f(y(t), t) - f(x(t), t)) \\
&= -10(y_1(t) - x_1(t))^2 + 11(y_1(t) - x_1(t))(y_2(t) - x_2(t)) - (y_2(t) - x_2(t))^2 \\
&\quad - 13.87(y_2(t) - x_2(t))(y_3(t) - x_3(t)) \\
&\quad - 10(y_1(t) - x_1(t))(\phi(y_1(t)) - \phi(x_1(t))) \\
&\le 2.7(y_1(t) - x_1(t))^2 - (y_2(t) - x_2(t))^2 + 11(y_1(t) - x_1(t))(y_2(t) - x_2(t)) \\
&\quad - 13.87(y_2(t) - x_2(t))(y_3(t) - x_3(t))
\end{aligned}$$

$$\le (2.7 + \frac{11\rho}{2})(y_1(t) - x_1(t))^2 + (-1 + \frac{11}{2\rho} + \frac{13.87\eta}{2})(y_2(t) - x_2(t))^2$$
$$+ \frac{13.87}{2\eta}(y_3(t) - x_3(t))^2$$
$$\le L_1(y(t) - x(t))^T\Gamma(y(t) - x(t)),$$

where $L_1 > 0$ is determined by choosing appropriate parameters $\rho, \eta > 0$.

Choosing $\rho = 1.4823$, $\eta = 0.6390$, one has $L_1 = 5.4263$. Then, $L = L_1 + (c_2k)/(2(1-\varepsilon)) + c_2/2 = 42.2098$, $-L/c_1 = -1.4070$.

In [Yu *et al.* (2009)], it was pointed out that it is better to use the high-degree pinning scheme when a small part of nodes is controlled. Here, from numerical calculation, orbits of $\lambda_{max}(A_2)$ as functions of the number of pinned nodes by high-degree (for matrix A), low-degree (for matrix A) and random pinning schemes are shown in Fig. 19.3, and we observe that one only needs 31, 24 and 9 nodes of network $\mathbb{Y}$ to realize the anti-synchronization between networks $\mathbb{X}$ and $\mathbb{Y}$ by using low-degree (for matrix A), random and high-degree (for matrix A) pinning schemes, respectively. Hence, it is better to use the high-degree pinning scheme in this case. Furthermore, it is found that we should use the low-degree pinning scheme when $-L/c_1 < -4$. Now, we apply adaptive feedback control to the first 9 most highly connected nodes of the non-delayed configuration matrix A based on (19.11), (19.12) and (19.13) with $q_{i_k} = 2$, $k = 1, \ldots, 9$. The evolution of anti-synchronization error $E(t)$ is illustrated in Fig. 19.4, which shows that the anti-synchronization between networks $\mathbb{X}$ and $\mathbb{Y}$ is achieved.

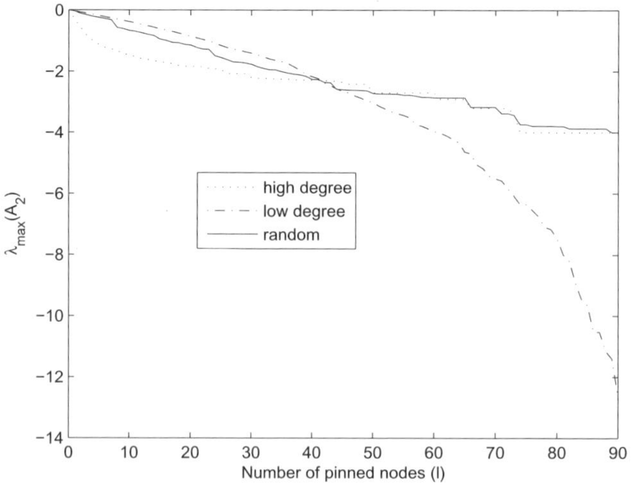

Fig. 19.3 Orbits of $\lambda_{max}(A_2)$ as functions of the number of pinned nodes by high-degree, low-degree and random pinning schemes.

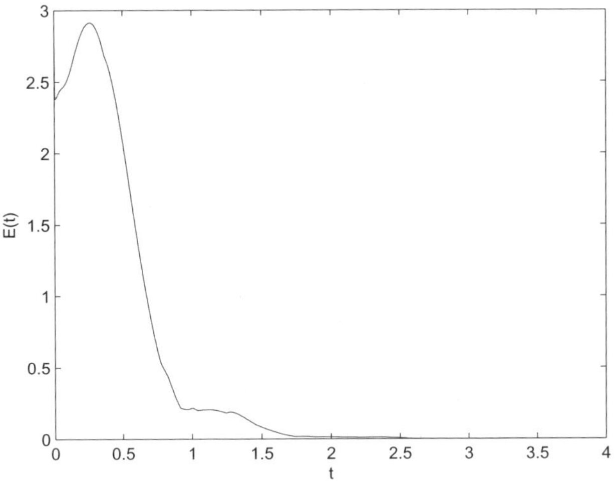

Fig. 19.4 Anti-synchronization error between networks $\mathbb{X}$ and $\mathbb{Y}$ by using high-degree pinning scheme with $l = 9$ and $q_{i_k} = 2,\ k = 1, \ldots, 9$.

19.4 Generalized synchronization between two networks

Consider the following driving-response (or master-slave) configurations [Wu *et al.* (2012)]:

$$\dot{x}_i = f(x_i(t)) + \mu \sum_{j=1}^{N_1} a_{ij} \Gamma_1 x_j,\ i = 1, \ldots, N_1, \tag{19.17}$$

$$\dot{y}_i = g(y_i(t)) + D(y_i, x_j) + \varepsilon \sum_{k=1}^{N_2} b_{ik} \Gamma_2 y_k, \\ i = 1, \ldots, N_2, j \in \{1, \ldots, N_1\}, \tag{19.18}$$

where $D(y_i, x_j) = \dot{x}_j - g(x_j) + (H - \partial g(x_j)/\partial x_j)(y_i - x_j)$. Here $x_i, y_i \in R^n$; $f, g : R^n \to R^n$ are continuously differentiable. $\mu, \varepsilon > 0$ are coupling strengths, Γ_1, Γ_2 are matrices (with order n) linking coupled variables. $A = (a_{ij})_{N_1 \times N_1}, B = (b_{ij})_{N_2 \times N_2}$ are respectively the inner connection matrices of the driving (or master) network $\mathbb{X}$ and response (or slave) network $\mathbb{Y}$. The matrix H is an arbitrary constant Hurwitz one (whose eigenvalues lie in open left semi-plane). The chosen interaction (19.17)–(19.18) is based on the open-plus-closed-loop (OPCL) method.

Definition 19.2. If there exist a transformation $\Phi : R^{n \times N_1} \to R^{n \times N_2}$, a manifold $M = \{(x, y) : y = \Phi(x)\}$, and a subset $N = N_x \times N_y \subset R^{n \times N_1} \times R^{n \times N_2}$ with $M \subset N$ such that all trajectories of (19.17)–(19.18) with initial conditions in the attractive basin N approach the manifold M as time goes to $+\infty$, then we say networks $\mathbb{X}$

and $\mathbb{Y}$ possess the property of generalized synchronization between $x \in R^{n\times N_1}$ and $y \in R^{n\times N_2}$.

19.4.1 *Generalized synchronization criterion*

Let $\phi_x^t : R^{n\times N_1} \to R^{n\times N_1}$ be the flow of network $\mathbb{X}$, $\phi^t = (\phi_x^t, \phi_y^t)$ be the flow of system (19.17)–(19.18) with $\phi_y^t : R^{n\times(N_1+N_2)} \to R^{n\times N_2}$. We find that the map Φ relates to the flow ϕ_y^t. To more accurately characterize the conditions of the occurrence of generalized synchronization for (19.17)–(19.18), the following criterion is presented.

Lemma 19.3. *Generalized synchronization occurs in* (19.17)–(19.18) *if and only if for all* $(x_0, y_0) \in N$ *the response network* $\mathbb{Y}$ *is uniformly asymptotically stable, i.e., for arbitrarily given* $x_0 \in N_x$, *and* $\forall y_{10}, y_{20} \in N_y$, $\lim\limits_{t\to+\infty} \|y(t, x_0, y_{10}) - y(t, x_0, y_{20})\| = 0$.

In the following, in order to study the asymptotic stability of the network $\mathbb{Y}$, we construct an auxiliary network $\mathbb{Z}$ as follows,

$$\dot{z}_i = g(z_i(t)) + D(z_i, x_j) + \varepsilon \sum_{k=1}^{N_2} b_{ik}\Gamma_2\, z_k, i = 1, \ldots, N_2, j \in \{1, \ldots, N_1\}, \quad (19.19)$$

where $D(z_i, x_j) = \dot{x}_j - g(x_j) + (H - \partial g(x_j)/\partial x_j)(z_i - x_j)$.

Letting $e_i = y_i - z_i$, and linearizing the error system around x_j, one has

$$\dot{e}_i = He_i + \varepsilon \sum_{k=1}^{N_2} b_{ik}\Gamma_2\, e_k, \;\; i = 1, \ldots, N_2. \quad (19.20)$$

If we set $e = (e_1, \ldots, e_{N_2}) \in R^{n\times N_2}$, then (19.20) can be rewritten in the compact form,

$$\dot{e} = He + \varepsilon\Gamma_2\, eB^T. \quad (19.21)$$

Then, the following lemma can be obtained.

Lemma 19.4. *Assume that* B *has* m_k *multiple eigenvalues* $\lambda_k \in R$ *where* $k = 1, \ldots, \ell_0$, $\lambda_k = \alpha_k + j\beta_k \in C$ $(\alpha_k, \beta_k \in R,\ \beta_k \neq 0)$ *where* $k = \ell_0 + 1, \ldots, \ell$, *and* $\sum_{k=1}^{\ell} m_k = N_2$. *If the real parts of all eigenvalues of* $H + \varepsilon\lambda_k\Gamma_2$ *(for* $k = 1, \ldots, \ell_0$*) are negative, and* $(H^T + H) + \varepsilon\alpha_k(\Gamma_2^T + \Gamma_2) < 0$ *for* $k = \ell_0 + 1, \ldots, \ell$, *then the zero solution to the matrix equation* (19.21) *is asymptotically stable.*

Combining Lemma 19.3 and Lemma 19.4, one has the following main result:

Theorem 19.4. *Assume that* B *has* m_k *multiple eigenvalues* $\lambda_k \in R$ *where* $k = 1, \ldots, \ell_0$, $\lambda_k = \alpha_k + j\beta_k \in C$ $(\alpha_k, \beta_k \in R,\ \beta_k \neq 0)$ *where* $k = \ell_0 + 1, \ldots, \ell$, *and* $\sum_{k=1}^{\ell} m_k = N_2$. *If the real parts of all eigenvalues of* $H + \varepsilon\lambda_k\Gamma_2$ *(for* $k = 1, \ldots, \ell_0$*) are negative, and* $(H^T + H) + \varepsilon\alpha_k(\Gamma_2^T + \Gamma_2) < 0$ *for* $k = \ell_0 + 1, \ldots, \ell$, *then generalized synchronization occurs in* (19.17)–(19.18).

19.4.2 *Numerical examples*

In the network $\mathbb{X}$, the dynamics at every node follows the Rössler system

$$\begin{aligned}\dot{x}_{i1} &= 2 + x_{i1}(x_{i2} - 4),\\ \dot{x}_{i2} &= -x_{i1} - x_{i3},\\ \dot{x}_{i3} &= x_{i2} + 0.45x_{i3},\ \ i = 1, \ldots, 10.\end{aligned} \tag{19.22}$$

The inner-coupling matrix is as follows,

$$A = \begin{pmatrix} -3 & 0 & 1 & 0 & 0 & 1 & 0 & 0 & 1 & 0\\ 0 & -4 & 1 & 0 & 0 & 1 & 0 & 1 & 0 & 1\\ 1 & 0 & -3 & 0 & 0 & 0 & 1 & 0 & 1 & 0\\ 0 & 1 & 0 & -4 & 1 & 0 & 1 & 0 & 0 & 1\\ 1 & 0 & 0 & 0 & -3 & 0 & 1 & 1 & 0 & 0\\ 1 & 1 & 0 & 1 & 0 & -5 & 0 & 1 & 1 & 0\\ 1 & 0 & 1 & 0 & 1 & 0 & -4 & 0 & 0 & 1\\ 0 & 1 & 0 & 1 & 0 & 1 & 0 & -4 & 0 & 1\\ 0 & 1 & 0 & 1 & 0 & 0 & 1 & 0 & -3 & 0\\ 1 & 0 & 1 & 0 & 0 & 1 & 1 & 0 & 0 & -4 \end{pmatrix}. \tag{19.23}$$

We may simply set $\Gamma_1 = \text{diag}(1, 0, 0), \mu = 1$.

In the network $\mathbb{Y}$, the dynamics of the individual nodes is described by the Lorenz system

$$\begin{aligned}\dot{y}_{i1} &= \sigma(y_{i2} - y_{i1}),\\ \dot{y}_{i2} &= \gamma y_{i1} - y_{i1}y_{i3} - y_{i2},\\ \dot{y}_{i3} &= y_{i1}y_{i2} - by_{i3},\ \ i = 1, \ldots, 12,\end{aligned} \tag{19.24}$$

where $\sigma = 10, \gamma = 28, b = 8/3$. The inner-coupling matrix is chosen as below,

$$B = \begin{pmatrix} 0 & 0 & 1 & 0 & 0 & 0 & 1 & 0 & 0 & 0 & 1 & 0\\ 0 & 0 & 1 & 0 & 1 & 0 & 1 & 0 & 0 & 0 & 0 & 0\\ 1 & 1 & 0 & 1 & 0 & 1 & 0 & 0 & 0 & 0 & 1 & 0\\ 0 & 0 & 1 & 0 & 1 & 0 & 0 & 0 & 1 & 0 & 0 & 0\\ 0 & 1 & 0 & 1 & 0 & 0 & 0 & 1 & 0 & 0 & 0 & 1\\ 0 & 0 & 1 & 0 & 0 & 0 & 0 & 1 & 0 & 1 & 0 & 0\\ 1 & 1 & 0 & 0 & 0 & 0 & 0 & 1 & 0 & 0 & 0 & 0\\ 0 & 0 & 0 & 0 & 1 & 1 & 1 & 0 & 1 & 0 & 0 & 0\\ 0 & 0 & 0 & 1 & 0 & 0 & 0 & 1 & 0 & 0 & 1 & 0\\ 0 & 0 & 0 & 0 & 0 & 1 & 0 & 0 & 0 & 0 & 0 & 0\\ 1 & 0 & 1 & 0 & 0 & 0 & 0 & 0 & 1 & 0 & 0 & 1\\ 0 & 0 & 0 & 0 & 1 & 0 & 0 & 0 & 0 & 0 & 1 & 0 \end{pmatrix}. \tag{19.25}$$

We may set $\Gamma_2 = \text{diag}(1, 1, 1)$, $j = 1$ in $D(y_i, x_j)$. The Hurwitz matrix H for the Lorenz system is

$$\begin{pmatrix} -\sigma & \sigma & 0\\ \gamma + p_1 & -1 & p_2\\ p_3 & p_4 & -b \end{pmatrix},$$

where $p_1, \ldots, p_4$ are parameters. A suitable choice of p_k is $p_1 < 1 - \gamma, p_2 = p_3 = p_4 = 0$. Here, let $p_1 = -29 < 1 - \gamma$ for the current simulation.

If $\varepsilon = 0.3$, it is easy to find that the conditions of Theorem 19.4 hold. In the numerical simulations throughout this subsection, the initial values are randomly chosen in the interval (0, 1). Figs. 19.5 and 19.6 show the simulation results of the driving-response networks with OPCL configurations. The evolution of state variables $x_i = (x_{i1}, x_{i2}, x_{i3})^T$ and $y_i = (y_{i1}, y_{i2}, y_{i3})^T$ are shown in Fig. 19.5 (a)–(c) and Fig. 19.6 (a)-(c), respectively. Obviously, networks $\mathbb{X}$ and $\mathbb{Y}$ do not reach complete synchronization.

Now we introduce an auxiliary network $\mathbb{Z}$ which is a replica of the response network $\mathbb{Y}$. Fig. 19.7 displays $e_i = (e_{i1}, e_{i2}, e_{i3})^T = (y_{i1} - z_{i1}, y_{i2} - z_{i2}, y_{i3} - z_{i3})^T$. The appearance of complete synchronization between $\mathbb{Y}$ and $\mathbb{Z}$ implies that generalized synchronization between $\mathbb{X}$ and $\mathbb{Y}$ of system (19.17)–(19.18) occurs.

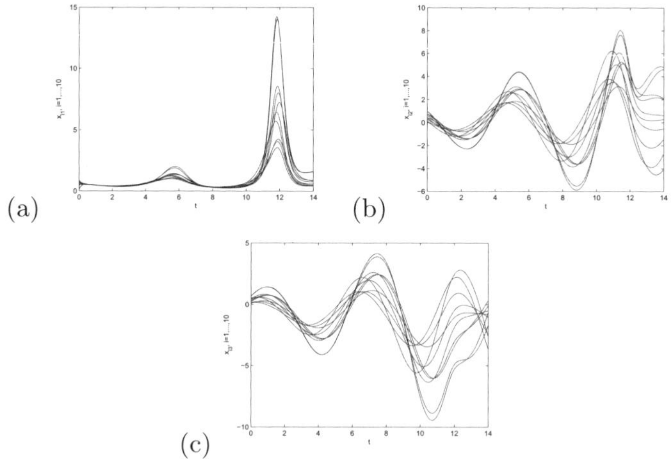

Fig. 19.5 Diagrams of the state variables in network $\mathbb{X}$.

19.5 Conclusion

In this chapter, anti-synchronization and generalized synchronization between two coupled networks have been presented. On condition that two networks are not ultimately connected and synchronization is necessary, we should design the controllers to realize it. Based on this principle, two adaptive controllers are given to realize anti-synchronization with nonlinear signal's connection and the inter-network actions. However, it is impossible to ensure synchronization by adding controllers to all nodes due to the complexity of the dynamical network. So researchers try to control a complex network by pinning part of nodes. And pinning controllers have

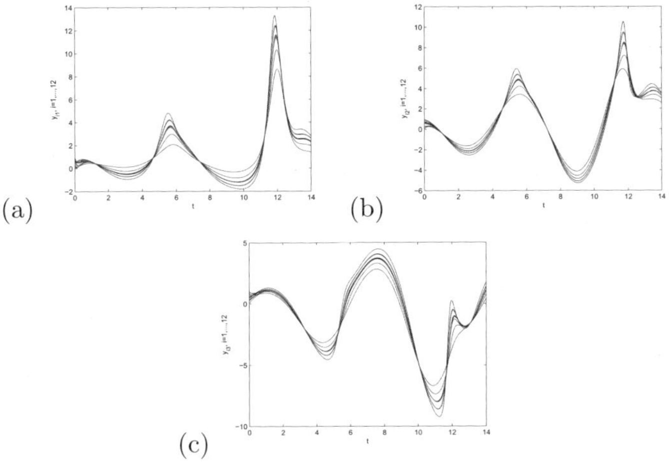

Fig. 19.6 Diagrams of the state variables in network $\mathbb{Y}$.

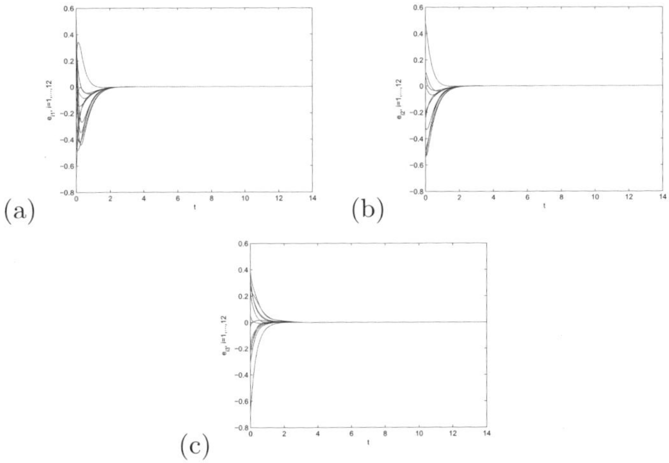

Fig. 19.7 Synchronization between networks $\mathbb{Y}$ and $\mathbb{Z}$. In (a)-(c), $(e_{i1}, e_{i2}, e_{i3})^T = (y_{i1} - z_{i1}, y_{i2} - z_{i2}, y_{i3} - z_{i3})^T$, $i = 1, \cdots, 12$.

been proposed to study anti-synchronization between two coupled complex networks with non-delayed and delayed coupling. Furthermore, the criterion for the occurrence of generalized synchronization in master-slave networks is introduced. The theoretical criterion is based on the uniform asymptotical stability of the response network, which can be verified by utilizing the Lyapunov stability theorem.

Bibliography

Albert, R. and Barabási, A. L. (2002). Statistic mechanics of complex networks, *Rev. Mod. Phys.* **74**, pp. 47–91.

Arenas, A., Diaz-Guilera, A., Kurths, J., Moreno, Y. and Zhou, C. S. (2008). Synchronization in complex networks, *Phys. Rep.* **469**, pp. 93–153.

Barabási, A. L. and Albert, R. (1999). Emergence of scaling in random networks, *Science* **286**, pp. 509–512.

Boccaletti, S., Latora, V., Moreno, Y., Chavez, M., and Hwang, D.-U. (2006). Complex networks: structure and dynamics, *Phys. Rep.* **424**, pp. 175–308.

Chen, J. R., Jiao, L. C., Wu, J. S. and Wang, X. H. (2009). Adaptive synchronization between two different complex networks with time-varying delay coupling, *Chin. Phys. Lett.* **26**, pp. 060505.

Chen, T. P., Liu, X. W. and Lu, W. L. (2007). Pinning complex networks by a single controller, *IEEE Trans. Circuits Syst. I* **54**, pp. 1317–1326.

Chen, M. and Zhou, D. (2006). Synchronization in uncertain complex netwroks, *Chaos* **16**, pp. 013101.

Chua, L. O., Itoh, M., Kocarev, L. and Eckert, K. (1993). Chaos synchronization in Chua's circuit, *J. Circ. Syst. Comput.* **3**, pp. 93–108.

Duan, Z. S., Chen, G. R. and Huang, L. (2007). Complex network synchronizability: Analysis and control, *Phys. Rev. E* **76**, pp. 056103.

Huang, D. (2006). Adaptive-feedback control algorithm, *Phys. Rev. E* **73**, pp. 066204.

Li, C. P., Sun, W. G. and Kurths, J. (2007). Synchronization between two coupled complex networks, *Phys. Rev. E* **76**, pp. 046204.

Li, C. P., Xu, C. X., Sun, W .G., Xu, J. and Kurths, J. (2009). Outer synchronization of coupled discrete-time networks, *Chaos* **19**, pp. 013106.

Li, Y., Liu, Z. R. and Zhang, J. B. (2008). Synchronization between different networks, *Chin. Phys. Lett.* **25**, pp. 874–877.

Li, X., Wang, X. F. and Chen, G. R. (2004). Pinning a complex dynamical network to its equilibrium, *IEEE Trans. Circuits Syst. I* **51**, pp. 2074–2087.

Newman, M. E. J. (2003). The structure and function of complex networks, *SIAM Review* **45**, pp. 167–256.

Shang, Y., Chen, M. Y. and Kurths, J. (2009). Generalized synchronization of complex networks, *Phys. Rev. E* **80**, pp. 027201.

Sorrentino, F. and Ott, E. (2008). Adaptive synchronization of dynamics on evolving complex networks, *Phys. Rev. Lett.* **100**, pp. 114101.

Sun, M., Zeng, C. Y. and Tian, L. X. (2009). Generalized projective synchronization between two complex networks with time-varying coupling delay, *Chin. Phys. Lett.*

26, pp. 010501.

Tang, H. W., Chen, L., Lu, J. A. and Tse, C. K. (2008). Adaptive synchronization between two complex networks with nonidentical topological structures, *Physica A* **387**, pp. 5623–5630.

Wang, G. J., Cao, J. D. and Lu, J. Q. (2009). Outer synchronization between two non-identical networks with circumstance noise, *Physica A* **389**, pp. 1480–1488.

Wang, X. F. and Chen, G. R. (2002). Synchronization in small-world dynamical networks, *Int. J. Bifurc. Chaos* **12**, pp. 187–192.

Wang, X. F. and Chen, G. R. (2002). Synchronization in scale free dynamical networks: robustness and fragility, *IEEE Trans. Circuits Syst. I* **49**, pp. 54–62.

Wang, X. F. and Chen, G. R. (2002). Pinning control of scale-free dynamical networks, *Physica A* **310**, pp. 521–531.

Watts, D. J. and Strogatz S. H. (1998). Collective dynamics of 'small-world' networks, *Nature* **393**, pp. 440–442.

Wu, Y. Q., Li, C. P., Wu, Y. J. and Kurths, J. (2012). Generalized synchronization between two different complex networks, *Commun. Nonlinear Sci. Numer. Simulat.* **17**, pp. 349–355.

Wu, Y. Q., Sun, W. G. and Li, S. S. (2011). Anti-synchronization between coupled networks with two active forms, *Commun. Theor. Phys.* **55**, pp. 835–840.

Wu, X. Q., Zheng, W. X. and Zhou, J. (2009). Generalized outer synchronization between complex dynamical networks, *Chaos* **19**, pp. 013109.

Yu, W. W., Chen, G. R. and Lü, J. H. (2009). On pinning synchronization of complex dynamical networks, *Automatica* **45**, pp. 429–435.

Zhang, G., Liu, Z. R. and Ma, Z. J. (2007). Synchronization of complex dynamical networks via impulsive control, *Chaos* **17**, pp. 043126.

Zheng, S., Bi, Q. S. and Cai, G. L. (2009). Adaptive projective synchronization in complex networks with time-varying coupling delay, *Phys. Lett. A* **373**, pp. 1553–1559.

Zhou, J., Lu, J. A. and Lü, J. H. (2006). Adaptive synchronization of an uncertain complex dynamical network, *IEEE Trans. Autom. Contr.* **51**, pp. 652–656.

Zhou. J., Xiang, L. and Liu, Z. R. (2007). Synchronization in complex delayed dynamical networks with impulsive effects, *Physica A* **384**, pp. 684–692.

Index